TODAY'S TECHNICIAN ™

Shop Manual for

Automotive Brake Systems

Fourth Edition

TODAY'S TECHNICIAN ™

Shop Manual for
Automotive Brake Systems

Fourth Edition

Clifton E. Owen
Griffin Technical College
Griffin, GA

THOMSON
DELMAR LEARNING

Australia · Canada · Mexico · Singapore · Spain · United Kingdom · United States

THOMSON

DELMAR LEARNING

Today's Technician: Shop Manual for Automotive Brake Systems, Fourth Edition
Clifton E. Owen

Vice President, Technology and Trades ABU:
David Garza

Director of Learning Solutions:
Sandy Clark

Managing Editor:
Larry Main

Senior Acquisitions Editor:
David Boelio

Product Manager:
Matthew Thouin

Marketing Director:
Deborah S. Yarnell

Marketing Manager:
Erin Coffin

Marketing Coordinator:
Patti Garrison

Director of Production:
Patty Stephan

Production Manager:
Stacy Masucci

Content Project Manager:
Barbara L. Diaz

Content Project Manager:
Cheri Plasse

Technology Project Manager:
Kevin Smith

Editorial Assistant:
Lauren Stone

Library of Congress Cataloging-in-Publication Data
Owen, Clifton E.
    Automotive brake systems /
    Clifton E. Owen. — 4th ed.
        p. cm.
    Includes index.
    ISBN 1-4180-6221-9
    1. Automobiles—Brakes. I. Title.
    TL269.O96 1997
    629.2'46—dc22
                                2007014992
ISBN-10: 1-4180-6221-9
ISBN-13: 9781418062217

NOTICE TO THE READER

CONTENTS

Photo Sequences

Job Sheets

PREFACE

Thanks to the support the Today's Technician™ series has received from those who teach automotive technology, Thomson Delmar Learning, the leader in automotive related textbooks, is able to live up to its promise to provide new editions of the series regularly. We have listened and responded to our critics and our fans and present this new updated and revised fourth edition. By revising our series regularly, we can and will respond to changes in the industry, changes in technology, changes in the certification process, and the ever-changing needs of those who teach automotive technology.

The Today's Technician™ series, by Thomson Delmar Learning, features textbooks that cover all mechanical and electrical systems of automobiles and light trucks (whereas the Heavy-duty Trucks portion of the series does the same for Heavy-duty vehicles). Principally, the individual titles correspond to the main areas of ASE (National Institute for Automotive Service Excellence) certification. Additional titles include remedial skills and theories common to all of the certification areas and advanced or specific subject areas that reflect the latest technological trends. Each text is divided into two volumes: a Classroom Manual and a Shop Manual.

Unlike yesterday's mechanic, the technician of today and for the future must know the underlying theory of all automotive systems and be able to service and maintain those systems. Dividing the material into two volumes provides the reader with the information needed to begin a successful career as an automotive technician without interrupting the learning process by mixing cognitive and performance learning objectives into one volume.

The design of Thomson Delmar Learning's Today's Technician™ series was based on features that are known to promote improved student learning. The design was further enhanced by a careful study of survey results, in which the respondents were asked to value particular features. Some of these features can be found in other textbooks, whereas others are unique to this series.

Each Classroom Manual contains the principles of operation for each system and subsystem. The Classroom Manual also contains discussions on design variations of key components used by the different vehicle manufacturers. This volume is organized to build on basic facts and theories. The primary objective of this volume is to allow the reader to gain an understanding of how each system and subsystem operates. This understanding is necessary to diagnose the complex automobiles of today and tomorrow. Although the basics contained in the Classroom Manual provide the knowledge needed for diagnostics, diagnostic procedures appear only in the Shop Manual. An understanding of the basics is also a requirement for competence in the skill areas covered in the Shop Manual.

A spiral-bound Shop Manual covers the "how-to's." This volume includes step-by-step instructions for diagnostic and repair procedures. Photo Sequences are used to illustrate some of the common service procedures. Other common procedures are listed and are accompanied by line drawings and photos that allow the reader to visualize and conceptualize the finest details of the procedure. This volume also contains the reasons for performing the procedures as well as when that particular service is appropriate.

The two volumes are designed to be used together and are arranged in corresponding chapters. Not only are the chapters in the volumes linked together, but also the contents of the chapters are linked. This linking of content is evidenced by marginal callouts that refer the reader to the chapter and page in which the same topic is addressed in the other volume. This feature is valuable to instructors. Without this feature, users of other two-volume textbooks must search the index or table of contents to locate supporting information in the other volume. This is not only cumbersome but also creates additional work for an instructor when planning the presentation of material and when making reading assignments. It is also valuable to the students; with page references, they also know exactly where to look for supportive information.

Both volumes contain clear and thoughtfully selected illustrations, many of which are original drawings or photos specially prepared for inclusion in this series. This means that the art is a vital part of each textbook and not merely inserted to increase the numbers of illustrations.

The page layout, used in the series, is designed to include information that would otherwise break up the flow of information presented to the reader. The main body of the text includes all of the "need-to-know" information and illustrations. In the wide side margins of each page are many of the special features of the series. Items that are truly "nice-to-know" information include simple examples of concepts just introduced in the text, explanations or definitions of terms that will not be defined in the Glossary, examples of common trade jargon used to describe a part or operation, and exceptions to the norm explained in the text. Many textbooks attempt to include this type of information and insert it in the main body of text; this tends to interrupt the thought process and cannot be pedagogically justified. By placing this information off to the side of the main text, the reader can select when to refer to it.

Highlights of This Edition—Classroom Manual

The text and figures of this edition are updated to show modern brake technology and applications. The Classroom Manual covers the complete mechanical-hydraulic automotive braking theories. It introduces the reader to the first generation of electric brake systems with a brief overview of brake-by-wire applications in the first chapter. The following chapters cover basic brake physics theories; discussion of newer components and materials, including a section on electric parking brakes, and many braking functions required for passenger cars and light trucks. The reader is introduced to fundamental information on trailer brakes and the DOT requirements for trailer brakes. In the trailer section electric, electrohydraulic, and surge-braking systems are discussed. A new Chapter 10 in this text entitled Electrical Braking Systems (EBS) simplifies the discussion on traditional antilock brake systems (ABS) while retaining the information for a complete understanding of ABS. This chapter guides the reader from ABS through traction control and the incorporation of those systems to vehicle stability systems. Electrohydraulic brakes are discussed in detail along with straightforward information on vehicle stability fundamentals. The Classroom Manual guides the reader from traditional hydraulic brakes to the brake system of the future.

Highlights of This Edition—Shop Manual

Safety information has been moved from the Classroom Manual to the first chapter of the Shop Manual. This places this critical subject next to the work to be accomplished. Chapter 2, Brake Service Tools and Equipment, covers basic tools with more information on brake special tools and equipment. Some of the safety information that is pertinent to a particular piece of equipment has been moved to this chapter so safety issues are presented just before operation of the equipment. Another major improvement is that the related system information is presented in Chapter 3 instead of the last chapter. This is more in accordance with standard brake operation and diagnosis procedures. Totally new information covers diagnosing electric parking brakes and electric braking systems. To clarify the diagnosis and repair procedures for electric braking two major ABS/TCS brands, Delphi DBC-7 and Teves Mark 20E, are used for discussion instead of the complete industry as in the last edition. This helps the reader better understand the technical diagnosing and repairing for all ABS/TCS. This edition of the Shop Manual will guide the student/technician through all the basic tasks in brake system repair.

Shop Manual

To stress the importance of safe work habits, the Shop Manual dedicates one full chapter to safety. Other important features of this manual include:

Performance Objectives

These objectives define the contents of the chapter and define what the student should have learned on completion of the chapter.

Although this textbook is not designed simply to prepare someone for the certification exams, it is organized around the ASE task list. These tasks are defined generically when the procedure is commonly followed and specifically when the procedure is unique for specific vehicle models. Imported and domestic model automobiles and light trucks are included in the procedures.

Tools Lists

Each chapter begins with a list of the Basic Tools needed to perform the tasks included in the chapter. Whenever a Special Tool is required to complete a task, it is listed in the margin next to the procedure.

Marginal Notes

Page numbers for cross-referencing appear in the margin. Some of the common terms used for components, and other bits of information, also appear in the margin. These marginal notes provide an understanding of the language of the trade and help when conversing with an experienced technician.

Photo Sequences

Many procedures are illustrated in detailed Photo Sequences. These detailed photographs show the students what to expect when they perform particular procedures. They also can provide a student a familiarity with a system or type of equipment that the school may not have.

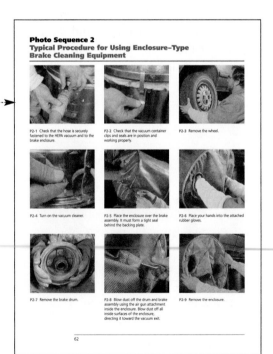

Service Tips

Whenever a special procedure is appropriate, it is described in the text. These tips are generally those things commonly done by experienced technicians.

Cautions and Warnings

Throughout the text, cautions are given to alert the reader to potentially hazardous materials or unsafe conditions. Warnings are also given to advise the student of things that can go wrong if instructions are not followed or if an unacceptable part or tool is used.

References to the Classroom Manual

Reference to the appropriate page in the Classroom Manual is given whenever necessary. Although the chapters of the two manuals are synchronized, material covered in other chapters of the Classroom Manual may be fundamental to the topic discussed in the Shop Manual.

Customer Care

This feature highlights those little things a technician can do or say to enhance customer relations.

Job Sheets

Located at the end of each chapter, the Job Sheets provide a format for students to perform procedures covered in the chapter. A reference to the ASE and NATEF Tasks addressed by the procedure is referenced on the Job Sheet.

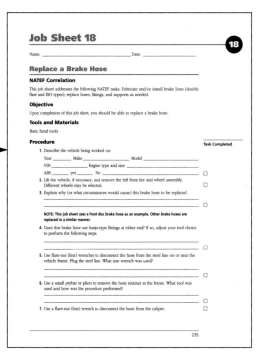

Job Sheet 18   18

Name: _____ Date: _____

Replace a Brake Hose

NATEF Correlation

This job sheet addresses the following NATEF tasks: Fabricate and/or install brake lines (double flare and ISO types); replace hoses, fittings, and supports as needed.

Objective

Upon completion of this job sheet, you should be able to replace a brake hose.

Tools and Materials

Basic hand tools

Procedure   Task Completed

1. Describe the vehicle being worked on:
   Year _____ Make _____ Model _____
   VIN _____ Engine type and size _____
   ABS _____ yes _____ No _____ ☐

2. Lift the vehicle, if necessary, and remove the left front tire and wheel assembly. Different wheels may be selected. ☐

3. Explain why (or what circumstances would cause) this brake hose to be replaced.
   _____ ☐

   NOTE: This job sheet uses a front disc brake hose as an example. Other brake hoses are replaced in a similar manner.

4. Does this brake hose use banjo-type fittings at either end? If so, adjust your tool choice to perform the following steps.
   _____ ☐

5. Use flare-nut (line) wrenches to disconnect the hose from the steel line on or near the vehicle frame. Plug the steel line. What size wrench was used?
   _____ ☐

6. Use a small prybar or pliers to remove the hose retainer at the frame. What tool was used and how was the procedure performed?
   _____ ☐

7. Use a flare-nut (line) wrench to disconnect the hose from the caliper. ☐

235

Case Studies

Case Studies concentrate on the ability to properly diagnose the systems. Chapters 3 through 10 end with a case study in which a vehicle has a problem, and the logic used by a technician to solve the problem is explained.

● **CASE STUDY**

The problem with a 1995 imported SUV was that the brakes would barely work first thing in the morning. After the vehicle was warmed up for a few minutes and driven about a half mile, the brakes started working normally. The owner suspected a stuck master cylinder, an ABS problem, or some similar "high-tech" cause. The tech working on the car took a simpler approach.

After the vehicle sat outside overnight, the technician started the engine and removed the vacuum line from the power brake booster. Engine speed jumped up several hundred rpm. The tech then shook the hose and blew through it several times. The engine speed then dropped back to normal.

On many imported vehicles, the vacuum check valve for the power booster is in line in the hose, not in the fitting on the booster. On this Isuzu, the valve was sticking open when cold. A few minutes of driving and some engine heat would get it working again. Replacing the hose with a new OEM unit that contained a new check valve solved the problem.

Terms to Know

Pressure differential   Spool valve   Vacuum suspended

ASE-Style Review Questions

1. *Technician A* says that if a vacuum booster is in good condition, starting the engine after the vacuum is exhausted should cause the brake pedal to drop slightly under foot pressure.
   Technician B says that the pedal should pulsate lightly after it drops.
   Who is correct?
   A. A only   C. Both A and B
   B. B only   D. Neither A nor B

2. The brake pedal of a high-mileage car is sluggish and moderately hard to apply:
   Technician A says that vacuum leaks or restrictions may exist between the engine and the booster.

Technician B says that a diesel-powered vehicle may have a hydraulic power brake booster.
Who is correct?
A. A only   C. Both A and B
B. B only   D. Neither A nor B

4. *Technician A* says that a defective vacuum check valve in the power brake booster may cause a hard-pedal problem.
   Technician B says that a vacuum check valve may not be installed in a vacuum booster.
   Who is correct?
   A. A only   C. Both A and B
   B. B only   D. Neither A nor B

valve in the booster is pressure from leaking akes are applied. valve is used to keep the booster when the

. Both A and B
. Neither A nor B

Terms to Know

Terms in this list can be found in the Glossary at the end of the manual.

ASE-Style Review Questions

Each chapter contains ASE-style review questions that reflect the performance objectives listed at the beginning of the chapter. These questions can be used to review the chapter as well as to prepare for the ASE certification exam.

Terms to Know

Aqueous	Discard dimension	Parallelism
Bearing end play	Heat checking	Rotor lateral runout
Bedding in	Loaded caliper	Vernier caliper
Burnishing	Nondirectional finish	
Cross-feed		

ASE-Style Review Questions

1. While servicing the front disc brakes on a FWD vehicle:
   Technician A determines the right wheel pad is worn and replaces both the right and left wheel pads.
   Technician B determines the pads are not worn but rotates their position to ensure even pad wear.
   Who is correct?
   A. A only   C. Both A and B
   B. B only   D. Neither A nor B

2. After servicing the disc brakes on a vehicle:
   Technician A reinstalls the wheel nuts using an impact wrench to ensure a tight fit.
   Technician B refills the master cylinder reservoirs to the proper level.
   Who is correct?
   A. A only   C. Both A and B
   B. B only   D. Neither A nor B

3. *Technician A* says loaded calipers are replacement calipers that come with pads and hardware already installed.
   Technician B says loaded calipers should be installed in axle sets.
   Who is correct?
   A. A only   C. Both A and B
   B. B only   D. Neither A nor B

4. *Technician A* says it is very important that the rotor surface be made nondirectional during refinishing.
   Technician B says rotors should always be refinished as part of routine disc brake service.
   Who is correct?
   A. A only   C. Both A and B
   B. B only   D. Neither A nor B

5. When performing disc brake work:
   Technician A works on one wheel at a time to avoid popping pistons out of the other caliper and to allow the other caliper to be used as a guide.

Technician B hangs the calipers from the brake hoses as a convenience to speed up the job.
Who is correct?
A. A only   C. Both A and B
B. B only   D. Neither A nor B

6. When measuring rotor runout:
   Technician A uses a micrometer.
   Technician B uses a dial indicator.
   Who is correct?
   A. A only   C. Both A and B
   B. B only   D. Neither A nor B

7. *Technician A* says that newly installed wheel bearings are self-adjusting and require no special adjustment.
   Technician B says that bearing assemblies are interchangeable with one another.
   Who is correct?
   A. A only   C. Both A and B
   B. B only   D. Neither A nor B

8. *Technician A* says that the minimum wear thickness of a rotor is the discard thickness of the rotor.
   Technician B says that the refinishing dimension is cast into a rotor.
   Who is correct?
   A. A only   C. Both A and B
   B. B only   D. Neither A nor B

9. When refinishing a rotor on a lathe, the rotor wobbles excessively:
   Technician A says that the lathe arbor may be bent.
   Technician B say that the mounting adapters of the lathe may be distorted.
   Who is correct?
   A. A only   C. Both A and B
   B. B only   D. Neither A nor B

ASE Practice Examination

A 50-question ASE practice exam, located in the Appendix, is included to test students on the content of the complete Shop Manual.

Appendix

ASE Practice Examination

1. A vehicle has a drift to the left as the brakes are applied:
   Technician A says that the left front disc brake is grabbing.
   Technician B says that the right rear drum brake is adjusted too loose.
   Who is correct?
   A. A only   C. Both A and B
   B. B only   D. Neither A nor B

2. The customer complains of rear wheel lockup during medium braking on a pickup truck with RWAL:
   Technician A says that the metering valve may be defective.
   Technician B says that a frame-mounted proportioning valve could be stuck.
   Who is correct?
   A. A only   C. Both A and B
   B. B only   D. Neither A nor B

3. Four-wheel, three-channel antilock brakes are being discussed:
   Technician A says that the system has four speed sensors.
   Technician B says that the two rear wheels are independently controlled.
   Who is correct?
   A. A only   C. Both A and B
   B. B only   D. Neither A nor B

4. Proportioning valves are being discussed:
   Technician A says that a bad valve may cause the rear brakes to engage too quickly during initial brake application.
   Technician B says that this valve prevents excessive pressure from reaching the front brakes during normal braking.
   Who is correct?
   A. A only   C. Both A and B
   B. B only   D. Neither A nor B

5. *Technician A* says that a jerk of the steering to the right only as the brakes are applied could indicate improper wheel alignment.
   Technician B says that brake fluid could cause the steering wheel to shake as the brakes are applied.
   Who is correct?
   A. A only   C. Both A and B
   B. B only   D. Neither A nor B

6. The brake pedal drops slowly to the floor after the vehicle has stopped and the brake is kept applied. The fluid level does not drop.
   Technician A says that the dump valve on the RABS could be leaking.
   Technician B says that the master cylinder could have an internal leak.
   Who is correct?
   A. A only   C. Both A and B
   B. B only   D. Neither A nor B

7. There is a heavy shuddering and vibration as the brakes are heavily applied on a dry road:
   Technician A says that the rotors are not parallel.
   Technician B says that the ABS is probably functioning.
   Who is correct?
   A. A only   C. Both A and B
   B. B only   D. Neither A nor B

8. ABSs are being discussed:
   Technician A says that a bent or damaged tone ring could cause the wheel to lock up.
   Technician B says that a damaged tone ring could cause the ABS to deactivate.
   Who is correct?
   A. A only   C. Both A and B
   B. B only   D. Neither A nor B

9. A vehicle with four-wheel antilock brakes has a problem with the right rear wheel locking:
   Technician A says that this could be caused by a bad speed sensor mounted at the wheel.
   Technician B says that the speed sensor mounted at the differential could cause this problem.
   Who is correct?
   A. A only   C. Both A and B
   B. B only   D. Neither A nor B

10. The brake pedal pulsates during braking:
    Technician A says that this should cause the ABS to deactivate.
    Technician B says that this could be caused by warped drums.
    Who is correct?
    A. A only   C. Both A and B
    B. B only   D. Neither A nor B

499

Classroom Manual

Features of this manual include:

Cognitive Objectives

These objectives define the contents of the chapter and define what the student should have learned on completion of the chapter.

Each topic is divided into small units to promote easier understanding and learning.

Related Systems: Tires, Wheels, Bearings, and Suspensions

CHAPTER 3

Upon completion and review of this chapter, you should be able to:

- Describe the basic kinds of tire construction and identify the most common construction method for modern tires.
- Identify and explain the various letters and numbers used in tire size designations and other tire specifications.
- Explain the basic effects of tire tread design on vehicle handling and braking.
- Explain the most important effects of tire design and condition on brake performance.
- Explain how wheel and tire run out and wheel rim width and offset affect braking.
- Identify the common types of wheel and axle bearings used on cars and light trucks.
- Identify the basic wheel alignment and steering angles.
- Explain how wheel alignment and steering angles can affect braking.
- Explain how the condition of steering and suspension parts can affect braking.

Introduction

The brake shoes or pads apply friction to the wheels, but it is the friction between the tires and the road that actually stops the car. Tire design, condition, and inflation pressure can affect braking, and attention to these factors often can solve braking problems.

The tires are mounted on wheels, which ride on bearings on steering knuckle spindles and axles. Steering and axle components, in turn, are supported by suspension struts and springs. Any of these components can create braking problems if they are not in proper working order. This chapter outlines the key relationships between brake systems and the related systems of wheels, tires, wheel bearings, and suspensions.

Tire Fundamentals

Brake systems are engineered in relation to many vehicle factors of weight, size, and performance. Among these factors are the construction, size, and tread design of the tires and the amount of traction or friction expected to be available between the tires and the road. For the best and most reliable brake performance, tires at all four wheels should be identical in construction, size, and tread pattern.

Carmakers' Recommendations

Most passenger cars and light trucks built since 1968 have a tire information placard on a door, a door pillar, or inside the glove compartment (Figure 3-1). The tire information placard lists the manufacturer's original equipment tire size and any recommended optional sizes. It also lists the recommended front and rear inflation pressures and maximum front and rear **gross vehicle weight rating (GVWR)**. Brake systems are engineered to work most efficiently with the tire sizes and pressures listed on the placard.

Marginal note: The gross vehicle weight is the weight of the vehicle plus driver, passenger(s), full fuel tank, and the amount of other material loaded onto the vehicle.

Shop Manual pages 86–93

Marginal note: Gross vehicle weight rating (GVWR) is the total weight of a vehicle plus its maximum rated payload.

41

References to the Shop Manual

Reference to the appropriate page in the Shop Manual is given whenever necessary. Although the chapters of the two manuals are synchronized, material covered in other chapters of the Shop Manual may be fundamental to the topic discussed in the Classroom Manual.

Hydraulic Principles and Brake System Engineering

Engineers must consider these principles of force, pressure, and motion to design a brake system for any vehicle that will give maximum stopping efficiency but still be easy to control. If the engineer chooses a master cylinder with relatively small piston area, the brake system can develop very high hydraulic pressure, but the pedal travel will be extreme. Moreover, if the master cylinder piston travel is not long enough, this high-pressure system will not move enough fluid to apply the large-area caliper pistons regardless of pressure. If, however, the engineer selects a large-area master cylinder piston, it can move a large volume of fluid but may not develop enough pressure to exert adequate braking force at the wheels.

The overall size relationships of master cylinder pistons, caliper pistons, and wheel cylinder pistons are balanced to achieve maximum braking force without grabbing or fading. Most brake systems with front discs and rear drums have large-diameter master cylinders (large piston area) to move enough fluid and a power booster to increase the input force.

Vacuum and Air Pressure Principles

Marginal note: Vacuum is generally considered to be air pressure lower than atmospheric pressure (a true vacuum is a complete absence of air).

Vacuum is another force used in most brake systems. Most power brake systems use vacuum to provide a power assist for the driver. Because the most significant use of atmospheric pressure and vacuum in a brake system is in the operation of a power booster, these principles are covered in Chapter 6 of this manual on power brake systems.

Electrical Principles

Many of the brake system components you will work on are controlled or powered by electricity. Examples are brake system warning lamps, stoplamp switches, brake fluid level sensors, and ABS components. Therefore, a basic understanding of some of the electrical principles, including amperage, voltage, and resistance, is needed.

Amperage, Voltage, and Resistance

Marginal note: Ampere (A) is the unit for measuring electric current where 1 ampere equals a current flow of 6.28×10^{18} electrons per second.

Marginal note: Voltage is the electromotive force that causes current to flow; the potential force that exists between two points when one is positively charged and the other is negatively charged.

Think of electricity in terms of the same principles that work in hydraulic systems. The flow of electricity through a circuit is similar to the flow of fluid through a hydraulic line. Current is the movement of free electrons, under pressure, in a conductor. A flow of current through a conductor requires a source of free electrons to supply the demand, just as fluid in a tank or reservoir is a source of flow in a hydraulic system. The rate of fluid moving through a line often is measured in gallons per minute. The rate of current flowing in a conductor is measured in amperes (A). One **ampere (A)** equals 6.28×10^{18} electrons passing a given point in a circuit per second.

Just as pressure is necessary to move fluid through hydraulic lines, there must be pressure to move electrons through a conductor. The pressure pushing the electrons through an electrical circuit is called the **voltage**, measured in **volts (V)**.

Friction between the walls of a hydraulic line and the fluid will cause some resistance to the flow of fluid. Similarly, some resistance to electron flow through a circuit is offered by any material. Electrical resistance is measured in **ohms (Ω)**.

To summarize the comparison of electrical and hydraulic systems, we can say:

- Voltage is the pressure (or electrical force) that moves electrons (current or amperes) through a wire just as pressure moves fluid through a pipe.
- Amperage, or current, is similar to the fluid flowing in a line.
- Electrical resistance is a load on the moving current that must be present to do any useful work, just as a hydraulic system must have the load of an output piston or motor to do work.

36

Marginal Notes

New terms are pulled out and defined. Common trade jargon also appears in the margin and gives some of the common terms used for components. These marginal notes allow the reader to speak and understand the language of the trade, especially when conversing with an experienced technician.

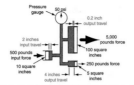

Figure 2-20 The output pistons' movement and their created force will be proportional to their size in relationship to the input piston size.

If the output piston is larger than the input piston, it exerts more force but travels a shorter distance. The opposite also is true. If the output piston is smaller than the input piston, it exerts less force but travels a longer distance. Apply the equation to the 5-square-inch output piston in Figure 2-20:

$$\frac{10 \text{ square inches (input piston)}}{5 \text{ square inches (output piston)}} = \frac{2}{1} \times 2 \text{ inches (input stroke} = 4.0 \text{ inches output motion)}$$

In this case, the smaller output piston applies only half the force of the input piston, but its stroke (motion) is twice as long.

This relationship of force, pressure, and motion in a brake system is shown when the force applied to the master cylinder pistons and the resulting brake force and piston movement at the wheels is considered. Wheel cylinder pistons move only a fraction of an inch to apply hundreds of pounds of force to the brake shoes, but the wheel cylinder piston travel is quite a bit less than the movement of the master cylinder piston. Disc brake caliper pistons move only a few thousandths of an inch but apply great force to the brake rotors.

A BIT OF HISTORY

All of the hydraulic principles that are applied in a brake system are based on the work of a seventeenth-century scientist named Blaise Pascal. Pascal's work is known as Pascal's law. Pascal's law says that pressure at any one point in a confined liquid is the same in every direction and applies equal force on equal areas.

One of the most important results of Pascal's work was the discovery that fluids can be used to increase force. Pascal was the first person to demonstrate the relationships of pressure, force, and motion and the inverse relationship of motion and force. On an automobile, Pascal's laws are applied not just to the brake system. These same hydraulic principles are at work in the hydraulic system of an automatic transmission and other systems. Pascal's laws are even at work in the movement of liquid fuel from a tank to the fuel-injection system on the engine.

A Bit of History

This feature gives the student a sense of the evolution of the automobile. This feature not only contains nice-to-know information, but also should spark some interest in the subject matter.

Summaries

Each chapter concludes with summary statements that contain the important topics of the chapter. These are designed to help the reader review the contents.

Summary

❏ Brake fluid specifications are defined by SAE Standard J1703 and FMVSS 116.
❏ Fluids are assigned DOT numbers: DOT 3, DOT 4, DOT 5, DOT 3/4, and DOT 5.1.
❏ Always use fluid with the DOT number recommended by any specific carmaker.
❏ Never use DOT 5 fluid in an ABS or mix with any other brake fluid.
❏ HSMO fluids are very rare and should never be used in brake systems designed for DOT fluids.
❏ The brake pedal assembly is a lever that increases pedal force to the master cylinder.
❏ The brake pedal lever is attached to a pushrod, which transmits force to the master cylinder pistons.
❏ The master cylinder has two main parts: a reservoir and a cylinder body.
❏ The reservoir can be a separate piece or cast as one piece with the cylinder.
❏ A dual-piston master cylinder has two separate pistons providing pressure for two independent hydraulic systems. Each of the two pistons in the master cylinder has a cup, a return spring, and a seal.
❏ During application, the piston and cup force fluid ahead of the piston to activate the brakes.
❏ During release, the return spring returns the piston.
❏ Fluid from the reservoir flows from the reservoir through the replenishing port around the piston cup.
❏ Excess fluid in front of the piston flows back into the reservoir through the vent ports.
❏ A front-to-rear split hydraulic system has two master cylinder circuits. One is connected to the front brakes and the other to the rear brakes.
❏ A diagonally split hydraulic system is one in which one master cylinder circuit is connected to the left front and right rear brakes and the other circuit is connected to the right front and left rear brakes.
❏ Quick take-up or fast-fill master cylinders have a step bore, which is a larger diameter bore for the rear section of the primary piston.
❏ Quick take-up master cylinders have a valve that provides rapid filling of the low-pressure spool area of the primary piston from the reservoir.
❏ Some ABS master cylinders have check valves in the heads of the pistons to reduce piston and pedal vibration and cup wear.

Terms to Know
Adjustable Pedal System (APS)
Cup seal
Diaphragm
Free play
Hydraulic system mineral oil (HSMO)
O-ring
Polyglycol
Quick take-up master cylinder
Quick take-up valve
Replenishing port
Reservoir
Residual pressure check valve
Vent port

Terms to Know

A list of new terms appears next to the Summary. Definitions for these terms can be found in the Glossary at the end of the manual.

Review Questions

Short-answer essay, fill in the blank, and multiple-choice questions follow each chapter. These questions are designed to accurately assess the student's competence in the stated objectives at the beginning of the chapter.

Review Questions

Short-Answer Essays

1. Identify and explain the systems of tire size designations and other tire specifications.
2. Describe the basic kinds of tire construction and identify the most common construction method for modern tires.
3. Explain the use of unidirectional tread pattern and its disadvantage compared to standard production tire tread.
4. Discuss run-flat tires.
5. List the basic adjustable wheel alignment and steering angles.
6. List the nonadjustable alignment and steering angles.
7. Identify the common types of wheel bearings used on cars and light trucks.
8. Explain how the condition of related systems may affect braking.
9. Identify the principal dimensions used to specify wheel size.
10. Explain the operational differences between WSB and PSB systems.

Fill in the Blanks

1. The steel beads around the rim and layers of cords or plies that are bonded together to give a tire its shape and strength are called the _____.
2. The layer of tire rubber that contacts the road and that contains a distinctive pattern is called the _____.
3. A tire with the cords in the body plies of the carcass running at an angle of 90 degrees to the steel beads in the inner rim of the carcass is a _____ _____ tire.
4. The _____ or the _____ is the percentage of tire cross-sectional height to cross-sectional width.
5. The most common system used to identify tires for passenger cars and many light trucks today is the _____ system.
6. On wet pavement, a slick tire will _____ on a layer of water trapped between the tread and the pavement.
7. The distance between the centerline of a rim and the mounting plane of the wheel is _____ _____.
8. The distance from the tire contact patch centerline to the point where the steering axis intersects the road is _____ _____.
9. The inward or outward tilt of the wheel measured from top to bottom and viewed from the front of the car is _____.
10. The backward or forward angle of the steering axis viewed from the side of the car is _____.

60

Reviewers

The author and publisher would like to extend special thanks to the following instructors for reviewing the draft manuscript:

Ron Chappell
Santa Fe Community College
Gainesville, FL

John Eichelberger
St. Philips College
San Antonio, TX

Raymond Karbowiak
University of Northwestern Ohio
Lima, OH

Jon D'Ambrosio
Mesa Community College
Mesa, AZ

Daniel Livingston
Jackson Community College
Jackson, MI

Darryl Malone
San Jacinto College
Pasadena, TX

Ruth Morrison
Southern Maine Community College
South Portland, ME

Les Peterson
Anoka Technical College
Anoka, MN

Anthony K. Rish
Gateway Community College
North Haven/New Haven, CT

Barry Stirn
University of Northwestern Ohio
Lima, OH

Christopher VanStavoren
Pennsylvania College of Technology
Williamsport, PA

Portions of materials contained herein have been reprinted with permission of General Motors Corporation.

Brake Safety

Upon completion and review of this chapter, you should be able to:

❏ Explain the purpose for government regulations of brake performance and standards.

❏ List the safety requirements for working with brake fluid.

❏ Describe the hazards of asbestos materials.

❏ Explain the safety concerns with solvents and other chemicals.

❏ Explain the general functions of the safety and environmental agencies of the United States and Canada.

❏ Discuss the principles of hazardous communications.

❏ Explain the need and methods for maintaining a safe working area.

❏ List and discuss some safety issues dealing with vehicle operation in the shop.

❏ Explain some of the commonsense rules for working with power equipment.

❏ Wear proper clothing and equipment in a shop.

❏ Discuss some of the safety concerns associated with antilock brake systems.

❏ Explain the first-aid step to remove chemicals from the eyes.

Introduction

Personal protection from injury involves not only what the technician is wearing, but also making and keeping the work area safe. The twofold advantage here is if one technician is protecting himself by wearing personal protection equipment *and* keeping the shop clean and safe, then all the other employees or visitors stand a good chance of avoiding accidents or injury. This chapter discusses those practices and equipment that will provide overall and personal safety.

Housekeeping

Good housekeeping is a safety issue. A cluttered shop is a dangerous shop. Each employee is responsible for keeping the work area and the rest of the shop clean and safe.

All surfaces must be kept clean, dry, and orderly. Any oil, coolant, or grease on the floor can cause slips that could result in injury. Use a commercial oil absorbent to clean up oil or brake fluid spills (Figure 1-1). Store dirty or oily rags in a sealed metal container to be disposed of properly. Keep all water off the floor; remember that water is a conductor of electricity. A serious shock hazard will result if a live wire falls into a puddle in which a person is standing.

When you raise a vehicle with a hand-operated jack, always set the car down on safety stands and remove the jack (Figure 1-2). Do not leave the jack handle sticking out from under the car where someone can trip over it.

Creepers also must be used and stored safely. When not in use, stand the creeper on end against a wall. Pushing it completely under the vehicle gets it out of the way, but it is easy to forget that it is there and to drive over it after the job is completed.

Air hoses and power extension cords should be neatly coiled and hung. Do not leave a tangled mess in walkways or on the shop floor.

Keep all exits open. A blocked exit violates fire codes and leaves the shop liable to legal action if people become trapped in a fire or dangerous situation. Memorize the route to the nearest exit in case of a fire or hazardous material spill.

Basic Tools

Safety glasses or goggles

Respirator

Vacuum with HEPA filter

Wet-clean system

Carbon monoxide vent system

Fire extinguisher(s)

Some oil dry or absorbent compounds have to be treated as hazardous waste after being used. They should not be thrown in the trash bin.

Figure 1-1 Use a commercial absorbent to soak up a spill.

Figure 1-2 Support a vehicle on safety stands such as these and move the jack out of the way.

Vehicle Operation

⚠️ **WARNING:** Use extra caution when moving a vehicle that requires brake repairs. The brakes may be poor or completely inoperative. Damage to the vehicle or shop or injury to yourself or others could result.

Test the brakes on the car to make sure they work before you start the engine. Push the car into the shop if it has a complete brake failure. After completing a brake repair and before moving the vehicle, always check the service brakes. There have been several small but embarrassing incidents where brakes were replaced but not seated. The first time the brakes were applied, there were no brakes.

Be very careful when driving a car in the shop. Be watchful of other workers or customers. Drive slowly and carefully, and get someone to act as a guide if visibility is blocked. Leave a window cranked down so instructions from someone can be heared outside the car.

Once the car is in the service area, place the automatic transmission shift lever in PARK. If the car has a manual transmission, put it in reverse gear with the engine off. Engage the parking brake by pulling the lever or setting the parking brake pedal.

The engine must often be operated in the shop to check for problems and to check your repairs. Several safety precautions should be followed when working on a running engine:

❏ Use wheel blocks to block the front and back of one of the wheels (Figure 1-3).
❏ Never get under a car when someone else is working on it or when the engine is running.
❏ Do not stand in front of or behind an automobile when the engine is running.
❏ Be careful of hot manifolds and moving engine parts if working under the hood.
❏ Many cars use electric cooling fans. Keep hands, tools, and test equipment clear of electric fans because they can start up at any time, even when the engine is not running.

Carbon Monoxide

Running an engine inside a shop can be very hazardous. Engine exhaust contains large amounts of **carbon monoxide,** a deadly gas that is odorless and colorless. Carbon monoxide poisoning begins with headaches and drowsiness. High exposure can lead to coma and death. Never run an engine in the shop without properly venting the exhaust fumes to the outside or to a dedicated ventilation system for exhaust gas (Figure 1-4) and make sure the ventilation system is working properly.

Housekeeping and Brake Dust

There are special tools and equipment designed to be used to collect and contain brake dust. This special equipment is discussed in detail in Chapter 2 of this manual, but some common sense should always be used when working on and around vehicles undergoing brake service.

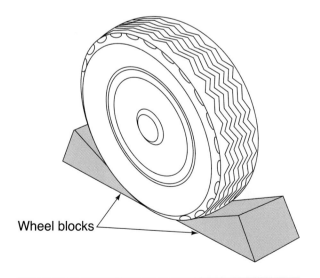

Wheel blocks

Figure 1-3 Block at least one wheel both in front and behind before raising the other end of the vehicle.

Figure 1-4 When running an engine in a shop, always connect the exhaust to the ventilation system.

CAUTION: Do not use compressed air to clean brake components. Brake dust will be present and can be blown into the eyes, embedded into the skin, and, at least, will contaminate the surrounding air. Use only authorized low-pressure washers or vacuum-cleaner-type equipment.

The first and probably most critical is to *never* use compressed air to blow dust from the braking components. This, obviously, moves and suspends the dust in the air. Use only the equipment or their equivalents listed in Chapter 3 to clean the brake components and surrounding area. A second commonsense rule is the wearing of safety glasses and gloves. As discussed earlier, brake fluid and cleaning solvents are hazardous materials and can cause injuries. If a vacuum cleaner is not available to clean the floor around the work area, mop the floor with water. When the mop is rinsed, the rinse water and the material it collects must be stored and treated as hazardous waste. This may seem to present some work problems, but like many things in the automotive repair business it must be done to protect the employees, the environment, and the community in general. Smokers or persons with some type of respiratory problems must be considered when dealing with brake dust. Even with so-called clean air, those individuals may suffer an extreme reaction to what we technicians consider everyday conditions. A technician should make every attempt to prevent the spread of brake dust while working on a vehicle.

Eye and Face Protection

The most frequent causes of eye injuries are flying objects, corrosive chemical splash, dangerous light rays, and poisonous gas or fumes. The most important point about eye injuries is that almost all of them are preventable.

WARNING: Grinds and cutting tools can be dangerous, even to a person not in the immediate area of the work. Ensure that the area is cleared of personnel as much as possible before metal-shaping work.

The best way to prevent eye injuries is to wear the correct type of eye protection (Figure 1-5). When you are performing jobs such as grinding metal, cutting metal, or driving a punch or chisel, your eyes are at risk from flying objects.

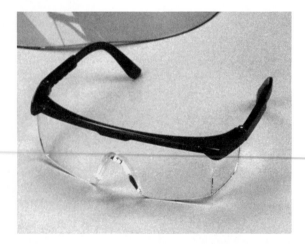

Figure 1-5 Occupational safety glasses provide protection from flying objects that ordinary eyeglasses do not.

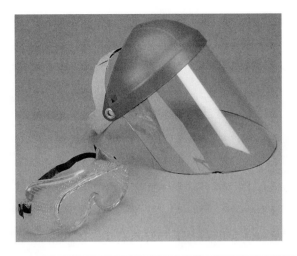

Figure 1-6 A face shield protects your entire face.

Occupational safety glasses (Figure 1-5) are the best protection against flying objects. These safety glasses are especially designed to provide the most protection. The glass or plastic lens provides maximum protection against an impact in the eye. The frames are constructed to prevent the lens from being pushed out of the frame during impact. They must have side shields to prevent objects from entering the eye from the side. They are available in prescriptions for people who need corrective lenses.

▲ **WARNING:** Wear occupational safety glasses when working in the shop, especially when performing any grinding or cutting operations. Ordinary prescription dress glasses are made to standards that provide impact protection, but the impact protection and the frame strength of dress glasses are much lower than occupational safety glasses.

The face shield (Figure 1-6) provides protection for the entire face and is a good choice when the danger is from flying objects or splashing liquids.

Goggles can be used for nearly every type of eye hazard, and they can be used over ordinary dress glasses. Goggles have another advantage over occupational safety glasses because they fit against the head, which allows them to distribute an impact better. Clear-cover goggles provide protection against flying objects or liquid splash. Some goggles have vents and baffles on top to prevent harmful vapors or fumes from getting into the eyes. When using goggles, do not overtighten the straps. They need be only taut enough to hold the goggles in place. As with all other clothing, they have to be worn for a while to become adapted to their weight and viewing area. When taking off goggles or a face shield, close your eyes. Small particles of sharp metal may have attached themselves to the outside of the goggles or face shield and may drop into your eyes.

Initial First Aid

Make sure you know the location and contents of the shop's first-aid kit. There should be eyewash solution or eyewash stations in the shop so you can rinse your eyes thoroughly if you get hydraulic fluid, battery acid, asbestos dust, or other irritants in them (Figure 1-7). See Photo Sequence 1 for details. After eyewashing, seek medical attention. Find out if there is a resident nurse in the shop or at the school, and locate the nurse's office. If there are specific first-aid rules in your school or shop, find out what they are and abide by them. In a school, a report is required for any injuries to a student.

Most shops and all schools require an accident report to be completed and filed.

Photo Sequence 1
Using Eye Wash

P1-1 Remove the eyewash bottle from the wall holder. The injured person may require assistance.

P1-2 Open the bottle. Attempt not to touch the mouth of the bottle once it is opened. The injured person may require assistance.

P1-3 Tilt the head back and over so the injured eye is lower that the other eye. The injured person may require assistance.

P1-4 Pour the water so the flow goes from the nose bridge, over the eye, and down the cheek. Keep both eyes open and looking upward during the flushing. The injured person may require assistance.

P1-5 If necessary, cover the injured eye with a sterile dressing and seek medical assistance. If the eye is covered, it is recommended that someone guide/transport the person to the neareast medical resource. A person loses depth perception and some vision when an eye patch is first used.

P1-6 The injured eye should be examined by a medical technician for injuries that may not be immeditely apparent.

If someone is overcome by carbon monoxide, move the person to fresh air immediately. Rinse burns immediately in cold water or apply an ice pack. To stop bleeding from a deep cut or puncture wound, apply pressure on or around the wound and get medical help. Never move someone you suspect has broken bones or a back injury unless the person is in danger from another hazard such as fire or carbon monoxide gas. Call for medical assistance.

Hand Protection

Hands are one of the most frequently injured parts of the body. This fact is not surprising when you think of how often we use our hands doing automotive repair. There are two parts to protecting

Figure 1-7 An eyewash solution will flush contaminants from your eyes.

the hands. One is to keep your hands out of dangerous areas. Rotating parts, such as the belts on the front of an engine, are hand danger areas. Make an effort to keep the hands out of those areas as much as possible.

The second part of hand safety is to wear hand protection when necessary. Special protective gloves are available for many jobs that require hand protection. There are heavy work gloves for metal working, rubber gloves for electrical shock protection, and nitrile gloves for handling used oil, brake fluid, and chemicals such as those used to clean parts. Always use the correct type of glove for the hand hazards in the work area.

Do not wear a wristwatch or rings while working. Watches can get caught in rotating machinery. Rings can get caught in machinery or provide a path for an electrical shock. Long hair can get caught in rotating machinery. Many bad injuries have been caused by the hair pulling the face into a rotating part. Always tie up long hair or wear a hat over it.

Always wear safety shoes in the shop. Safety shoes have metal or fiberglass protection over the toe to prevent an injury if a heavy object falls on your foot. Safety shoes should at least have oil-resistant soles that grip slippery floors better than casual dress shoes.

Lifting and Carrying

The back is one of the most often injured parts of the body. The most frequent kind of back injury at work is caused by improper lifting. Not all back injuries are caused by lifting too much weight but by lifting relatively small, light objects. The problem occurs while lifting the object and twisting the body or lifting when the load is unbalanced. Most back injuries can be prevented by following these ten simple rules:

1. Do not lift any heavy object by yourself. Get someone to share the load or get some equipment such as a chain hoist to do the lifting.
2. Study the load before you attempt to lift it. Use your head before you use your back.

If you lose control of a lifted object, do not attempt to catch it. Step back and let the object drop.

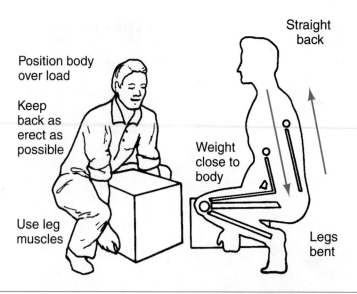

Figure 1-8 Keep your back straight and bend your legs to lift heavy objects safely.

3. Place your body close to the object as shown in Figure 1-8. Keep your legs close to the load and positioned for good balance.

4. Bend your legs, not your back.

5. Get a strong grip on the object with your hands.

6. Lift with your legs, keeping your back as straight as possible.

7. Keep the load close to your body as you lift it up.

8. Keep a tight grip on the object and do not try to change your grip while lifting.

9. Do not twist your body to change direction. Move your feet in the new direction.

10. When you are ready to set the load down, do not bend forward. Keep the load close to your body and lower it by bending your legs. When placing the object on a shelf, place the edge of the load on the surface of the shelf and slide it forward. When setting an object on the floor, lower it by bending your knees and keeping your back straight. Bending forward strains your back muscles.

Having the body out of position can lead to painful injury even if nothing is being lifted. The most common muscle sprain or injury happens when the person is lifting a small weight but the body is twisted off center.

Brake System Safety Regulations

In the United States, brake systems are regulated by Part 571 of the **Federal Motor Vehicle Safety Standards (FMVSS).** These regulations are established and enforced by the U.S. **Department of Transportation (DOT).** The standards that relate to brake systems are:

❏ FMVSS 105   Hydraulic Brake Systems
❏ FMVSS 106   Brake Hoses
❏ FMVSS 108   Lamps, Reflective Devices, and Associated Equipment
❏ FMVSS 116   Motor Vehicle Brake Fluids
❏ FMVSS 121   Air Brake Systems
❏ FMVSS 122   Motorcycle Brake Systems
❏ FMVSS 211   Wheel Nuts, Wheel Discs, and Hub Caps

Many U.S. states and Canadian provinces also have regulations that govern the brake's safety, condition, and operation. Several of the federal standards apply to specific components included in this text. General performance requirements for service brakes and parking brake systems are governed by FMVSS 105. This standard became effective in 1967, was revised significantly in 1976, and has undergone several smaller changes since then. FMVSS 105 spells out the "requirements for hydraulic service brake and associated parking brake systems to ensure safe braking performance under normal and emergency conditions . . . for passenger cars, multipurpose passenger vehicles, trucks, and buses with hydraulic service brakes."

FMVSS 105 does not prescribe the design of brake systems; it establishes brake performance requirements. By so doing, however, it also establishes the baseline for system safety. The standard regulates four major features of brake systems: instrument panel warning lamps, the fluid reservoir and its labeling, automatic adjustment, and mechanically operated friction parking brakes.

Although FMVSS does not dictate brake system hardware and design, one of its first major effects that car owners saw was the introduction of dual-chamber master cylinders and split hydraulic systems on 1967 model-year cars. Also, the increased performance requirements in the 1976 revision made it impractical to use drum brakes on the front wheels of cars. The standard did not specify front disc brakes, but discs were the most practical way to meet the performance requirements.

Brake systems are not designed just to meet minimum legal standards, however. They are designed in relation to the performance and intended use of a vehicle. Trucks have larger brakes than passenger cars, for example, to stop a vehicle with a heavier payload. A high-performance car will have high-performance brakes, but an economy compact car will not. Every vehicle has a brake system that meets motor vehicle safety requirements and matches the performance capabilities and intended use of that vehicle. Thus, brake systems reflect both safety regulations and sound engineering practices.

Brake Performance Test

The brake performance test of FMVSS 105 defines the minimum requirements for the hydraulic brake system on any vehicle driven on the highway. The technician should know a bit about the performance test, not for the sake of being able to quote government regulations, but because parts of the test define the kind of performance a brake system should deliver after the vehicle is serviced.

The brake performance test is divided into eighteen stages and begins with a new set of brakes on a test vehicle. The first stage of testing is discussed as an example of the other seventeen. The first stage is to install the test instruments on the vehicle and verify that they operate correctly. The vehicle then goes through what is called the "first effectiveness test." This test is performed with fresh brake linings before they have had a chance to burnish in. The vehicle makes six stops from 30 mph and six stops from 60 mph. At least one of the stops from 30 mph must be made in 57 feet or less, and one stop from 60 mph must be in 216 feet or less (Figure 1-9). These stopping distances and stopping distances in other stages of the test are absolute requirements for any vehicle of any size and weight. Remember that FMVSS 105 defines minimum brake performance. It is up to the engineers to design the vehicle and the brake system to meet the performance standards.

Complete United States brake testing procedures and standards can be found on the Web at http://www.nhtsa.dot.gov/cars/rules/import/FMVSS/index.html. At the same Web location, information can be found for tires, wheels, and other vehicle safety equipment. The Canadian government has a similar site at http://www.Nasdpts.org/documents/PubSBMTCMiniGuide04.pdf.

While the governments set the minimal rules and regulations for design and manufacturing, the technician should understand the ramifications if a brake system is not returned to its

Department of Transportation (DOT) is the U.S. government executive department that establishes and enforces safety regulations for motor vehicles and for federal highway safety and oversees, inspects, and regulates all interstate transportation including road, rail, and water facilities; commercial operators training/certification; and commercial vehicles. They are assisted by state-funded transportation departments.

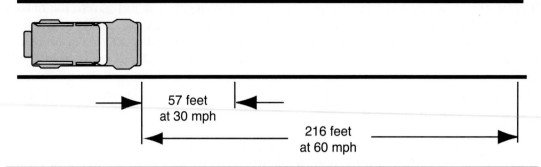

Figure 1-9 One of the eighteen stages of the brake performance test in FMVSS 105 requires one stop from 30 mph in 57 feet or less and one stop from 60 mph in 216 feet or less.

designed capability. Failure to follow correct repair procedures could cause a vehicle accident, resulting in damage, injuries, and lawsuits. It takes less time to do it right, rather than to take a shortcut that saves time and labor in the short term but may result in much greater loss of time and money later.

Brake Service Laws and Regulations

After new vehicles are first sold, the responsibility for maintaining safe brake operation falls on the vehicle owners. The owners, in turn, rely on the service technician to keep the brakes in proper operating condition. Many states and provinces have laws that govern brake system operation and brake service.

Some states require periodic vehicle safety inspection, either every year or every 2 years. These safety inspections usually include at least an inspection of brake components. Some also include dynamic stopping tests, done on a brake system analyzer or on a measured course. If a vehicle fails any part of the safety inspection, its registration cannot be renewed until all defects are fixed. Some states require that a vehicle that has failed a brake test or inspection or that has been cited for unsafe brakes by a police officer can only be repaired at a state-authorized repair facility.

In addition, some states, provinces, counties, or cities have regulations for the licensing or certification of brake service technicians. Some areas conduct their own certification programs; others rely on Automotive Service Excellence (ASE) certification in brake service. ASE is a non-profit organization that technically certifies automotive technicians with a series of standardized written tests. Automotive business leaders, technicians, and educators select and write the test questions.

Working in an area that has brake service regulations, the technician will find that safety is not only good common sense, it is good business. Service technicians who pass all certification requirements for brake systems will get more of the service business, have more secure employment, and earn higher wages. In addition, any technician who provides high-quality brake service can take satisfaction in knowing that he or she is contributing to driving safety.

Brake Warnings and Cautions

At the beginning of many manufacturers' service manuals and at appropriate points throughout the manuals are various cautions and warnings to alert the technician to some dangers inherent in brake repair. Some of the most common manufacturers' warnings and cautions are paraphrased and listed in Table 1-1. Within this text and the *Classroom Manual,* there are also warnings and

Table 1-1 A Sample of Brake Warnings and Cautions. Other Components and Subsystems of the Brake System Will also Have Specific Alert Messages Similar to Those Listed Here.

Pertaining to brake fluids:

 CAUTION: Brake fluid is corrosive to body finish. Do not allow fluid to spill onto the paint or components. The fluid will damage the finish and possibly damage some components.

WARNING: Brake fluid can damage the eyes and skin. Wear safety glasses and chemical-resistant gloves when handling brake fluid. Damage to the eyes or skin can be caused by direct contact with brake fluids.

WARNING: Never mix different types of brake fluids unless specifically authorized by the vehicle manufacturer. Mixing different types of brake fluid may result in a loss of braking ability and cause damage or injury.

Pertaining to disc brake calipers:

WARNING: Do not hang the caliper from its brake hose. Damage to the hose could occur that may result in poor braking ability. Damage or injuries could result.

CAUTION: Do not use a sharp object to remove the caliper seal. Scratches or nicks could prevent proper sealing around the piston. Damage to the caliper bore or the piston could result.

cautions pertaining to servicing brake systems. To prevent damage both to the vehicle and vehicle equipment, and possible injury, it is imperative that the technician adhere to the information contained in the alert messages. As further chapters are read and studied, the reader will become more conversant with the warnings and cautions and why they are necessary.

Asbestos Health Issues

One of the greatest safety concerns in any shop doing brake service is personnel exposure to **asbestos** dust. Exposure to asbestos was a greater problem in automotive service many years ago than it is today. The importance of avoiding asbestos exposure and the concern for asbestos safety have not decreased in the slightest, however.

Asbestos is a silicate compound that is very resistant to heat and corrosion. Its excellent heat dissipation abilities and coefficient of friction make it ideal for automotive friction materials such as clutch and brake linings. Unfortunately, asbestos has other characteristics that make it an extreme health hazard.

Asbestos contains millions of small, linked fibers that give it both strength and flexibility. Because asbestos does not deteriorate or decompose naturally, inhaling asbestos fibers lodges them in the respiratory passages and the lungs. Once inhaled, these fibers are in place forever. Even moderate quantities of inhaled asbestos fibers can lead to serious diseases. The most serious are asbestosis and lung cancer.

Asbestosis is a progressive lung disease caused by asbestos fibers continually lodging in the lungs and inflaming the lung air sacs. The inflammation of **asbestosis** can heal, but it leaves scar tissue in the lungs that thickens the air sacs and makes it increasingly more difficult for oxygen to enter the bloodstream. Over a period of years, breathing becomes increasingly more difficult. Once started, asbestosis is irreversible. Lung cancer is the most deadly of any asbestos-related disease. Asbestos exposure combined with other respiratory irritations, such as tobacco smoke, can accelerate the development of cancer and produce more severe effects. It is possible for a person to develop both asbestosis and lung cancer from severe asbestos exposure. Heavy exposure to asbestos also can lead to other cancers of the respiratory and digestive systems.

Asbestos Control Laws and Regulations

Occupational Safety and Health Administration (OSHA) is a division of the U.S. Department of Labor that establishes and enforces workplace safety regulations.

Regulations of the U.S. **Occupational Safety and Health Administration (OSHA)** control asbestos exposure and handling of materials that contain asbestos. OSHA regulations state that fibers of 5 microns or larger are hazardous. These regulations further say that no worker can be exposed to more than 0.1 fiber per cubic centimeter of air during an 8-hour period. That is an extremely small exposure to an extremely small amount of material. These low exposure limits can be maintained in a brake service shop, however, through the proper use of brake cleaning equipment and respiratory safety devices.

A   B I T   O F   H I S T O R Y

The Occupational Safety and Health Act (OSHA), which regulates workers safety, was passed into federal law in 1970.

A respirator designed specifically for protection against asbestos inhalation is your best personal protection. The respirator shown in Figure 1-10 is approved by the National Institute for Occupational Safety and Health (NIOSH) and has replaceable filters for maximum protection. A brake dust vacuum cleaning enclosure (Figure 1-11) and a brake washing system will keep asbestos dust within safe limits for the entire shop area.

Along with OSHA, the U.S. **Environmental Protection Agency (EPA)** regulates some aspects of asbestos safety. EPA regulations are concerned primarily with handling and disposal of asbestos waste. These regulations state that any waste material containing more than 1 percent asbestos must be disposed of by rigidly controlled methods that do not endanger public health.

Environmental Protection Agency (EPA) is the U.S. government executive department that establishes and enforces regulations to protect and preserve the physical environment, through the control of hazardous materials and waste, including landfills. It is best known for regulations relating to air quality. Many times OSHA and EPA authority overlap in large incidents.

Technicians' concern with asbestos safety does not end with prescribed cleaning of brake systems and respiratory safety. They must dispose of cleaning residue according to EPA regulations. Because brake dust may contain more than 1 percent asbestos, any vacuum cleaner bags, filters, and cloths used to wipe up brake dust must be sealed in double plastic bags or a similar nonpermeable container. The bag or container must then be labeled with an asbestos exposure warning, similar to the following:

CAUTION: HAZARDOUS MATERIALS

THIS CONTAINER HOLDS ASBESTOS FIBERS. AVOID CREATING DUST WHEN MOVING OR OPENING. BREATHING PROTECTION SHOULD BE WORN WHEN UNSEALING/SEALING CONTAINER. ASBESTOS FIBERS ARE HAZARDOUS AND CAN CAUSE CANCER AND LUNG DISEASE.

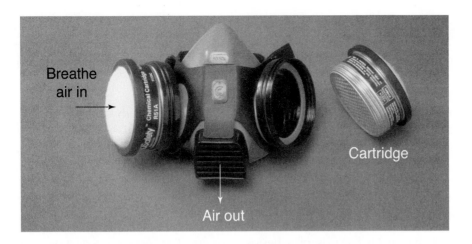

Figure 1-10 This NIOSH-approved filter-type respirator is ideal for brake work.

Figure 1-11 This full-enclosure asbestos vacuum system traps brake dust and helps keep the shop's air free of dust.

AUTHOR'S NOTE: All containers in which you store hazardous material or waste must be labeled as to contents. The text can be handwritten or printed with other methods as long as it cannot be easily wiped off or will not fade during the time the container is used for this purpose.

In most areas of the country, it is acceptable to turn over properly contained asbestos residue to local sanitation agencies for burial in a landfill. This eliminates the hazard of airborne fibers. Local asbestos disposal regulations may vary, however, and some may require additional special handling. It is the technician's responsibility to know the local regulations and to ensure that they are observed in the workplace.

Additional Respiratory Safety

Concern for the health hazards of asbestos exposure has led to a reduction of its use in automotive components. For many years, asbestos has been eliminated from the brake linings of new cars and light trucks sold in North America and from replacement brake linings made in the United States and Canada. These restrictions do not apply, however, to replacement brake linings manufactured outside North America and imported into the United States or Canada. Furthermore, the asbestos content of imported brake linings may not be identified clearly on packaging.

As asbestos content was reduced in brake linings, other materials took its place. Today, many brake linings are made primarily of organic or semimetallic compounds. As with asbestos, however, these materials wear and create airborne dust. Semimetallic brake linings, for example, may contain copper or iron compounds, and these materials become part of the brake dust. Although exposure to these metals may not be as hazardous as asbestos exposure, it cannot be good for a person to inhale copper or iron dust.

For all of these reasons, proper use of brake cleaning equipment and respiratory safety devices is as important today as it has ever been. Do not think for a moment that the reduced use of asbestos in automotive materials has reduced the requirements for safe material handling. You must take the proper steps to protect yourself and create a safe work environment.

A BIT OF HISTORY

Health concerns were one good reason to remove asbestos from brake linings, but there was an equally good engineering reason. Modern brake friction materials work better than asbestos. Asbestos was still common in some friction materials in the early 1990s, but higher temperatures of smaller disc brakes caused the asbestos pads to wear faster than was acceptable. Moreover, even the best asbestos material will start to glaze at temperatures as low as 250°F (122°C). Modern semimetallic and organic linings are safer not only from a health standpoint, but they also provide better braking performance than asbestos did.

Chemical Safety

Asbestos is not the only hazardous material found in auto service facilities. Solvents, cleaners, brake fluids, gasoline, oils, and other chemicals all present hazards if not handled properly. They may be flammable, emit harmful vapors, or be irritating to the eyes or skin.

Brake Cleaning Solvents

One reason liquid solvents were developed for brake cleaning was to reduce the hazard of blowing off brake assemblies with compressed air and creating clouds of airborne fibers and dust. Wetting the dirt and dust residue on the brakes with solvent keeps the toxic materials out of the air.

Always work with cleaning solvents in a well-ventilated area that is free of sparks or flames. The fumes from aerosol cleaners and open part washers are heavier than air and will settle to the lower part of the work area such as below floor-level dynamometers and alignment pits. Solvent vapors may also be harmful if inhaled, particularly in large quantities for prolonged periods. If necessary, use a respirator to prevent inhaling the vapors. Chlorinated hydrocarbon solvents may be absorbed through the skin with toxic effects. Always wear gloves when using any cleaning solvent. The best first aid for skin and eye contamination is flushing with large amounts of water and contacting medical personnel. Inhalation exposure requires quick removal to clean air and medical attention.

Although they are a lesser health hazard than asbestos, various cleaning solvents used on brake systems must be handled with specific precautions. Among the most significant from a safety standpoint are those that contain **chlorinated hydrocarbon solvents** such as **1,1,1-trichloroethane, trichloroethylene,** and **tetrachloroethylene** (Figure 1-12). These are all colorless solvents with a strong odor of ether or chloroform. The vapors from these solvents can cause drowsiness or loss of consciousness. Very high levels of exposure, even for a short time, may be fatal. Although these hydrocarbon solvents are not flammable, they decompose when exposed to flame and release toxic gases such as **phosgene,** carbon monoxide, and hydrogen chloride.

This family of chlorinated hydrocarbon solvents reacts in the atmosphere and depletes the Earth's ozone layer. Their manufacture has been restricted since January 1, 1996. Other solvents such as hexane, heptane, and xylene are replacing chlorinated hydrocarbons in brake cleaners (Figure 1-13). Hexane and heptane are flammable, however, so all fire safety precautions must be observed when using these solvents.

Causes and Effects of Chemical Poisoning

A person may be exposed to chemical health hazards in three ways: by ingestion, by inhalation, and by contact with the skin. Material safety data sheets (MSDS), discussed in more detail in

Most new automotive cleaning solvents no longer contain chlorine. Chlorine is suspected of causing damage to the ozone layer and is banned from common use by the EPA.

Chlorinated hydrocarbon solvents are a class of chemical compounds that contain various combinations of hydrogen, carbon, and chlorine atoms.

Trichloroethane is a chlorinated cleaning solvent often used in aerosol brake cleaner.

Trichloroethylene is a chlorinated toxic cleaning solvent often used in aerosol brake cleaner and as an insecticide fumigant.

Phosgene is a poisonous gas that is formed when certain other gases are exposed to flame; it is also known as mustard gas, the principal poison gas used in World War I.

Figure 1-12 This aerosol brake cleaner contains tetrachloroethylene. You should know and practice the safe use of all solvents in the shop.

Figure 1-13 Nonchlorinated cleaning solvents also require specific handling and safety precautions.

subsequent paragraphs, describe any poisoning hazards and how to counteract poisonous effects. An MSDS for every solvent used in the shop should be readily available to every worker.

Obviously, swallowing any solvent—even soap—can be hazardous, but this does not happen very often. Solvents can be ingested, however, by a smoker who lights a cigarette while working with the solvent. Solvents must always be handled carefully and kept in properly labeled containers. When not in use, the containers must be stored away from untrained personnel or children.

Contact with solvents occurs most often through inhalation or absorption through your skin. Inhalation probably is the more serious and has the more immediate effect. Absorption can be just as dangerous, however, and its effects may not be noticeable for several days after exposure.

Current OSHA standards for exposure to airborne trichloroethylene say that more than 100 parts per million (ppm) in the air during 8 hours is dangerous. To give you an idea of how small the allowable exposure is, 100 ppm equal 0.0001 percent.

WARNING: Always wear nitrile gloves when working with chemicals. Exposure can lead to skin injuries and, sometimes, ingestion through the skin into the bloodstream. Serious injury could result.

There is no current standard for physical contact with these solvents, but the immediate effect is the removal of natural skin oils, which causes drying of the skin and redness and irritation. Prolonged skin contact with solvent can have the same effects as inhalation.

Exposure to chlorinated hydrocarbons and other solvents by any means can cause nausea, drowsiness, headache, dizziness, and eventually unconsciousness. Prolonged exposure can lead to liver and kidney damage.

Safety and Envirnomental Agencies

Environment Protection Agency

The Environment Protection Agency, or EPA, is a federal agency charged with instituting and enforcing regulations that assist in protecting the environment. It was formed in the early 1970s to reduce air pollution caused by vehicle and manufacturing emissions. Inherent within that charter was the control and disposal of waste products from almost all businesses, including the local automotive repair shop and individuals. The main issue with the EPA is the storage and disposal of hazardous waste from major manufacturers, plants, the local garbage dump, and everything in between. Although its formation met with much resistance, the results some 40 years later are cleaner air and less ground and water pollution. Unless something changes, the agency will be in operation for the foreseeable future. The EPA's Web site is http://www.epa.gov.

Occupational Safety and Health Administration (OSHA)

OSHA was formed to help protect employees and, ultimately, employers. It has the legal authority to inspect businesses and ensure that working areas are safe for the employees. Some safety concerns of utmost interest are the control of chemicals within the workplace, the equipment/facility in which to store or use those chemicals, the equipment and tools used within the facility, and the general working environment. It should be noted that since the formation of OSHA, accidents resulting from unsafe working environments have been reduced, with an increase in production associated with lowered loss of man-hours and fewer accidents. A suggested Web site for OSHA is http://www.osha.gov/SLTC/index.html.

Environmental Canada

Environmental Canada is the Canadian version of the United States EPA. It has requirements that relate to Canada's more northern environment and citizens. Within its organization are subagencies, such as the Canadian Environmental Assessment Agency, that may not be directly related to subagencies of the United States EPA. As far as the automotive industry is concerned, however, the legal and environmental control requirements are almost exactly the same. Section 7 of the Canadian Environmental Protection Act specifically covers the Canadian automotive industry. The Web site best suited for information on this agency is http://www.ceaa.gc.ca/ppp/index_e.htm#1.

Canadian Center for Occupational Health and Safety (CCOHS)

The **Canadian Center for Occupational Health and Safety (CCOHS)** is similar to the United States OSHA with a similar mandate, responsibility, and authority. It performs inspections, determines administrative fines, may file criminal charges, and directs training programs in much the same manner as the United States OSHA does. The Web site is http://www.ccohs.ca/html.

It should be noted that each of the four agencies listed operates "over border" because many pollutants tend to cross borders. Automotive manufacturing, vehicle repair, and vehicle operation are shared by the United States and Canada and many associated problems are the result of actions in one country affecting the environment of its neighbor. Each of the listed Web sites has a large amount of information pertaining to almost any environmental and safety issue.

Hazardous Communications

Each of the agencies noted in the last section enforces what are known as right-to-know laws or hazardous communications. Basically, right-to-know requires the employer to notify employees of dangerous materials that are housed or used on-site. They also require the initial training of new employees; annual (or more often) refresher training of all employers; and employer-designated personnel with specific authority to train, maintain records, and, in some instances, act as first responders to fires or accidents. Of direct interest to all employees are the three main informational documents pertaining to on-site chemicals.

Important information about such materials is contained in **material safety data sheets (MSDS),** which are multiple-page information sheets (Figure 1-14). The MSDS is issued by the manufacturer of the material; and it provides detailed information on hazardous materials, including dangerous ingredients, corrosiveness, reactivity, toxicity, fire and explosion data, health hazards, spill and leak procedures, and special precautions. Federal law requires that an MSDS be available for each hazardous material in the workplace. They are sometimes posted in the shop or available in the office. An employee must have access to all MSDS documents pertaining to his or her work area.

The MSDS often states recommended uses for the material and lists specific handling instructions and safety precautions that must be observed. Emergency treatments for accidental ingestion, inhalation, and eye and skin contact are given when applicable. Guidelines for cleaning up spills or responding to other emergencies are included. The Canadian equivalent to the MSDS is the **Workplace Hazardous Materials Information Sheet.**

The employer is responsible for obtaining all MSDS for the hazardous materials in the shop and for making this information available to all employees. The employer must also provide formal training on the safe handling of all hazardous materials and must update this training yearly.

Containers storing potentially hazardous materials must be properly labeled with regard to health, fire, reactivity, and handling hazards (Figure 1-15). The simplest way to ensure compliance is to keep materials in their original containers. If a chemical is moved into another container, it is the responsibility of the shop to see that the container is the proper type and is correctly labeled. Do not use materials in unmarked containers. They may not be what they appear to be, or they may be contaminated.

Every employer also must maintain documentation on all hazardous materials used in the shop. The employer must provide proof of training programs, keep records of all accidents or spills, and satisfy all employee requests to review MSDS. Even if a hazardous material is phased out of use, the MSDS must be kept on file for 30 years. OSHA and other regulatory agencies are quite serious when it comes to employee safety and hazardous materials. You should be too.

During the workday, a technician may use any number of materials that can be hazardous. For example, there are solvents, brake cleaners, and brake fluids. The containers for these and all other hazardous materials must have a label that should be read before using them (see Figure 1-15).

Figure 1-16 shows a typical label. The label must identify the hazardous chemicals in the product and tell what the specific hazards are. For example, the label would tell the technician that the material might be poisonous or flammable and list what precautions should be taken. There might be a warning to wear eye protection or to use the material in a well-ventilated area. First-aid information is also provided on the label.

Unlabeled materials can be very dangerous. Many people have been injured when they did not know what was in a container. There may be times when you put material from a labeled container into another container. Always make a label for the new container that describes the contents. You may not be the only person to use the container or material.

MATERIAL SAFETY DATA SHEET

CRC Industries, Inc. • 885 Louis Drive • Warminster, PA 18974 • (215) 674-4300

PRODUCT NAME CLEAN-R-CARB (AEROSOL)   #-MSDS05079
PRODUCT- 5079,5079T,5081,5081T
(Page 1 of 2)

1. INGREDIENTS

	CAS #	ACGIH TLV	OSHA PEL	OTHER LIMITS	%
Acetone	67-64-1	750 ppm	750 ppm		2-5
Xylene	1330-20-7	100 ppm	100 ppm		68-75
2-Butoxy Ethanol	111-76-2	25 ppm	25 ppm	(skin)	3-5
Methanol	67-56-1	200 ppm	200 ppm		3-5
Detergent	-	NA	NA		0-1
Propane	74-98-6	NA	1000 ppm		10-20
Isobutane	75-28-5	NA	NA	1000ppm	10-20

2. PHYSICAL DATA : (without propellent)

Specific Gravity : 0.865   Vapor Pressure   : ND
                                  % Volatile   : > 99
Boiling Point   : 176°F initial   Evaporation Rate : Moderately fast
Freezing Point : ND   Vapor Density   : ND
Appearance and Odor: pH: NA
    A clear colorless liquid, aromatic odor

Solubility   :   Partially soluble in water.

3. FIRE AND EXPLOSION DATA

Flashpoint : -40 F Method : TCC
Flammable Limits   : propellent   LEL:1.8   UEL:9.5
Extinguishing Media : CO2, dry chemical, foam
Unusual Hazards   : Aerosol cans may explode when heated above 120°F.

4. REACTIVITY AND STABILITY

Stability   : Stable
Hazardous decomposition products
                    : C_{O2}, carbon monoxide (thermal)

Materials to avoid : Strong oxidizing agents and sources of ignition.

5. PROTECTION INFORMATION

Ventilation :   Use mechanical means to insure vapor conc. is
                    below TLV.

Respiratory :   Use self-contained breathing apparatus above TLV.

    Gloves : Solvent resistant   Eye & Face : Safety glasses
Other Protective Equipment: Not normally required for aerosol product usage.

Figure 1-14 The MSDS for any chemical lists physical and chemical properties and all necessary safety information. (Courtesy of CRC Industries, Inc.)

Many of the waste materials from shop use are also considered hazardous (Figure 1-17). Dirty solvent, used engine coolant, used batteries, used engine oil, and vacuum cleaner bags with brake dust are just a few examples of shop hazardous waste. Never throw these materials in the trash or pour them down a drain. They could end up in a place where they could injure

Figure 1-15 Chemical storage cabinets must be labeled as to contents and fire hazards.

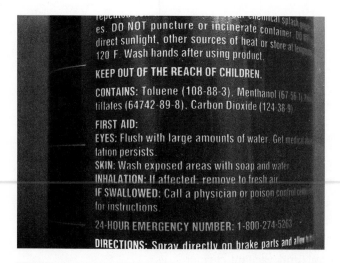

repeated ... chemical splash
es. **DO NOT** puncture or incinerate container. DO ...
direct sunlight, other sources of heat or store at ...
120 F. Wash hands after using product.

KEEP OUT OF THE REACH OF CHILDREN.

CONTAINS: Toluene (108-88-3), Menthanol (67-56-1) ...
tillates (64742-89-8), Carbon Dioxide (124-38-9)

FIRST AID:
EYES: Flush with large amounts of water. Get medical ...
tation persists.
SKIN: Wash exposed areas with soap and water.
INHALATION: If affected, remove to fresh air
IF SWALLOWED: Call a physician or poison control cen...
for instructions.

24-HOUR EMERGENCY NUMBER: 1-800-274-5263

DIRECTIONS: Spray directly on brake parts and allow ...

Figure 1-16 The label on a can of brake fluid lists hazards, warnings, and first-aid information.

Oily Rags

Figure 1-17 Hazardous waste materials must be stored in clearly labeled safety containers until they can be disposed of properly.

someone. Federal laws regulate how hazardous waste materials should be handled. Automotive shops usually have contracts with companies that pick up these materials and dispose of them properly.

Handling of Hazardous Waste

When the shop is finished using a hazardous material, it becomes hazardous waste. The EPA defines hazardous waste as solid or liquid materials that have one or more of the following characteristics:

- ❏ *Ignitability.* This characteristic applies to liquids with flash points below 140°F or solids that can spontaneously ignite.
- ❏ *Corrosivity.* Materials that dissolve metals or other materials or burn the skin on contact are considered corrosive.
- ❏ *Reactivity.* Reactive materials include those that react violently with water or other materials. They may release cyanide gas, hydrogen sulfide gas, or similar gases when exposed to low-pH acid solutions. They may also generate toxic or flammable vapors.
- ❏ ***Extraction Procedures* (EP)** toxicity. Materials that leach one or more heavy metals in concentrations greater than 100 times primary drinking water standard concentrations are considered toxic.

A complete list of hazardous wastes may be found at the EPA or CCOHS Web sites. When handling any hazardous waste material, always wear the safety equipment specified in the MSDS. In many cases, this includes full eye protection, chemical-resistant gloves, and a respirator (Figure 1-18).

Figure 1-18 Wear proper safety equipment when handling hazardous materials such as cleaning solvents.

Cleaning Equipment Safety

Parts cleaning is an important part of any brake repair job. Be careful when using solvents. Most are toxic, caustic, and flammable. Avoid placing your hands in solvent; wear protective gloves, if necessary. Read all manufacturer's precautions and instructions and material safety data sheets (MSDS) before using.

Do not use gasoline to clean components. This practice is very dangerous. Gasoline vaporizes at such a rate that it can form a flammable mixture with air at temperatures as low as −50°F. Gasoline also is dangerous if it gets on your skin because the chemicals in gasoline can be absorbed through the skin and get into your body.

Small cleaning jobs are often done with aerosol cleaners. These spray cans contain chemicals that break down dirt and grease and allow them to be removed. Do not throw empty spray containers in the trash without punching a hole in the side. Always read the warnings on the can and follow them. Wear eye protection, proper gloves, and a shop coat to prevent exposure to the skin or eyes. Always do the cleaning in a well-ventilated area.

Many of the solvents used in solvent cleaning tanks are flammable. Be careful to prevent an open flame around the solvent tank. Never mix solvents. One could vaporize and act as a fuse to ignite the others.

Wear neoprene gloves when washing parts. Some solvents can be absorbed through the skin and into your body. This is especially true if you have a cut on your hand. Do not blow compressed air on your hands if they get wet with solvent, as this can cause the solvent to go through your skin.

Wipe up spilled solvents promptly, and store all rags in closed, properly marked metal containers. Store all solvents either in their original containers or in approved, properly labeled containers. Finally, when using a commercial parts washer, be sure to close the lid when you are finished.

Antilock Brake Hydraulic Pressure Safety

Many ABSs generate extremely high brake fluid pressures that range from 2,000 to 3,000 psi. Failure to fully depressurize the hydraulic accumulator of an ABS before servicing any part of the system could cause severe personal injury from high-pressure brake fluid escaping from a service connection. Follow the exact shop manual procedure for the vehicle being serviced. A typical repressurizing procedure follows, with complete details in Chapter 10.

1. Disconnect the negative (−), or ground, battery cable (Figure 1-19A).
2. Be sure the ignition key is off (Figure 1-19B).

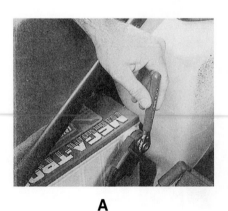
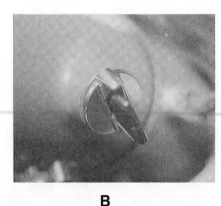
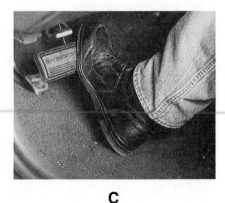

A   B   C

Figure 1-19 To relieve high ABS pressures, (A) disconnect the battery negative cable, (B) be sure the ignition is off, and (C) pump the brake pedal twenty-five to fifty times until you feel a definite increase in pedal pressure.

3. Pump the brake pedal at least 25 to 50 times, using about 50 pounds of pedal force (Figure 1-19C).

4. Continue pumping until you feel a definite increase in pedal pressure. Pump the pedal a few more times to ensure complete relief of hydraulic pressure from the system.

5. Proceed with system service.

Air Bag Safety

⚠️ **WARNING:** Most late-model cars and light trucks have supplemental inflatable restraint systems (SIRSs), known as air bags. To avoid accidental deployment of the air bag and possible injury or vehicle damage, always disconnect the battery ground (negative) cable, then the positive battery cable and wait a minimum of 10 minutes before working near any of the impact sensors, steering column, or instrument panel. Do not use any powered electrical test equipment on any of the air bag system wires or tamper with them in any way unless specifically directed by your instructor or supervisor. Do not use memory saver devices.

Most vehicles built since the early 1990s have a **supplemental inflatable restraint system (SIRS),** more commonly called an air bag. This system is designed to protect the driver and other passengers from injury in case of a collision. The system consists of an air bag module in the center of the steering wheel, another in the right side of the instrument panel, and possibly others in the side panels and headrests.

When working on brake system components under the instrument panel or near any of the air bag sensors or actuators, it is a good idea to deactivate the air bag system as described in the warning above. Exact procedures may vary from one vehicle to another, so consult the specific vehicle service manual for details.

Automotive manufacturers installed side and headrest air bags in the 2000 and later model vehicles. Some side bags are in the doors, whereas others are in the side of the seat backrest. They are protection during a side impact. The headrest bags are designed to reduce head and neck injuries during a collision from the rear. It is an accepted fact that SIRSs of this type can be dangerous to automotive and emergency technicians. The newest SIRSs are disarmed in a manner similar to that for driver and passenger bags. Always consult the service manual before beginning work in or around any SIRS components.

Fire Control

There are four general classifications of fires and a type of fire extinguisher to match the burning materials (Figure 1-20). Each class of fire is matched with a type of fire extinguisher containing the best material for controlling or extinguishing that fire. The automotive repair shop is normally in danger of fire from fuel, mostly gasoline, or from electrical fires. Electrical fires can sometimes be easily extinguished by disconnecting the battery, but do not go in harm's way trying to do this. Fuel fires will continue to burn as long as there is fuel. One thing not to use on fuel fires is spraying water. That will only spread the fuel and the fire. A Class B or a multiple-purpose fire extinguisher is the best tool for stopping a fuel fire. Most automotive shops have multiple-purpose-type extinguishers because they will work on different types of fire.

	Class of Fire	Typical Fuel Involved	Type of Extinguisher
Class **A** Fires (green)	**For Ordinary Combustibles** Put out a class A fire by lowering its temperature or by coating the burning combustibles.	Wood Paper Cloth Rubber Plastics Rubbish Upholstery	Water*[1] Foam* Multipurpose dry chemical[4]
Class **B** Fires (red)	**For Flammable Liquids** Put out a class B fire by smothering it. Use an extinguisher that gives a blanketing, flame-interrupting effect; cover the whole flaming liquid surface.	Gasoline Oil Grease Paint Lighter fluid	Foam* Carbon dixoide[5] Halogenated agent[6] Standard dry chemical[2] Purple K dry chemical[3] Multipurpose dry chemical[4]
Class **C** Fires (blue)	**For Electrical Equipment** Put out a class C fire by shutting off power as quickly as possible and by always using a nonconducting extinguishing agent to prevent electric shock.	Motors Appliances Wiring Fuse boxes Switchboards	Carbon dioxide[5] Halogenated agent[6] Standard dry chemical[2] Purple K dry chemical[3] Multipurpose dry chemical[4]
Class **D** Fires (yellow)	**For Combustible Metals** Put out a class D fire of metal chips, turnings, or shavings by smothering or coating with a specially designed extinguishing agent.	Aluminum Magnesium Potassium Sodium Titanium Zirconium	Dry powder extinguishers and agents only

*Cartridge-operated water, foam, and soda-acid types of extinguishers are no longer manufactured. These extinguishers should be removed from service when they become due for their next hydrostatic pressure test.

Notes:

(1) Freezes in low temperatures unless treated with antifreeze solution, usually weighs over 20 pounds, and is heavier than any other extinguisher mentioned.

(2) Also called ordinary or regular dry chemical (sodium bicarbonate).

(3) Has the greatest initial fire-stopping power of the extinguishers mentioned for class B fires. Be sure to clean residue immediately after using the extinguisher so sprayed surfaces will not be damaged (potassium bicarbonate).

(4) The only extinguishers that fight A, B, and C classes of fires. However, they should not be used on fires in liquefied fat or oil of appreciable depth. Be sure to clean residue immediately after using the extinguisher so sprayed surfaces will not be damaged (ammonium phosphates).

(5) Use with caution in unventilated, confined spaces.

(6) May cause injury to the operator if the extinguishing agent (a gas) or the gases produced when the agent is applied to a fire is inhaled.

Figure 1-20 Class B- and C-type fires present the greatest fire concern in an automotive shop. A multiple-purpose fire extinguisher will work on each type.

The first thing that should be done when a fire is discovered is to sound the alarm, then locate and remove the extinguisher from its mount. Using a fire extinguisher is fairly simple provided that the employer and employee have done their routine checks. Each fire extinguisher in the shop must have a tag where the date and time of inspection have been completed. This inspection is performed and the tag initialed each month. Usually the local fire marshal will conduct an annual inspection visit of each facility, and this is one of the things that will be checked. Before placing the fire extinguisher into action, check the small gauge near the handle. The needle should be in the green zone. If it is not, then the extinguisher is no longer charged and will not function.

Exercise extreme caution when fighting a fire. If at any point it appears out of control, immediately evacuate the building or area and allow the first responders to control the situation.

Figure 1-21 ASE-certified technicians may wear these shoulder patches.

Technician Training and Certifications

Technician training can start as early as the early teens, helping family or friends repair personal vehicles. In many ways, this is one of the best ways to start a career in automotive service. Back in the good old days, before electronics, a person who was known for working on his or her personal vehicle and keeping it operational could get a job in almost any automotive repair center. With the highly sophisticated vehicles of today, that backyard experience does not count for much with today's service managers. It is almost imperative that any person desiring to be an automotive service technician receive formal training. This training may start in high school and continue through a postsecondary technical school or college. A diploma or degree from a postsecondary school at least will get an applicant a job, but the training will not stop there. All dealerships and most independent shops will require additional training throughout the technician's career. Before selecting a postsecondary automotive program, check out its job placement program. A job placement in the 90th percentile level means the employers served by that program trust the content of the program and the instructors. Based on this trust won over many years, they will hire graduates or near-graduates, knowing that the new employees have the will and knowledge to be successful.

Another key to success, or at least to proving success, are certification programs for the technician. The most well known for automotive technicians is the Automotive Institute for Excellence (ASE), Figure 1-21. This is a nonprofit organization that conducts semiannual written tests on the eight system areas of the vehicle. More information can be gained through its Web site at http://www.asecert.org. ASE has a subagency named the National Automotive Technician Education Foundation (NATEF). NATEF certifies automotive training programs ranging from high school to postsecondary and manufacturer-specific schools. Another thing to consider during postsecondary program selection may be: Is it NATEF certified?

Terms to Know

Asbestos

Asbestosis

Canadian Center for Occupational Health and Safety (CCOHS)

Carbon monoxide

Chlorinated hydrocarbon solvents

Department of Transportation (DOT)

Environmental Canada

Environmental Protection Agency (EPA)

Extraction procedures (EP)

Federal Motor Vehicle Safety Standards (FMVSS)

Material safety data sheet (MSDS)

Occupational Safety and Health Administration (OSHA)

Phosgene

Supplemental inflatable restraint system (SIRS)

Tetrachloroethylene

Trichloroethane

Trichloroethylene

Workplace Hazardous Materials Information Sheet

ASE-Style Review Questions

1. Which of the following items is the LEAST preferred method to clean the floor of oil or dust?
   A. using a straw broom
   B. wiping with a rag
   C. using a vacuum cleaner
   D. using a damp mop

2. *Technician A* says that the Workplace Hazardous Materials Information Sheet is published by the Canadian government. *Technician B* says that the MSDS is published by the chemical manufacturer. Who is correct?
   A. A only
   B. B only
   C. Both A and B
   D. Neither A nor B

3. Wearing jewelry around the shop is being discussed:
   Technician A says that jewelry can become dangerous when working on electrical systems.
   Technician B says that jewelry is dangerous when working around moving machinery.
   Who is correct?
   A. A only
   B. B only
   C. Both A and B
   D. Neither A nor B

4. All of the following are usually part of the firefighting procedures in an automotive shop EXCEPT
   A. selecting a Class B or multiple-purpose fire extinguisher.
   B. using a fire extinguisher with the gauge needle in the green.
   C. moving the spray back and forth at the base of the fire.
   D. remaining about 6 to 8 feet downwind of the fire.

5. An air hose is found to have a bulge in its outer lining. This means
   A. the hose is in danger of blowing out.
   B. the hose is good to use if there is no air escaping.
   C. the hose is no good and should be disconnected from the compressed air source.
   D. both A and C.

6. Trichloroethylene, 1,1,1-trichloroethane, and tetrachloroethylene are classified as what type of solvents?
   A. emulsifying soaps
   B. chlorinated hydrocarbons
   C. fluorocarbons
   D. polyglycol solvents

7. A brake system is being disassembled and cleaned for service:
   Technician A says to use a full-enclosure vacuum unit to keep dust out of the air. *Technician B* says to wear a breathing respirator for personal safety. Who is correct?
   A. A only
   B. B only
   C. Both A and B
   D. Neither A nor B

8. Chemical poisoning can occur by any of the following EXCEPT
   A. inhalation.
   B. refraction.
   C. absorption.
   D. ingestion.

9. Legal responsibility for safe operation of a brake system on a vehicle in use rests with the
   A. vehicle owner.
   B. service technician.
   C. vehicle manufacturer.
   D. federal government.

10. *Technician A* says that skin contact with brake fluid is not harmful. *Technician B* says that brake fluid can harm the vehicle's finish. Who is correct?
    A. A only
    B. B only
    C. Both A and B
    D. Neither A nor B

Brake Service Tools and Equipment

Upon completion and review of this chapter, you should be able to:

❏ List the basic units of measure for length and volume in the metric and U.S. customary systems.

❏ Identify and describe the purpose and use of hand tools commonly found in a service technician's toolbox.

❏ Identify and describe the purpose and use of special tools used for brake service.

❏ Identify and describe the purpose and use of commonly used power tools.

❏ Identify and use the major measuring tools and instruments used in brake service work.

❏ Identify and use the electrical test tools used in ABS diagnostic work.

❏ Explain how to measure with both an inside and outside micrometer.

❏ Describe asbestos containment equipment commonly used in the shop.

Fasteners

Although not truly tools, nuts and bolts and other fasteners are part of everyday automotive work; and you must understand their characteristics and grading system. Figure 2-1 identifies U.S. customary and metric fastener dimensions and terminology. Figure 2-2 shows the Society of Automotive Engineers (SAE) grade and strength markings for bolts used in automotive and other industrial assemblies. Metric fastener strength ratings are indicated by numbers embossed in the head of the bolt or nut. The most common metric fastener grades for automotive use are 9.8 and 10.9. Service manuals also list important fastener information. Using an incorrect fastener or a fastener of poor quality can result in dangerous failures and personal injury.

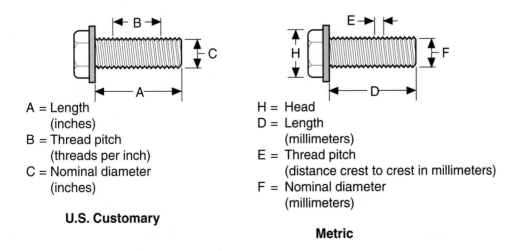

A = Length
    (inches)
B = Thread pitch
    (threads per inch)
C = Nominal diameter
    (inches)

U.S. Customary

H = Head
D = Length
    (millimeters)
E = Thread pitch
    (distance crest to crest in millimeters)
F = Nominal diameter
    (millimeters)

Metric

Figure 2-1 The basic dimensions of an automotive fastener are length, diameter, and thread.

SAE Grade Markings	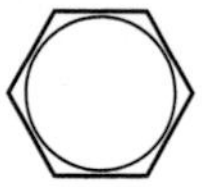			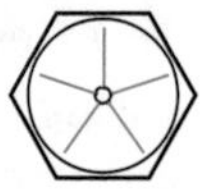	
DEFINITION	No lines: unmarked indeterminate quality SAE grades 0-1-2	3 Lines: common commercial quality automotive and AN bolts SAE grade 5	4 Lines; medium commercial quality automotive and AN bolts SAE grade 6	5 Lines: rarely used SAE grade 7	6 Lines: best commercial quality NAS and aircraft screws SAE grade 8
MATERIAL	Low carbon steel	Medium carbon steel, tempered	Medium carbon steel, quenched and tempered	Medium carbon alloy steel	Medium carbon alloy steel, quenched and tempered
TENSILE STRENGTH	65,000 psi	120,000 psi	140,000 psi	140,000 psi	150,000 psi

Figure 2-2 Lines on bolt heads indicate the relative strength of SAE inch-sized fasteners.

CAUTION: Some U.S. customary and metric fasteners may appear to be exactly the same and to require the same wrench. An example is a fastener requiring a ½-inch wrench and a fastener requiring a 13-mm wrench. Although the wrench may appear to fit, it does not fit tight enough to prevent damage to the fastener's head.

Measuring Systems

International System of Units or metric system is the modern international metric system used by the automotive industry and other industries.

Two different measurement systems are currently used in the United States and Canada: the U.S. customary system and the **International System of Units or metric system.** Note the speedometer markings on your personal vehicle. They are in miles per hour and kilometers per hour. The dual markings are part of an ongoing education of the American public to accept and use the international standard of measurement. Domestic auto manufacturers adopted the metric system for many fasteners and specifications about 25 years ago, but many component dimensions and specifications, as well as fastener sizes, are still based on the customary inch system. Most vehicles imported from Europe and Asia are built to metric standards, so the imported vehicles will require metric wrenches and sockets. Fortunately, vehicle specifications and tightening torque are normally listed in both metric and customary units, so tools such as micrometers and torque wrenches can be based on either system.

U.S. Customary System

Linear means in a straight line.

In the U.S. customary system, the basic unit of **linear** measurement is the inch. The inch can be divided into fractions, such as those used to designate wrench or socket sizes (¼, 5⁄16, ½, 9⁄16, and so on).

For component size and tolerance measurements, the inch is commonly divided into tenths, hundredths, thousandths, and even ten-thousandths. When an inch is divided in this way, the measurement is written with a decimal point. A single digit to the right of the decimal point indicates tenths of an inch (0.1 in.). Tenths of an inch can be further divided by ten into hundredths of an inch, which is written using two digits to the right of the decimal point (0.01 in.). The division after hundredths is thousandths (0.001 in.), followed by ten-thousandths (0.0001 in.).

The following paragraphs summarize metric units of linear measurement, as well as metric volume, weight, pressure, torque, and temperature measurement units that you may encounter in brake service.

Metric System

The metric system has a basic unit for every kind of measurement. The most common metric units used in automotive service are:

❏ Linear measurement: the meter (m)
❏ Volume measurement: the cubic meter (m^3)
❏ Weight: the kilogram (kg)
❏ Pressure: the pascal (Pa)
❏ Torque: the newton-meter (Nm), or the older units of kilogram-meters or kilogram-centimeters (kg-m or kg-cm)
❏ Temperature: degrees Celsius (Technically, the basis for metric temperature units is the kelvin, but it is not used for everyday measurements.)

The basic metric units are called **stem units.** For large measurements, any metric stem unit is multiplied one or more times by 10. For small measurements, any stem unit is divided one or more times by 10. Units larger than the stem units are called **multiples;** units smaller than the stem units are called **submultiples.** Multiples and submultiples are indicated by prefixes written or abbreviated in front of the stem units. The most common metric prefixes for automotive measurements are listed in Table 2-1:

A **stem unit** is any metric unit to which a prefix can be added to indicate larger or smaller measurements to some power of 10.

Multiple refers to the metric measurement unit that is larger than the stem unit through multiplying by a power of 10.

Submultiple is the metric measurement unit that is smaller than the stem unit through dividing by a power of 10.

Table 2-1 METRIC PREFIXES FOR COMMON AUTOMOTIVE MEASUREMENTS

MULTIPLE	PREFIX	ABBREVIATION
× 1,000	kilo	k
× 1,000,000	mega	M

SUBMULTIPLE	PREFIX	ABBREVIATION
× 0.01	centi	c
× 0.001	milli	m
× 0.000001	micro	μ

CAUTION: Never use a metric wrench or socket on a U.S. customary bolt or nut or an inch-sized wrench on a metric fastener. The wrench or socket will always be slightly oversized and will slip off or damage the fastener.

Linear. The metric base unit for linear measurement is the meter, which is about 39 inches (Figure 2-3). For distance measurement on the road, the common unit is the kilometer (1,000),

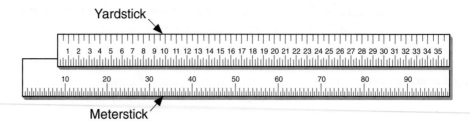

Yardstick

Meterstick

Figure 2-3 A meter is approximately 39⅜ inches.

which is abbreviated km. For smaller measurements, the meter is divided by 10 into submultiples. The most common measurements for automotive work are:

❑ one-hundredth of a meter (0.01 m), or a centimeter (cm)
❑ one-thousandth of a meter (0.001 m), or a millimeter (mm)

The millimeter is the most common metric measurement used in automotive work. Metric wrenches are sized by millimeter: for example, 10 mm, 12 mm, and so on. The exact conversion factors for kilometers, centimeters, and millimeters into customary units are:

❑ kilometers (km) × 0.621377 = miles
❑ kilometers (km) = 1.6093 × miles
❑ centimeters (cm) × 0.3937 = inches
❑ centimeters (cm) = 2.540 × inches
❑ millimeters (mm) × 0.03937 = inches
❑ millimeters (mm) = 25.400 × inches

The volume or displacement of an engine may be given as cubic inches (ci), cubic centimeters (cc), or liters. The most common are liters and cubic inches.

Volume. Although the cubic meter (m³) is the base metric unit for volume, we are more accustomed to the common unit used for fluid volume: the liter. The liter is not an official metric unit, but it is recognized as a volume of 1 cubic decimeter (¹⁄₁₀ m³) for liquids. The liter also is often divided into submultiples of one-thousandth, or cubic centimeters. The official abbreviation for the cubic centimeters is cm³, but the auto industry has used "cc" for so long that either is acceptable. The cubic centimeter also equals 1 milliliter (ml), which you may see occasionally in fluid measurements. The common conversion factors for liters into quarts and gallons (Figure 2-4) are:

❑ liter (l) × 0.946 = quart
❑ liter (l) = 1.057 × quart
❑ liter (l) × 0.26418 = gallon
❑ liter (l) = 3.7854 × gallon

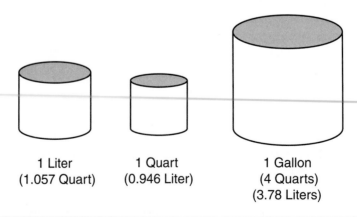

1 Liter
(1.057 Quart)

1 Quart
(0.946 Liter)

1 Gallon
(4 Quarts)
(3.78 Liters)

Figure 2-4 A liter is slightly larger in volume than a U.S. quart.

Weight. The metric base unit for weight is the kilogram (kg), but the kilogram itself is a multiple of a stem unit: the gram. A gram is a very small unit, so the kilogram—1,000 grams—is common in automotive work. The exact conversion factors for kilograms into pounds are:

❑ kilogram (kg) × 2.2046 = pounds
❑ kilogram (kg) = 0.4536 × pounds

Pressure. Many different units are used for pressure measurement, but the customary pressure unit most used in automotive service is pounds-per-square-inch (psi). Older metric specifications usually also were written in units of **force** divided by area, such as kilograms-per-square-centimeter (kg-cm^2). The SI metric pressure unit, however, is the pascal (Pa). The basic pascal unit is very small, so pressure is usually expressed in thousands of pascal, or kilopascals (kPa). The conversion factors are:

❑ kilopascals (kPa) × 0.145 = psi
❑ kilopascals (kPa) = 6.895 × psi

Torque. Customary units for torque measurement are compound units of force times distance: foot-pounds and inch-pounds. Since 1971, the metric base unit has been the newton-meter (Nm), but you may see older metric specifications given in kilogram-meters (kg-m) or kilogram-centimeters (kg-cm). The conversion factors for newton-meters to foot-pounds and inch-pounds are:

❑ newton-meter (Nm) × 0.737 = foot-pounds
❑ newton-meter (Nm) = 1.356 × foot-pounds
❑ newton-meter (Nm) × 0.089 = inch-pounds
❑ newton-meter (Nm) = 0.113 × inch-pounds

Temperature. The zero-point of the metric Celsius temperature scale is the freezing point of water. The boiling point of water is 100 degrees Celsius (100°C). In the customary Fahrenheit scale, the freezing and boiling points of water are 32°F and 212°F, respectively. Therefore, the conversion factors require both subtraction and multiplication or division:

❑ Celsius (°C) = °F − 32 × 5/9, or
❑ Celsius (°C) = °F − 32 ÷ 1.8
❑ Fahrenheit (°F) = (1.8 × °C) + 32

Force is the power working against resistance to cause motion.

The measurement, pascal, is derived from Blaise Pascal, who researched and formulated the theories of hydraulics.

Most torque wrenches used in the United States are calibrated in foot-pounds or inch-pounds. Many have U.S. customary and metric markings.

Measuring Tools

Many of the procedures for brake service require exact measurements of parts and clearances. Accurate measurements require precision measuring tools capable of measuring to the thousandth of an inch and smaller. For example, the acceptable lateral run out of a brake rotor may be as little as 0.004 inch. The minimum acceptable thickness variation may be as small as 0.0005 inch. Tools for measuring such small increments are delicate instruments and should be handled with great care. Never strike, pry, drop, or force these tools. Clean them before and after every use.

As with all tools, measuring tools should be used only for the purposes for which they were designed. Some instruments are not accurate enough for very precise measurements; others are too accurate to be practical for less critical measurements.

All measuring tools should be calibrated or checked periodically against known-good equipment or tool standards. This ensures that they work properly and give accurate measurements.

Many different measuring devices are used by automotive technicians. The following sections cover those commonly used to service brake systems.

Steel Ruler or Tape Measure

A steel ruler or tape measure is needed to measure brake pedal travel. Although a drum micrometer (described next) is necessary to accurately measure drum diameter, a ruler or tape measure can be used for a quick check.

Micrometers

Micrometers are used to measure the outside diameter of an object or the inside diameter of a bore or drum. An outside micrometer has a fixed anvil and a movable spindle in a C-shaped frame (Figure 2-5). Figure 2-5 also shows an inside micrometer such as you would use to measure a cylinder bore. Measurement gradations are on the fixed sleeve or body and on the movable thimble.

Any micrometer has an accurately ground screw thread that is rotated in a fixed nut to change the distance between two measuring surfaces. Turning the thimble of an outside micrometer moves the spindle in and out toward the anvil. The screw thread of a decimal-inch micrometer has 40 threads per inch. One revolution of the thimble moves the spindle $1/40$, or 0.025 inch. Similarly the lines on the fixed sleeve correspond to the micrometer screw threads. Each line represents $1/40$, or 0.025 inch. Every fourth line is numbered to indicate hundreds of thousandths (Figure 2-6). Line 1 equals 0.100; line 2 equals 0.200; line 3 equals 0.300, and so on. The beveled edge of the movable thimble has twenty-five lines, each representing 0.001. Turning the thimble from one line to another moves the spindle in or out 0.001 inch.

To measure a small object with an outside micrometer, open the jaws and slip the object between the spindle and the anvil. Hold the object against the anvil and lightly turn the thimble so that the object just fits between the spindle and the anvil (Figure 2-7). Do not overtighten the thimble. The object should slide between spindle and anvil with very slight resistance.

To read a micrometer graduated in thousandths of an inch, multiply the number of vertical divisions on the spindle by 0.025 and then add the number of thousandths shown by the line on the thimble that matches with the horizontal line on the spindle. Figure 2-8 shows the following measurement:

Line 1 on the spindle equals 0.100.

Three more lines on the spindle equal 3 × 0.025 or 0.075.

Line 3 on the thimble coincides with the horizontal line on the spindle, which equals
3 × 0.001 or 0.003.

The micrometer reading is 0.178.

Some micrometers have even finer gradations on the sleeve to give readings in ten-thousandths of an inch. These gradations are ten division marks on the sleeve that occupy the same space as

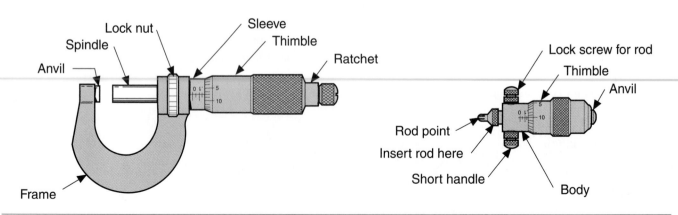

Figure 2-5 Major parts of an outside micrometer (left) and an inside micrometer (right).

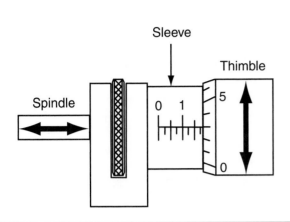

Figure 2-6 Readings on a micrometer graduated in thousandths of an inch.

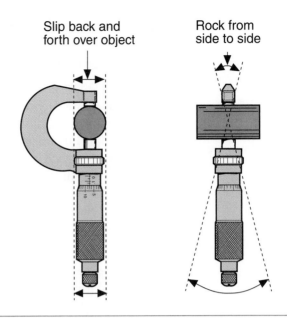

Figure 2-7 Slide the micrometer over the part being measured so that it drags just lightly. Rock the micrometer slightly and close the thimble gently.

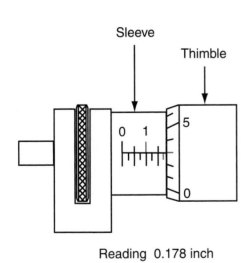

Reading 0.178 inch

Figure 2-8 Reading a micrometer graduated in thousandths of an inch (0.178 in.).

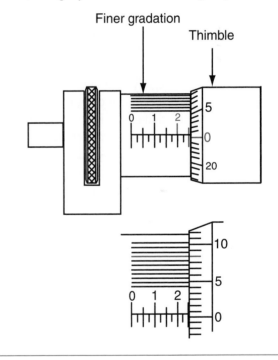

Figure 2-9 Readings on a micrometer graduated in ten-thousandths of an inch (0.2507 in.).

nine divisions on the thimble (Figure 2-9). The difference between one space on the fixed sleeve and one space on the movable thimble is 0.1 of a thimble division, or 0.0001.

To read this micrometer, first take the reading in thousandths and then see which line on the fine scale of the sleeve matches a line on the thimble. If it is line 1, add 0.0001 to the thimble reading; if it is line 2, add 0.0002, and so on. Figure 2-9 shows the following measurement:

Line 2 on the sleeve equals 0.200.

Two more lines on the spindle equal 2 × 0.025 or 0.050.

The longitudinal line on the sleeve is between the 0 and 1 on the thimble, indicating that a ten-thousandth measurement must be read from the vernier.

Line 7 on the vernier coincides with a line on the thimble: 7 × 0.0001 or 0.0007.

The micrometer reading is 0.2507.

Metric micrometers have a spindle screw thread pitch of 0.5 mm so that each full turn of the thimble moves the spindle 0.5 mm. The longitudinal line on the sleeve is graduated from 0 to 25 mm and marked in 1.0 mm and 0.5 mm increments (Figure 2-10). The beveled edge of the thimble has fifty divisions, and every fifth line is numbered. A full turn of the thimble moves the spindle 0.5 mm, so each thimble gradation equals $\frac{1}{50}$ of 0.5, or 0.01 mm.

To read a metric micrometer, add the mm reading from the sleeve to the reading in hundredths of a mm on the thimble. Figure 2-10 shows the following measurement:

Line 5 on the spindle equals 5.0 mm.

One more half-mm line on the spindle equals 0.05 mm.

Line 28 on the thimble is on the longitudinal line of the spindle (each line is 0.01 mm): 28 × 0.01 = 0.28 mm.

The micrometer reading is 5.78 mm.

Outside micrometers come in many different sizes, but the measurement range of the spindle is usually just 1 inch or 25 mm. That is, a 4-inch micrometer measures from 3 to 4 inches. Special micrometers are made for measuring brake rotor thickness. The throats on these micrometers are deeper than on standard micrometers to allow for the diameter of the rotor. The anvil and spindle of a rotor micrometer are usually pointed to allow measurements at the deepest points of any grooves in the rotor.

To measure a large object such as a brake rotor, slide the micrometer over the object and lightly tighten the thimble while continuing to slide the micrometer across the object. When you feel a light drag, rock the micrometer slightly to be sure it is contacting the object squarely. Then tighten the locknut (if equipped) and take the reading.

A micrometer is a precision instrument. Observe these precautions when using one:

❏ Be sure the anvil is clean before measuring any item.
❏ Always hold the part being measured squarely between the anvil and the spindle.
❏ Do not overtighten the thimble, which could damage the micrometer or the part being measured, or both.
❏ If the micrometer has a torque-limiting ratchet, use it to tighten the thimble.

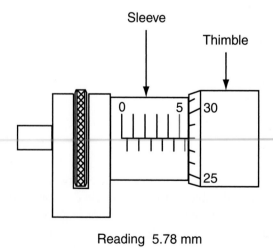

Reading 5.78 mm

Figure 2-10 Reading a micrometer graduated in hundredths of a millimeter (5.78 mm).

Depth Gauges

A depth gauge is used to check the lining thickness of drum brake shoes (Figure 2-11). A depth gauge is graduated and read like a micrometer. Both customary inch and metric models are available.

Drum Micrometer

A drum micrometer is a single-purpose instrument, used to measure the inside diameter of a brake drum (Figure 2-12). A drum micrometer has two movable arms on a shaft. One arm has a precision dial indicator; the other arm has an outside anvil that fits against the inside of the drum. In use, the arms are secured on the shaft by lock screws that fit into grooves every $\frac{1}{8}$ inch (0.125 in.) on the shaft. The dial indicator is graduated in 0.005-inch increments.

Some of the older drum micrometers can be difficult to set up and read. The newest drum micrometers available are electronic digital types.

Figure 2-11 Use a micrometer depth gauge to measure lining thickness precisely.

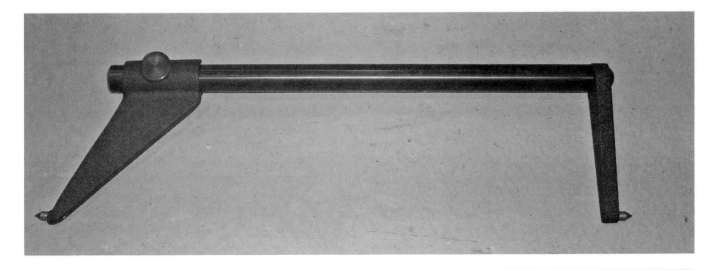

Figure 2-12 An electronic disc rotor micrometer is easy to use, is accurate, and can, at the press of a button, convert from USC to metric or vice versa.

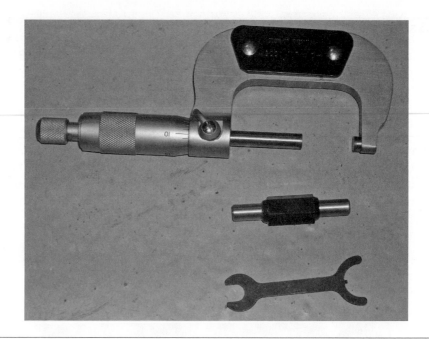

Figure 2-13 The tool in the center is a micrometer standard used to check the accuracy of outside micrometers and dial calipers.

To use a manual drum micrometer, loosen the lock screws and move the arms along the shaft so that the micrometer can fit inside the drum. Then extend the arms on the shaft until they align with the increments that indicate the nominal size of the drum. To measure a 10.5-inch drum, set one arm at the 10-inch mark on the shaft and the other arm at the other 10-inch mark, plus four 0.125-inch marks (0.500 in.) for the 10.5-inch drum. Then place the micrometer inside the drum and rock it gently on the drum surface until the highest reading is obtained. If the dial indicator reads 0.019 inch, for example, add this reading to the 10.5-inch measurement for a total diameter of 10.519 inches. Always take six to eight measurements around the drum circumference to check for drum distortion.

SERVICE TIP: Handle precision measuring tools gently and store them properly. Micrometers should be checked for accuracy before use and at specified intervals. Standards are used to measure the tool's accuracy (Figure 2-13).

Metric drum micrometers work the same way except that the shaft is graduated in 1 cm major increments, and the lock screws fit in notches every 2 mm.

Electronic drum micrometers are available at a reasonable cost. They are much easier to use and are extremely accurate.

Vernier Calipers

Vernier calipers (Figure 2-14) can take both inside and outside measurements, and vernier depth and height gauges are made with scales for measurements to 0.001 inch or 0.02 mm.

On a twenty-five-division vernier caliper that measures thousandths of an inch, the bar is graduated in forty units of 0.025 inch. Each fourth gradation shows tenths of an inch. The twenty-five divisions on the movable vernier plate occupy the same space as twenty-four divisions on the bar. Since one division on the bar equals 0.025 inch, twenty-four divisions equal 0.600 inch (24 × 0.025 = 0.600). The twenty-five divisions on the movable plate also equal 0.600 inch, so each division equals ¹⁄₂₅ × 0.600, or 0.024 inch. Therefore, the difference between each mark on the bar and on the plate is 0.001 inch.

Figure 2-14 The vernier caliper can make inside and outside measurements.

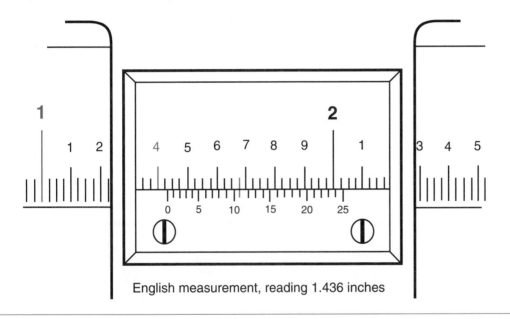

English measurement, reading 1.436 inches

Figure 2-15 Typical decimal-inch measurement on a twenty-five-division vernier scale.

To read a twenty-five-division **vernier scale,** count the number of inches, tenths (0.100), and fortieths (0.025) that the 0 mark on the plate is from the 0 mark on the bar. Then add the number of thousandths shown by the line on the plate that exactly matches a line on the bar. The example in Figure 2-15 shows a reading of 1.000 inch, plus 0.400, plus 0.025, or 1.425 inches. The eleventh line on the plate matches a line on the bar, so add 0.011 to 1.425 for an exact reading of 1.436 inches.

On a twenty-five-division metric vernier, the gradations on the movable plate are ⅕₀, or 0.02 mm. The bar is graduated in centimeters, millimeters, and one-half (0.5) millimeters. The example in Figure 2-16 shows a reading of 4.00 cm (40.0 mm), plus 1.00 mm, plus 0.50 mm, or 41.50 mm.

Dial Indicators

You can use various kinds of dial indicators to measure hole depth, surface smoothness, and out-of-round and runout of cylinders and rotating parts. You will usually use them to check

A **vernier scale** is a fine auxiliary scale that indicates fractional parts of a larger scale.

Dial indicators are the most common measurement device used with automotive disc brakes.

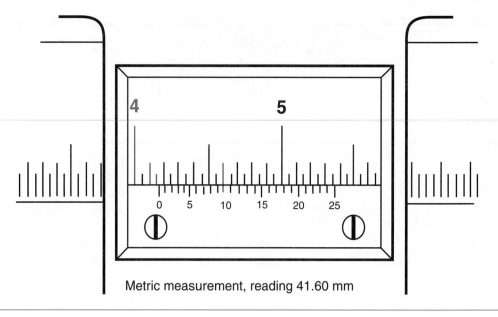

Metric measurement, reading 41.60 mm

Figure 2-16 Typical metric measurement on a twenty-five-division vernier scale.

measurement variations rather than single linear measurements. In brake service, the most common use of a dial indicator is to measure the runout in a disc brake rotor (Figure 2-17).

Dial indicators have a plunger and a precision gear mechanism to turn a pointer on a dial. Most are graduated in thousandths (0.001) or ten-thousandths (0.0001) of an inch, or hundredths (0.01) or thousandths (0.001) of a millimeter. Continuous dial indicators read in one direction from the zero point (Figure 2-18). Balanced dial indicators read in both directions, plus and minus, from zero (Figure 2-18). Dial indicators come with various mounting fixtures to hold them for different jobs. On most indicators, the dial is turned to align the 0 point with the pointer and then locked in place before starting a measurement. Pointer movement then indicates the variation from zero.

Figure 2-17 Use a dial indicator to measure rotor runout.

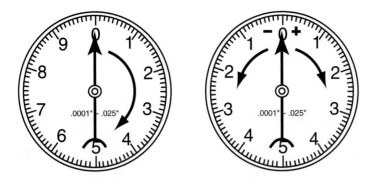

Figure 2-18 Dial indicators are either continuous reading (left) or balanced reading (right).

Feeler Gauges

A feeler gauge is a thin strip of metal of a known thickness. A feeler gauge set is a collection of these strips, each with a different thickness (Figure 2-19). A steel feeler gauge set usually contains strips of 0.002-inch to 0.010-inch thicknesses (in increments of 0.001 inch) and strips of 0.012-inch to 0.024-inch thicknesses (in increments of 0.002 inch). Feeler gauges are used to check piston-to-wheel cylinder bore, drum to brake shoe, and other types of small clearances down to the thousandths of an inch. Metric feeler gauges are usually graduated in increments of 0.05 mm.

Brake Shoe Adjusting Gauge (Calipers)

A brake shoe adjusting gauge is an inside-outside measuring device (Figure 2-20). This gauge is often called a brake shoe caliper. During drum brake service, the inside part of the gauge is placed inside a newly surfaced drum and expanded to fit the drum diameter. The lock screw is then tightened and the gauge moved to the brake shoes installed on the backing plate. The brake shoes are then adjusted until the outside part of the gauge just slips over them. This provides a rough adjustment of the brake shoes. Final adjustment must still be done after the drum is installed, but the brake shoe gauge makes the job faster.

The brake shoe adjusting gauge is particularly helpful on the brake systems that require the drum to be removed for brake adjustment.

Pressure Tools

The brake system can be diagnosed using pressure testers or gauges. This is especially true for ABSs in which a failure in one valve, line, or other hydraulic can result in a diagnostic trouble code (DTC) being set. The fault may be traced down by using brake pressure gauges tapped into various places and comparing the pressure generated against the specifications or against like components such as one wheel brake against another. Figure 2-21 shows a typical kit sold as

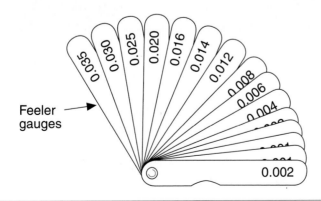

Figure 2-19 Feeler gauges are available in both decimal-inch and metric gradations.

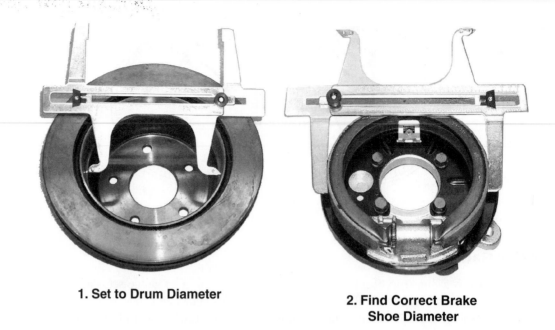

1. Set to Drum Diameter

2. Find Correct Brake Shoe Diameter

Figure 2-20 A brake shoe adjusting gauge is used for preliminary shoe adjustment. In this illustration, it is the only adjustment for the GM truck parking brake.

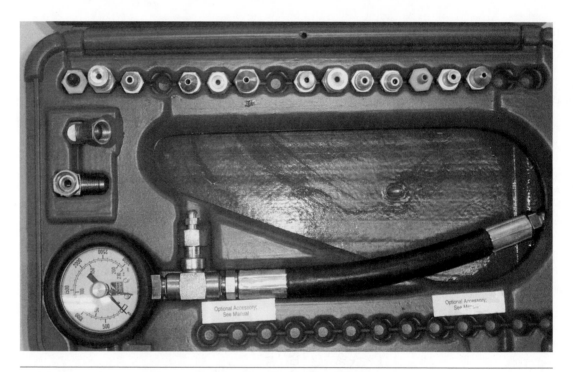

Figure 2-21 Typical kit sold as a master cylinder pressure gauge set.

a master cylinder pressure gauge set. It can be used in areas of the system other than the master cylinder if a very high pressure gauge is included with the kit and the appropriate adapter is available. This type of kit may be most valuable on comeback jobs in which the problem is hard to trace. Connecting the gauge(s) into a master cylinder outlet immediately determines if the cylinder is generating the same pressure by each piston. The same connection can isolate a bad proportioning valve built into or directly attached to the master cylinder. Moving the gauge(s) to a frame-mounted proportioning valve can help determine if that valve is functioning properly.

Locating the connection at the caliper end of the two front brake hoses can help the technician diagnose the proper application or release of the fluid to each wheel. If the pressure in one wheel fails to increase with the matching wheel, then there is probably a blockage in the hose. A blockage may also be identified if the two pressures do not decrease the same amount when the brakes are released.

> ☑ **SERVICE TIP:** A noncontact pyrometer is an infrared thermometer that you simply aim at an object to read its temperature. You can use a pyrometer to check brake temperatures on a car with a brake-pull problem. Drive the car and stop it with moderately hard braking effort and immediately measure drum or rotor temperature from front to rear or side to side. If rotor or drum temperatures differ significantly, caliper pistons may be sticking or a problem may exist with the pads or shoes. Temperature changes are a result of pressure and force applied.

> ⚠ **WARNING:** Ensure that the connection of the gauge to the system is tight and leak-free. Wear safety glasses when using pressure gauges for diagnosing and testing. Apply just enough force to the brake pedal to charge the system and check for leaks. If leak-free, apply normal pedal pressure. When possible, do not get in the same plane as the gauge/system connection. Brake systems generate high, dangerous pressures and leaking fluid could be injected into the flesh or eye over a distance of several feet.

Selection, Storage, and Care of Tools

Brake system service requires a wide variety of tools. Many of these tools are the common hand and power tools used in all types of automotive service. Other tools are more specialized and are used only for specific brake system testing and repair. Measuring tools are particularly important in brake service work. Brake service involves the precision measurement of rotors, brake drums, and other components where measurements as small as one ten-thousandth of an inch (0.0001 in.) can determine the serviceability of a part.

To work quickly, safely, and efficiently, you must have the right tools for the job. Professional service technicians use quality tools and keep them clean, organized, and within reach. Any tool should feel comfortable and balanced in your hand, and the tool's finish should be smooth for easy cleaning.

A roll cabinet (Figure 2-22), tool chest, and smaller tote tray are standard items used to store and carry your tool set. Store cutting tools such as files, chisels, and drills in separate drawers to avoid damaging the cutting edges. Keep the most frequently used tools at hand, and keep tool sets such as wrenches, sockets, and drill bits together. Leaving a tool on a workbench is the first step in losing it. Make it a habit to return frequently used tools, such as screwdrivers, to your tool chest or tote tray after each use. Keep delicate measuring tools, such as micrometers, in their protective cases and store in a clean, dry area. After each use, clean the tool using a lightly oiled lint-free cloth.

The best method of tracking tools is the use of tool control equipment. The equipment may be a cheap socket rail and wrench holder to expensive full-blown versions sold by tool manufacturers with perfect cutouts for each tool in the box. Most technicians find something in the middle to control their tools.

Common Hand Tools

Both basic and specialized brake tools are discussed in the following paragraphs.

Wrenches

To work on late-model brake systems, you should have complete sets of both inch-sized and metric wrenches. Most older domestic cars are built with bolts and nuts that require inch-sized

Common tools are usually the technician's personal set. Many technicians are buying expensive diagnostic tools and equipment formerly purchased by the shop.

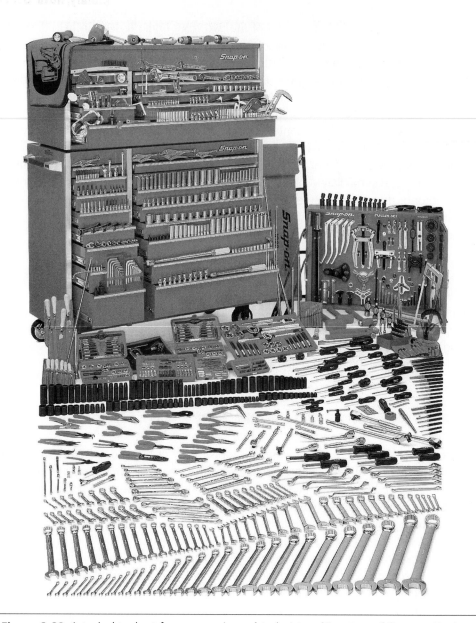

Figure 2-22 A typical tool set for an experienced technician. (Courtesy of Snap-on Tools Company)

wrenches. Inch-sized wrenches are most commonly sized in increments of $\frac{1}{16}$ of an inch. Metric wrenches are sized in increments of 1 mm.

Wrenches are either boxed or open ended. A box wrench (Figure 2-23) completely encircles a nut or bolt head. It is less likely to slip and cause damage or injury. But when clearances are very tight, it may not be possible to place a box wrench around the nut or bolt.

Open-end wrenches solve the problem of tight clearances. They have open, squared ends and grasp only two of the nut's four or six flats (Figure 2-24). Open-end wrenches are more likely to slip than box wrenches but may be the only alternative in a tight spot. An open-end wrench is normally the best tool for turning the nut down before final tightening or for holding a bolt head.

To reduce cost and storage space, you may want to have a set of combination wrenches that have an open-end wrench on one end and a box wrench on the other. Both ends are sized the same and can be used interchangeably on the same nut or bolt.

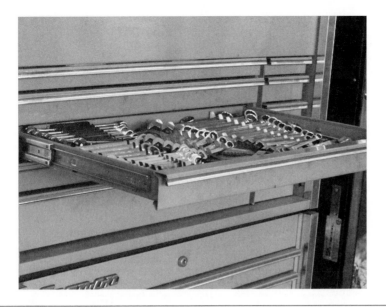

Figure 2-23 Assorted box-end wrenches.

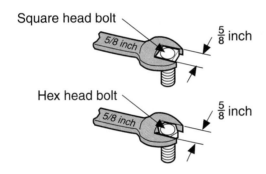

Figure 2-24 An open-end wrench grips only two flanks (sides) of a nut or bolt head.

Flare-Nut (Line) and Bleeder Screw Wrenches

Flare-nut wrenches (Figure 2-25) should be used to loosen or tighten brake line or tubing fittings. Using open-end wrenches on these fittings will tend to round the corners of the nut, which are typically made of soft metal and can distort easily. Flare-nut wrenches surround the nut and provide a better grip on the fitting. A section is cut out so that the wrench can be slipped around the brake line and dropped over the flare nut.

Special bleeder valve wrenches often are used to open bleeder screws (Figure 2-26). Bleeder valve wrenches are small, six-point box wrenches with strangely offset handles for access to bleeder screws in awkward locations. The six-point box end grips the screw more securely than a twelve-point box wrench can and avoids damage to the screw.

CAUTION: Do not use an open-end wrench on a bleeder screw. The awkward locations of many bleeder screws almost ensure that an open-end wrench will slip and round off the screw shoulders. Avoid using a twelve-point box wrench on a bleeder screw; use a six-point bleeder valve wrench whenever possible.

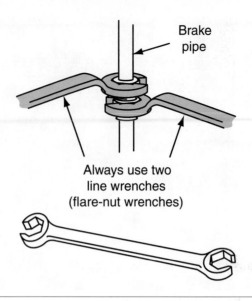

Brake
pipe

Always use two
line wrenches
(flare-nut wrenches)

Figure 2-25 Use flare-nut (line) wrenches, not open-end wrenches, to loosen tube fittings.

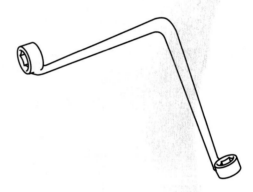

Figure 2-26 Brake bleeder valve wrenches are bent to get into the awkward locations of many bleeder screws.

Torque Wrenches

Torque wrenches measure the torque, or twisting force, applied to a fastener. Many fasteners such as caliper mounting bolts, bleeder screws, and brake hose-to-caliper fasteners must be tightened to a torque specification that is expressed in foot-pounds or newton-meters. Some bleeder screws may require an inch-pound torque wrench.

SERVICE TIP: Torque wrenches must be handled carefully and must be properly stored. Torque wrenches should be backed down (set) to zero before storing. The wrench's calibration can be knocked off by rough handling. Torque wrenches should be recalibrated about every 12 months.

Torque wrenches are available with ¼-inch, ⅜-inch, and ½-inch drives. Common torque wrench designs include the dial type (Figure 2-27), the beam and breakover types (Figure 2-28), and the electronic digital-readout type. These all have a scale that measures turning effort as the fastener is tightened. With a breakover torque wrench, the desired torque is dialed in. The wrench then makes an audible click when you have reached the correct force. For accurate torque

Figure 2-27 Dial-type torque wrench.

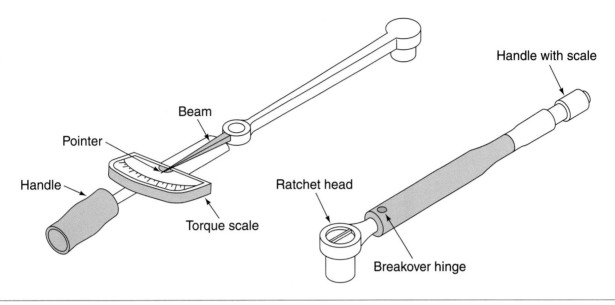

Figure 2-28 Beam-type torque wrench (left) and breakover torque wrench (right).

readings, the threads on the bolt and hole must be clean and dry. Whenever possible, pull on the wrench rather than pushing on it.

> **CAUTION:** Use a torque wrench for tightening purposes only. Do not use this tool to break nuts or bolts loose. Some torque wrenches will break if used to break fasteners loose.

Special Brake Tools

Many tools are designed for a specific purpose. Car manufacturers and specialty tool companies work together to design and manufacture special tools required to repair cars. Most special tools for brake service are listed in carmakers' service manuals. Special tools needed for routine brake service procedures are discussed in this *Shop Manual.* Following are brief descriptions of some of the special tools used in brake service.

Hold-Down Spring and Return Spring Tools

Brake shoe return springs used on drum brakes are very strong and require special tools for removal and installation. Although mass retailers offer some brake tools, they are designed for the do-it-yourself vehicle owner. The tools are excellent for that use but tend to fail quickly under typical shop operations. Most return spring tools have special sockets and hooks to release and install the spring ends. Some are built like pliers (Figure 2-29).

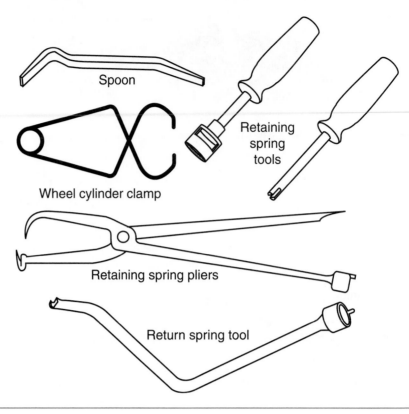

Figure 2-29 Special brake tools include adjusting spoons for drum brakes (top), wheel cylinder clamps (top center), brake spring pliers (bottom center), and return spring tools (bottom).

Hold-down springs for brake shoes are much lighter than return springs, and many such springs can be released and installed by hand. A hold-down spring tool looks like a cross between a screwdriver and a nut driver. A specially shaped end grips and rotates the spring retaining washer.

Drum Brake Adjusting Tools

Although almost all drum brakes built for more than 30 years have some kind of self-adjuster, the brake shoes still require an initial adjustment after they are installed. The star wheel adjusters of many drum brakes can be adjusted with a flat-blade screwdriver. Brake adjusting spoons (see Figure 2-29) and wire hooks designed for this specific purpose can make the job faster and easier, however.

Boot Drivers, Rings, and Pliers

Dust boots attach between the caliper bodies and pistons of disc brakes to keep dirt and moisture out of the caliper bores. A special driver (Figure 2-30) is used to install a dust boot with a metal ring that fits tightly on the caliper body. The circular driver is centered on the boot placed against the caliper and then hit with a hammer to drive the boot into place. Other kinds of dust boots fit into a groove in the caliper bore before the piston is installed. Special rings or pliers (see Figure 2-30) are then needed to expand the opening in the dust boot and let the piston slide through it for installation.

Caliper Piston Removal Tools

A caliper piston can usually be slid or twisted out of its bore by hand. Rust and corrosion (especially where road salt is used in the winter) can make piston removal difficult. One simple tool that helps with the job is a set of special pliers that grip the inside of the piston and let you move it by hand with more force (Figure 2-31). These special pliers work well on pistons that are only mildly stuck.

Dust boot installer

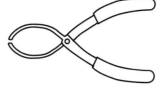
Dust boot pliers

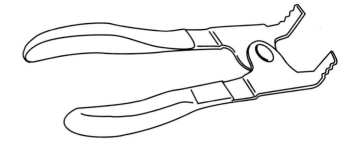

Figure 2-30 This driver and special pliers are used to install dust boots on calipers.

Figure 2-31 These special pliers are used to remove pistons from calipers.

For a severely stuck caliper piston, a hydraulic piston remover can be used. This tool requires that the caliper be removed from the car and installed in a holding fixture. A hydraulic line is connected to the caliper inlet and a hand-operated pump is used to apply up to 1,000 psi of pressure to loosen the piston. Because of the danger of spraying brake fluid, always wear eye protection when using this equipment.

There are some special ratchet-type tools used to compress the caliper piston back into its bore (Figure 2-32). This is necessary to provide room for new pads to fit over the rotor. The tool can be used to compress front pistons, but it is especially good on rear disc calipers because many of them ratchet outward as their self-adjusting mechanism.

Chapter 7 on disc brake service contains specific procedures and safety precautions for using compressed air to remove a caliper piston. Follow the instructions and observe the required precautions in Chapter 7 to prevent injury and damage to vehicle parts.

Brake Cylinder Hones

Cylinder hones (Figure 2-33) are used to clean light rust, corrosion, pits, and built-up residue from the bores of master cylinders, wheel cylinders, and calipers. A hone can be a very useful— sometimes necessary—tool when you have to overhaul a cylinder. A hone removes material, makes the bore bigger, and must be used accordingly. A hone will not, however, save a cylinder with severe rust or corrosion.

Another kind of hone is the **brush hone.** It has abrasive balls attached to flexible metal brushes that are, in turn, mounted on the hone's flexible shaft. In use, centrifugal force moves the abrasive balls outward against the cylinder walls; tension adjustment is not required. A brush hone provides a superior surface finish and is less likely to remove too much metal as is a stone hone.

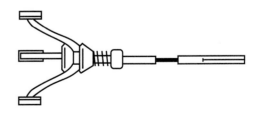

Figure 2-32 This tool is very useful for winding (screwing) in the caliper on a rear disc caliper. It may also be used to push the piston in on the standard front disc calipers.

Figure 2-33 Adjustable brake cylinder hone.

Ideally, hones are used only on larger brake components. Generally, a damaged caliper, wheel cylinder, or master cylinder on a light vehicle is replaced rather than rebuilt. Larger vehicles, having much larger components, can be honed and rebuilt cheaper. It is desirable to use only the brush-type hone because less material is removed. The stone-type hones should be used only when absolutely necessary, and, even then, the technician must consider the ramifications of making the cylinder bore and pistons larger. Even a small change of a few thousandths of an inch in bore and piston diameters will alter the vehicle's braking characteristics to some extent. It is recommended that no more than one pass be made in a cast-iron brake component. Aluminium brake bores should never be honed. Either kind of hone must be lubricated with brake fluid during use. Do not use a petroleum-based lubricant. After honing a cylinder, flush it thoroughly with denatured alcohol to remove all abrasives and dirt.

Brake Pedal Effort Gauge

The brake pedal effort, or force, gauge (Figure 2-34) is used more and more often to test and service modern brake systems. Some carmakers specify the use of this gauge when checking brake pedal travel and effort or when adjusting the parking brakes.

The assembly has a hydraulic plunger and gauge that measures the force applied by your foot in pounds or in metric newtons. The gauge assembly is mounted on a steel bracket that can be attached to a brake pedal.

Tubing Tools

The rigid brake lines or pipes of the hydraulic system are made of steel tubing to withstand high pressure and to resist damage from vibration, corrosion, and work hardening. Although copper is flexible and easy to work with, copper tubing cannot be used for brake lines, because it cannot withstand high pressure and is susceptible to breaking due to vibration.

CAUTION: Do not attempt to hand bend a pipe or tubing. In most cases the bend will crimp and weaken the pipe/tubing material, split the material, or block the pipe/tubing.

Rigid brake lines often can be purchased in preformed lengths to fit specific locations on specific vehicles. Straight brake lines can be purchased in many lengths and several diameters and

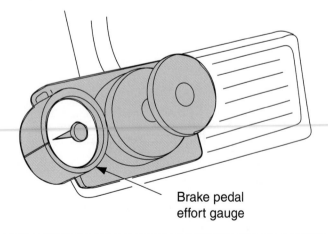

Figure 2-34 A brake pedal effort gauge measures the force applied to the brake pedal.

Brake pedal
effort gauge

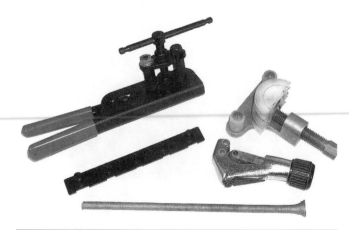

Figure 2-35 A tubing cutter, tubing bender, and flaring tools are used to create brake lines.

bent to fit specific vehicle locations. Even with prefabricated lines available, there may be occasions to cut and bend steel lines and form flared ends for installation. The common tools should include:

❑ A tubing cutter and reamer
❑ Tubing bender
❑ A double flaring tool for SAE flares
❑ An ISO flaring tool for European-style ISO flares

A **tubing cutter** is used to cut a pipe or hose at a flat angle. A pipe **reamer** is used to smooth the interior where the cutter broke through the metal. A **tubing bender** is a collection of interchangeable curved sections used to bend a pipe to the correct radii without crimping. The curved sections must fit the pipe's outside diameter. Figure 2-35 shows some of these tubing tools. Their uses are explained in detail in Chapter 5 of this manual, along with the subjects of forming and installing brake lines.

Power Tools

Power tools, whether electric or pneumatic (air), save time and energy. Stationary power tools such as bench grinders, brake lathes, and air compressors are part of the shop equipment. The next section outlines some of the necessary inspection and use of power tools.

General Safety Guidelines

Portable power tools are powered by electricity or air (pneumatic). Never operate a power tool without being properly trained in its use. Always use a tool for the job it was designed to do and keep all guards in place and in working order. Follow these general safety practices with all power tools:

1. Always wear the proper eye and ear protection. Gloves and a face shield are required when operating air chisels or air hammers.

⚠ **WARNING:** Do not use electric power tools while standing on a damp floor or ground. Even a properly grounded tool could cause a shock if an easier path to ground is available to the electrical current.

2. All electrical tools, unless they are the double-insulated type, must be grounded. Replace damaged or frayed power cords (Figure 2-36), and do not use a two-prong electrical adapter to plug in a three-prong, grounded piece of equipment. Never use a grounded piece of equipment that has the third ground plug removed.

3. Never try to make adjustments to, lubricate, or clean a power tool while it is running or plugged in. Keep all guards in place and in working order.

4. Be sure pneumatic tools and lines are attached properly.

5. Turn off and unplug all portable power tools when not in use, and return all equipment to its proper place.

When operating stationary power tools such as brake lathes and grinders:

1. Always wear eye and ear protection.

2. Make sure that all others are clear of the power tool before turning on the power.

3. Be sure all machine safety guards are in the correct position before you start the machine.

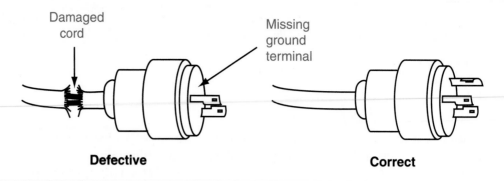

Damaged cord

Missing ground terminal

Defective

Correct

Figure 2-36 Replace damaged power cords, and do not use an electrical plug with the third grounding terminal removed.

4. Start your own power tool and remain with it until you have turned it off and it has come to a complete stop.

5. Stay clear of power tools being operated by others.

6. Do not talk to or distract someone who is using a power tool.

7. Do not operate any machine without receiving instructions on the correct operating procedures. Read the owner's manual to learn the proper applications of a tool and its limitations. Make sure all guards and shields are in place. Remove all key and adjusting wrenches before turning the tool on.

8. Give the machinery your full attention. Do not look away or talk to fellow students or workers. Keep the work area clean and well lighted. Never work in a damp or wet location.

9. Do not abuse electrical cords by yanking them from receptacles or running over them with vehicles or equipment.

10. Inspect equipment for any defects before using it. Make all adjustments before turning on the power. Whenever safety devices are removed to make adjustments, change blades or adapters, or make repairs, turn off and unplug the tool. Lock and tag the main switch, or keep the disconnected power cord in view at all times.

11. Always wait for the machine to reach full operating speed before applying work.

12. Do not leave the power tool until it has come to a complete stop. Allow a minimum of 4 inches between your hands and any blades, cutters, or other moving parts. Do not overreach; be sure to keep the proper footing and balance at all times.

Compressed Air Safety

Compressed air is used in most shops to inflate tires, operate air tools, and blow dry parts. An air compressor in the shop sends air under pressure through air lines to flexible air hoses. Badly worn air hoses will burst under pressure. The fittings on the ends of an air hose are attached with special crimped connectors that can withstand several hundred pounds of pressure. Air hose fittings should never be attached with hose clamps. Replace worn hoses to prevent accidents.

The discharge of high-pressure air through a blowgun nozzle can injure your face or body. Never pull the trigger when the gun is pointed directly toward anyone.

Be sure that an air nozzle used for cleaning has a safety tip that limits air outlet pressure to 30 pounds per square inch (psi) when the nozzle tip is blocked or dead ended (Figure 2-37). This type of nozzle is commonly referred to as an OSHA-approved nozzle. When you use an air nozzle to clean parts, direct the airflow away from yourself and other personnel. Never use an air nozzle to dust off your clothing or hair.

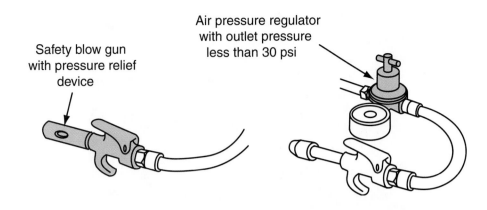

Safety blow gun with pressure relief device

Air pressure regulator with outlet pressure less than 30 psi

Figure 2-37 The nozzle to the left is the most common in an automotive shop. It is also known as an OSHA air nozzle.

 CAUTION: When cleaning and drying bearings of any type with compressed air, ensure that the bearing cannot spin. The bearing can spin at high speed and disintegrate, causing damage.

Wear safety glasses when working with pneumatic equipment. Do not look into the discharge nozzle while trying to find out if the nozzle is clogged. Check hose connections before turning on the air. When turning air on or off, hold the air hose nozzle to prevent it from whipping.

WARNING: Never blow brake dust off a part with compressed air. The dust contains particles that could enter the lungs.

WARNING: Make sure you know how to operate a power tool before using it. Carelessness or mishandling of power tools can cause serious injury.

Impact Wrenches

An impact wrench (Figure 2-38) hammers or impacts a nut or bolt to loosen or tighten it. Removing wheel nuts from wheels is a common job for an impact wrench. Light-duty impact wrenches are available in three drive sizes: ¼ inch, ⅜ inch, and ½ inch; and two heavy-duty sizes: ¾ inch and 1 inch.

 CAUTION: Do not use an impact wrench to tighten critical fasteners or parts that may be damaged by the hammering force of a wrench.

Sockets for use with an impact wrench are special heavy, hardened sockets to withstand the blows of the impact hammer. Sockets for an ordinary hand ratchet or breaker bar must not be used with an impact wrench because they may shatter and cause damage or injury. Many technicians use torque sticks to install wheel assemblies quickly. Do not use torque sticks on other assemblies or to remove fasteners.

Air Ratchet

Air ratchets are used for general disassembly or reassembly work. Because they turn sockets without a jarring impact force, air ratchets can be used on most parts and with ordinary sockets. Air ratchets typically have a ⅜-inch drive (Figure 2-39).

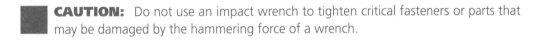

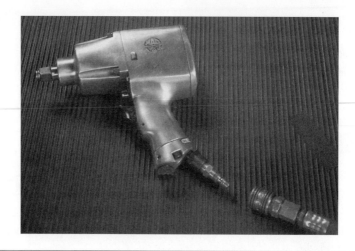

Figure 2-38 An impact wrench is a high-torque power wrench that loosens and removes fasteners by hammering them.

Figure 2-39 An air ratchet speeds fastener removal and installation.

Air ratchets are not torque sensitive. After snugging a fastener down with an air ratchet, a torque wrench must be used to set the final fastener tightness.

WARNING: Always maintain control of air ratchets. When a fastener begins to torque down, the ratchet will attempt to rotate itself instead of the fastener. Fingers can be mashed or cut and damage to the component or fastener may occur.

Brake Lathes

Brake lathes are special power tools used only for brake service. They are used to turn and resurface brake rotors and drums (Figure 2-40). Turning involves cutting away very small amounts of metal to restore the surface of the rotor or drum. The traditional brake lathe is an assembly mounted on a stand or workbench. This so-called bench lathe requires that the drum or rotor be removed from the vehicle and mounted on the lathe for service. No person should ever operate a power tool without proper training. Training may consist of reading a pamphlet or attending formal classes. When assigned to use a machine notify the supervisor or instructor if training has not been received.

As the drum or rotor is turned on the lathe spindle, a carbide steel cutting bit is passed over the drum or rotor friction surface to remove a small amount of metal. The cutting bit is mounted rigidly on a lathe fixture for precise control as it passes across the friction surface.

Obviously a brake drum must be removed from its axle or spindle to turn it on a lathe. The friction surface of a rotor is exposed, however, when the wheel, tire, and caliper are

Figure 2-40 A typical bench brake lathe for machining (turning) drums and rotors.

removed. Then it is possible to apply a cutting tool to the rotor friction surface without removing the rotor from the car. With the universal adoption of disc brakes, on-car brake lathes were developed for rotor service.

An on-car lathe is bolted to the vehicle suspension or mounted on a rigid stand to provide a stable mounting point for the cutting tool (Figure 2-41). The rotor may be turned either by

Figure 2-41 A typical on-car lathe for turning disc brake rotors.

the vehicle engine and drive train (for a front-wheel-drive car) or by an electric motor and drive attachment on the lathe. As the rotor is turned, the lathe cutting tool is moved across both surfaces of the rotor to refinish it. An on-car lathe not only has the obvious advantage of speed, it rotates the rotor on the vehicle wheel bearings and hub so that these sources of run out or wobble are compensated for during the refinishing operation.

Most drum and rotor lathes include attachments for applying a final surface finish to the rotor or for grinding hard spots on drums. Some much older bench lathes also may have a shoe arcing attachment for arc grinding the linings of brake shoes, but such attachments have become very rare. Shoe arcing was a necessary operation years ago, but concerns about asbestos dust led brake lining manufacturers to develop preground shoes that do not require arcing. Whenever possible, it is good practice to buy preground shoes. They not only speed up the brake job, they eliminate brake dust hazards due to shoe arcing. Chapter 7 covers disc brake rotors servicing and machining and Chapter 8 covers drum brakes.

Lifting Tools

Lifting tools are necessary for most brake service procedures and are usually provided by the shop. Correct operating and safety procedures should always be followed when using lifting tools.

Jacks

Jacks are used to raise a vehicle off the ground. Two basic kinds of jacks are found in most shops: hydraulic and pneumatic. The most popular jack is the hydraulic floor jack, which is classified by the weight it can lift: $1\frac{1}{2}$, 2, $2\frac{1}{2}$ tons, or more. A hydraulic floor jack is operated by pumping the handle up and down. A pneumatic portable floor jack operates on compressed air from the shop air supply.

Safety Stands

Whenever a vehicle is raised by a jack, it must be supported by safety stands (Figure 2-42). Never work under a car with only a jack supporting it; always use safety stands. The hydraulic seals of a floor jack can let go and allow the vehicle to drop.

Check in the vehicle service manuals for the proper locations for positioning the jack and the safety stands. Always follow the guidelines for each vehicle.

Figure 2-42 Always use safety stands to support a car after it has been raised with a floor jack.

Hoist Safety

Most brake service requires that you raise the vehicle off the floor. If an automobile were to fall from the hoist, the person working underneath would be seriously injured. Great care must be taken in using hoists or lifts. Always make sure the vehicle is properly placed on the hoist (lift) before raising it and always use the safety locks to prevent the lift from coming down unexpectedly.

Hoists use electricity, hydraulics, or compressed air for power. There are several styles of hoists. The single-post lift is simple to operate and can be adjusted quickly for any size vehicle by swinging the four large pads under the vehicle frame. The large center post, however, makes working on the middle of the vehicle difficult. To work on the center of the vehicle, a twin-post lift can be used. Its two posts lift the vehicle by its two sides.

 WARNING: Refer to a shop manual before lifting any vehicle to identify the proper locations for lifting. Failure to use the correct lifting points is dangerous and may cause damage or injury.

CAUTION: When using any lift for the first time, learn the control operation by raising and lowering it without a vehicle in place.

To ensure safe operation of a hydraulic lift, follow these ten guidelines:

1. Inspect each lift daily.
2. Never operate a lift if it does not work properly or if it has broken or damaged parts.
3. Never overload a lift. The rate capacity of the lift is listed on the manufacturer's nameplate. Never exceed that rating.
4. Always make sure the vehicle is properly positioned before making the lift. Refer to a service manual before lifting any vehicle to identify the proper lifting points (Figure 2-43).
5. Never raise a vehicle with someone in it.
6. Always keep the lift clean and clear of obstructions.
7. Before moving a vehicle over a lift, position the arms and supports to allow for free movement of the vehicle over the lift. Never drive over or hit the lift arms, adapters, or supports because this may damage the vehicle or the lift.
8. Carefully load the vehicle onto the lift and align the lift arms and contact pads with the specified lift points on the vehicle. Raise the lift until it barely supports the vehicle, then check the contact area of the lift.
9. Always lock the lift into position while working under the raised vehicle.
10. Before lowering the lift, make sure all tools and other equipment are removed from under the vehicle. Also make sure no one is standing under or near the vehicle as it descends.

Hydraulic Lifts

When used correctly, a hydraulic lift (hoist) is the safest, most convenient lifting tool. It allows you to raise the vehicle high enough to work comfortably and quickly. There are several styles of lifts. A single-post lift is simple to operate and may be adjusted quickly for any size vehicle by swinging the four large pads under the automobile frame. However, the large center post makes working on the middle of the vehicle difficult. To work on the center of the vehicle, double-post lift (Figure 2-44) can be used. Its two posts lift the vehicle by its two sides.

All lifts or hoists have automatic locking devices. After lifting the vehicle to the correct working height set the locks before beginning work.

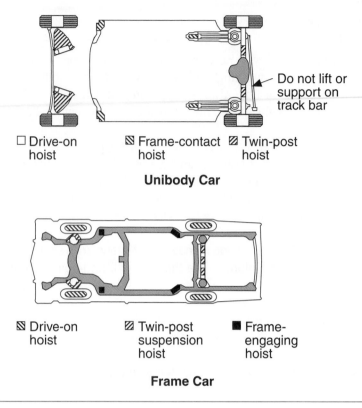

□ Drive-on hoist   ▨ Frame-contact hoist   ▨ Twin-post hoist

Do not lift or support on track bar

Unibody Car

▨ Drive-on hoist   ▨ Twin-post suspension hoist   ■ Frame-engaging hoist

Frame Car

Figure 2-43 If there is any doubt about the vehicle's lift points, consult the service manual.

Figure 2-44 A twin-post hoist provides good access to undercar components as well as to the brakes.

Various safety features prevent a hydraulic lift from dropping if a seal does leak or if air pressure is lost. Before lifting a vehicle, make sure the vehicle is properly placed on the lift and always use the safety locks to prevent the lift from coming down unexpectedly.

Always refer to a shop manual before lifting any vehicle to identify the proper locations for lifting. When using any lift for the first time, learn the control operation by raising and lowering it without a vehicle in place. Review the lift safety instructions in Chapter 1 of this *Shop Manual.*

Pressure Bleeders

Pressure bleeding is a fast and efficient way to bleed a brake system for two reasons. First, the master cylinder does not have to be refilled several times; and second, the job can be done by one person.

A pressure bleeder is a tank separated into two sections by a flexible diaphragm. The top section is filled with brake fluid. Compressed air is fed into the bottom section, and as the air pushes on the diaphragm, the brake fluid above it also is pressurized. A pressure bleeder is normally pressurized to about 30 psi. Higher pressures should be avoided because fluid may be forced rapidly through the hydraulic system, causing a swirling or surging action. This, in turn, may actually create air pockets in the lines and valves and make bleeding difficult.

A supply hose runs from the top of the tank to the master cylinder. The hose is connected to the master cylinder by an adapter fitting that fits over the reservoir, taking the place of the reservoir cap. These adapters exist in different shapes for the different types of reservoirs, including the plastic reservoirs on some of the newer vehicles.

The pressurized brake fluid flows into the master cylinder and out through the brake lines, quickly forcing air out of the lines.

Because most brake fluids (except for silicone) tend to absorb moisture from the air, always keep containers tightly capped. It is better to buy smaller containers of brake fluid and keep them sealed until needed. Taking these steps to minimize water in the brake fluid will help reduce corrosion and keep the brake fluid boiling point high throughout the hydraulic system.

Another type of bleeder uses a hand-operated pump either to inject fluid into the system or to suck fluid from the system. The equipment known as the Phoenix Injector can be used to bleed all wheels or individual wheels as required (Figure 2-45). The unit comes in a basic kit with additional adaptors available as accessories.

⚠️ **WARNING:** Do not use the same pressure bleeder for silicone-based and glycol-based brake fluids. The wrong type of fluid could be introduced into a system and cause extensive damage and possible injury.

Other tools used in brake bleeding operations include a large rubber syringe, used to remove fluid from the master cylinder on some systems; master cylinder bleeder tubes, used to return fluid to the master cylinder reservoir from the outlet ports during bench bleeding; and assorted line and port plugs, used to close lines and valves temporarily during service and keep out dirt and moisture.

■ **CAUTION:** Do not depress the brake pedal when pressure brake bleeding equipment is being used. This could damage seals and piston cups within the brake system.

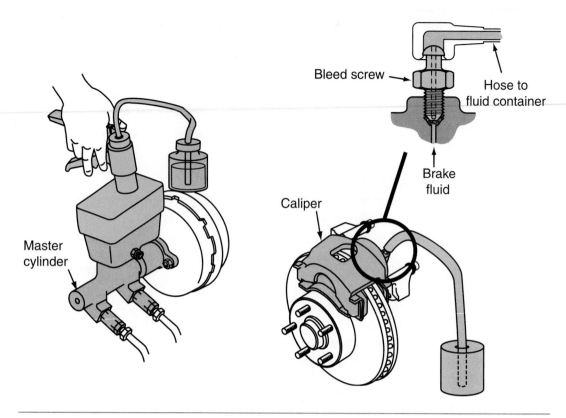

Figure 2-45 A Phoenix Injector™ set up to inject fluid through the brake system. (Courtesy of Phoenix Systems LLC)

Cleaning Equipment and Containment Systems

The following systems and methods are used to safely contain brake dust in the workplace.

Negative-Pressure Enclosure and HEPA Vacuum Systems

In a negative-pressure enclosure, brake system cleaning and inspection are performed inside a tightly sealed protective enclosure that covers and contains the brake assembly (Figure 2-46). The enclosure prevents the release of asbestos fibers into the air.

The enclosure is designed so you can clearly see the work in progress. It has impermeable sleeves and gloves that let you perform the brake cleaning and inspection. Examine the condition of the enclosure and its sleeves before beginning work. Inspect the enclosure for leaks and a tight seal.

A **high-efficiency particulate air (HEPA) filter** vacuum keeps the enclosure under negative pressure as work is done. Because particles cannot escape the enclosure, compressed air can be used to remove dust, dirt, and potential asbestos fibers from brake parts. The HEPA vacuum also can be used to loosen the asbestos-containing residue from the brake parts. Once the asbestos is loose, draw it out of the enclosure with the vacuum port. The dust is then trapped in the vacuum cleaner filter.

When the vacuum cleaner filter is full, spray it with a fine mist of water, then remove it and place it immediately in an impermeable container. Label the container as follows:

A **high-efficiency particulate air (HEPA) filter** is the filter that removes the smallest particulates from the air.

A HEPA filter can clean large volumes of air.

Figure 2-46 This negative-pressure enclosure is used to contain brake dust during cleaning.

> This container holds asbestos fiber. Avoid creating dust when removing it. Asbestos fibers can cause cancer and lung disease.

Asbestos waste must be collected, recycled, and disposed of in sealed impermeable bags or other closed, impermeable containers. Any spills or release of asbestos-containing waste material from inside the enclosure or vacuum hose or vacuum filter should be cleaned up immediately using vacuuming or wet-cleaning methods. Review the asbestos safety instructions in Chapter 1 of this *Shop Manual.*

Low-Pressure Wet-Cleaning Systems

Low-pressure wet-cleaning systems wash dirt from the brake assembly and catch the contaminated cleaning agent in a basin (Figure 2-47). The reservoir contains water with an organic solvent or wetting agent. To prevent any asbestos-containing brake dust from becoming airborne, control the flow of liquid so that the brake assembly is gently flooded.

You can use the cleaning liquid to wet a brake drum and backing plate before the drum is removed. After the drum is removed, thoroughly wet the wheel hub and the back of the assembly to suppress dust. Wash the brake backing plate, shoes, and brake parts used to attach the brake shoes before removing the old shoes.

Figure 2-47 Shown is a heated, low-pressure aqueous brake washer. It has a filter to trap the brake dust. The filter must be treated as hazardous waste when it is filled. (Courtesy of R&D Automotive Industrial)

Some wet-cleaning equipment uses a filter. When the filter is full, first spray it with a fine mist of water; then remove the filter and place it in an impermeable container. Label and dispose of the container as described earlier.

Wet-Cleaning Tools and Equipment

The wet-cleaning method of containing asbestos dust is the simplest and easiest to use, but it must be done correctly to provide protection. First, thoroughly wet the brake parts using a spray bottle, hose nozzle, or other implement that creates a fine mist of water or cleaning solution. Once the components are completely wet, wipe them clean with a cloth.

Place the cloth in a correctly labeled, impermeable container and properly dispose of it. The cloth can also be professionally laundered by a service equipped to handle asbestos-laden materials and can then be reused.

The use of wet-cleaning tools is probably the most common brake cleaner used in the typical shop. It is cheap and easy to use and the waste can be stored fairly easily and safely.

✓ **SERVICE TIP:** Do not be tempted to remove fingerprints and small grease spots from the friction material on brake pads or shoes with aerosol brake cleaner. Aerosol cleaners are petroleum products and will contaminate the friction materials. You can remove small dirt spots by lightly sanding them with medium sandpaper and wiping them off with denatured alcohol and a clean cloth.

Vacuum Cleaning Equipment

Several types of vacuum cleaning systems are available to control asbestos in the shop. The vacuum system must have a HEPA filter to handle asbestos dust (Figure 2-48). A general-purpose shop vacuum is not an acceptable substitute for a special brake vacuum cleaner with a HEPA filter. After vacuum cleaning, wipe any remaining dust from components with a damp cloth.

Because they contain asbestos fibers, the vacuum cleaner bags and cloths used in asbestos cleanup are classified as hazardous material. Such hazardous material must be disposed of in accordance with OSHA regulations. Always wear your respirator when removing vacuum cleaner bags or handling asbestos-contaminated waste. Seal the cleaner bags and cloths in heavy plastic bags. Label and dispose of the container as described previously.

Cleaning with Enclosure Equipment

Photo Sequence 2 shows how to use enclosure equipment to clean a drum brake assembly.

Figure 2-48 Always use a vacuum cleaner with a HEPA filter to clean the shop. This cleaner also works well in cleaning brake parts.

Photo Sequence 2
Typical Procedure for Using Enclosure–Type Brake Cleaning Equipment

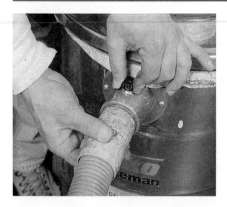

P2-1 Check that the hose is securely fastened to the HEPA vacuum and to the brake enclosure.

P2-2 Check that the vacuum container clips and seals are in position and working properly.

P2-3 Remove the wheel.

P2-4 Turn on the vacuum cleaner.

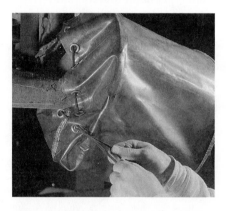

P2-5 Place the enclosure over the brake assembly. It must form a tight seal behind the backing plate.

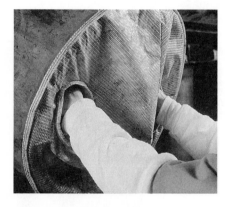

P2-6 Place your hands into the attached rubber gloves.

P2-7 Remove the brake drum.

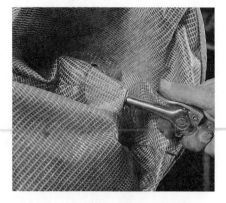

P2-8 Blow dust off the drum and brake assembly using the air gun attachment inside the enclosure. Blow dust off all inside surfaces of the enclosure, directing it toward the vacuum exit.

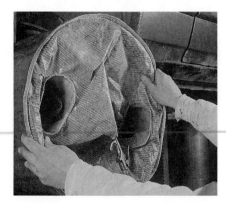

P2-9 Remove the enclosure.

Photo Sequence 2
Typical Procedure for Using Enclosure–Type Brake Cleaning Equipment (continued)

P2-10 Turn off the vacuum. Then proceed with the drum brake inspection.

P2-11 When the vacuum cleaner filter is full, spray it with a fine mist of water, then remove and place it immediately in an impermeable container. Label the cleaner as explained in the text.

Cleaning Equipment Safety

Parts cleaning is an important part of any brake repair job. You must often clean parts to find problems and measure for wear. Brake parts cleaning must always be done with approved equipment because of the danger of asbestos exposure.

Be careful when using solvents. Most are toxic, caustic, and flammable. Avoid placing your hands in solvent; wear protective gloves, if necessary. Read all manufacturer's precautions and instructions and MSDS before using.

Do not use gasoline to clean components. This practice is very dangerous. Gasoline vaporizes at such a rate that it can form a flammable mixture with air at temperatures as low as –50°F. Gasoline also is dangerous if it gets on your skin because the chemicals in gasoline can be absorbed through the skin and get into your body.

Small cleaning jobs are often done with aerosol cleaners. These spray cans contain chemicals that break down dirt and grease and allow them to be removed. Do not throw empty spray containers in the trash without punching a hole in the side. If the EPA inspector finds an undamaged can in the trash, she will test it for pressure. A few hours in the warm sun will build enough pressure in the can to get a warning or a citation from the inspector. Always read the warnings on the can and follow them. Wear eye protection, proper gloves, and a shop coat to prevent exposure to your skin or eyes. Always do your cleaning in a well-ventilated area.

Many of the solvents used in solvent cleaning tanks are flammable. Be careful to prevent an open flame around the solvent tank. Never mix solvents. One could vaporize and act as a fuse to ignite the others.

Wear neoprene gloves when washing parts. Some solvents can be absorbed through the skin and into your body. This is especially true if you have a cut on your hand. Do not blow compressed air on your hands if they get wet with solvent, as this can cause the solvent to go through your skin.

Wipe up spilled solvents promptly, and store all rags in closed, properly marked metal containers. Store all solvents either in their original containers or in approved, properly labeled containers. Finally, when using a commercial parts washer, be sure to close the lid when you are finished.

SERVICE TIP: After resurfacing disc brake rotors, some technicians like to clean those rotors with aerosol brake cleaner. This type of solvent does not remove small particles of iron and abrasive material as well as detergent and hot water do, however, and it is not recommended by most carmakers. Washing the rotors with soap and water is the preferred cleaning method. Besides, brake cleaner is more expensive than soap and water.

Brake Lubricants

Special lubricants are used in brake service to aid assembly and to help prevent corrosion and mechanical seizure.

Assembly Fluid

As with brake fluid, brake assembly lubricants should be stored in tightly capped, dry containers.

If you overhaul hydraulic parts such as cylinders and calipers, assembly fluid will help to install pistons past seals or to install piston seals into bores. You can use brake fluid as an assembly lubricant, but special assembly fluid has a higher viscosity than brake fluid. The higher viscosity provides better lubrication and lets the fluid remain on the part until it is installed and put into service.

Brake Grease

Brake grease is applied to the shoe support pads on backing plates and to moving caliper parts on disc brakes. Brake grease has a melting point above 500°F (260°C) and contains solid lubricants that will remain in place even if the temperature rises above the grease melting point. Brake grease often can be identified by its light color (almost white). Wheel bearing grease or chassis grease should not be used in place of brake grease because when the grease is heated by brake operation it may run onto the pads and linings and ruin them.

SERVICE TIP: Ford Motor Company specifically prohibits the use of any petroleum-based brake grease on all of its cars and light trucks built since the early 1980s. Petroleum-based grease may damage the EPDM rubber parts used in Ford's disc brakes. Ford specifies silicone grease that meets Ford specification ESE-M1C171-A for all disc and drum brakes on its vehicles.

Rubber Lubricants

Some master cylinder overhaul kits contain a small package of nonpetroleum grease for application to the rubber boot of the cylinder where the pushrod enters. This special grease should be used only on rubber boots of cylinders. Do not use it in place of assembly fluid or brake grease.

Wheel Bearing Grease

Brake service often includes repacking front or rear wheel bearings. Although some multipurpose greases can be used for both chassis lubrication and wheel bearings, you must be sure that whatever grease you use is identified as suitable for wheel bearing lubrication.

All greases are made from oils blended with thickening agents so that the grease will stick to the surfaces to be lubricated. Greases are identified by the National Lubricating Grease Institute (NLGI) number and by the kinds of thickeners and additives that the grease contains. The higher the NLGI number, the higher viscosity (thicker) the grease is. Almost all wheel bearing and chassis greases are NLGI number 2 greases. Most wheel bearing grease uses a lithium-based thickener for temperature resistance. Molybdenum disulfide is another common additive that improves the antiseize properties of wheel bearing grease.

There is no way to tell what kind of grease was used previously on a wheel bearing. Therefore, wipe away all old grease thoroughly when repacking wheel bearings and use only new grease that meets the vehicle maker's specifications for wheel bearing service.

Electronic Test Equipment

With the increasing use of ABSs, electronic controls have become part of the brake systems you service. For troubleshooting these systems, you need two basic pieces of electrical test equipment—a circuit tester and a digital multimeter.

● **CUSTOMER CARE:** One of those little things that irritate customers during repair is having to reset the radio, power seat, CD player, and other electronic devices because the vehicle battery was disconnected. Before disconnecting the battery, connect a memory-saver battery so all of the memories are preserved. If working in an air bag area, advise the customer that you will note the radio and some pre-sets, but ones such as the power seat memory will be lost. Most customers will understand and appreciate your thoughtfulness.

Circuit Testers

Circuit testers, or test lights, are used to identify shorted and open electrical circuits.

Probe Light. The most common test light, or probe light, looks like an ice pick. Its handle is transparent and contains a light bulb. A probe extends from one end of the handle and a wire and clip extend from the other end. When the clip is attached to a ground and the probe is touched to a live connector, the bulb in the handle lights up. If the bulb does not light, voltage is not available at the connector (Figure 2-49). The test light also can be reversed by connecting its wire to a voltage source and touching the probe to a possible ground. If the bulb lights, the circuit is complete and you have found an electrical ground.

■ **CAUTION:** Do not use a probe light to test the low-current circuits of an electronic control system. The incandescent lamp bulb may draw too much current and damage the electronic integrated circuits. Always test electronic circuits with a high-impedance voltmeter.

LED Test Light. A test light made with a light-emitting diode (LED) provides higher resistance than an incandescent lamp and is suitable for use on an electronic circuit (Figure 2-50). An LED test light is used in the same way as an incandescent probe light.

An auto-ranging meter is usually the best choice for an automotive technician because of its ease of use.

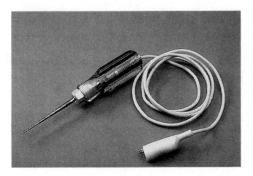
Probe Light

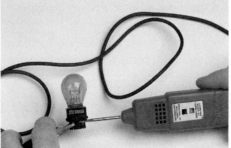
Continuity Light

Figure 2-49 A test light (probe light) and a self-powered test light (continuity tester) are common electrical troubleshooting tools.

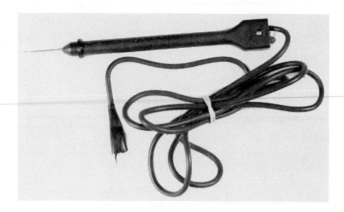

Figure 2-50 LED test lights such as this one are recommended for use on low-power electronic circuits instead of a probe light.

Self-Powered Test Light. A self-powered test light is called a continuity tester. It is used with the power off in the circuit being tested. It looks like a probe light, except that it has a small internal battery. When the clip is attached to one terminal of a component and the probe is touched to the other, the bulb will light if the circuit is complete. If an open circuit exists, the bulb will not light.

Multimeters

The multimeter is one of the most versatile tools for diagnosing electrical and electronic systems. The most common multimeter is the volt, ohm, milliamp meter (Figure 2-51). Such a meter is referred to variously as a volt-ohm-milliamp (VOM) meter, a digital volt-ohmmeter (DVOM), or a **digital multimeter (DMM).** Regardless of the name, this type of multimeter tests voltage, resistance, and current.

As the name indicates, multimeters are multifunctional. Most measure direct current (dc) and alternating current (ac) amperes (A), volts (V), and ohms (Ω). More advanced multimeters may also test engine revolutions per minute (rpm), ignition dwell, diode connections, distributor conditions, frequency, and even temperature.

Multimeters are available with either **digital** or **analog** displays. Many electronic tests require very precise voltage measurement. Digital multimeters provide this accuracy by measuring volts, ohms, or amperes in tenths, hundredths, or thousandths of a unit. Several test ranges are usually provided for each of these functions. Some meters have multiple test ranges that must be manually selected. Others are **auto ranging.** A typical multimeter may have eighteen test ranges. It may measure dc voltage readings as small as $\frac{1}{10}$ millivolt (0.0001 volt) or as high as 1,000 volts.

Analog meters use a needle and scale to display readings and are not as precise as digital meters. Another problem with analog meters is their low internal resistance (input impedance). The low input impedance allows too much current to flow through circuits; therefore, analog meters should not be used on electronic devices.

Digital meters have a **high impedance**—high input resistance—of at least 10 megohms (10 million ohms). Metered voltage for resistance tests is well below 5 volts, reducing the risk of damage to sensitive components and computer circuits.

Scan Tools

Scan tools (Figure 2-52) are essential for servicing most modern ABSs. The **scan tool** plugs into the brake system control computer and allows you to read stored or current DTCs and system

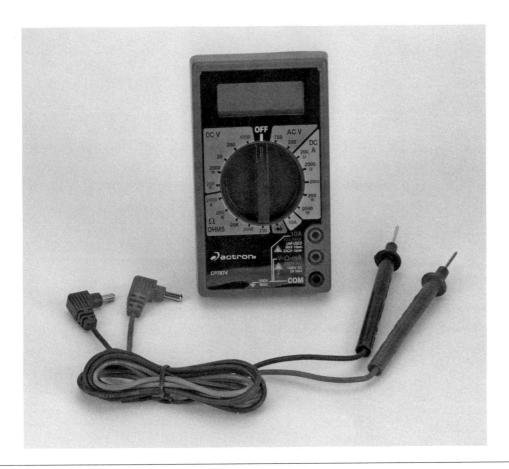

Figure 2-51 A digital multimeter (DMM) should always be used to check electronic circuits.

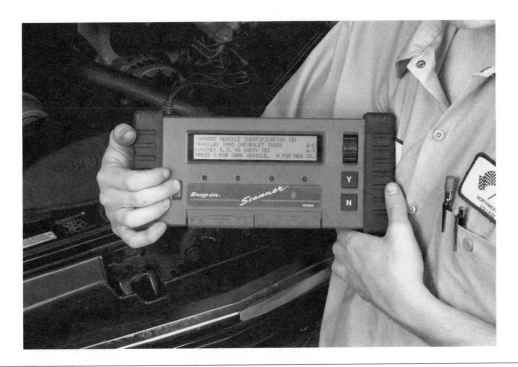

Figure 2-52 Scan tools are essential for complete testing of ABS control systems.

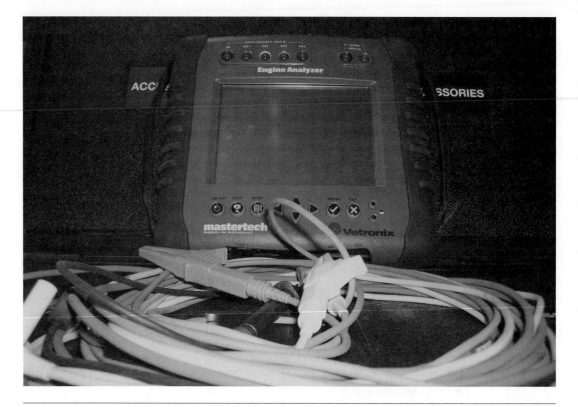

Figure 2-53 A four-trace DSO can graph and store various electronic operations by measuring volts, amperes, and resistance all at the same time for comparison data.

operating data that help pinpoint system problems. Many scan tools also can activate system functions to test individual components or to record a snapshot of the system in operation during a test drive.

Graphing Meters

Graphing meters are commonly known as scopes or oscilloscopes. They are electronic diagnostic instruments that can graph the flow of electrical values and come in different sizes and capabilities. The **digital storage oscilloscope (DSO)** was at one time the most elaborate, most expensive, and a large-size unit. The typical DSO is shown in Figure 2-53. It is a small hand-held unit capable of displaying four traces or graphs at once. Each trace can be attached to a different electrical device so comparisons can be viewed simultaneously. Similar DSOs are available from most of the major vehicle tool vendors.

Electrical Principles

ABSs rely on electronic actuation and control, but modern brake systems also use simple electrical circuits for basic system status checks and safety functions. For example, an electrical circuit is energized if the parking brake is applied or if fluid level in the master cylinder reservoir falls below a certain point. An electrical circuit may also be energized if a pressure difference develops in the hydraulic brake lines. All of these circuits are tied into one or more brake system warning lamps on the instrument panel. When a problem exists, the warning lamp lights to alert the driver.

Some vehicles with electronic instrument panels flash a written message instead of lighting a warning lamp. However, the basic circuit operation is essentially the same between the instrument panel and the warning circuit switch.

Brake stoplamp circuits to the rear stoplamps and the center high-mounted brake lamp are energized when the brake pedal is depressed. These lamps warn other motorists that the brakes on the vehicle in front of them have been applied.

Although most circuits involved in brake stoplamp and warning lamp operation are simple circuits, the increased use of electrical equipment on modern vehicles may make them more difficult to find and trace. Always work from the service manual electric schematics and follow the principles of electrical troubleshooting outlined later.

In a vehicle electrical system, electric power flows from a power source to a load device and then back to the source of power to form a complete circuit. In addition to the power source and loads, most automotive circuits contain circuit control and protection components (Figure 2-54).

The vehicle battery and alternator are the power sources that provide power for all electric circuits during startup or alternator failure. During normal operation the alternator assumes that responsibility of providing the vehicle with electrical power. Circuit protection devices include items such as fuses, circuit breakers, and fusible links. They provide overload protection for the circuit. Circuit controllers such as switches or relays are used to control the power within a circuit. They open and close the circuit. Circuit loads may be lamps, motors, or solenoids.

Electric circuits can be series circuits, parallel circuits, or series-parallel circuits.

A **parallel circuit** is an electric circuit in which all the loads are connected to form more than one current path with the power source.

A **series-parallel circuit** is an electric circuit in which some loads are in parallel with each other and one or more loads are in series with the power source and with the parallel branches.

Series Circuit

In a **series circuit,** the electrical load devices are connected to form one current path to and from the power source. Series circuit voltage is shared by all the components proportionally to the resistance of each. Current flows through every component in series and remains constant throughout the circuit (Figure 2-55).

Parallel Circuit

In a **parallel circuit,** the electrical loads are connected to form more than one current path to and from the power source. Voltage is equal for each parallel current path, and current varies proportionally to the resistance of each parallel branch (Figure 2-56).

Series-Parallel Circuits

A **series-parallel circuit** consists of some loads in parallel with each other and one or more loads in series with the power source and with the parallel branches (Figure 2-57).

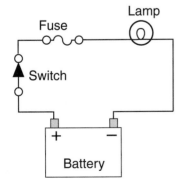

Figure 2-54 An electrical circuit consists of a power source (battery), a control device (switch), circuit protection (fuse), and one or more load devices (lamp).

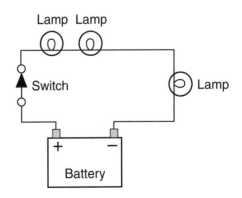

Figure 2-55 A series circuit has the loads connected one after another with the power source.

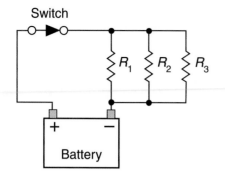

Figure 2-56 A parallel circuit has several alternative (parallel) loads connected to the power source.

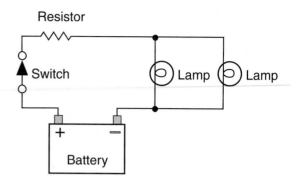

Figure 2-57 A series-parallel circuit has one or more loads in series with the parallel branches and with the power source.

Circuit Testing

Refer to Chapter 4 of this *Shop Manual* for electrical diagnostic procedures and instructions for using electrical test equipment.

Service Manuals

Good service manuals, paper or computerized, are the best single all-around tool a technician can use. This applies to new technicians on their first day as well as technicians nearing retirement.

Service manuals are essential for safe, complete brake service. They are needed to obtain specifications on torque values and critical measurements such as drum and rotor discard limits. Manuals also provide drawings and photographs that show where and how to perform service procedures on a particular vehicle. Special tools or instruments also are listed and illustrated when they are required. Precautions are given to prevent injury or damage to parts.

Most automobile manufacturers publish a service manual, or set of manuals, for each model and year of their cars and trucks. These manuals provide the best and most complete information for those vehicles. Several independent publishers reproduce carmakers' information in aftermarket manuals and information systems. This information is available both in traditional printed books and in computerized information systems. Brake parts manufacturers also often publish service manuals and service literature independently of vehicle manufacturers.

The most common service manual today is the computerized version. Paper manuals are expensive and take up a lot of storage space, a lot of different ones are required, and they are easily damaged. Computerized manuals offer a quick, easy method to extract specific data and usually need only a computer terminal with Internet connection. The Internet data bank is updated daily in most cases. The major vendors are Snap-on, Mitchell, and All Data. Each offers a version of the same thing: a comprehensive data bank covering almost all years and models of vehicles back to about 1983. In addition, the more sophisticated systems can display labor times and parts. Some even offer Internet order capability so the technician can determine what part is needed and immediately order it online from a local parts vendor. Most can be tied into the service repair order and accounting programs, so, in theory at least, there is no paperwork involved. Everything from the opening of the repair order through the payment by the customer is done via the shop's computer internal network.

Although the manuals from different publishers vary in presentation and arrangement of topics, all service manuals are easy to use after you become familiar with their organization. Most manuals are divided into several sections, each of which covers different systems of the vehicle. Each section includes diagnostic, service, and overhaul procedures and has an index indicating specific areas of information.

Repair Order

The repair order (Figure 2-58) is a legal document and must be completed on a vehicle being repaired on the premises or on the road under the auspices of the business license. There are a few items that the technician/service manager should insist on with regard to a repair order. If the work is being persormed free or at a reduced rate for a person or institution, make sure the order

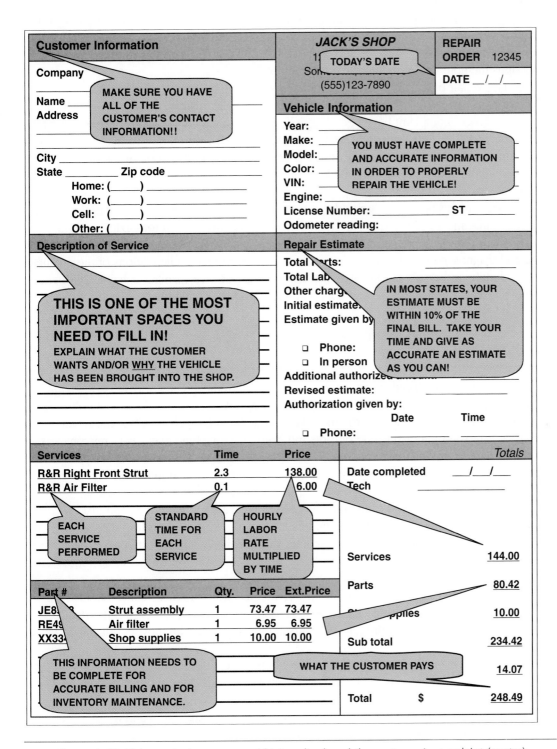

Figure 2-58 This repair shows a recent history (top) and the customer's complaint (center). (Courtesy of Spalding Lincoln/Mercury/Mazada/Dodge)

is completed in the same manner as a routine "for-hire" order. The only difference should be the cost of the repair, usually paid by the vehicle owner. A completely "free" repair still legally obligates the shop and employees the same as a regular cutomer-pay work order. The second occurrence is as rare as the free work, but both happen now and then. If during the course of the repair something is found to need service and the customer refuses to approve the work, make sure it is stated on the work order and the customer signs that she refuses the service. This is especially critical if the refused service pertains to some safety aspects of vehicle operation such as bald tires or completely worn brakes. It is one thing for a customer to refuse an oil change, which may cause the shop to buy an engine. But refusal to replace the bald tire, which causes a death or injury and the shop being sued for all costs, is an entirely different matter. Ensure that the customer signs the note indicating the repair was refused.

Usually repair orders are opened, completed, and paid in a routine manner. A typical repair order that has been started is shown in Figure 2–58. The repair history, the vehicle identification, and the customer's complaint have been inserted. At this point the technician would research the shop's database of related **Technical Service Bulletins (TSB),** warranties, recalls, and, possible, repair and service diagnosis. After the repair is accomplished, parts, labor, and the total cost of the repair will be added. A copy will be kept on file at the shop and a copy will be given to the customer.

Throughout this book, you are told to refer to the appropriate service manual to find the correct procedures and specifications. Although the brake systems of all automobiles work in much the same way, there are many variations in design, particularly for ABSs. Each design has its own repair and diagnostic procedures. You must always follow the recommendations of the manufacturer to identify and repair problems. Make using vehicle service manuals a habit. The benefits include increased productivity, less rework, and safer working conditions.

Summary

❏ Fasteners must be replaced with the same type fastener removed.
❏ Measuring systems may be metric or U.S. Customary (USC).
❏ The USC unit of measurement is the inch. The metric unit of measurement is the meter.
❏ Precision measuring tools include outside and inside micrometers, rule, depth micrometers, and rotor and drum micrometers.
❏ Rotor and drum micrometers are specially designed to make measuring rotors and drums quick and accurate.
❏ Vernier calipers and dial indicators may be used to measure rotors, drums, and inside measurement of calipers and wheel cylinders.
❏ Brake adjusting tools provide a quick accurate way to initially adjust drum brakes before installing the drum.
❏ Hand tools should always be stored clean after each job.
❏ Common hand tools for brake repair are the same ones used in other vehicle repairs.
❏ Flare-nut or line wrenches and bleeder wrenches are critical hand tools for brake repairs.
❏ Special brake tools include spring tools, adjusting tools, boot/seal removers/drivers, and hones.
❏ Tubing tools are required to cut and form steel brake lines.
❏ Power tools used in brake repairs are typically the same ones used in other automotive repairs.
❏ Brake lathes are power-driven machining tools used to resurface rotors and drums.
❏ Brake lathes may be bench-mounted or portable on-the-car types.
❏ Lifting tools include typical automotive lifts and floor jacks.

- Pressure bleeders make one-person brake bleeding easier and help prevent running the master cylinder during the bleeding process.
- Pressure bleeders create a pressure within the master cylinder reservoir that forces brake fluid past the piston cups.
- Vacuum bleeders apply a vacuum (suction) at the output end of the brake hydraulic system to pull fluid and air from the system.
- Brake cleaning systems are designed to create minimum dust and to capture the brake residue from cleaning.
- Brake cleaning systems may be completely sealed or use low-pressure solvent and a capture container.
- Brake lubricants are specifically formulated to resist heat, dust, and brake fluid contamination.
- Scan tools are used to diagnose some electrical/electronic faults with the brake system, particularly antilock brakes.
- DSOs show graphs or traces of electrical values as they actually occur.
- The principles of electricity within a brake system include current, voltage, resistance, conductors, power source, and load.
- Brake repair or informational data may be on paper or in computerized format.

Terms to Know

Analog	High impedance	Series-parallel circuit
Auto ranging	International System of Units or metric system	Stem unit
Brush hone	Linear	Submultiple
Cylinder hone	Multiple	Technical Service Bulletin (TSB)
Digital	Parallel circuit	Tubing bender
Digital storage oscilloscope (DSO)	Reamer	Tubing cutter
Digital multimeter (DMM)	Scan tool	Vernier scale
High-efficiency particulate air (HEPA) filter	Series circuit	

ASE-Style Review Questions

1. *Technician A* always uses flare-nut wrenches to disconnect brake line fittings.
   Technician B uses open-end wrenches for the same job.
   Who is correct?
   A. A only
   B. B only
   C. Both A and B
   D. Neither A nor B

2. *Technician A* states that the only danger in using a metric wrench on an inch-sized bolt is skinned knuckles if the wrench slips.
   Technician B says that the head of the bolt may become rounded as well.
   Who is correct?
   A. A only
   B. B only
   C. Both A and B
   D. Neither A nor B

3. The use of torque wrenches is being discussed:
   Technician A says that the bolt and hole must be clean to get an accurate torque reading.

Technician B states that it is safer to pull the torque wrench toward you rather than push it.
Who is correct?

A. A only   **C.** Both A and B
B. B only   **D.** Neither A nor B

4. Impact wrenches are being discussed:
Technician A says that they are commonly used to remove wheel lug nuts.
Technician B says that they should not be used to tighten fasteners on critical parts.
Who is correct?

A. A only   **C.** Both A and B
B. B only   **D.** Neither A nor B

5. *Technician A* says that a dial indicator is an ideal tool for measuring disc brake rotor thickness.
Technician B says that a dial indicator often is used to measure brake lining thickness.
Who is correct?

A. A only   **C.** Both A and B
B. B only   **D.** Neither A nor B

6. *Technician A* says that using a pressure bleeder is the fastest way to bleed a brake system.
Technician B says that with a pressure bleeder, one technician can do the job.
Who is correct?

A. A only   **C.** Both A and B
B. B only   **D.** Neither A nor B

7. Cleaning asbestos from a brake system is being discussed:
Technician A says that an air gun can be used to blow dust off if the worker is wearing a HEPA-type respirator.

Technician B says that an air gun can be used only if the work area is enclosed in a negative-pressure containment enclosure.
Who is correct?

A. A only   **C.** Both A and B
B. B only   **D.** Neither A nor B

8. *Technician A* says that circuit continuity can be tested with a self-powered test light.
Technician B says that digital multimeters are preferred over analog meters when working with electronic control systems.
Who is correct?

A. A only   **C.** Both A and B
B. B only   **D.** Neither A nor B

9. *Technician A* refers to the vehicle service manual for specific torque values when installing caliper guide pins.
Technician B refers to the service manual to find the proper lift points of the vehicle.
Who is correct?

A. A only   **C.** Both A and B
B. B only   **D.** Neither A nor B

10. *Technician A* says that using a high-pressure spray hose is a preferred way to remove asbestos from brake parts.
Technician B says that all component washing should be at low pressure or done by hand.
Who is correct?

A. A only   **C.** Both A and B
B. B only   **D.** Neither A nor B

Job Sheet 1

Name: _____ Date: _____

Linear Measurement Practice

NATEF Correlation

This job sheet addresses the following NATEF task: Measure brake pedal height; determine necessary action.

Upon completion of this job sheet, you should be able to make measurements using a rule and a micrometer.

Tools and Materials

Vehicle

Rule

Micrometer

Describe the Vehicle Being Worked On

Year _____ Make _____ Model _____

VIN _____ Engine type and size _____

Procedure

NOTE TO INSTRUCTORS: Other devices may be used to check the student's skill with measuring instruments.

1. Consult with the instructor for the use of a vehicle or other device to be measured with a rule.

> ⚠ **WARNING:** Ensure that wheel blocks are placed in front of and behind at least one wheel and the parking brake is set. Damage or injury could occur if the vehicle should move when the clutch is disengaged, or the automatic transmission is not in "PARK."

2. Use a rule to determine the following brake pedal heights. If vehicle has a manual transmission, the clutch pedal can also be used for the same measurements for additional practice.

Engine off

Brake pedal topmost height from floor _____

Brake pedal height at brake engagement (free play) _____

Brake pedal height at full brake application _____

If power assisted, pump pedal several times and measure pedal height from floor with full brake application _____

If power assist, start the engine.

Task Completed

☐

Brake pedal topmost height from floor _____

Brake pedal height at brake engagement (free play) _____

Brake pedal height at full brake application _____

Optional. Based on your training to this point, have you determined if a defect exists as shown by your measurements and observations? If so, what? Wrong guesses count as two (2) answers.

_____

_____

_____

3. Measure a brake rotor or similar device for thickness using a brake micrometer. Specify the measurement units and then convert to other measurement units, that is, metric to U.S. customary. Make the measurements at least six different points around the device.

1 _____ 2 _____ 3 _____ 4 _____ 5 _____ 6 _____

7 _____ 8 _____ 9 _____ 10 _____ 11 _____ 12 _____

Optional. Based on your training to this point, have you determined if a defect exists as shown by your measurements and observation? If so, what? Wrong guesses count as two (2) answers.

_____

_____

_____
Task Completed

4. INSTRUCTORS: Additional space is provided for more measurements if deemed necessary.

1 _____ 2 _____ 3 _____ 4 _____ 5 _____ 6 _____

7 _____ 8 _____ 9 _____ 10 _____ 11 _____ 12 _____

1 _____ 2 _____ 3 _____ 4 _____ 5 _____ 6 _____

7 _____ 8 _____ 9 _____ 10 _____ 11 _____ 12 _____

☐

5. Clean the area and store the tools.

Problems Encountered _____

_____

Instructor's Response _____

_____

Job Sheet 2

Name: _____ Date: _____

Electrical Measurement Practice

NATEF Correlation

This job sheet addresses the following NATEF task: Demonstrate the proper use of a digital multimeter (DMM) during diagnosis of electrical circuit problems.

Upon completion of this job sheet, you should be able to make measurements using a digital multimeter.

Tools and Materials

Vehicle

Digital multimeter

Describe the Vehicle Being Worked On

Year _____ Make _____ Model _____

VIN _____ Engine type and size _____

Procedure

NOTE TO INSTRUCTORS: Other electrical systems or devices may be used for measurements. The ones selected for this exercise are for convenience on the vehicle.

1. Raise and support the hood. Locate and describe the battery if visible.

Task Completed

CCA _____ Cleanliness/general condition _____

_____ ☐

2. Make the following measurements with the multimeter.

Battery voltage (red lead + post, black lead – post) _____V

Voltage drop, positive side (red lead + battery cable terminal, black + post) _____V

Voltage drop, negative side (red lead – post, black lead – cable terminal) _____V

Voltage drop, post to case (red lead + cable terminal, black to case) (measure at several points on top of case) _____V _____V _____V.

3. Based on your training to this point and your observations, is the battery charged and serviceable? _____

4. Select the headlight with the most accessible electrical connection or consult the instructor for a different circuit.

5. Disconnect the headlight. ☐

6. How many and what colors are the wires leading to the connection? _____ _____

_____

☐   **7.** Connect the multimeter's black lead to a good ground.

☐   **8.** Use the multimeter to determine which of the wires is grounded.

☐   **9.** Set the multimeter to ohms or resistance.

☐   **10.** Ensure that the power to the lights is off.

11. With the black lead connected to ground, touch the red lead to each of the conductors in the wiring connector. Record the meter's reading at each conductor.

Wire 1 _____ Wire 2 _____ Wire 3 _____

Wire 4 _____ Wire 5 _____ Wire 6 _____

12. Based on the readings, which is the grounded wire? _____

☐   **13.** Move the black lead to the ground wire cavity at the connector end. Check the connection by placing the red lead on a ground. The reading should be the same as that recorded in step 12.

☐   **14.** If the connection is correct, leave the black wire in place.

☐   **15.** Switch the multimeter to volts DC.

☐   **16.** Turn on the power to the headlights.

17. With the black connected to the ground wire, probe each of the other cavities for voltage. Record the results.

Wire 1 _____ Wire 2 _____ Wire 3 _____

18. Based on your measurement, which wire(s) is(are) the feed wire(s) to the low-beam filament on this lamp? _____

_____

19. Switch the headlights to high beam and repeat step 17.

Wire 1 _____ Wire 2 _____ Wire 3 _____

20. Based on your measurement, which wire(s) is(are) the feed wire(s) to the high-beam filament on this lamp? _____

_____

☐   **21.** Turn off the power, disconnect and store the multimeter, and reconnect the headlights. Test the headlights and front running lights for operation.

☐   **22.** Clean the area and store the tools.

Problems Encountered   _____

_____

Instructor's Response   _____

_____

_____

Job Sheet 3

3

Name: _____ Date: _____

Vehicle Service Data

NATEF Correlation

This job sheet addresses the following NATEF task: Remove, clean (using proper safety procedures), inspect, and measure brake drums; determine necessary action.

Upon completion of this job sheet, you should be able to collect and use specific vehicle data.

Tools and Materials

Vehicle or brake drum(s) (number instructor choice)

Brake drum caliper (manual or electronic)

Service manual (paper or computerized)

Describe the Vehicle Being Worked On

Year _____ Make _____ Model _____

VIN _____ Engine type and size _____

Procedure

Task Completed

1. Use paper or computerized service manuals to collect the following service information.

    Drum discard diameter _____

    Drum discard diameter _____

    Drum discard diameter _____

    Drum discard diameter _____ ☐

2. Use the brake drum diameter to measure each drum's diameter.

    Drum diameter _____

    Drum diameter _____

    Drum diameter _____

    Drum diameter _____ ☐

3. Compare the measurements against the specifications. Determine if the drum(s) are serviceable.

    Drum diameter   Yes _____ No _____

    Drum diameter   Yes _____ No _____

    Drum diameter   Yes _____ No _____

    Drum diameter   Yes _____ No _____ ☐

4. Of the serviceable drum(s), state the amount of metal that can be removed during machining and still retain serviceability.

Drum _____

Drum _____

Drum _____

Drum _____

☐

Problems Encountered _____

_____

Instructor's Response _____

_____

_____

Job Sheet 4

4

Name: _____ Date: _____

Opening a Service Repair Order

NATEF Correlation

This job sheet addresses the following NATEF task: Complete work order to include customer information, vehicle identifying information, customer concern, related service history, cause, and correction.

NOTE TO INSTRUCTORS: A simple repair order is included on the following page for the use of this job sheet and others as desired. Copy and/or adjust as desired. Customer and vehicle information may be changed as desired.

Upon completion of this job sheet, you will be able to open or start a repair order based on supplied customer and vehicle information.

Tools and Materials

Service manual (paper or computerized)

Procedure

    1. Open the repair order using the following information.

Task Completed

       Customer: Joe Somebody

                    111 Run Everywhere Road

                    Anywhere, Any State

                    Phone: 111-111-1111

                    Cell: 222-222-2222

       Vehicle: 1996 Chrysler Town and Country. 3.3L

                    VIN 1FA3345ZA1C12456 (Fictitious)

                    Mileage   105336 miles

       Customer complaint:

       Customer smells gasoline when exiting the vehicle.   ☐

       No service history

    2. Complete the repair order.   ☐

Problems Encountered _____

_____

Instructor's Response _____

SERVICE REPAIR ORDER

CUSTOMER _____ DATE _____

PHONE _____ CELL _____

ADDRESS

_____

_____

VEHICLE VIN _____

MAKE _____ YEAR _____ MODEL _____ ENGINE _____

MILEAGE IN _____ MILEAGE OUT _____

PROBLEM/SYMPTOM

_____

_____

_____

_____

SERVICE HISTORY

_____

_____

RECALL _____ YES _____ NO _____ RECALL NUMBER _____

TECHNICAL SERVICE BULLETIN

_____

_____

SPECIFICATIONS RELATING TO REPORTED PROBLEM(S)

_____

_____

_____

DIAGNOSIS

_____

_____

_____

REPAIR ACTION RECOMMENDED

_____

_____

REPLACEMENT PARTS REQUIRED

_____

_____

_____

LABOR TIME REQUIRED (ACTUAL OR GUIDE)

_____

Problems Encountered _____

_____

Instructor's Response _____

_____

_____

Related Systems Service

Upon completion and review of this chapter, you should be able to:

❏ Isolate brake problems from those originating in other systems.

❏ Check and inspect tires and inflate them to proper pressures.

❏ Install tires and wheels and torque wheel nuts or bolts to specifications.

❏ Check radial and lateral runout of wheels and tires and correct if possible.

❏ Troubleshoot braking problems related to wheel bearings.

❏ Remove, clean, inspect, lubricate, reinstall, and adjust tapered roller wheel bearings.

❏ Inspect steering suspension and wheel alignment for possible causes of braking problems.

Isolating Brake Problems

There are times when what appears to be a brake problem is actually a problem caused by a defect in a related system. Problems of this type are usually caused by tires or alignment. However, damaged or worn components in the steering or suspension may create problems in several systems. Experienced technicians can quickly locate the root cause using their experience and a good working knowledge of the systems involved. The first step is to listen to what the customer is saying. Ask questions on when, what, where, and what conditions. The second step is to road test the vehicle if the problem is not obvious.

Before road testing the vehicle, conduct a visible inspection of the outside of the vehicle and perform some basic operational checks inside the vehicle. Make the following checks:

❏ Do the tires appear to be properly inflated?
❏ Do all tires have acceptable tread?
❏ Check the steering wheel for looseness or free play.
❏ Check the brake pedal with the engine off and operating.
❏ Does the vehicle appear to be safe for a road test?

Conduct the road test on a smooth level road with minimum traffic if possible. Listen, feel, and observe for any of the following:

❏ Noise, binding, or jerking of the steering wheel
❏ Pulling or drifting to one side while the vehicle is moving
❏ Pulling or drifting as the brakes are applied lightly and then firmly
❏ Movement of the steering wheel or brake pedal during braking
❏ Noise when the brakes are applied

After completing the road test, use the following guidelines to separate actual brake problems from those caused by the related systems:

Symptom	Cause in order of probability
Steering wheel movement/binding	Steering mechanism, wheel bearing
Pull/drift to one side, brake off	Tire inflation, alignment, dragging brake shoes/pads, wheel bearing
Pulling/drifting, brake applied	Brake inoperative one side, excessive positive camber
Steering wheel/brake pedal movement	Brake rotor/drum, wheel bearing, tire tread
Noise, brakes applied	Brakes pads/shoe, wheel bearing

Basic Tools

Basic technician's
  tool set
Service manual

Special Tools

Tire gauge

Although the guidelines are general in nature, they provide a starting point for the diagnosis. Usually when a customer complains of a brake problem, it *is* a brake problem, but the technician must be aware of contributing causes that affect braking action. One of the most common causes is a simple lack of maintenance by the owner.

Tire and Wheel Service

Tire and wheel problems can reduce the effectiveness of even the best brake service and, in some cases, affect the operation of the brakes themselves. The tire and wheel problems that can affect brake performance the most are tire condition, inflation pressure, and tire and wheel installation methods. The following sections explain the inspection and service procedures for troubleshooting and fixing brake problems associated with wheels and tires.

Tire and Wheel Inspection

Tires that are badly worn, mismatched in size or tread condition, or incorrectly inflated can cause brake problems. Simple inspection and checking with a pressure gauge can eliminate many of these problems.

Tire Inflation. Improper inflation is the most common tire problem, and underinflation is far more common than overinflation. Do not rely on your eye to tell you if a tire is inflated properly; use a pressure gauge. Overinflation and underinflation both will change the tire contact patch and affect brake performance. Unequal inflation from side to side can cause brake pull.

Several myths and misunderstandings work against proper tire maintenance. One myth that has persisted for decades is that the sidewalls of radial ply tires are supposed to bulge. That may have been the case in the late 1960s when radials were built with higher, 78- and 80-series profiles and ran at 32-psi pressure. Today, with higher-pressure 50-, 60-, and 70-series low-profile radials, that old myth is just an excuse to ignore proper inflation pressure. A low-profile radial ply tire can be 10 to 20 pounds underinflated before it will show a noticeable bulge in its sidewall.

Another equally old myth is that radial tires should only be switched front to rear on the same side of the car. Manufacturing technology in the 1960s and 1970s was not what it is today. Tire molds were not as precise, and radial plies were known to acquire a rotational set or direction. In the late 1960s and early 1970s, many carmakers did recommend keeping radials continuously on the same side of the car. That has not been the case for years, however, except for high-performance directional tires. Follow the rotation patterns in current tire service manuals and owner's manuals (Figure 3-1).

> ✔ **SERVICE TIP:** At times a pull to one side may develop after a tire rotation. This may be caused by lack of previous rotation or by some type of tread problem; that is, the tire ran underinflated. Swap the two front tires side to side and road test the vehicle. In most instances, the pull will be corrected. In a few instances, it may be necessary to "unrotate" the tires or install two new tires on the front.

Problems caused by underinflation and lack of rotation can be compounded on tires with a mud-and-snow tread design. When a tread block hits the pavement during braking, it has the same effect as a pencil eraser hitting a sheet of paper. The rubber wears at an angle. If the tires are not rotated to reverse the direction of wear, the wear becomes excessive and leads to chunking. Underinflation increases this kind of wear because the rubber tends to dig into the pavement more.

Mismatched Tires and Tire Wear. Equal braking performance at each wheel requires that the tires at all four wheels be the same type, size, and condition. Tires of different size, construction, or tread design will have different traction characteristics.

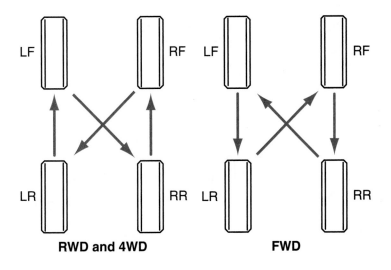

RWD and 4WD   **FWD**

Figure 3-1 These tire rotation patterns are recommended by most carmakers and by the Rubber Manufacturers Association.

Unequal tire diameters on the same axle can cause the vehicle to pull during braking and to wander under other driving conditions. Tire diameter can be measured when troubleshooting brake problems by placing a straightedge horizontally across the wheel and tire at the wheel center point. Then place smaller straightedges across the tread, perpendicular to the tire sidewall. A better, more accurate method is to measure the circumference of the tire, inflated and loaded, using a tape measure (Figure 3-2).

The best way to eliminate tire differences as a cause of brake problems is to mount identical tires on all four wheels. If this is not possible, tires of the same size and tread condition should be mounted on each axle.

All modern car and light truck tires have tread wear indicators that appear as continuous bars across the tread when the tread wears down to the last $\frac{2}{32}$ ($\frac{1}{16}$) inch (Figure 3-3). When a tread wear indicator appears across two or more adjacent grooves, the tire should be replaced. You also can measure tread depth with a tire tread gauge. Although $\frac{2}{32}$ inch is the minimum safe tread depth, a tire should be replaced before it wears to this point to ensure vehicle safety and proper braking performance. Tire wear can be due to several causes (Figure 3-4).

Incorrectly sized tires can cause engine, vehicle, and ABS problems unless all of the sensors are replaced with ones to match tire size.

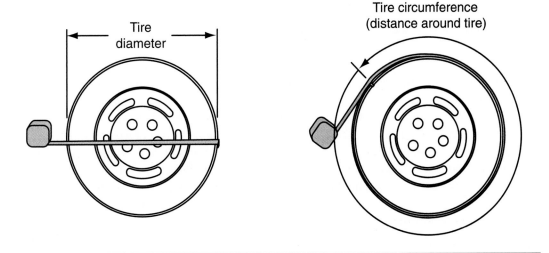

Figure 3-2 A method to determine the rolling circumference of an inflated and loaded tire.

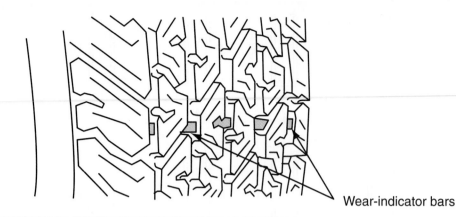

Wear-indicator bars

Figure 3-3 When wear-indicator bars appear across several tread grooves, the tire is ready for replacement.

Conditions	Rapid wear at shoulders	Rapid wear at center	Cracked treads	Wear on one edge	Feathered edge	Bald spots	Scalloped wear
Effect	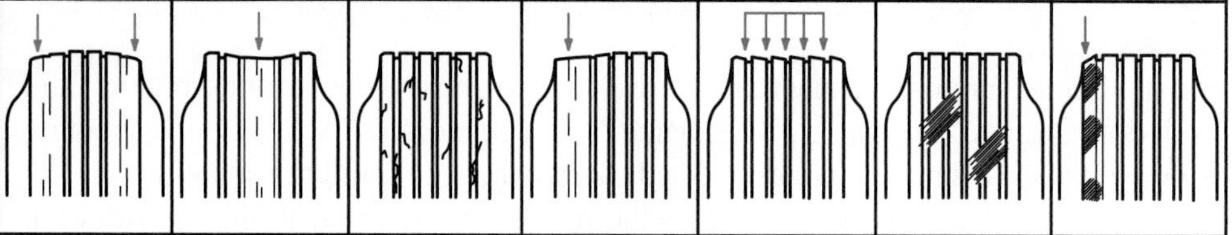						
Causes	Underinflation or lack of rotation	Overinflation or lack of rotation	Underinflation or excessive speed	Excessive camber	Incorrect toe	Unbalanced wheel Or tire defect	Lack of rotation of tires or worn or out-of-alignment suspension
Corrections	Adjust pressure to specifications. When tires are cool, rotate tires.			Adjust camber to specs	Adjust toe to specs	Dynamic or static balance wheels	Rotate tires and inspect suspension

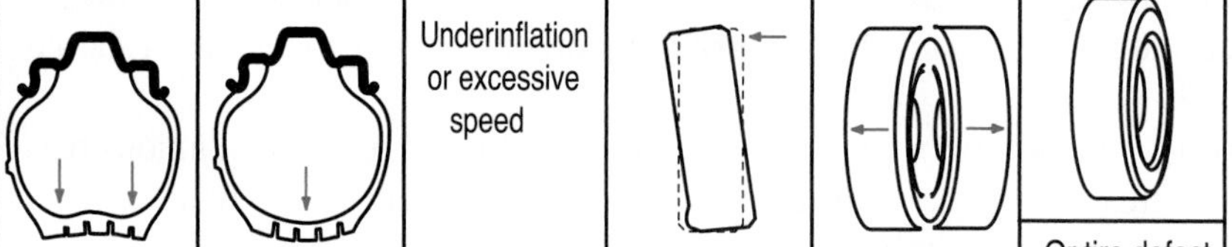

Figure 3-4 Tire wear may be caused by several problems related to the suspension and steering system and lack of preventive maintenance.

Uneven braking also can result from variations in tread pattern and condition from side to side and from front to rear. Dry and wet pavement will affect braking performance based on tread condition. A slick tire may have a high coefficient of friction on dry pavement and good stopping power. On wet pavement, the same slick tire will **hydroplane** on a layer of water trapped between the tread and the pavement. Hydroplaning occurs when the tire is lifted from the road surface by the water on the road. Consider that a water skier skis on top of the water, but he or she cannot ski *through* the water. Because liquid cannot be compressed, the water on the road acts the same as the water acts on waterskis, lifting the tire and eliminating friction. It must be remembered that new tires with good tread may also hydroplane if the tread is not formed to shed or direct water from under the tire. The traction and braking performance are then the opposite of what they are on dry pavement.

Tire and Wheel Installation

Two factors of tire and wheel installation that affect braking are the wheel nut or bolt torque and the sequence in which the wheel nuts or bolts are tightened. If wheel nuts or bolts are overtightened or tightened unevenly or in the wrong sequence, the brake drum or rotor can be distorted. This distortion creates runout in the drum or rotor that, in turn, causes pedal pulsation and uneven brake application.

☑ **SERVICE TIP:** There is a special tool available that will reduce cleaning time and efficiency. It is a drill-driven wire brush that fits over the wheel stud and cleans the hub and/or rotor area directly around the stud. It is also used to clean the remainder area of the hub and/or rotor.

To install a tire and wheel, be sure that the mounting surfaces of the wheel and the brake drum or rotor are free of dirt, rust, burrs, or any condition that could prevent proper installation and create wheel runout. Clean the wheel, drum, or rotor with a wire brush if necessary. Lift the wheel straight onto the drum or rotor, being careful not to damage the threads on the studs. Then install the nuts or bolts. Do not use any lubricant or antiseize compound on wheel nut or bolt threads. Lubricants and antiseize compounds will increase the torque applied to the nuts or bolts and result in overtightening.

Every vehicle manufacturer specifies torque values for wheel nuts and bolts. For most passenger cars, these range from 50 to 100 foot-pounds. Specifications for individual vehicles can be found in tire service manuals and owner's manuals. Again, these are dry specifications, without any lubricant or antiseize compound.

An impact wrench can be used to remove a wheel from a vehicle, but should not be used to reinstall the wheel unless a torque stick is used. Wheel nuts and bolts should always be installed with a torque wrench or with a specified torque stick used with an impact wrench (Figure 3-5). A torque stick is an extension for an impact wrench that includes the correct size socket for the wheel nut and that acts as a torsion bar to limit the torque applied by the impact wrench. Torque

Special Tools

Lift or jack with stands

Impact tools

Torque wrench

Special Tools

Lift or jack with stands

Dial indicator

Torque stick

Figure 3-5 A torque stick can be used with an impact wrench for safe wheel installation. Never use a torque stick to loosen fasteners.

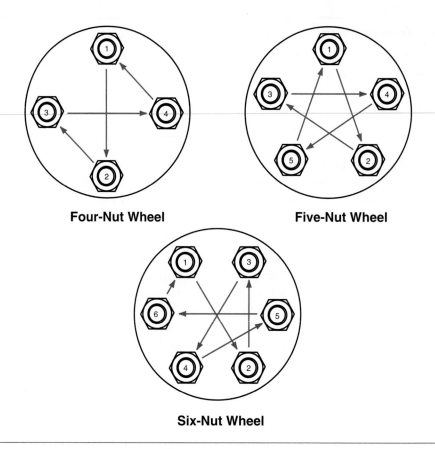

Four-Nut Wheel

Five-Nut Wheel

Six-Nut Wheel

Figure 3-6 Wheel nut tightening patterns.

sticks come in sets that contain several sticks with specific torque ratings. Do not use a torque stick to remove a wheel nut or bolt.

> **CAUTION:** A torque stick that is used to remove fasteners can shatter. The splinters can embed in the shin or eyes and cause serious injury. Never use a torque stick to loosen fasteners.

The sequence in which the wheel nuts are tightened is as important as the torque value. All tightening sequences are crisscross patterns in which the nuts or bolts are tightened alternately from one side of the wheel to the other (Figure 3-6). Of equal importance, the nuts or bolts should not be tightened to the final torque value in one step. They should be tightened in three or four stages to one-half, three-quarters, and full torque. Then they should be retightened to the final value a second time in the specified sequence.

Checking Tire and Wheel Runout

Tire and wheel runout can be factors in troubleshooting braking performance. Excessive radial runout can prevent a tire from maintaining uniform traction, which can cause braking force to vary as the tire rotates. Excessive lateral runout can cause the contact patch to shift as the tire rotates and also can contribute to uneven braking.

When you measure radial or lateral runout, measure at the tire first because this will indicate the total runout of both the wheel and the tire. If runout at the tire is out of limits, then measure at the wheel to isolate the problem to either the wheel or the tire or to tolerances that have accumulated in one direction during mounting.

SERVICE TIP: If you are troubleshooting a low-pedal complaint on a RWD vehicle with rear disc brakes, do not overlook rear axle runout as a possible cause. Excessive lateral runout in an axle flange can become rotor runout that knocks the caliper piston farther back in its bore. The pedal then must travel farther to take up the extra pad-to-rotor clearance. This was a common problem on old Corvettes with Delco Moraine fixed-caliper rear discs.

Before measuring runout, be sure the wheel bearings are not loose. Some technicians prefer to tighten the bearings temporarily while measuring runout or to measure runout with the tire and wheel mounted on a wheel balancer to eliminate the wheel bearings as a cause of runout. If you measure runout with the wheel and tire mounted on the vehicle, be sure the wheel nuts are tightened properly as explained previously. Finally, drive the vehicle for several miles to warm the tires and eliminate any flat spots that may have developed from sitting stationary.

A special dial indicator with a small roller on the end is the best tool to use for measuring tire runout. The roller keeps the indicator from catching on the tread or uneven spots on the wheel or tire. Measure radial runout first (Figure 3-7), then lateral runout (Figure 3-8), as follows:

1. Raise the vehicle on a hoist and support it safely so that the wheels can turn freely.

✔ **SERVICE TIP:** Place masking tape, aligned with the center tread, around the circumference of the tire. The tape will help prevent the tip of the dial indicator from hanging on the tread and provide a more accurate reading.

2. Position the pointer of the dial indicator against the center tire tread where it can contact a uniform section of tread around the circumference (see Figure 3-7A).

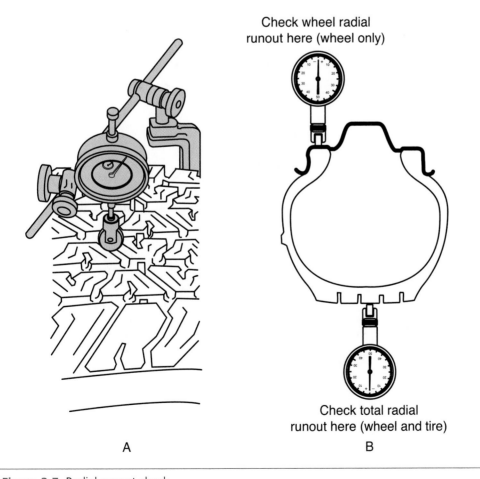

Check wheel radial
runout here (wheel only)

Check total radial
runout here (wheel and tire)

A   B

Figure 3-7 Radial runout check.

3. Rotate the tire until you get the lowest reading on the dial indicator. Then set the indicator to zero.

4. Rotate the tire until you get the highest reading on the dial indicator. This is the total radial runout of the tire and the wheel, and it should not exceed 0.060 inch (1.5 mm). Mark the location on the tire. If radial runout is excessive, proceed to check the runout of the wheel alone. If radial runout is within limits, proceed to step 8 and check lateral runout.

5. To check the radial runout of the wheel alone, reposition the dial indicator pointer to the inside of the rim flange (see Figure 3-7B).

6. Rotate the wheel until you get the lowest reading on the dial indicator. Then set the indicator to zero.

7. Rotate the wheel until you get the highest reading on the dial indicator. This is the radial runout of the wheel, and it should not exceed 0.040 inch (1.0 mm) for a steel wheel or 0.030 inch (0.75 mm) for an alloy wheel. Mark the location on the wheel.

8. Measure lateral runout of the wheel and tire by repositioning the dial indicator plunger to the tire sidewall, just below the wear rib on the sidewall (see Figure 3-8A). Do not place the pointer against the wear rib because any damage or wear on this area will cause a false measurement.

9. Rotate the tire until you get the lowest reading on the dial indicator. Then set the indicator to zero.

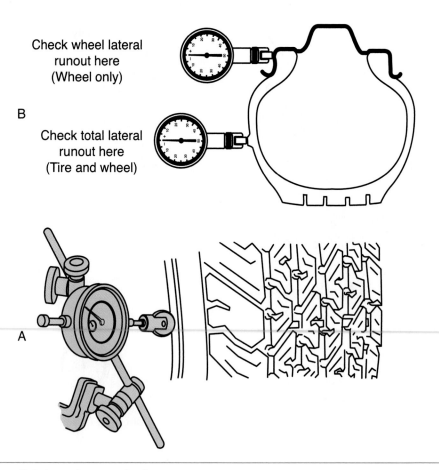

Figure 3-8 Lateral runout check.

10. Rotate the tire until you get the highest reading on the dial indicator. This is the total lateral runout of the tire and the wheel, and it should not exceed 0.080 inch (2.0 mm). Mark the location on the tire. If lateral runout is excessive, proceed to check the runout of the wheel alone.

11. To check the lateral runout of the wheel alone, reposition the dial indicator pointer to the side of the rim flange as shown in (Figure 3-8B).

12. Rotate the wheel until you get the lowest reading on the dial indicator. Then set the indicator to zero.

13. Rotate the wheel until you get the highest reading on the dial indicator. This is the lateral runout of the wheel, and it should not exceed 0.045 inch (1.15 mm) for a steel wheel or 0.030 inch (0.75 mm) for an alloy wheel. Mark the location on the wheel.

Radial runout is more often a cause of braking problems than is lateral runout, but neither should be overlooked. Runout often can be reduced, if not eliminated altogether, by repositioning the tire on the rim. When measuring runout, mark the locations of maximum radial and lateral runout for both the wheel and the tire. You can reduce excessive runout by remounting the tire so that the points of maximum runout are opposite each other. Rebalancing will be required if the tire is remounted on the wheel. Repositioning the tire and wheel assembly on the brake drum or rotor also may help to reduce runout.

Radial runout also may be reduced by shaving the tire on a tire truing machine. Some tires also may have abnormally stiff sections in their sidewalls and areas of the tread. These stiff sections keep a tire from rolling concentrically and uniformly when loaded. Although this condition is hard to detect when vehicle weight is unloaded from the tire and wheel, some wheel balancers have the capability to load the tire and identify hard spots and stiff areas.

☑ **SERVICE TIP:** The sealed, preloaded, zero-clearance front wheel bearings on late-model cars require little or no toe-in. This lack of toe-in is great to reduce rolling resistance and improve fuel economy, but bearings with no end play cannot absorb small amounts of hub and rotor runout. This makes rotor runout and thickness measurements very critical on these cars.

Tapered Roller Bearing Service

The wheels and axles of late-model cars and light trucks are supported by tapered roller bearings, straight roller bearings, or ball bearings. Straight roller bearings and ball bearings do not require periodic service or adjustment. They are replaced only when defective, and the procedures are outside the scope of a brake service text. Details on this service may be found in Thomson Delmar Learning's *Today's Technician Automotive Suspension and Steering.*

Some vehicles have sealed double-row ball bearings or tapered roller bearings in an assembly that includes the wheel hub, the bearing races, and a mounting flange. They may be used at the front drive axles on FWD cars or on nondriving wheels. These are nonserviceable, nonadjustable assemblies that must be replaced if damaged or defective.

Tapered roller bearings are used on the front or rear nondrive wheels of many vehicles. Tapered roller bearings must be cleaned and repacked with grease periodically and then adjusted (Figure 3-9). These bearing services are usually part of a complete brake job. If the bearings are too tight, the hub and brake assembly may overheat with accompanying problems that include brake fade. If wheel bearings are too loose with too much end play, wheel runout may be excessive.

Tapered Roller Bearing Troubleshooting

Bearings rarely fail suddenly. Rather, they deteriorate slowly from dirt, lack of lubrication, and improper adjustment. Bearing wear and failure are almost always accompanied by noise.

Certain bearing services are usually included as part of the brake service.

Classroom Manual
pages 50–52

Figure 3-9 Although many technicians prefer to use hand packing, this bearing packer is easy to use and less messy.

Start diagnosing possible wheel bearing problems by raising the vehicle on a hoist so that the wheels can turn freely. Rotate the wheel and listen to the sounds of brake drag and bearing rotation. If the brakes are dragging slightly, they will produce a high-pitched swishing sound. Bearing noise will be a lower-pitched rumble. Low-level bearing noise and brake drag are normal. Bearing sounds should be uniform through the wheel revolutions. An uneven rumble or a grinding sound indicates possible bearing problems.

If a bearing is unusually noisy, roughness can usually be felt as the wheel is rotated. Hold onto the wheel; listen for a low-frequency rumbling sound and feel for rough and uneven rotation. Compare left and right wheels to determine whether a problem may exist.

While rotating the wheel, try to move it in and out on the spindle and note the amount of axial movement. Worn or damaged bearings or bearings that need adjustment will have a noticeable amount of end play. Also grasp the top and bottom of the tire and try to wobble the wheel and tire back and forth. There should be little or no wobble in a properly adjusted bearing that is in good condition.

Measure bearing end play precisely with a dial indicator placed against the wheel hub (Figure 3-10). Set the indicator to zero, move the wheel in and out on the spindle, and note the reading. Tapered roller bearings can have 0.001 inch to 0.005 inch (0.025 mm to 0.127 mm) of end play. Check the vehicle manufacturer's specifications for the exact amount of allowable end play.

Finally, inspect the wheel and the brake drum or rotor for grease being expelled through a leaking seal. Bearing grease can contaminate brake linings, and a leaking grease seal can let dirt into the bearing. A leaking seal must be replaced, but the bearings also must be cleaned, inspected, and repacked to be sure they have not been damaged.

SERVICE TIP: You can use a noncontact pyrometer or infrared thermometer to solve a wheel bearing problem on a car with sealed wheel bearings. Drive the car for several miles at highway speeds and immediately use the pyrometer to measure wheel temperatures from side to side. If one wheel is hotter than the other, take a close look at the sealed wheel bearing on that side. The same kind of test can help you pinpoint constant velocity (CV) joint problems on FWD cars and dragging brakes on any vehicle.

Bearing Service Guidelines

The procedures to remove, clean, lubricate (repack), and install tapered roller bearings are basically the same for servicing a used bearing or installing a new one. The outer races or cups are

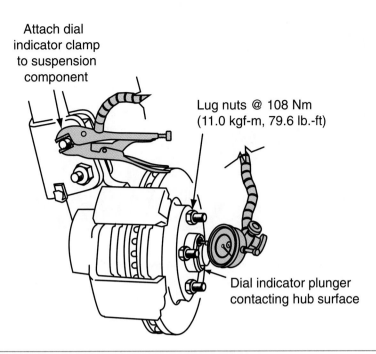

Attach dial indicator clamp to suspension component

Lug nuts @ 108 Nm (11.0 kgf-m, 79.6 lb.-ft)

Dial indicator plunger contacting hub surface

Figure 3-10 Notice that the lug nuts are on and torqued. Also note where the plunger of the dial indicator is placed.

pressed into the hub and must be driven or pressed out when installing a new bearing. A bearing and its outer race must be replaced as a set. Installing a new bearing in a used race will cause an uneven wear pattern that can lead to premature bearing wear or failure.

When installing a new bearing, leave the bearing in its protective packaging until you are ready to lubricate it. Be sure your hands are clean before handling a bearing. Some technicians prefer to apply a light film of motor oil to their fingers before handling a bearing. This prevents acid from their fingers from damaging the bearing.

Whether repacking and reinstalling a used bearing or installing a new one, always install a new grease seal. Seals wear with age and can be damaged during bearing removal. A new seal ensures that bearing grease stays in the hub where it belongs, not on the brake linings. A new seal also ensures that dirt and moisture stay out of the bearing.

Tapered Roller Bearing Removal

With drum brakes, you can leave the tire and wheel on the drum to remove the bearing if you wish. The drum and bearing are easier to handle with the weight of the wheel and tire removed, however. With disc brakes, the wheel and tire must be removed for access to remove the caliper (Figure 3-11). Remove inner and outer tapered roller bearings as follows:

1. Raise the vehicle on a hoist and support it safely so that the wheels can be removed.

2. Remove the wheel cover, if so equipped, and remove the wheel and tire from the drum or rotor.

3. If the axle has disc brakes, remove the brake caliper as explained in Chapter 7 of this *Shop Manual* and suspend it out of the way. Do not support the caliper by the brake hose.

4. Pull the dust cap from the center of the hub to expose the bearing nut (Figure 3-12).

5. Remove the cotter pin and nut lock.

Special Tools

Lift or jack with stands

Impact tools

Cotter pin puller

Seal puller

Brass punch

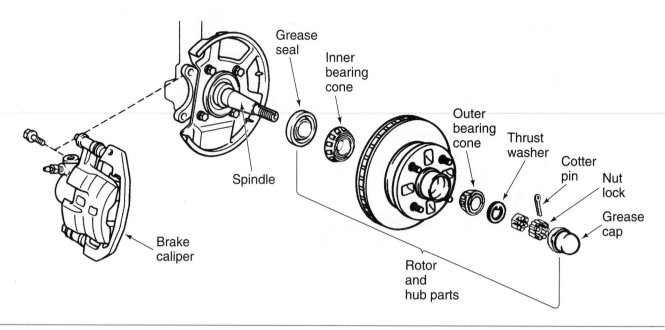

Figure 3-11 Typical hub and tapered roller bearing assembly.

6. Loosen and remove the adjusting nut.

7. Lift the outer bearing and thrust washer from the hub and set them aside for cleaning.

8. Place the thumb of one hand over the thrust washer outboard of the outer bearing and carefully slide the drum or rotor off the spindle. Support the drum or rotor with both hands so that it does not drag on the spindle as you remove it. If a brake drum catches on the shoes, reinstall the adjusting nut and back off the brake adjustment as explained in Chapter 8 of this *Shop Manual*. Then remove the drum.

9. Set the drum or rotor on a clean work surface with the inboard side down.

10. Remove the inner bearing and grease seal in one of the following ways (see Photo Sequence 3):

Figure 3-12 Removing the grease cap using grease cap pliers.

SERVICE TIP: There is an easy way to remove the inner bearing and seal on hubs with an internal bore large enough to let the adjusting nut slide through. Use this method only if the bearing is being replaced. After removing the outer bearing and thrust washer, screw the adjust nut onto the spindle several turns. Grip the hub, slide it toward the outer end of the spindle, then jerk it outward and downward. The adjusting nut will pull the inner bearing and seal from the hub. Remember, use this method only if the bearing is being replaced. Regardless of the reason for removing the seal, it must be replaced once removed.

    A. Turn the drum or rotor over, hook the claw of a seal puller against the inside of the grease seal, and lever the seal out of the hub. A large screwdriver can also be used to pry the seal out of the hub.

    B. Raise the drum or rotor on suitable blocks so that the inside of the hub is 1 inch to 2 inches off the bench. Then insert a large, nonmetallic drift punch through the outer bearing diameter and place it against the inner race of the inner bearing (Figure 3-13). Tap the bearing at several locations around its circumference until the bearing and seal are driven from the hub.

11. Set the inner bearing aside for cleaning or replacement.

12. If a new bearing is to be installed, remove the race by inserting a large drift punch through the opposite side of the hub and placing it against the inner edge of the race. Strike the punch with a hammer at several locations around the circumference of the race until the race is driven from the hub. A bearing race can be removed from a hub using a press and suitable adapters.

Tapered Roller Bearing Cleaning and Inspection

After the bearings are removed from the hub, wipe the old grease off the spindle and out of the center cavity of the hub with clean, lint-free shop cloths or paper towels. Examine the old grease

Special Tools

Parts washer

Cloths

Figure 3-13 Removing a bearing race from a hub.

Photo Sequence 3
Removing and Installing a Bearing Race

P3-1 Shown is the inner bearing race. Note the depth of the race inside the hub.

P3-2 Flip the hub over. Position the punch through the cavity and against the edge of the race.

P3-3 Strike the punch with medium force. Move the punch tip around the edge of the race to help keep the race even in the bore during removal.

P3-4 Observe the area where the race was removed from and ensure there is no damage from the punch or from the race spinning in its seat.

P3-5 Fit the race in as far as possible with hand force. Align the driver over and into the race.

P3-6 Use moderate hammer force to drive the race in place. When the race bottoms in the cavity, it will be noticeable because a different sound will come from the hammer and the driver.

P3-7 Check to ensure the race is completely in place. The space between the race and outer edge of the hub should be close to where the old race was positioned.

for metal particles from the bearings or their races. Any sign of metal particles in the grease is a clue to inspect the bearings very closely for wear and damage. Also inspect the old grease for dirt, rust, and signs of moisture that could indicate a leaking grease seal.

If the outer races for the bearings are still installed in the hub, wipe old grease off them with a clean, lint-free shop cloth and inspect them closely for wear or damage (Figure 3-14). Also wipe as much of the old grease as possible from the bearings with a clean, lint-free shop cloth or paper towel and inspect them similarly for wear or damage. Inspect the grease removed from the bearings for metal particles, rust, and dirt. Turn the rollers in the cage and listen and feel for roughness.

- ❏ *Galling.* **Galling** is indicated by metal transfer or smears on the ends of the rollers and is caused by overloading, lubricant failure, or overheating. Overheating is usually the result of adjustment that is too tight.

- ❏ *Etching.* **Etching** is a condition that results in a grayish-black bearing surface and is caused by insufficient or incorrect lubricant.

- ❏ *Brinelling.* **Brinelling** occurs when the surface is broken down or indented and is caused by impact loading, usually because the adjustment is too loose.

- ❏ *Abrasive wear.* Abrasive wear results in scratched rollers and is caused by dirty or contaminated bearing lubricant.

Clean the bearings in a parts washer, using clean petroleum-based solvent. Rotate the bearings in the cage as you clean them and remove old grease with a stiff-bristled brush. Wash each bearing separately and keep them sorted to be sure they are reinstalled in the same hubs from which they were removed.

CAUTION: Do not spin the bearings with compressed air during cleaning. High-speed rotation of an unlubricated bearing will damage it. Compressed air also can dislodge bearings from the cage and cause additional damage or injury.

After cleaning, flush all of the cleaning solvent from the bearings with a non-petroleum-based brake-cleaning solvent. Let the bearings dry in the air for a few minutes and then blow any remaining solvent from the bearings and cages with low-pressure compressed air. Direct the air through the bearing from side to side, along the axis of the rollers. Do not spin the bearings with compressed air. All solvent must be removed from the bearing in order for the new solvent to adhere to the bearing components. Carefully inspect the bearings after cleaning. Turn each roller completely around and check the surface. Carefully inspect each of the races also. Figure 3-15 shows a good bearing and some common bearing problems.

If the roller or race shows signs of any of the conditions shown in Figure 3-14 or if unsure of the bearing condition in any way, replace the bearing. A replacement bearing contains inner and outer races and the rollers in their cage. Always compare the old parts with their replacements to be sure the items are correct.

Tapered Roller Bearing Lubrication

Tapered roller bearings must be lubricated or packed with grease made especially for wheel bearing use. Disc brakes cause wheel bearings to run at higher temperatures than do drum brakes, and many wheel bearing greases are labeled for use with disc brakes. Always use the type of grease specified by the vehicle manufacturer.

Because wheel bearing grease must operate at higher temperatures than chassis grease, it contains oils and other lubricants that have been made particularly heavy with thickening agents.

Special Tools

Bearing packer

Tapered Roller Bearing Diagnosis

Consider the following factors when diagnosing bearing condition:
1. General condition of all parts during disassembly and inspection.
2. Classify the failure with the aid of the illustrations.
3. Determine the cause.
4. Make all repairs following recommended procedures.

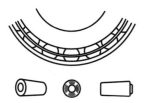

Abrasive Step Wear

Pattern on roller ends caused by fine abrasives. Clean all parts and housing; check seals and bearings and replace if leaking, rough, or noisy.

Galling

Metal smears on roller ends due to overheating, lubricant failure, or overload. Replace bearing, check seals, and check for proper lubrication.

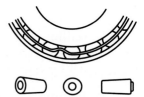

Bent Cage

Cage damaged due to improper handling or tool usage. Replace bearing.

Abrasive Roller Wear

Pattern on races and rollers caused by fine abrasives. Clean all parts and housings; check seals and bearings and replace if leaking, rough, or noisy.

Etching

Bearing surfaces appear gray or grayish black in color with related etching away of material, usually at roller spacing. Replace bearings, check seals, and check for proper lubrication.

Bent Cage

Cage damaged due to improper handling or tool usage. Replace bearing.

Indentations

Surface depressions on race and rollers caused by hard particles of foreign material. Clean all parts and housings. Check seals and replace bearings if rough or noisy.

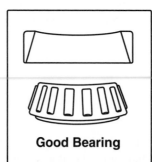

Good Bearing

Misalignment

Outer race misalignment due to foreign object. Clean related parts and replace bearing. Make sure races are properly sealed.

Figure 3-14 Good and bad bearings.

Figure 3-15 A good bearing and race (right) and a badly worn bearing and race (left).

Figure 3-16 A typical wheel bearing packer.

Hand packing is still the preferred packing technique among many technicians because the new grease can be seen moving through the bearing.

Different grease manufacturers use different thickening agents and other additives, so it is best to avoid mixing greases by removing as much of the old grease as possible before repacking a bearing.

Tapered roller bearings can be packed with grease most efficiently and thoroughly with a **bearing packer** (Figure 3-16). This packer uses a grease gun to insert the grease. Place the bearing in the packer with the taper pointing down and screw the cone down on the bearing. Apply a hand-operated grease gun to the fitting on the center shaft of the packer and force grease into the bearing

Figure 3-17 Packing bearings by hand.

and out around the rollers in the cage. Do not use a pneumatic grease gun to avoid spraying grease out of the bearing. Some bearing packers contain a supply of grease. Pushing on the cone handle forces the bearing down into the packer and lubricant in the tool is forced into the bearing.

If a bearing packer is unavailable, force grease by hand through the large end of the bearing and around the rollers. Rotate the bearing several times while packing it by hand to ensure that grease is forced completely around each roller (Figure 3-17). Finish by spreading an even film of grease around the outside of all the rollers. After packing the bearings, set them aside on a clean sheet of lint-free paper until you are ready to install them.

Before installing the bearings and the hub, inspect the spindle for wear or damage. Remove any burrs with a fine file or emery cloth. Spread a light coating of bearing grease around the entire spindle. Also spread fresh grease inside the hub to a level just below the bearing races. Applying grease to the spindle and hub, even to nonbearing surfaces, will protect against rust and other moisture damage. Grease on the bearing surfaces of the spindle also lets the inner bearing races creep slightly on the spindle and distribute the bearing load uniformly. Be careful to keep grease off the braking surfaces of the drum or rotor.

☑ **SERVICE TIP:** There is a myth that grease must completely fill the interior of the hub and the dust cap. This is wrong because the grease will expand when it heats up during normal vehicle operation. Excessive grease will force past the seal and once the seal is violated, a permanent leak will be present. A film of grease should be applied to the bearing races and only about one-quarter of the dust cup should be filled.

Tapered Roller Bearing Installation and Adjustment

Special Tools

Seal driver

Bearing driver

Bearing adjustment procedures are the same whether a used bearing is being repacked and reinstalled or a bearing is being replaced. If you are replacing a bearing, however, you must install a new race in the hub. Install and adjust tapered roller bearings as follows:

1. If the old bearing race has been removed from the drum, install a new one as follows:

   A. Apply a thin film of grease to the back of the race and to the bore in the hub to aid installation.

   B. Position the race squarely at the end of the bore and use a **bearing driver** (Figure 3-18) or a suitably sized socket and a hammer to drive the race into the hub. But care must be taken to ensure the socket fits the outer edge of the race.

   C. Be sure the tool contacts the outer edge of the race squarely.

   D. Listen for a change in the hammer sound as the race is seated in the hub.

2. Place the drum or rotor with the outer side down on a workbench.

3. Apply a light coat of grease to the outer race for the inner bearing. Then insert the freshly packed bearing into the race.

4. Place a new grease seal into the inner bore of the hub with the lip pointing inward (Figure 3-19).

Figure 3-18 Bearing driver used to install a new bearing race.

Figure 3-19 The grease seal lip must face inward in the hub when installed.

5. Drive the seal into place with a seal driver until the outer surface of the seal is flush with the hub (Figure 3-20).

6. Apply a thin film of bearing grease to the lip of the seal to protect it during installation on the spindle.

7. Turn the hub over and apply a light coat of grease to the outer race for the outer bearing. Then insert the freshly packed bearing into the race.

8. Install the rotor or drum carefully onto the spindle in a straight line and be careful not to hit the inner bearing on the spindle threads.

9. Support the rotor or drum with one hand and install the outer bearing and thrust washer into the hub.

10. Install the bearing adjusting nut finger tight against the thrust washer (Figure 3-21).

Figure 3-20 Install a new grease seal with the proper size driver.

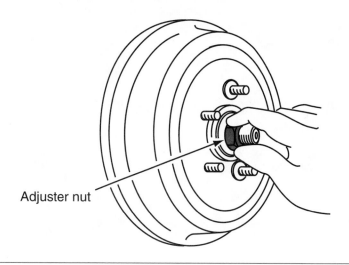

Adjuster nut

Figure 3-21 Install the spindle nut finger tight.

11. Rotate the drum or rotor by hand and lightly tighten the adjusting nut by hand to seat the bearings. Then adjust the bearings by one of the following methods (Figure 3-22):

   A. Rotate the drum or rotor and lightly snug up the adjusting nut with a wrench to seat the bearings. Then back off the nut one-quarter to one-half turn or until it is just loose while continuing to rotate the drum or rotor. Tighten the nut by hand to a snug fit and lock it as described later. This is a method used by experienced technicians. Beginners should use method B or C.

   B. Rotate the drum or rotor and tighten the adjusting nut with a torque wrench to the torque specified by the carmaker, which is usually 12 ft.-lb to 25 ft.-lb. Then back off the nut one-third turn and retorque it to the value specified by the carmaker while continuing to rotate the drum or rotor. Final torque is usually 10 to 15 in.-lb. Lock the adjusting nut as described below.

1. Hand spin the wheel.

2. Tighten the nut to 16 Nm (12 ft.-lb) to fully seat the bearings—this overcomes any burrs on threads.

3. Back off the nut until just loose.

Bend end of cotter pin legs flat against nut. Cut off extra length.

4. Hand "snug up" the nut.

5. Loosen the nut until a hole in the spindle lines up with a slot in the nut. Insert cotter pin.

6. When the bearing is properly adjusted there will be from 0.03 to 13 mm (0.001" to 0.005") of end play.

Figure 3-22 Typical bearing adjustment specifications.

C. To adjust bearing end play with a dial indicator, rotate the drum or rotor and tighten the adjusting nut with a torque wrench to 12 ft.-lb to 25 ft.-lb. Then back off the nut one-quarter to one-half turn or until it is just loose. Mount a dial indicator on the drum or rotor with its pointer against the spindle. Move the drum or rotor in and out and note the indicator reading. Turn the adjusting nut as necessary to obtain the specified end play, which is usually 0.001 inch to 0.005 inch (0.025 mm to 0.125 mm). Lock the adjusting nut as described below.

12. Install the nut lock over the top of the bearing adjusting nut so that the slots in the nut lock align with the cotter pin hole in the spindle (Figure 3-23).

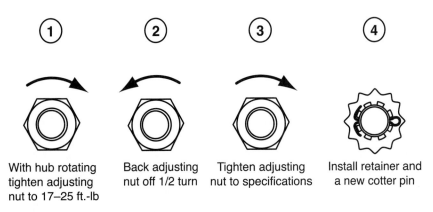

Figure 3-23 Typical spindle nut, nut lock, and cotter pin installation.

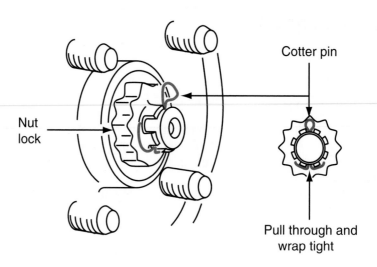

Figure 3-24 Wrap the ends of the cotter pin as shown here or cut them off to provide clearance for the grease cap.

13. Install a new cotter pin through the spindle and nut lock and bend its ends to secure it. Reinstall the dust cap in the hub (Figure 3-24).

14. If the axle has drum brakes that were backed off to allow removal of the drum, readjust the brakes as explained in Chapter 8 of this *Shop Manual*.

15. If the axle has disc brakes, reinstall the caliper as explained in Chapter 7 of this *Shop Manual*.

16. Reinstall the wheel and tire and lower the vehicle to the ground.

Photo Sequence 4 outlines the key points of bearing adjustment and provides additional details.

Wheel Alignment, Steering, and Suspension Inspection

Classroom Manual
pages 52–59

Problems in the steering and suspension systems that affect braking usually are related to a pull to one side or the other. And braking problems related to steering, suspension, and wheel alignment usually involve front-end components. Components that are damaged, worn, or otherwise loose make it hard to maintain directional stability. When braking forces are applied to loose parts, the pulling reaction often becomes more apparent.

Repairing steering and suspension problems and correcting wheel alignment are beyond the scope of brake service, but brake system troubleshooting often includes inspection of these other undercar systems.

For a thorough inspection, raise the vehicle on a hoist so that the wheels hang freely. Before lifting the vehicle, however, inspect the suspension as it sits on the shop floor. Look for any noticeable sagging or uneven ride height from side to side or front to rear. This condition often is a sign of broken or weak springs or other suspension parts. Examine the front and rear wheels as the vehicle weight is on them and look for any apparently extreme camber angles or toe angles. Look at each wheel by itself and compare right to left sides, front and rear.

Raise the vehicle on a hoist so that the wheels hang freely and inspect the tie rod ends, shock absorbers, and ball joints for looseness. Also check steering idler arms and pitman arms for

Photo Sequence 4
Typical Procedure for Adjusting Tapered Roller Bearings

P4-1 Always make sure the car is positioned safely on a lift before working on the vehicle.

P4-2 Remove the dust cap from the wheel hub.

P4-3 Remove the cotter pin and nut lock from the bearing adjusting nut.

P4-4 Tighten the bearing adjusting nut to 17 to 25 foot-pounds.

P4-5 Loosen the bearing adjusting nut one-half turn.

P4-6 Rotate the wheel while tightening the bearing adjusting nut to specification.

P4-7 Position the adjusting nut lock over the adjusting nut so the slots are aligned with the holes in the nut and spindle.

P4-8 Install a new cotter pin and bend the ends around the retainer flange.

P4-9 Install the dust cap and be sure the hub rotates freely.

looseness and worn or damaged bushings. Inspect shock absorbers and struts for fluid leakage and check all the mounting fasteners for looseness.

Grab the front tires by the treads and try to move the toe in and out. There should be no noticeable movement. Then try to move the steering linkage up and down while watching the front wheel toe angle. Toe angle should not change and there should be no noticeable looseness in the steering linkage.

Loose steering linkage can make it impossible to adjust the toe angle. Looseness in the linkage can cause enough toe angle change during acceleration and braking to cause a pull. Similarly, loose control arm or strut bushings may cause caster or camber angles to change during acceleration, braking, and cornering, which can cause or contribute to a brake pull.

Special Tools

Jack

Pry bar

Excessive ball joint wear is sometimes felt as a bump or heard as a popping sound as the brakes are applied.

A **wear-indicating ball joint** is a ball joint with a visual indicator to show the amount of wear on the joint.

☑ **SERVICE TIP:** When troubleshooting a problem of steering shimmy while braking on a car with four-wheel disc brakes, do not overlook the rear rotors as a possible cause. Severe runout and rearend shimmy can be felt through the steering wheel. If the front rotors check out within specs for runout and parallelism, take some close measurements on the rear rotors. Install good sealing plugs at the proportioning valve rear brake outlets or at the ends of the line or hose leading to the rear brake calipers or cylinders. This prevents fluid from entering the rear brakes. Do not clamp brake hoses. Clamping could damage the interior lining of the hose. Test stop the vehicle. If the steering shimmy goes away, suspect the rear rotors as the cause.

Ball Joint Inspection

Many cars have **wear-indicating ball joints.** These ball joints must be inspected with the vehicle weight resting on the joints. Some vehicles have ball joints on which the grease fitting recedes into the housing as the joint wears. Replace the ball joint if the shoulder of the fitting is flush with or receded into the ball joint housing (Figure 3-25). On other wear-indicating ball joints, try to wiggle the fitting by hand. If the fitting moves at all, replace the ball joint.

If a vehicle does not have wear-indicating ball joints, check the joints with the vehicle weight removed from them. Most of the wear will occur on the load-carrying joint, which must be unloaded to check it for looseness. If the spring is between the frame and a lower control arm, raise the vehicle on a jack and support the lower control arm as far outboard as possible. If the

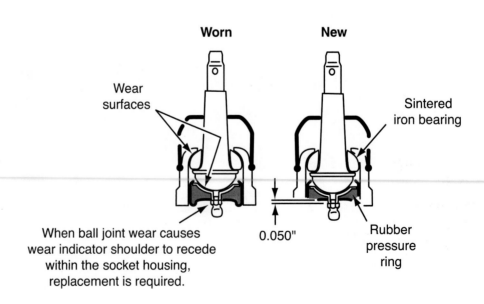

Worn   New

Wear surfaces

Sintered iron bearing

When ball joint wear causes wear indicator shoulder to recede within the socket housing, replacement is required.

0.050"

Rubber pressure ring

Figure 3-25 Typical wear-indicating ball joint.

spring is between the frame and an upper control arm or if the vehicle has Macpherson struts, raise the vehicle on a jack and support the vehicle inboard of the lower control arm.

With the vehicle supported as required, grasp the tire at the top and bottom and try to rock it in and out. Watch the ball joint, or have an assistant watch it, for signs of radial (in-and-out) play. Then place a pry bar under the tire and try to lever it up and down while watching for axial (up-and-down) play. If radial or axial play of the load-carrying ball joint exceeds specifications, replace the ball joint. Replace a non-load-carrying ball joint if it is loose in any way.

Wheel Alignment Inspection

If inspection of the steering and suspension reveals any looseness or if ball joints are replaced, it is a very good idea to check wheel alignment on an alignment rack, using alignment equipment. Pay close attention to the caster and camber variations from side to side.

Front wheel caster and camber are often used to keep a vehicle from pulling to one side or the other. Some carmakers specify different caster angles for right and left front wheels to compensate for a pull caused by the crown of the road. Road crown causes a pull to the right so if different caster angles are used from right to left, the right front wheel will have more positive caster to create a slight caster lead and compensate to the left. Different camber angles from right to left also can compensate for road crown pull, but because of the effects on tire wear, camber variations are not used as often as caster variations.

Braking problems can arise when caster or camber angles are significantly different from right to left. In such cases, the vehicle will tend to pull toward the side with the more negative caster or the more positive camber. A pull caused by extreme caster or camber variations may be more noticeable under braking than under acceleration or cruising because of the torque reaction forces imposed on the wheel while braking. Excessive vehicle setback also may contribute to brake pull problems. Setback cannot be adjusted. It is set up by the vehicle designers and manufacturing process.

When troubleshooting a complaint of brake pull, do not overlook the caster and camber angles and the setback measurement. If the wheel alignment may be contributing to a brake problem, realignment to manufacturer's specifications is a good investment.

> ☑ **SERVICE TIP:** When you are troubleshooting a brake-pull problem and cannot pinpoint the cause, try driving the car in reverse and applying the brakes. If the car pulls in opposite directions while braking in forward and reverse (for example, pulls left when going forward and pulls right in reverse), look closely for loose or broken suspension and steering components such as ball joints and tie rod ends.

CASE STUDY

Most race cars are set up with the right rear tire larger than the left rear tire. On a sprint car, the size differential is really obvious; but even with tires of the same size specification, manufacturing variations create slight size differences. Racers use this to advantage to improve traction in corners, particularly on oval tracks. The racers call this selective tire sizing "stagger." What works well on the race track, however, does not always work well on the highway.

The Cadillac Fleetwood Brougham had some weird drivability symptoms that included an accelerator pedal that became hard to press at about 45 mph to 50 mph. The cruise control would shut off all by itself at about the same speed. Along with an ABS, the car also had a TCS, and the tech working on the problem thought the TCS might be involved.

She road tested the car with a scan tool connected and monitored the wheel speeds on the ABS and TCS data stream. As the car accelerated, the left rear wheel gradually went faster than the other three wheels and faster than the car speedometer. At 45 mph to 50 mph (the problem speed), the left rear wheel was 3 mph to 4 mph faster.

The tech questioned the car owner and learned that the left rear tire had been replaced recently. The original tire suffered a sidewall blowout. The other three tires on the car had only about 20,000 miles on them, so they were not replaced. It seemed strange but, yes, just a 3 mph or 4 mph difference in driving wheel speed caused the TCS to kick in at moderate cruising speed in a straight line. Rotating the new tire to the front eliminated the tire stagger and solved the problem.

Terms to Know

Bearing driver	Etching	Wear-indicating ball joint
Bearing packer	Galling	
Brinelling	Hydroplane	

ASE-Style Review Questions

1. *Technician A* says that wheel nuts should be installed with an impact wrench set at its highest torque setting.
   Technician B says that a torque stick can be used with an impact wrench to install wheels.
   Who is correct?
   A. A only   **C.** Both A and B
   B. B only   **D.** Neither A nor B

2. *Technician A* says that radial runout may cause a wheel to wobble.
   Technician B says that lateral runout causes a wheel to hop or move up and down.
   Who is correct?
   A. A only   **C.** Both A and B
   B. B only   **D.** Neither A nor B

3. *Technician A* says that wheel alignment can correct wheel runout.
   Technician B says that wheel runout and total runout should both be measured radially and laterally.
   Who is correct?
   A. A only   **C.** Both A and B
   B. B only   **D.** Neither A nor B

4. *Technician A* says that tires should always be inflated to the pressure listed on the sidewall.
   Technician B says that air should be removed from a tire to reduce hot inflation pressure.
   Who is correct?
   A. A only   **C.** Both A and B
   B. B only   **D.** Neither A nor B

5. *Technician A* says that radial runout is measured at the center of the tire tread.
   Technician B says that lateral runout is measured at the sidewall, just inboard of the tread.
   Who is correct?
   A. A only   **C.** Both A and B
   B. B only   **D.** Neither A nor B

6. *Technician A* says that incorrect caster can cause the center of a tire tread to wear more than the outside edges.
   Technician B says that incorrect camber can cause one side of a tire tread to wear more than the other.
   Who is correct?
   A. A only   **C.** Both A and B
   B. B only   **D.** Neither A nor B

7. *Technician A* says that tapered roller wheel bearings should be installed with a 15 foot-pound preload.
   Technician B says that straight roller bearings require a 15 foot-pound preload.
   Who is correct?
   A. A only   **C.** Both A and B
   B. B only   **D.** Neither A nor B

8. *Technician A* says that tapered bearings should be lubricated with bearing grease before installation.
   Technician B says that sealed ball bearings should be lubricated with 10W30 motor oil before installation.
   Who is correct?
   A. A only   **C.** Both A and B
   B. B only   **D.** Neither A nor B

9. *Technician A* says that wear-indicating ball joints must be inspected in an unloaded condition. *Technician B* says that wear-indicating ball joints are used on most independent rear suspension systems. Who is correct?
   - **A.** A only
   - **B.** B only
   - **C.** Both A and B
   - **D.** Neither A nor B

10. *Technician A* always cleans and greases a wheel bearing seal before reinstalling it. *Technician B* says that a lip seal should be installed with the lip facing outward. Who is correct?
    - **A.** A only
    - **B.** B only
    - **C.** Both A and B
    - **D.** Neither A nor B

ASE Challenge Questions

1. *Technician A* says that a tire with low pressure may cause the vehicle to pull more to one side with the brakes off than with them on. *Technician B* says that tire treads could cause activation of the ABS even under moderate braking. Who is correct?
   - **A.** A only
   - **B.** B only
   - **C.** Both A and B
   - **D.** Neither A nor B

2. Wheel bearings are being discussed: *Technician A* says that repacking tapered wheel bearings is usually part of the brake service. *Technician B* says that the bearings in the hub of a FWD vehicle should be repacked when the disc pads are replaced. Who is correct?
   - **A.** A only
   - **B.** B only
   - **C.** Both A and B
   - **D.** Neither A nor B

3. *Technician A* says that a bad wheel seal could cause the brakes to grab or lock up. *Technician B* says that differential fluid may reduce the braking effect of the rear brakes. Who is correct?
   - **A.** A only
   - **B.** B only
   - **C.** Both A and B
   - **D.** Neither A nor B

4. *Technician A* says that excessive radial runout of a wheel assembly may cause brake pedal pulsation during braking. *Technician B* says that improper camber will not affect braking but may affect steering. Who is correct?
   - **A.** A only
   - **B.** B only
   - **C.** Both A and B
   - **D.** Neither A nor B

5. *Technician A* says that a bumping sound as the brakes are being applied may indicate a bad joint or worn or missing suspension bushing. *Technician B* says that a too loose wheel bearing could cause excess pad and rotor wear and vibration when the vehicle is braked. Who is correct?
   - **A.** A only
   - **B.** B only
   - **C.** Both A and B
   - **D.** Neither A nor B

Job Sheet 5

5

Name: _____ Date: _____

Remove, Repack, and Install a Wheel Bearing

NATEF Correlation

This job sheet addresses the following NATEF tasks: Remove, clean, inspect, repack, and install wheel bearings and replace seals; install hub and adjust wheel bearings; replace wheel bearing and race.

Upon completion and review of this job sheet, you should be able to remove, repack, and install a wheel bearing and seal.

Tools and Materials

Service manual

Impact tools

Bearing packer

Seal remover

Seal driver

Lift or jacks with stands

Hub with inner bearing and seal installed

Blocking materials

Parts washer

Describe the Vehicle Being Worked On

Year _____ Make _____ Model _____

VIN _____ Engine type and size _____

ABS _____ yes _____ no _____ If yes, type _____

NOTE: This job sheet assumes the hub is removed from the vehicle and the outer bearing removed.

Procedure

Task Completed

1. Place the hub with the inner side facing up. If necessary, use blocks so clearance is available between the hub and bench. ☐

2. Use the seal remover to force the seal from its cavity. ☐

3. Remove the inner bearing. ☐

4. Wash the bearing and hub cavity using a parts washer. ☐

▲ **WARNING:** Do not allow the bearing to spin when using compressed air as a drying agent. Serious injury could result if the bearing comes apart.

5. Use reduced-pressure compressed air to dry the bearing. Use a catch basin to collect the cleaner and waste.

6. Inspect the bearings, races, and hub for any damage or abnormal wear.

   Results _____

   _____

   _____

☐

7. Place the bearing, small end down, into the bearing packer.

☐

8. Force the grease into the bearing until clean ribbons of grease are visible at the top of the rollers.

☐

9. Smear a light coat of grease on the outer sides of the rollers.

☐

10. Install the bearing, small end first, into the cleaned and dried hub cavity.

☐

11. Place the grease seal, lip first, over the cavity.

☐

12. Hold the seal in place with the seal driver while using the hammer to drive the seal in place.

☐

13. Drive the seal in until it is flush with the edge of the hub.

14. Use your fingers to rotate the bearing within the hub. If it moves freely, the task is complete.

   Results _____

   _____

   _____

Problems Encountered _____

_____

Instructor's Response _____

_____

_____

Job Sheet 6

6

Name: _____ Date: _____

Inspect the Tires for Abnormal Wear

NATEF Correlation

This job sheet addresses the following NATEF task: Diagnose tire wear patterns; determine necessary action. (Suspension and Steering task)

Upon completion and review of this job sheet, you should be able to inspect the tires for abnormal wear.

Tools and Materials

Service manual

Tires

Describe the Vehicle Being Worked On

Year _____ Make _____ Model _____

VIN _____ Engine type and size _____

ABS _____ yes _____ no _____ If yes, type _____

Procedure

Task Completed

1. Record the tires' specification data.

   Size _____

   Size _____

   Size _____

   Size _____

   Maximum inflation _____

2. Inspect each tire and record the type of wear.

   Tire 1 _____

   _____ ☐

   Tire 2 _____

   _____ ☐

   Tire 3 _____

   _____ ☐

   Tire 4 _____

   _____ ☐

3. Based on the inspection, the *Classroom Manual, Shop Manual,* and the service manual, determine the probable cause of any abnormal wear.

Results _____

_____

☐

_____

4. Recommendation _____

_____

_____

☐

_____

5. Explain how the abnormal wear may affect braking. _____

_____

_____

_____

☐

_____

6. What would you expect the customer complaint to be based on your inspection? _____

_____

☐

_____

Problems Encountered _____

_____

Instructor's Response _____

_____

Job Sheet 7

Name: _____ Date: _____

Remove and Install a Wheel Assembly on a Vehicle

NATEF Correlation

This job sheet addresses the following NATEF task: Install wheel, torque lug nuts, and make final checks and adjustments.

Objective

Upon completion of this job sheet, you will be able to remove and install a wheel assembly on a vehicle.

Tools and Materials

Service manual, paper or computerized
Impact wrench with impact socket
Torque stick (socket) or torque wrench (recommended)
Lift or jack and safety stands

Procedure

Task Completed

1. Determine the following information from the service manual.

   Vehicle Make _____ Model _____ Year _____

   Lug nut torque _____

   ABS Cautions, if equipped _____

   _____ ☐

2. Lift the vehicle until the tire is free of the floor. ☐

3. Inspect the hub cap or center piece to determine if it is held on by the lug nuts. If not, use a hub cap hammer or wide blade screwdriver to remove the hub cap or center piece. ☐

4. Select the correct size impact socket and fit it to the impact wrench. Ensure the impact wrench is set for the correct rotation direction. ☐

5. Remove the lug nuts in a staggered manner. When the last lug nut is set to be removed, place one hand at the top of the wheel assembly to prevent the assembly from flipping from the lug nut studs. ☐

6. Grip the tire at about the 3 and 9 o'clock position and lowered the assembly to the floor. ☐

   ▮ **CAUTION:** Bend at the knees, not the back, as the assembly is lowered. Keep the back as straight as possible during this maneuver.

7. Inspect the lug nut stud threads for any damage. If damage is apparent, consult with the instructor. ☐

☐

8. Place the wheel assembly near and below the vehicle hub. Grip the tire at about the 3 and 9 o'clock position and lift the assembly to the hub and slide it into the studs. Balance the assembly on the studs with one hand until at least one lug nut is started onto the stud.

NOTE: If the center piece or hub cap is secured by the lug nuts, place the cap or center in place before screwing on the first lug nut.

CAUTION: Bend at the knees, not the back, as the assembly is lifted. Keep the back as straight as possible during this maneuver.

☐

9. Screw each of the lug nuts several turns onto the studs.

☐

10. Use a ratchet and socket or hand to screw each of the lug nuts as far as possible onto the studs.

☐

11. Torque to specifications. Torque the lug nuts in a diamond pattern. The torque can be applied in two ways.

A ½-drive impact wrench with the correct torque stick (socket) may tighten the lug nuts, but the recommended method is to use a standard torque wrench.

WARNING: Ensure the impact wrench is strong enough to create the correct torque and the compressed air is of sufficient pressure. A typical torque stick requires a wrench capable of 250 foot-pounds of torque at 120 psi of air pressure. Use of weak or insufficient wrenches or low air pressure will prevent the correct torque being applied to the lug nuts. This could lead to the loss of a wheel during vehicle operation and cause damage and/or injury.

☐

12. With the lug nut torqued and hub cap or center piece in place, lower the vehicle to the floor.

Problems Encountered _____

_____

Instructor's Response _____

_____

Master Cylinder and Brake Fluid Service

Upon completion and review of this chapter, you should be able to:

❏ Perform a safe brake system test drive.

❏ Diagnose problems in the brake pedal linkage and repair as necessary. Adjust pedal free play to manufacturer's specifications.

❏ Diagnose poor stopping, brake drag, or hard pedal caused by master cylinder problems and perform needed repairs.

❏ Check the master cylinder fluid level and fill as necessary. Analyze the condition of a vehicle's brake fluid from its appearance.

❏ Inspect a master cylinder for leaks and defects.

❏ Test a master cylinder for leakage and air entrapment and determine needed repairs.

❏ Remove and replace a master cylinder and bench bleed the master cylinder before installation.

❏ Overhaul a master cylinder.

❏ Locate the hydraulic bleeding sequence and instructions for a specific vehicle in a service manual.

❏ Bleed and flush the brake hydraulic system.

Brake System Road Test

To operate safely, the master cylinder and other hydraulic components of a brake system must work properly. Leaks in the master cylinder or brake lines can rob the system of pressure and cause dangerous operating conditions, which is why the master cylinder and hydraulic system must be inspected whenever the brake pads or linings are changed or when a customer complains of poor braking. Any problems must be corrected immediately.

Check for the following conditions that can cause poor brake performance:

❏ *Tire Problems.* Worn, mismatched, or underinflated or overinflated tires cause unequal braking.

❏ *Unequal Vehicle Loading.* A heavily loaded vehicle requires more braking power. If the load is unequal from front to back or side to side, the brakes may grab or pull to one side.

❏ *Wheel Misalignment.* Wheels that are out of alignment may cause problems that appear to be related to the brakes. For example, tires with excessively unequal camber or caster settings pull to one side.

⚠ **WARNING:** Road test a vehicle under safe conditions and while obeying all traffic laws. Do not attempt any maneuvers that could jeopardize vehicle control. Failure to adhere to this precaution could lead to serious personal injury and vehicle damage. Refer to Chapter 3 for a preoperation inspection checklist before moving onto a roadway.

If the tires are in good shape and the wheel alignment and vehicle loading do not appear to be the problem, proceed with a brake system road test. Follow these guidelines when road testing a vehicle for brake problems:

❏ Test drive the vehicle on a dry, clean, relatively smooth roadway or parking lot. Roads that are wet or slick or that have loose gravel surfaces will not allow all wheels to grip the road equally. In many cases, loose gravel roads may cause the ABS to function. Rough roads can cause the wheels to bounce and lose contact with the road surface.

Basic Tools

Basic technician's tool set

Clean shop towel

Flare nut wrench

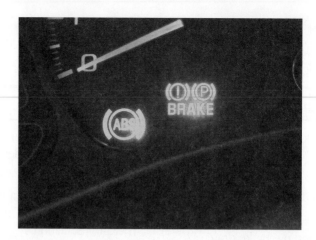

Figure 4-1 The BRAKE warning lamp should light when the ignition switch is in the start position.

❑ Avoid crowned roadways. They can throw the weight of the vehicle to one side, which will give an inaccurate indication of brake performance.
❑ First test the vehicle at low speeds. Use both light and fairly heavy pedal pressure. If the system can safely handle it, test the vehicle at higher speeds. Avoid locking the brakes and skidding the tires.

The brake light is red in color and is usually one of the largest warning lights on the instrument panel.

Check the brake warning lamp on the instrument panel. It should light when the ignition switch is in the start position and go off when the ignition returns to the run position with engine running (Figure 4-1).

If the brake warning lamp stays on when the engine is running, verify that the parking brake is fully released. If it is, the problem may be a low brake fluid level in the master cylinder. Some vehicles have a separate master cylinder fluid level warning lamp. If either warning lamp remains on, check the fluid level in the master cylinder reservoir.

Listen for unusual brake noise during the test drive. Are there squeals or grinding? Do the brakes grab or pull to one side? Does the brake pedal feel spongy or hard when applied? Do the brakes release promptly when the brake pedal is released?

Brake Pedal Mechanical Check

Classroom Manual
pages 68–72

Special Tool

Coworker

Checking the brake pedal mechanical operation is an important part of brake troubleshooting. Whether you do it as part of the brake system road test or during a system leak test, check these points of pedal operation:

❑ Check for friction and noise by pressing and releasing the brake pedal several times (with the engine running for power brakes). Be sure the pedal moves smoothly and returns with no lag or noise.

❑ Move the brake pedal from side to side. Excessive side movement indicates worn pedal mounting parts.

❑ Check stoplamp operation by depressing and releasing the brake pedal several times. Have a coworker check that the lamps light each time the pedal is pressed and go off each time it is released (Figure 4-2) including the third or center—high-mounted—stoplight.

Figure 4-2 Checking stoplamp operation.

Pedal Travel and Force Test

Air in the hydraulic system causes most low-pedal problems, and bleeding the system usually solves the problems. Low pedal also can be caused by a leak in the hydraulic system, incorrect pushrod length adjustment, a service brake that is out of adjustment, worn brake shoes, or a brake shoe adjuster that is not working.

SERVICE TIP: Before starting any diagnosis, refer to the vehicle's service history if available. Note any recent history pertaining to this repair order, for example, brake pedal low. A recent brake repair may point the way to a quick, accurate diagnosis.

When a given amount of force is applied to the pedal, brake pedal travel must not exceed a specified maximum distance. This maximum travel specification is normally about 2.5 inches (64 mm) when 100 pounds (445 N) of force is applied. The exact specifications can be found in the vehicle service manual.

Failure to exhaust brake boost pressure will result in an incorrect pedal travel or force measurement. Use a brake pedal effort gauge to measure force applied to the pedal with these five procedures:

1. Turn off the engine. On vehicles with vacuum assist, pump the pedal until all reserve vacuum is exhausted from the booster.

2. Install the brake pedal effort gauge on the brake pedal (Figure 4-3).

3. Hook the lip of the tape measure over the top edge of the brake pedal and measure the distance from the pedal to the steering wheel rim (Figure 4-4). You can use a yardstick on some vehicles in place of a tape measure.

4. Apply the brake pedal until the specified test force registers on the brake pedal effort gauge (Figure 4-5).

5. Note the change in pedal position on the tape measure or yardstick. The increased distance should not exceed the maximum specification listed in the vehicle service manual. If it does, look for a leak in the hydraulic system and check pushrod adjustment. Worn shoes, bad shoe adjusters, or a poorly adjusted parking brake also can cause excessive pedal travel.

Classroom Manual
pages 68–72

Special Tools

Brake pedal effort gauge
Tape measure
Service manual

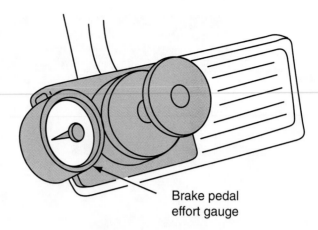

Brake pedal
effort gauge

Figure 4-3 Install the brake pedal effort gauge on the brake pedal.

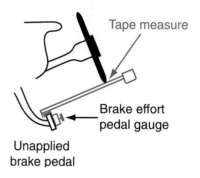

Tape measure

Brake effort
pedal gauge

Unapplied
brake pedal

Figure 4-4 Use a tape measure or a yardstick to measure the distance from the pedal to the steering wheel.

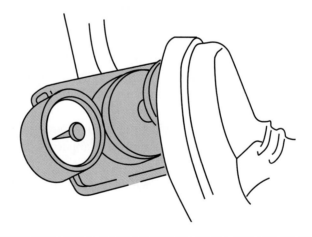

Figure 4-5 Apply the specified amount of pedal force.

Pedal Free Play Inspection and Adjustment

Classroom Manual
pages 68–72

Brake pedal free play is the clearance between the brake pedal or booster pushrod and the primary piston in the master cylinder. A specific amount of free play must exist so that the primary piston is not partially applied when the pedal is released and so that pedal travel is not excessive. Free play at the primary piston is usually only a small fraction of an inch or a few millimeters. The pedal ratio multiplies this free play to about ⅛ inch to ¼ inch at the pedal (Figure 4-6).

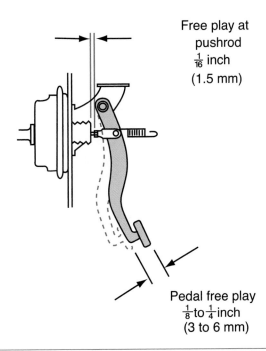

Free play at
pushrod
$\frac{1}{16}$ inch
(1.5 mm)

Pedal free play
$\frac{1}{8}$ to $\frac{1}{4}$ inch
(3 to 6 mm)

Figure 4-6 The pedal ratio multiplies free play at the master cylinder several times at the pedal.

✓ **SERVICE TIP:** Some vehicles require an adjustment of the booster pushrod when the booster is replaced. Some require the booster pushrod be adjusted before adjusting the pedal pushrod, or vice versa. It is important to check both pushrods if the booster is replaced. The manufacturer will have specific instructions in its service manual.

Too much free play causes the pedal to travel too far before moving the pistons far enough to develop full pressure in the master cylinder. Excessive free play can severely reduce braking performance and create an unsafe condition.

Too little free play causes the pedal to maintain contact with the primary piston. This can cause the piston cup to block the vent port and maintain pressure in the lines when the pedal is released. Unreleased pressure can cause the brakes to drag, overheat, fade, and wear prematurely. Additionally, with the piston cup blocking the vent port, each stroke of the pedal draws fluid from the low-pressure area behind the piston into the high-pressure area in front of the piston as the pedal is released. Eventually, enough fluid pressure can accumulate ahead of the piston to lock the brakes.

Check pedal free play by pumping the brake pedal with the engine off to exhaust vacuum in the booster. Place a ruler against the car floor, in line with the arc of pedal travel. Then press the pedal by hand and measure the amount of travel before looseness in the linkage—the free play—is taken up. Measure at the top or bottom of the pedal, whichever provides the most accurate view. Refer to vehicle specifications for the exact amount of free play to be provided.

Adjust the free play by lengthening or shortening the pushrod. On most vehicles, loosen the locknut on the pushrod at the pedal and rotate the pushrod while rechecking free play measurement. Tighten the locknut when adjustment is correct.

 AUTHOR'S NOTE: The following procedure is based on a 2000–2006 Honda S2000. Other vehicles have similar procedures.

Adjusting Pedal Height

One method to adjust the brake pedal height and free play follows. Disconnect and loosen the brake pedal position switch until it is no longer touching the brake pedal lever (Figure 4-7, A and B). Gain clear access to the floorboard by lifting the carpet and the insulator (Figure 4-8, C). Measure

Special Tools

Tape measure or rule

Service manual

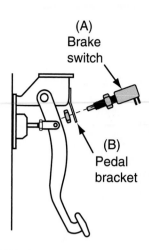

Figure 4-7 Remove the pedal position switch or stoplamp switch from the pedal bracket.

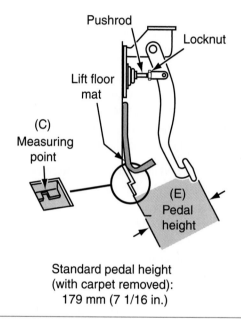

Standard pedal height
(with carpet removed):
179 mm (7 1/16 in.)

Figure 4-8 Remove the floor mat and a portion of the carpet to gain clear access to the floorboard.

the pedal height, E in Figure 4-8, from the right center of the brake pad to the cleared floorboard. In the case of this Honda, the pedal height should be 179 mm (7 $\frac{1}{16}$ inches). If necessary to adjust the pedal height, loosen the locknuts, and turn the pushrod to obtain the correct measurement (Figure 4-9). With the correct height obtained, hold the pushrod in place while tightening the locknut to 15 N · m (11 ft.-lb).

Install the brake pedal position switch until its plunger is against the pedal lever and completely pushed into the switch (Figure 4-10). Unscrew the switch until there is 0.3 mm (0.01 inch) between the switch's threaded end and the pedal pad. Connect the switch to its electrical harness. Have an assistant check the brake lights as the brake pedal is depressed and released.

CAUTION: If the switch is not adjusted correctly, the brakes will drag. This may cause heat problems with the friction materials and poor braking performance.

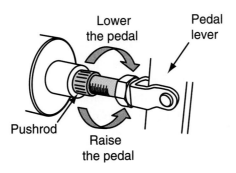

Figure 4-9 Loosen the locknut and turn the pushrod to make the rod longer or shorter depending on the movement needed.

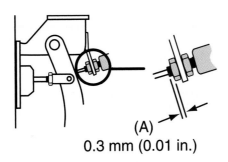

Figure 4-10 Turn the switch within its locknut until the proper clearance is obtained. The clearance on this switch should be 0.3 mm (0.01 inch) at point A.

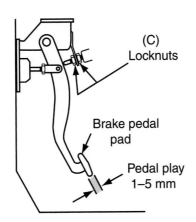

Figure 4-11 Check the pedal free play. If adjustment is needed, turn C until the proper free play is achieved. Check the stoplamp's operation.

Adjusting Pedal Free Play

Using the same Honda vehicle as the example, the pedal free play is checked and adjusted in the following manner. The engine should be off. Push on the brake by hand while measuring the distance the pedal travels before a stiff resistance is felt. This measurement is taken at the brake pedal foot pad and should be 1 mm to 5 mm ($\frac{1}{16}$ inch to $\frac{3}{16}$ inch) (Figure 4-11). If necessary, adjust the free play by loosening the locknut on the brake pedal switch and turning the switch in the appropriate direction until the free play is correct. Do not forget to tighten the locknut after the adjustment is made and recheck the free play after the locknut is tightened.

If the car has a mechanical stoplamp switch on the brake pedal linkage, check switch operation and adjust it if necessary after adjusting pedal free play.

Adjusting the Stoplamp Switch

 SERVICE TIP: At one time, a stoplamp switch could be adjusted by warping its mount to get the plunger lined up. However, today's stoplamp switches are usually multi-functional units with up to four or five different internal switches or contacts that serve many computer systems. They must be adjusted according to the service manual, and the technician must be concerned with not just the stoplights but all the other circuits involved. Some stoplamp switches can be accessed easily, whereas others may require some ingenuity. Study the service manual and then inspect the area around the switch to discover the best method to access the switch.

AUTHOR'S NOTE: The following procedure is based on a 2005 DaimlerChrysler 300 Series and Magnum vehicle.

Use a brake pedal depressor to hold the brake pedal down (check the alignment machine for a depressor). Rotate the stoplamp switch approximately 30 degrees counterclockwise and pull rearward on the switch. It should separate from its mount (Figure 4-12). Using hand force only, pull the switch plunger out to its fully extended position. Low clicks should be heard as the plunger ratchets out.

Ensure the brake pedal is down as far as it will go and is firmly held in place. Align the switch's index key to the notch in the bracket and push the switch into place. Rotate the switch about 30 degrees clockwise until it locks.

CAUTION: Do not release the brake pedal by pulling the depressor out and letting the pedal slam up to its stop. The stoplamp switch will not adjust properly and may be damaged.

Apply foot force to the brake pedal and remove the pedal depressor. Allow the pedal to gently raise until it stops. Using gentle hand force, pull up on the brake pedal until it stops moving. This will ratchet the plunger to the correct position. The switch adjustment is initially checked by having an assistant observe the brake lights as the brake pedal is depressed and released. However, the final check requires a road test on a road where the cruise control can be safely used. During the road test, engage the cruise control at a safe speed. Once the system is stabilized, depress the brake slightly. The cruise control should turn off. If not, then the switch must be checked and readjusted as needed.

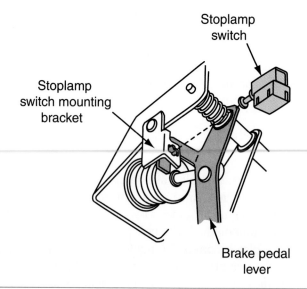

Figure 4-12 Pull the switch plunger all the way out before installation. The pedal should be locked down and not released until the switch is installed.

● **CUSTOMER CARE:** A customer's only contact, literally, with the brake system in his or her car is through the brake pedal. Customers tend to judge brake performance by "pedal feel." It is always a good idea to evaluate the feel and action of the brake pedal before starting any brake job. Then when you deliver the finished job, pedal feel should be noticeably improved. The biggest cause of spongy or low brake pedal action is air in the system, so careful bleeding of the system will do a lot to ensure customer confidence.

Adjustable Pedal Systems (APS) Service (Based on a 2005 Jeep Grand Cherokee)

Removal. Use the dash-mounted switch to move the pedals, accelerator, and brake to their most forward or down position if possible. Disconnect the battery by removing the negative cable and fastening it to one side. Remove the brake booster.

Within the engine compartment, locate and remove the two intermediate shaft bearing nuts at the bulkhead where the steering column comes through (Figure 4-13). Remove the pinch bolt where the lower shaft connects to the steering gear. Disconnect the lower steering shaft from the column.

▲ **WARNING:** Wait a sufficient time for the air bags to disarm before starting work in the passenger compartment. Serious injury or death could result if an air bag functions when personnel are working around the steering column.

Lock the steering wheel in place if not already done. Remove any lower dash panels or other components to gain access to the lower portion of the steering column. Disconnect the pedal motor electrical connection and the accelerator pedal cable. Remove the two nuts that connect the accelerator pedal sled to the brake pedal sled (Figure 4-14). Loosen the two top brake pedal sled nuts. Remove the pedal assembly. The assembly includes the brake and accelerator pedals (Figure 4-15).

Installation. As usual the installation is the reverse of the removal. Some important points that may be overlooked during installation are:

❏ Torque all fasteners to specifications.

❏ Connect the accelerator cable and the APS electrical connection.

❏ Adjust the pushrod as needed.

❏ Reconnect the vacuum line to the booster.

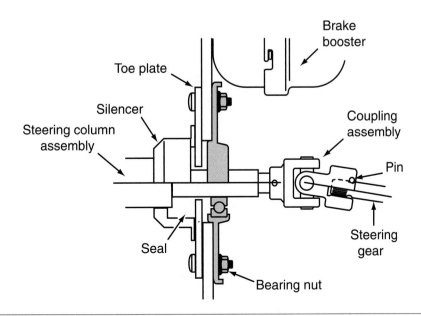

Figure 4-13 The bearing nuts are located on the firewall below the booster area.

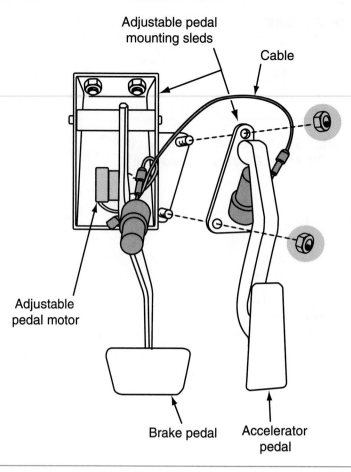

Figure 4-14 Disconnect the electrical connection and remove the two nuts to separate the two sleds. Loosen the two top nuts on the brake pedal sled.

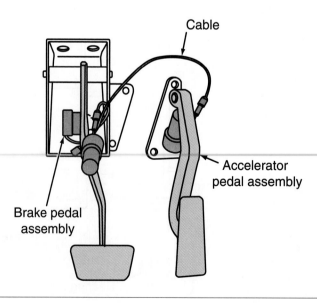

Figure 4-15 Remove the pedal assembly.

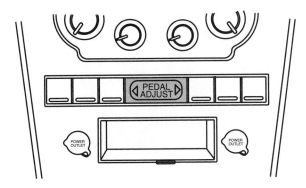

Figure 4-16 This APS switch is located in the lower instrument panel.

ABS Switch Service

The switch is located in what DaimlerChrysler calls a lower instrument panel switch pod (Figure 4-16). There is more than one switch in this pod and, if any single switch, including the pedal switch, is damaged or inoperative, the entire pod is replaced. The procedure to remove and install the pod requires some disassembly of the instrument panel. Although this may seem to be a straightforward operation, sometimes removal of the instrument panel involves some ingenuity as well as following the instructions. It is recommended that the service manual directions be consulted before attempting to replace the APS switch.

Brake Fluid Precautions

The specifications of all automotive brake fluids are defined by SAE Standard J1703 and Federal Motor Vehicle Safety Standard (FMVSS) 116. Fluids classified according to FMVSS 116 are assigned U.S. Department of Transportation (DOT) numbers: DOT 3, DOT 4, and DOT 5. Basically, the higher the DOT number, the more rigorous the specifications for the fluid. Review Chapter 4 of the *Classroom Manual* for a detailed explanation of brake fluid specifications.

Classroom Manual
pages 63–68

Choosing the right fluid for a specific vehicle is not the simple idea that if DOT 3 is good, DOT 4 must be better, and DOT 5 better still. Domestic carmakers all specify DOT 3 fluid for their vehicles, but Ford calls for a heavy-duty variation that meets the basic specifications for DOT 3 but has the higher boiling point of DOT 4. Import manufacturers are about equally divided between DOT 3 and DOT 4.

DOT 5 fluids are all silicone based because only silicone fluid can meet the DOT 5 specifications. However, no vehicle manufacturer recommends DOT 5 fluid for use in its brake systems, particularly in antilock brake systems. Although all three fluid grades are compatible in certain aspects, they do not combine if mixed together in a system. DOT 5 in particular should never be mixed with or used to replace other types of brake fluids. Therefore, the best general rule is to use the fluid type recommended by the carmaker and do not mix fluid types in a system.

The DOT rating is found on the brake fluid container (Figure 4-17). The vehicle service manual and owner's manual specify what rating is correct for the car. Do not use a brake fluid with a lower DOT rating than specified by the manufacturer. The lower-rated fluid could boil and cause a loss of brake effectiveness.

DOT 3/4 and DOT 5.1 synthetic brake fluids should be stored in the same manner as other brake fluids and can be mixed or used to replace DOT 3 or DOT 4 fluids. Refer to the manufacturer or service manual to see if synthetic fluids can be used in the vehicle being repaired.

✓ **SERVICE TIP:** If the vehicle is still under manufacturer warranty or covered by an extended warranty, check with the warranty writer or agent to determine if the use of synthetic brake fluids violates the warranty terms. Some warranties are very specific about repair or replacement parts and fluids.

Figure 4-17 The brake fluid DOT number is found on every brake fluid container.

CAUTION: DOT 5.1 and DOT 5.1 long-life brake fluids are *not* silicone-based fluids. Do *not* mix or replace DOT 5 fluid with DOT 5.1 or DOT 5.1 long-life fluids. Damage to the brake system and possible injury could occur due to damage to brake system components.

WARNING: Brake fluid can cause permanent eye damage. Always wear eye protection when handling brake fluid. If you get brake fluid in your eye, see a doctor immediately.

CAUTION: When working with brake fluid, do not contaminate it with petroleum-based fluids, water, or any other liquid. Keep dirt, dust, or any other solid contaminant away from the fluid. Contaminated fluid may cause system failure.

WARNING: Brake fluid will irritate your skin. If fluid gets on your skin, wash the area thoroughly with soap and water.

CAUTION: Never mix glycol-based and silicone-based fluids because the mixture could cause a loss of brake efficiency and possible injury.

CAUTION: Always store brake fluid in clean, dry containers. Brake fluid is hygroscopic; it will attract moisture and must be kept away from dampness in a tightly sealed container. When water enters brake fluid, it lowers the boiling point. Never reuse brake fluid.

CAUTION: DOT 3 and DOT 4 *polyglycol* fluids have a very short storage life. As soon as a container of DOT 3 or DOT 4 fluid is opened, it should be used completely because it immediately starts to absorb moisture from the air.

CAUTION: Do not spill glycol-based brake fluid on painted surfaces. Glycol-based fluids damage a painted surface. Always flush any spilled fluid immediately with cold water.

WARNING: Brake fluid is a toxic and hazardous material. Dispose of used brake fluid in accordance with local regulations and EPA guidelines. Do not pour used brake fluid down a wastewater drain or mix it with other chemicals awaiting disposal.

Using DOT 5 Silicone Fluid

DOT 5 silicone fluid does not absorb water and has a very high boiling point. It is noncorrosive to hydraulic system components, and it does not damage paint as does polyglycol fluid. DOT 5 fluid has other characteristics that are not so beneficial.

DOT 5 silicone fluid has a lower specific gravity than polyglycol fluid. If the two types are mixed, they do not blend; the silicone fluid separates and floats on top of the polyglycol fluid.

Therefore, if a customer wants silicone fluid in his vehicle, all the polyglycol fluid must be completely flushed out. The best time to convert to silicone fluid is during a complete brake system overhaul.

Silicone fluid compresses slightly under pressure, which can cause a slightly spongy brake pedal feel. Silicone fluid also attracts and retains air more than polyglycol fluid does, which makes brake bleeding harder; it tends to outgas slightly just below its boiling point; and it tends to aerate from prolonged vibration. DOT 5 fluid has other problems with seal wear and water accumulation and separation in the system. All of these factors mean that DOT 5 silicone fluid should *never be used* in an ABS.

The best practice is to use a single, high-quality brand of brake fluid of the DOT type specified for a particular vehicle. Avoid mixing fluids whenever possible.

WARNING: DOT 5 should never be mixed with or used to replace DOT 3, DOT 4, or DOT 3/4, or DOT 5.1 fluids because of their chemical noncompatability. They will not mix and silicone can damage seals designed for polyglycol liquids.

CAUTION: DOT 5.1 and DOT 5.1 long-life brake fluids are *not* silicone-based fluids. Do *not* mix or replace DOT 5 fluid with DOT 5.1 or DOT 5.1 long-life fluids. Damage to the brake system and possible injury could occur due to damage to brake system components.

SERVICE TIP: Two good reasons for periodic flushing and refilling of the brake hydraulic system are: (1) Flushing the system and refilling with fresh fluid keeps sediment out of ABS valves. (2) Flushing and refilling the system also keeps that sediment out of the self-adjusting parking brake mechanisms on rear-wheel disc brakes.

Master Cylinder Fluid Service

Brake fluid service procedures are the most basic—but among the most important—brake system services. The following paragraphs provide instructions for checking the master cylinder fluid level and adding fluid to the system. Later sections of this chapter contain procedures for bench bleeding a master cylinder. Complete system bleeding instructions and fluid flushing and filling instructions are discussed later.

Classroom Manual
pages 63–68, 72–74

Checking Master Cylinder Fluid Level

Master cylinder fluid level and fluid condition should be inspected at least twice a year as part of a vehicle preventive maintenance schedule. If the car has a translucent fluid reservoir, general fluid level can be checked every time the motor oil is checked or changed.

Although normal brake lining wear causes a slight drop in fluid level, an abnormally low level in either chamber—especially an empty reservoir—usually means that there is a leak in the system.

When you check the fluid in the master cylinder, you are checking two things. First, be sure that the reservoir is filled to the correct level. A two-piece master cylinder with a plastic reservoir usually has graduated markings to indicate the correct fluid level (Figure 4-18). The markings may be on the outside of the reservoir if the reservoir is translucent, or they may be inside if the reservoir is opaque. Fill the reservoir to the FULL mark or equivalent.

If the master cylinder has a one-piece cast body with an integral reservoir, fluid level may not be marked in the reservoir. In this case, fill the reservoir to ¼ inch (6 mm) from the top. If the reservoir is mounted at an angle on the master cylinder, measure fluid level at the point closest to the reservoir rim (Figure 4-19). Some composite master cylinders have opaque plastic reservoirs (Figure 4-20). Remove the reservoir caps or covers to check fluid level as you would for a one-piece iron reservoir.

Special Tools

Brake fluid
Cloths

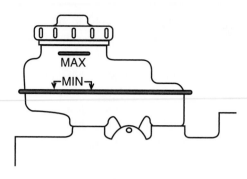

Figure 4-18 Most translucent reservoirs have markings for the minimum and maximum fluid levels (arrows).

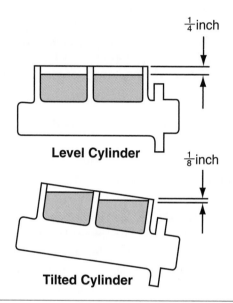

Figure 4-19 Check fluid level at the point closest to the reservoir rim if the master cylinder is tilted. Fluid level is usually higher for a tilted reservoir.

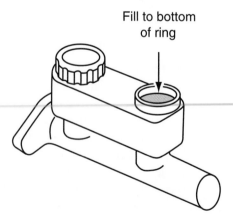

Figure 4-20 Check fluid level in an opaque plastic reservoir as you would for a cast-iron reservoir.

Photo Sequence 5
Typical Procedure for Filling a Master Cylinder Reservoir

 CAUTION: Be careful to avoid spraying brake fluid. To protect your face, never bend directly over the reservoir.

On some antilock brake systems, you must depressurize the system before adding brake fluid. When depressurized, fluid level may rise slightly, giving a more accurate level reading.

P5-1 Thoroughly clean the reservoir cover before removing it to prevent dirt from entering the reservoir body.

P5-2 Remove the reservoir cover and the diaphragm.

P5-3 Check that the vent hole in the reservoir cover is open.

P5-4 Inspect the diaphragm for holes, tears, and other damage. Replace as needed.

P5-5 Check the brake fluid level and its appearance. The fluid level should be within ¼ inch of the top of the reservoir or at the reservoir's (FULL or MAX) level marking. The fluid should be clean, with no rust or other contamination. Note the level of fluid in each section of the reservoir for diagnostic purposes.

P5-6 Fill the reservoir with the recommended brake fluid. The wrong type of brake fluid, contaminated fluid, water, or mineral oil may cause the brake fluid to boil or the rubber components in the system to deteriorate.

Do not overfill any reservoir and use only the type of fluid specified by the vehicle maker. For 90 percent of the vehicles, this will be DOT 3 or DOT 4 fluid. Photo Sequence 5 shows the procedure for filling the master cylinder reservoir.

The second thing you look for when you check brake fluid is contamination. Most DOT 3 and DOT 4 fluids are clear or light amber when fresh, and, ideally, fluid in service should retain most of its original appearance. Fluid in good condition should be clear and transparent, although some darkening is allowable. Any of the following conditions may indicate the need for flushing and refilling the system or for more serious service:

Figure 4-21 Test strips for determining if the brake fluid is contaminated.

❏ *Cloudy fluid.* Cloudiness usually indicates moisture contamination.
❏ *Dark brown or murky fluid, not transparent.* Very dark fluid usually indicates excessive contamination by rust and dirt.
❏ *Layering or separation.* These conditions indicate a mixture of two fluids that have not blended together. The contaminating fluid may be oil or some other petroleum-based product, or it may be DOT 5 silicone fluid.
❏ *Layering or separation accompanied by cloudy or murky color and deteriorated rubber parts.* This condition almost always indicates oil contamination. In this case, the system must be flushed thoroughly and all seals and other rubber parts replaced.

Layering or separation of fluids occurs when a lightweight fluid separates and floats on top of heavier fluids. Mix oil and water and the oil will sink to the bottom.

✓ **SERVICE TIP:** Phoenix Systems offers a set of test strips to check the contamination (Figure 4-21). This strip is not a moisture or pH test but tests for copper instead. A strip is dipped into the fluid in the reservoir and the color change is compared to a supplied test chart. This is almost a fail-safe method of field testing the brake fluid and making a service recommendation to the customer. Check http://www.brakestrips.net for more information.

You may also see the following conditions when checking fluid in the master cylinder. These conditions, by themselves do not always indicate a need for master cylinder service:

❏ Unequal fluid levels in the master cylinder reservoir chambers on front disc and rear drum systems may result as fluid moves from the reservoir into the calipers to compensate for normal lining wear. Fill both chambers to the full marks.
❏ A slight squirt of brake fluid from one or both master cylinder reservoir chambers when the brake pedal is applied is normal. It is caused by fluid moving through the reservoir vent ports as the master cylinder pistons move forward in the bore.
❏ Light fluid turbulence in the reservoir when the brake pedal is released is the result of brake fluid returning to the master cylinder after the brakes have been released.
❏ A slight trace of brake fluid on the booster shell below the master cylinder mounting flange is normal. It results from the lubricating action of the master cylinder wiping seal.

● **CUSTOMER CARE:** Remind your customers to check the brake fluid level in their master cylinder reservoirs periodically, particularly on older vehicles that may develop slow leaks. Leaking fluid is a sign of trouble. Emphasize that service is needed whenever this problem is noticed.

Checking Boiling Point and Water Content

You can use a simple instrument called a **refractometer** for a quick check of brake fluid water content and boiling point. A refractometer works by measuring the way light is bent or refracted as it passes through a liquid. A refractometer works on the principle that different liquids have different specific gravities, and liquids of different specific gravities will refract or bend light waves at different angles. As brake fluid absorbs water, its **specific gravity** changes. Water refracts light in a specific way. A brake fluid refractometer is graduated so that the specific gravity is correlated to the water content of the fluid and the resulting change in boiling point.

To use the refractometer, place a couple drops of fluid on the sample window, close the cover, point the instrument toward a light source, and focus the eyepiece. Light shining through the fluid sample will refract or bend and indicate a line across the measuring scale.

✔️ **SERVICE TIP:** Wagner Brake Products sells a set of test strips used to indicate the amount of moisture content in DOT 3 or DOT 4 brake fluid. The product is called Wet Check and works in less than 1 minute. Dip the end of the strip into the brake fluid and watch the color. The shade of color indicates the amount of moisture (water) in the fluid. Snap-on Tools has an instrument that will boil the water and alert the technician to contaminated fluid.

SAE tests have shown that the average 1-year-old car contains 2 percent water in its brake fluid. This occurs just through the normal hygroscopic attraction of polyglycol fluids for water. As little as 4 percent water content can cut the boiling point of DOT 3 fluid in half.

Checking brake fluid with a refractometer can often be the deciding point on whether complete flushing and refilling of the hydraulic system should be recommended.

Checking ABS Fluid Level

Some older ABSs have an electrohydraulic booster that shares the reservoir with the master cylinder. Some require the system accumulator to be charged, whereas others require it to be discharged before checking fluid level. Teves supplies ABS components to several vehicle manufacturers, including Ford and General Motors. However, the fluid level may be checked in different manners. Using the Ford and General Motors versions of the Teves ABS, the different checks are discussed here.

Ford

Most Ford-installed Teves ABSs require the accumulator to be charged before checking fluid level. Turn the ignition on and pump the brake pedal until the pump motor runs. When the pump stops, the accumulator is charged and an accurate check of the fluid level can be made. The fluid level should be at the FULL mark. This system uses DOT 3 brake fluid.

General Motors

This Teves ABS requires the accumulator to be discharged. Turn the ignition key to off and pump the brake pedal at least thirty times to discharge the accumulator. The pedal will eventually become hard to push. Check the fluid to ensure the level is at the FULL mark. Top off as needed with General Motors specified brake fluid. Turn the ignition key to run and allow the pump to recharge the accumulator.

⚠️ **WARNING:** Never open or loosen a brake line serving the ABS controls or modules without ensuring that the pressure has been released. Opening a pressurized line could result in injury.

Special Tools

Refractometer
Service manual

A **refractometer** is a test instrument that measures the deflection or bending of a beam of light.

Specific gravity is the weight of a volume of any liquid divided by the weight of an equal volume of water at equal temperature and pressure.

Master Cylinder Test and Inspection

Classroom Manual
pages 63–66, 74–77

Special Tools

Brake fluid

Service manual

Coworker

Rule

Paint missing from the booster or vehicle finish below the driver's end of the master cylinder indicates an external leak. The master cylinder should be replaced.

Check for cracks in the master cylinder housing. Look for drops of brake fluid around the master cylinder. A slight dampness in the area surrounding the master cylinder is normal and is usually no reason for concern. However, if a reservoir chamber is cracked, it may be completely empty and the surrounding area may be dry. This is because the fluid drained very quickly and has had time to evaporate or wash away. But with only one-half of the brake system operational, the BRAKE warning lamp should be lit and a test drive should reveal the loss of braking power.

Refill the master cylinder reservoir section that is empty and apply the brakes several times. Wait 5 to 10 minutes and check for leakage or fluid level drop in the reservoir. Although brake pedal response and reservoir fluid levels are strong indicators of problems with the master cylinder or hydraulic system, other tests can be performed to help pinpoint the problem.

Hydraulic System Inspection

If the master cylinder does not appear to be leaking, raise the vehicle on a lift and inspect all brake lines, hoses, and connections (Figure 4-22). Look for brake fluid on the floor under the vehicle and at the wheels. Brake lines must not be kinked, dented, or otherwise damaged, and there should be no leakage. Brake hoses should be flexible and free of leaks, cuts, cracks, and bulges.

Drum brake backing plates and disc brake calipers should be free of brake fluid and grease. Any parts attached to them should be tight.

Testing the Hydraulic System for Trapped Air

The following test requires a helper to pump the brake pedal while you observe the reaction of fluid in the master cylinder (Figure 4-23):

1. Top off the master cylinder reservoir with fresh brake fluid.

2. Loosely place the gasket and cover on the master cylinder. Do not tighten the cover.

   CAUTION: This test may result in brake fluid bubbling or spraying out of the master cylinder reservoir. Wear safety goggles. Cover the master cylinder reservoirs with clear plastic wrap or other suitable cover to keep brake fluid off the paint.

3. Have your helper rapidly pump the brake pedal approximately twenty times. The brake pedal should be held down on the last application.

4. Remove the master cylinder cover and observe the fluid in the reservoirs.

A   B   C

Figure 4-22 Inspect the hydraulic system for leaks at these and other points: (A) brake lines and hoses, (B) drum brake backing plates, and (C) disc brake calipers.

Air may be indicated by spongy pedal, low pedal, or bottoming pedal.

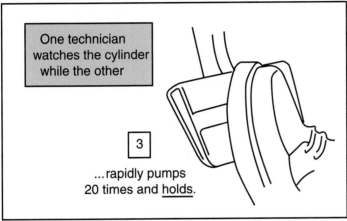

1. Check fluid level, replenish if necessary.

2. Replace cap loosely atop cylinder.

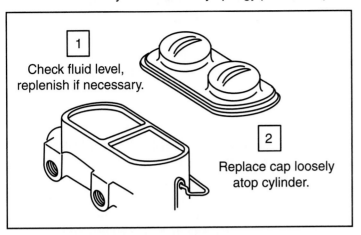

One technician watches the cylinder while the other

3. ...rapidly pumps 20 times and holds.

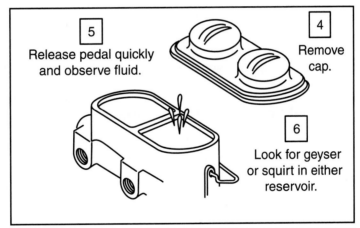

5. Release pedal quickly and observe fluid.

4. Remove cap.

6. Look for geyser or squirt in either reservoir.

RESULTS:

Geyser from reservoir:
Indicates air trapped in the system. It is compressed by pumping and causes a squirt when released. (If pedal is low, rear brake misadjustment can also cause a geyser.)

Action:
Bleed the affected system or systems. If one reservoir only, you need not bleed the other.

Figure 4-23 This quick test will help to determine which half of the brake system may have air trapped in it.

5. Have your helper release the brake pedal quickly. If air is trapped in the system, pumping the brake pedal will compress it. When the brake pedal is released, the compressed air will expand. This will push brake fluid back into the master cylinder with enough force to produce bubbling or even a small geyser.

6. If air is found in the system, bleed the brake system following the service manual recommended sequence.

External Leak Test

Hydraulic brake system leaks can be internal or external. Most internal leaks are actually fluid bypassing the cups in the master cylinder. If the cups lose their ability to seal the pistons, brake fluid leaks past the cups and the pistons cannot develop system pressure.

Internal and external rubber parts wear with use or can deteriorate with age or fluid contamination. Moisture or dirt in the hydraulic system can cause corrosion or deposits to form in the bore, resulting in the wear of the cylinder bore or its parts. Although internal leaks do not cause a loss of brake fluid, they can result in a loss of brake performance. This internal leakage, or fluid bypassing back to the reservoir, is the cause of many soft-pedal or no-pedal complaints and can be hard to pinpoint.

When external leaks occur, the system loses fluid. External leaks are caused by cracks or breaks in master cylinder reservoirs, loose system connections, damaged seals, or leaking brake lines or hoses. Check for a brake fluid leak as follows:

1. Run the engine at idle with the transmission in neutral.

2. Depress the brake pedal and hold it down with a constant foot pressure. The pedal should remain firm, and the foot pad should be at least 2 inches from the floor for manual brakes and about 1 inch for power brakes (Figure 4-24) without an ABS.

3. Hold the pedal depressed with medium foot pressure for about 15 seconds to make sure that the pedal does not drop under steady pressure. If the pedal drops under steady pressure, the master cylinder or a brake line or hose may be leaking. Inspect the system as described later in this chapter.

If your inspection reveals no external leakage, but the pedal still drops under steady pressure, check for fluid bypassing the piston cups inside the master cylinder as explained under Internal Leak Test (Fluid Bypass Test).

Internal Leak Test (Fluid Bypass Test)

If the primary piston cup seal is leaking, the fluid will bypass the seal and move between the vent and replenishing ports for that reservoir or, in some cases, between reservoirs.

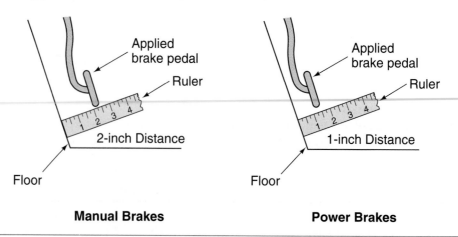

Figure 4-24 With the brakes applied, the distance from the pedal to the floor should be about 2 inches for manual brakes or 1 inch for power brakes.

If no sign of external leakage exists, but the BRAKE warning lamp is lit, the master cylinder may be bypassing or losing pressure internally. Another sign of internal leakage, or bypassing, is a fluid level that rises slightly in one or both reservoirs when pressure is held on the brake pedal. Test for fluid bypassing the piston cups as follows:

1. Remove the master cylinder cover and be sure the reservoirs are at least half full.

2. Watch the fluid levels in the reservoirs while a helper slowly presses the brake pedal and then quickly releases it.

3. If fluid level rises slightly under steady pressure, the piston cups are probably leaking. Fluid level rising in one reservoir and falling in the other as the brake pedal is pressed and released also can indicate that fluid is bypassing the piston cups.

4. Replace or rebuild the master cylinder if it is bypassing fluid internally.

Another quick test for internal leakage or bypassing is to hold pressure on the brake pedal for about 1 minute. If the pedal drops but no sign of external leakage exists, fluid is probably bypassing the piston cups.

> **✓ SERVICE TIP:** The most common customer complaint that indicates an internal master cylinder leak is: "The brakes stop well, but while I am sitting at a stoplight the pedal slowly drops." To the technician this means that the fluid ahead of one of the master cylinder pistons is leaking past the piston seal and returning to the reservoir. Double-check before replacing the master cylinder.

This is a quick, accurate test that can be used after a master cylinder or booster replacement to confirm pushrod placement.

Test for Open Vent Ports

Test for open vent ports in the master cylinder as follows:

1. Remove the cover from the master cylinder. While a helper pumps the brake pedal, observe the fluid reservoirs. You should see a small ripple or geyser in the reservoirs as the brakes are applied.

2. If you see no turbulence, loosen the bolts securing the master cylinder to the vacuum booster about ⅛ inch to ¼ inch and pull the cylinder forward, away from the booster. Hold it in this position and repeat step 1.

3. If turbulence (indicating compensation) now occurs, adjust the brake pedal pushrod length. If turbulence still does not occur, replace the master cylinder. Turbulence can be seen only in the front (secondary) reservoir of a quick take-up master cylinder.

If turbulence does not occur in the master cylinder during the preceding steps, the pistons are probably restricting the vent ports, meaning that the pushrod is not allowing the pistons to return to the fully released positions. Adjusting the brake pedal pushrod at its connection to the pedal lever (as instructed in step 3) often fixes the problem on a manual brake system (one without power assist). More often, however, the output pushrod of the power brake booster requires adjustment. Refer to Chapter 6, Power Brake Service, for booster pushrod adjustment instructions.

Quick Take-Up Valve Test

The quick take-up valve is used in quick take-up master cylinders with low-drag disc brakes to provide a high volume of fluid on the first pedal stroke. This action takes up the slack in low-drag caliper pistons.

No direct test method exists for a quick take-up valve, but excessive pedal travel on the first stroke may indicate that fluid is bypassing the valve. If this symptom exists, check for a damaged or unseated valve.

If the pedal returns slowly when the brakes are released, the quick take-up valve may be clogged so that fluid flow from the cylinder to the reservoir is delayed.

Removing a Non-ABS Master Cylinder

Classroom Manual
pages 74–84

Special Tools

Hand Tools

Catch basin

Cloths

After exhausting the booster, the brake pedal should become very firm even with a malfunctioning master cylinder.

Remove the master cylinder from the vehicle as follows:

1. Disconnect the battery ground (negative) cable.

2. Relieve any residual vacuum by pumping the pedal fifteen to twenty times until you feel a change in pedal effort.

3. Use a shop towel to remove any loose dirt or grease around the master cylinder that could work its way into open lines or the vacuum booster unit.

4. Unplug the electrical connector from the fluid level sensor, if equipped.

5. Place a container under the master cylinder to catch any brake fluid that leaks from the outlet ports when the lines are disconnected.

6. Use a flare-nut wrench to disconnect each brake line fitting from the master cylinder. Plug the end of each line as it is disconnected to keep dirt out of the line and to prevent excessive brake fluid loss.

7. Remove the nuts attaching the master cylinder to the vacuum booster unit (Figure 4-25).

CAUTION: Do not drip brake fluid onto the vehicle paint while removing the master cylinder. Paint damage will occur.

8. Lift the master cylinder out of the vehicle. It may be necessary to insert a small prybar between the booster and the master cylinder to free the master cylinder. Before removing the master cylinder from some vehicles, the proportioning valve must be slid off of the master cylinder mounting studs. On some vehicles, the vacuum valve from the booster must be removed, and the pressure warning switch connector must be disconnected before removing the master cylinder. In vehicles that have manual (nonpower) master cylinders, the pushrod must be disconnected from the brake pedal before the master cylinder can be removed.

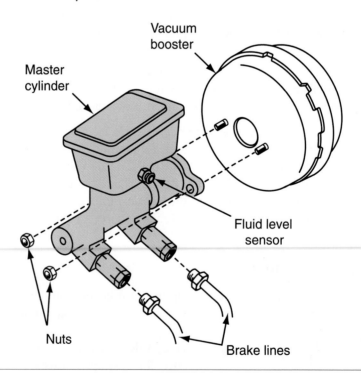

Figure 4-25 Remove the mounting nuts and disconnect the brake lines and any electrical connectors to remove the master cylinder.

9. Clean the master cylinder and vacuum booster contact surfaces with a clean shop towel.

For many vehicles with an ABS, the master cylinder is part of the ABS hydraulic modulator and master cylinder assembly. This type of ABS assembly is removed as a unit, and the master cylinder is then separated from the modulator. See Chapter 10 for more information on removing the modulator and master cylinder assembly.

Master Cylinder Reservoir Removal and Replacement

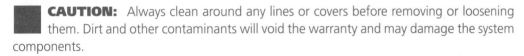

 SERVICE TIP: Some newer reservoirs are held to the master cylinder with a fastener (Figure 4-26). The fastener is usually below and to one side of the reservoir. Before trying to pry or pull the reservoir from the cylinder, ensure that this fastener(s) is removed or damage to the reservoir may occur. Remember to reinstall the fastener(s) after the reservoir is installed again.

To remove a plastic reservoir without damaging it, secure the master cylinder in a vise. Clamp on the metal cylinder body flange to avoid damaging the cylinder body. Insert a prybar between the reservoir and cylinder body and push the reservoir body away from the cylinder (Figure 4-27). When the reservoir is free, remove and discard the rubber grommets that seal the reservoir to the cylinder body. Make sure the reservoir is not cracked or deformed. Replace it if it is.

⚠ **WARNING:** Never use the old grommets when replacing the master cylinder reservoir. A leak may occur and cause a loss or a severe reduction in braking.

■ **CAUTION:** Always clean around any lines or covers before removing or loosening them. Dirt and other contaminants will void the warranty and may damage the system components.

If the reservoir is serviceable, clean it with denatured alcohol and dry it with clean, unlubricated compressed air. Using clean brake fluid, lubricate the new grommets and the bayonets on the bottom of the reservoir. To reinstall the reservoir, place the reservoir top down on

Classroom Manual
pages 72–74

Special Tools

Small prybar

Vise

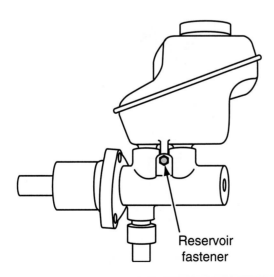

Reservoir fastener

Figure 4-26 Before cracking the reservoir, make sure there is no fastener of some sort holding it in place.

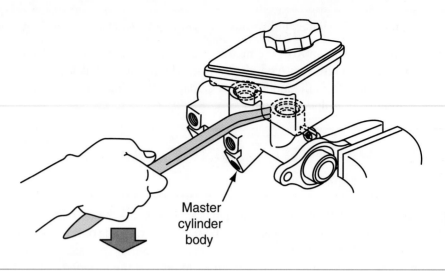

Figure 4-27 Carefully pry the reservoir off the master cylinder with a prybar or large screwdriver.

a hard, flat surface, such as a workbench. Start the cylinder body onto the reservoir at an angle, working the lip of the reservoir bayonets completely through the grommets until seated. Using a steady downward force and a smooth rocking motion, press the cylinder body onto the reservoir (Figure 4-28).

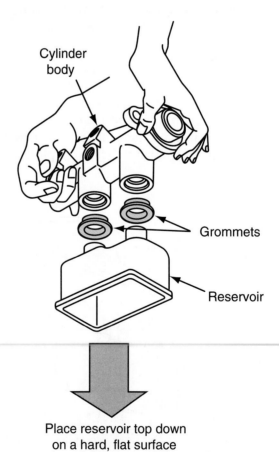

Place reservoir top down
on a hard, flat surface

Figure 4-28 Install new grommets on the cylinder and carefully push the reservoir onto the cylinder. Lubricate the grommets with brake fluid to aid assembly.

Rebuilding the Master Cylinder

Rebuilding a master cylinder involves the following major steps:

- ❏ Draining the reservoirs
- ❏ Completely disassembling the unit
- ❏ Inspecting all parts, including the cylinder bore
- ❏ Replacing all rubber seals (cups) and O-rings
- ❏ Reassembling the unit with new seals and O-rings

■ **CAUTION:** Do not clean master cylinder parts with gasoline, kerosene, solvent, or other petroleum products. Damage to rubber seals and O-rings will result.

The rebuild kit for the master cylinder usually contains all replaceable seals, O-rings, and retainer clips (Figure 4-29). Other components such as the reservoir body, external valves, and fluid sensors are replaced only if they are faulty. See Chapter 5, Hydraulic Line, Valve, and Switch Service, for details on testing these sensors and valves.

The information presented is just for general information because master cylinders on passenger cars and light trucks are almost never rebuilt. Usually it is cheaper and much faster to replace the master cylinder. However, there are times when it may be cheaper to rebuild the master cylinder, such as on some medium trucks and equipment. The process is fairly simple: remove, clean, disassemble, inspect, and clean the interior and parts and reassemble with new components. Like all rebuilding operations, use a clean work area and follow the instructions in the service manual.

Classroom Manual
pages 74–84

Special Tools

Stone or brush hone

Brake fluid

Electric drill

Service manual

The rebuilding bench or table must be kept clean and a catch basin must be readily available during rebuilding of the master cylinder.

Bench Bleeding Master Cylinders

■ **CAUTION:** Do not remove protective shipping seals, covers, or plugs before preparing to bleed the new master cylinder. Dirt and other contaminants may enter and damage the system components or void warranty.

To remove all air from a new or rebuilt master cylinder, bench bleed it before installing it on the vehicle. **Bench bleeding** reduces the possibility of air getting into the brake lines. Proper bench

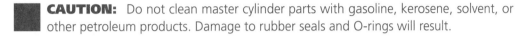

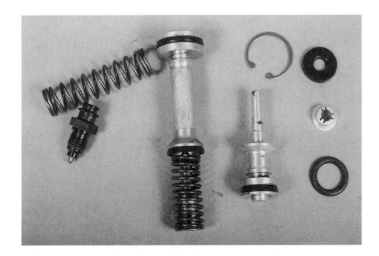

Figure 4-29 Typical master cylinder rebuild kit.

Classroom Manual
pages 71–74

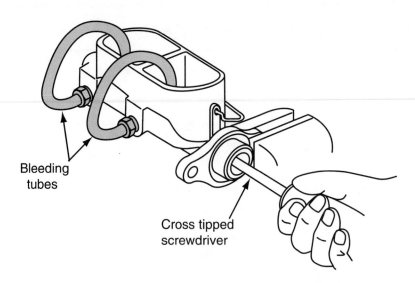

Figure 4-30 Bench bleeding a master cylinder with bleeding tubes.

Bleeding
tubes

Cross tipped
screwdriver

bleeding is particularly important with dual-piston cylinders, tandem chamber reservoirs, and master cylinders that mount on an angle other than horizontal.

> ⚠️ **WARNING:** Avoid spraying brake fluid or making it bubble violently during the bench-bleeding procedure. Do not hold your face directly above the reservoirs. Wear safety goggles or a face shield.

Bench bleeding involves mounting the master cylinder in a vise and forcing all air out of the unit. The most popular bench-bleeding technique involves installing lengths of tubing to the cylinder ports and feeding them back into the reservoir (Figure 4-30). Bench-bleeding kits are available that contain assorted fittings and tubing that will fit most vehicles. You can also make your own bleeding kit using brake lines or hoses and fittings. Always make sure the tube nuts are tightened securely to guard against air being drawn into the master cylinder on the return stroke.

Pump the cylinder pistons manually to recirculate fluid back to the reservoir in a closed loop until all air bubbles to the surface. The procedure for bench bleeding the master cylinder is shown in Photo Sequence 6.

Master Cylinder Bench Bleeding by Syringe

Another bench-bleeding technique for the master cylinder uses a special bleeding syringe to draw fluid out of the reservoir, remove air from it, and inject the fluid back into the unit (Figure 4-31):

> 📋 **AUTHOR'S NOTE:** Figure 4-31 shows the bleeding process using a syringe. The old-fashioned syringe depicted has been replaced with tools sold by several manufacturers that do the same job in a manner similar to the one shown but that is a lot cleaner and more efficient.

1. Plug the outlet ports of the master cylinder. Carefully mount it in a vise with the pushrod end slightly elevated (Figure 4-31, Item 1). Do not clamp the cylinder by the bore or exert pressure on a plastic reservoir.

2. Pour brake fluid into the master cylinder until it is half full.

3. Remove a plug from one outlet port so you can use the syringe to draw fluid out of the cylinder.

Photo Sequence 6
Typical Procedure for Bench Bleeding a Master Cylinder

P6-1 Mount the master cylinder firmly in a vise, but do not apply excessive pressure to the casting. Position the master cylinder so the bore is horizontal. When possible, clamp the master cylinder by its mounting lug.

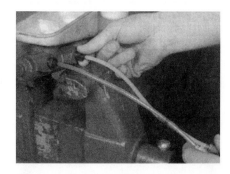

P6-2 Connect short lengths of tubing to the outlet ports, making sure the connections are tight.

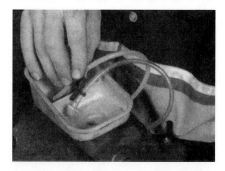

P6-3 Bend the tubing lines so that the ends are in each chamber of the master cylinder reservoir.

P6-4 Fill the reservoirs with fresh brake fluid until the level is above the ends of the tubes.

P6-5 Using a wooden dowel or a Phillips screwdriver, slowly push on the master cylinder pistons until both are completely bottomed out in their bore.

P6-6 Watch for bubbles to appear at the tube ends immersed in the fluid. Slowly release the cylinder piston and allow it to return to its original position. On quick take-up master cylinders, wait 15 seconds before pushing in the piston again. On other units, repeat the stroke as soon as the piston returns to its original position. Slow piston return is normal for some master cylinders.

P6-7 Pump the cylinder piston until no bubbles appear in the fluid. Light tapping on the body may help to dislodge trapped air bubbles.

P6-8 Remove the tubes from the outlet ports and plug the openings with temporary plugs or your fingers. Keep the ports covered until you install the master cylinder on the vehicle.

P6-9 Install the master cylinder on the vehicle. Attach the lines, but do not tighten the tube connections.

Photo Sequence 6
Typical Procedure for Bench Bleeding a Master Cylinder (continued)

P6-10 Place rags under the fittings to absorb any fluid that is expelled. Have an assistant slowly depress the brake pedal several times to force out any air that might be trapped in the connections. Loosen the nuts slightly before each pedal depression and tighten them before releasing the pedal.

P6-11 When there are no air bubbles in the fluid, tighten the connections to the manufacturer's specifications. Make sure the master cylinder reservoirs are adequately filled with brake fluid.

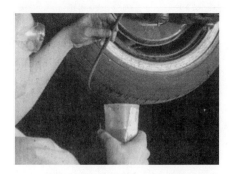

P6-12 After reinstalling the master cylinder, bleed the entire brake system on the vehicle.

4. Depress the syringe plunger completely and place its rubber tip firmly against the outlet port to seal it.

5. Slowly pull back on the plunger to draw fluid out of the cylinder. Fill the syringe body about one-half full (Figure 4-31, Item 2).

6. Point the tip of the syringe upward. Slowly depress the plunger until all air is expelled (Figure 4-31, Item 3).

7. Place the tip of the syringe (with fluid) firmly against the same outlet and slowly depress the plunger to inject the fluid back into the cylinder (Figure 4-31, Item 4). Air bubbles should appear in the reservoir.

8. When these bubbles stop, remove the syringe and plug the outlet. Repeat this procedure
at the other outlet. Plug all outlets tightly.

9. With the pushrod end tilted downward slightly, reclamp the master cylinder in the vise.

10. Slowly slide the master cylinder pushrod back and forth about ⅛ inch until you see no air bubbles in the reservoirs (Figure 4-31, Item 5).

11. Remount the master cylinder with the pushrod end up (Figure 4-31, Item 6). Fill the syringe with brake fluid and expel the air as in step 6.

12. Remove one outlet plug at a time and repeat step 6 and step 7. The master cylinder is now completely bled.

✔ **SERVICE TIP:** Do not rush to condemn the master cylinder as the cause of a low brake pedal. Bench bleeding the master cylinder before installation and careful bleeding of the entire system are important to ensure proper pedal height.

If the pedal stays low after thorough bleeding of the master cylinder and the system, check the adjustment of the rear drum brakes, if equipped. Inoperative self-adjusters and poor adjustment of the rear brakes when serviced are leading causes of low-pedal problems.

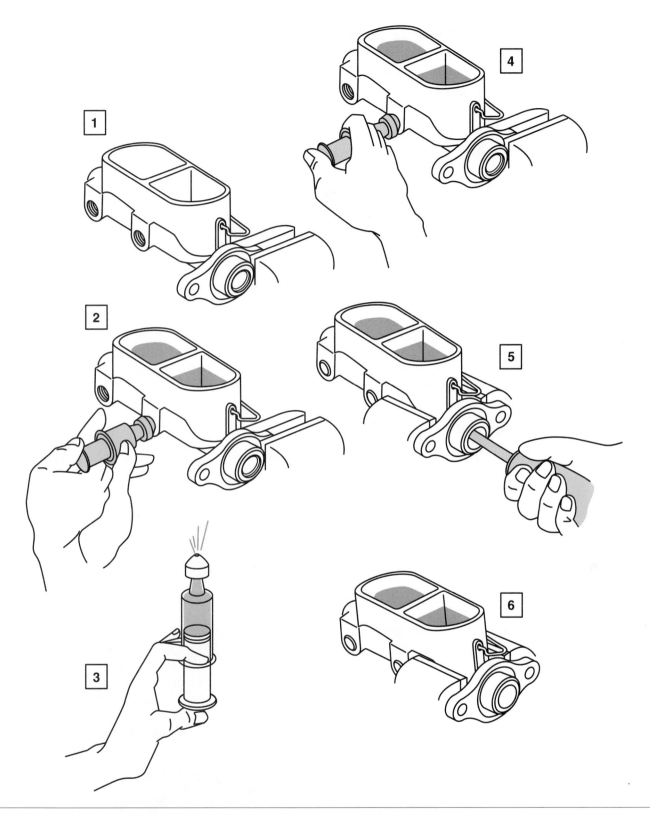

Figure 4-31 You can bench bleed a master cylinder with a special syringe as shown in these six steps. When possible, clamp the cylinder by its mounting lug.

Installing a Non-ABS Master Cylinder

Classroom Manual
pages 74–81

Special Tool

Hand tools

After bench bleeding a master cylinder, install it as follows:

1. Install the master cylinder onto the vacuum booster studs.
2. Start each brake line fitting into the master cylinder port but do not tighten.
3. Install the retaining nuts and torque to specifications.
4. Unplug each outlet port and use a flare-nut wrench to install the brake line fitting. Tighten securely (Figure 4-32).
5. Reconnect the wiring harness connector to the brake fluid level sensor connector, if equipped.
6. Reconnect the battery ground (negative) cable.
7. Bleed the system.

CAUTION: Ensure that the fitting is not cross-threaded when reconnecting. This could damage the fitting, the component, or both.

If the brake system is a manual system without a vacuum booster, install the master cylinder on the mounting studs on the engine compartment bulkhead and pass the pushrod through the bulkhead opening. Install the mounting nuts loosely to hold the cylinder in place but let it move around on the mounting studs. Connect the pushrod to the brake pedal inside the vehicle (Figure 4-33) and then tighten the cylinder mounting nuts securely.

Pushrod Adjustment

Proper adjustment of the master cylinder pushrod is essential for safe and correct brake operation. If the pushrod is too long, the master cylinder piston will restrict the vent ports. This can prevent hydraulic pressure from being released and can result in brake drag.

If the pushrod is too short, the brake pedal will be low and the pedal stroke length will be reduced, which can result in a loss of braking power. When the brakes are applied with a short pushrod, groaning noises may be heard from the vacuum booster.

If a problem with pushrod length or adjustment is suspected, perform the "Test for Open Vent Ports" presented earlier in this chapter. If fluid does not spurt from the vent ports when the

Figure 4-32 Use tubing wrenches to connect the brake line fittings to the master cylinder.

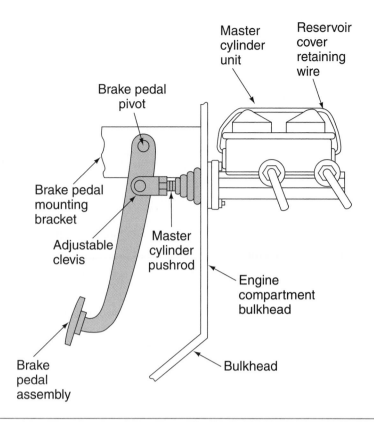

Figure 4-33 The pushrod connection to the brake pedal arm is similar for both manual and power brakes.

brake pedal is released, the pushrod may be holding the master cylinder pistons in positions that partially restrict the ports.

In a vacuum power brake system, the pushrod is part of the booster and is matched to the booster during assembly. It is normally adjusted only when the vacuum booster or the master cylinder is serviced. Vacuum booster pushrod length is usually checked with a gauge. Because pushrod length is most often checked and adjusted as part of vacuum booster overhaul or replacement, procedures are included in Chapter 6, Power Brake Service.

Checking and adjusting brake pedal free play as explained earlier in this chapter also will help to ensure proper master cylinder piston travel.

Master Cylinder Bleeding on the Vehicle

Whenever possible, bleed the master cylinder on the bench before installing it on the vehicle. Bleeding the cylinder after installing it on the car removes any final air bubbles in the outlet ports or air that may enter the fittings when they are connected.

Fluid pressure to bleed the master cylinder can be supplied by a pressure bleeder or by having a helper press the brake pedal. Because the master cylinder is the highest point of the hydraulic system and because air rises, bleed the master cylinder before bleeding any of the wheel brakes. The following procedures explain how to bleed a master cylinder with and without bleeder screws.

 WARNING: Brake fluid can be irritating to the skin and eyes. In case of contact, wash skin with soap and water or rinse eyes thoroughly with water.

Special Tools

Hand tools
Cloths

On-Vehicle Bleeding Without Bleeder Screws

Most master cylinders do not have bleeder screws, but you can remove any trapped air at the outlet ports by loosening the line fittings and applying fluid pressure:

1. Discharge the vacuum from the power booster, if equipped, by pumping the pedal with the engine off until it becomes hard.
2. Fill the master cylinder with fresh brake fluid and ensure that it stays at least half full during the bleeding procedure.
3. Using a flare-nut wrench, loosen the forward or highest line fitting on the master cylinder outlet ports (Figure 4-34).
4. Apply fluid pressure with a pressure bleeder or by having a helper press and hold the brake pedal.
5. Hold a clean rag or a container under the fitting to catch fluid that escapes.
6. While maintaining fluid pressure, tighten the fitting.
7. Repeat step 3 through step 6 until no air escapes from the fitting along with the fluid. Then repeat these steps at each remaining fitting on the master cylinder, working from the highest to the lowest.

On-Vehicle Bleeding with Bleeder Screws

Some master cylinders have bleeder screws. These usually are cylinders that mount level on the vehicle. You will need a flare-nut wrench, a length of clear plastic tubing that fits the end of the bleeder screws, and a container partly filled with fresh brake fluid. Proceed as follows:

1. Discharge the vacuum from the power booster, if equipped, by pumping the pedal with the engine off until it becomes hard.

Figure 4-34 Loosen one of the two brake lines as the brake pedal is being depressed. Tighten it back before the pedal is released. Do one line at a time.

2. Fill the master cylinder with fresh brake fluid and ensure that it stays at least half full during the bleeding procedure.

3. Attach the plastic tubing to the end of the forward bleeder screw on the master cylinder and submerge the other end of the tubing in the container of fresh fluid.

4. Using a flare-nut wrench, loosen the bleeder screw on the master cylinder and apply fluid pressure with a pressure bleeder or by having a helper press and hold the brake pedal.

5. While maintaining fluid pressure, tighten the bleeder screw.

6. Repeat step 3 through step 5 until no air escapes from the bleeder screw along with the fluid. Then repeat these steps at each remaining bleeder screw on the master cylinder.

CAUTION: Prevent brake fluid from coming in contact with the vehicle's finish. Brake fluid damages paint and finish immediately on contact. If fluid contacts the finish, wash area thoroughly with running water and soap if possible.

SERVICE TIP: Brakes dragging? Check the fluid level in the master cylinder. Overfilling the reservoir can prevent fluid from returning completely from the brake lines. The excess fluid in the lines then maintains slight pressure on the brakes and causes drag and overheating. The fluid level should not be above the MAX line in the reservoir.

SERVICE TIP: Bleeding the brakes or master cylinder that are mounted at an angle can be best done by lifting the rear of the vehicle until the master cylinder is horizontal or as near as possible. This will help prevent air bubbles from being trapped in the forward end of the master cylinder.

Hydraulic System Bleeding

Brake bleeding is the process of removing air from the hydraulic lines by opening a bleeder port at each wheel and sometimes elsewhere in the system. These ports are sealed with **bleeder screws** that are opened to allow fluid and air to escape the system (Figure 4-35). Bleeder screws are located at high points throughout the brake system (Figure 4-36). A bleeder screw is normally installed in each drum brake wheel cylinder, in each disc brake caliper, next to the outlet ports of some master cylinders, and on some combination valves. Some bleeder screws have a threaded passage and a protective dust cap screw that must be removed before a drain hose can be installed on the bleeder screw.

Classroom Manual
pages 63–68, 84–86

SERVICE TIP: Russel Performance Products makes a special tool used for one-person brake bleeding. The Speed Bleeder has a built-in check valve and replaces the standard brake bleeder screw. Loosen the Speed Bleeder one quarter turn and pump the brake pedal. When that wheel is bled, close the bleeder and move to the next wheel. Do not remove the Speed Bleeder. The customer gets the tool along with the brake repairs.

If air is trapped in the system, the brake pedal will be low and feel spongy when first applied. Rapidly pumping the pedal several times will compress much of the air and cause the pedal to rise and become more firm. As soon as pressure is released from the pedal, the air will expand again and the pedal will return to its spongy and low condition. Upward bends in the brake tubing as it is routed through the vehicle chassis also can trap air in the system (Figure 4-37).

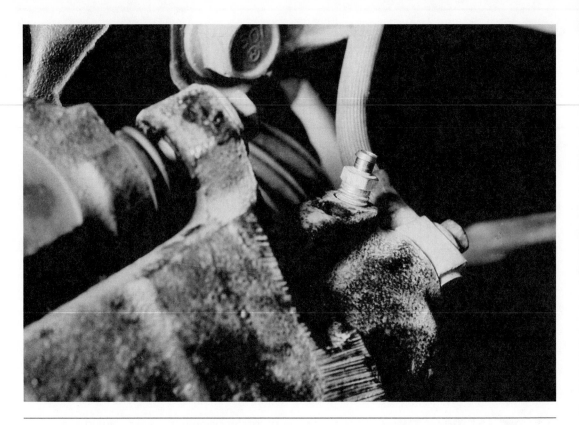

Figure 4-35 Bleeder screws, such as this one on a disc brake caliper, are also installed on wheel cylinders and some master cylinders and valves.

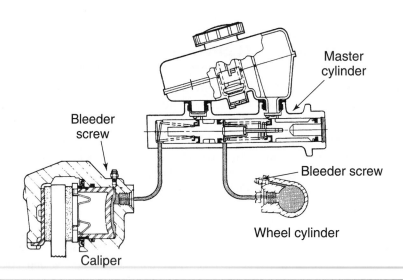

Figure 4-36 Bleeder screws are located at the highest points on calipers and wheel cylinders.

This condition makes bleeding the system even more important and often more difficult. All of the air must be removed from the system to ensure proper brake operation and pedal feel.

The brake pedal symptoms of a system that contains air can be similar to the pedal symptoms caused by drum brakes that need adjustment. A low pedal caused by misadjusted drum brakes, however, will become firm after two or three applications and will not have the spongy

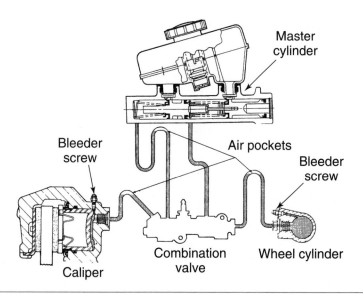

Figure 4-37 Bends in brake hoses and tubing can trap air and make bleeding more difficult. All air must be removed for proper operation.

feeling caused by air in the system. You can use the following air-entrapment test to help determine if there is air in the system.

Air-Entrapment Test

WARNING: Wear safety glasses to protect your eyes during the air-entrapment test. Also protect the vehicle paint from possible brake fluid splatters.

Remove the master cylinder cover and be sure the reservoirs are filled to the proper level. Hold the cover and gasket against the reservoir top but do not secure the clamp or screws. Then have an assistant pump the brake pedal ten to twenty times rapidly and maintain pressure after the last pedal application.

Remove the reservoir cover and have the assistant quickly release pedal pressure. Watch for a squirt of brake fluid from the reservoirs. If air is compressed in the system, it will force fluid back through the compensating ports faster than normal and cause fluid to squirt in the reservoir. If a fluid squirt appears in one side of the reservoir but not the other, that side of the split hydraulic system contains the trapped air.

 SERVICE TIP: When troubleshooting a problem of brake drag, overheating, or premature wear, connect a pressure gauge to the bleeder port on each caliper and wheel cylinder in turn. When the brakes are applied, pressure should rise quickly and drop just as quickly when the brakes are released. Some drum brakes may retain about 10 psi in the lines to hold the cup seals against the wheel cylinder bores. Other than that, any residual pressure in one brake line usually indicates a restricted tube or hose. Inspect all the brake lines for that wheel closely.

Overall Brake Bleeding Sequences

Bleeding may be performed on all or just part of the hydraulic system. As explained earlier in this chapter *Shop Manual*, a master cylinder usually is bled on the bench before it is installed on the car. Then, final bleeding can be performed at the master cylinder fittings or bleeder screws to remove any air trapped in the connections. If brake pedal and master cylinder operation is normal after these procedures, it may not be necessary to bleed the wheel brakes.

Depending on where the hydraulic system was opened to air, bleeding may be needed at all four wheels and the master cylinder; or it might be needed only at two wheels. If all fittings are disconnected from a dual-reservoir master cylinder, or if air was introduced into the system through low fluid level in both cylinder reservoirs, bleeding is required at the master cylinder and all four wheels.

If the tubes for one set of wheels are disconnected at the master cylinder, or if air entered through a low fluid level in only the reservoir for that set of wheels, only those wheels and lines need to be bled. If there is any doubt about air in the system, however, the entire system should be bled.

CAUTION: On some vehicles, it is possible to install a right-hand caliper on the left-side wheel and vice versa. If you inadvertently do this, the bleeder screw will be located at the bottom of the caliper bore. This low position makes it impossible to bleed all air out of the system.

Air rises, certainly, so if an entire brake system is to be bled, it is bled starting at the highest point in the system. An overall brake bleeding sequence would be:

❑ Master cylinder
❑ Combination valve
❑ Wheel cylinders and brake calipers
❑ Height-sensing proportioning valve

CAUTION: Do not spill any brake fluid on the car finish. If you do, flush it immediately with water. Clean dirt away from the master cylinder reservoir cap or cover before opening it. Do not let contamination enter the hydraulic system.

Whether bleeding the complete system or just part of it, the first step is to fill the master cylinder with fresh fluid. Being careful not to let dirt or other contamination enter the reservoir, remove the cap or cover and fill it to the specified level with fresh fluid. If fluid in the reservoir falls below the level of the compensating and replenishing ports, air will enter the system. If this happens, bleeding steps must be repeated to ensure that all air is removed.

General practice with any type of bleeding is to slip a hose over the end of the bleeder screw. Place the free end of the hose into a jar half filled with brake fluid. Always keep the end of the hose submerged in brake fluid. This prevents air from being drawn back into the system and lets you observe when air bubbles stop flowing from the bleeder to indicate that air has been removed from that portion of the system.

Freeing a Frozen Bleeder Screw

When opening or removing a bleeder screw, use a special six-sided bleeder screw box wrench of the correct size. The shoulders of a bleeder screw are easily rounded if the wrong tool is used. Do not exert too much force on the screw. It is possible to break off the small screw in the housing. Fixing this problem requires drilling and tapping so avoid it by working carefully. If drilling and tapping are not possible, the entire cylinder or caliper must be replaced.

Before performing cylinder or caliper service, always check the bleeder screw to see if it can be loosened. If the caliper or cylinder is serviceable but the bleeder screw is frozen, one of the following methods may free it.

Use a special wrench to exert pressure on the screw by striking it with a hammer. The shock to the screw from the hammer blows and the tension against it may free the screw. Apply a few drops of penetrating oil around the screw threads. Let the oil soak in for a few minutes and try again to turn the screw.

WARNING: Never apply heat to a closed brake system. The fluid could boil, generating great pressure and rupturing the weakest point. Serious injury could occur. If heat must be used, remove the pistons and brake lines so the heat can dissipate through the openings.

As a last resort, use a welding torch to heat the housing around the screw to a dull red. When the caliper or cylinder body is hot, apply pressure to the screw with the correct size wrench. When using this method, be sure to remove the rubber and plastic parts in the caliper or cylinder to prevent heat damage. This method is best done with the caliper or cylinder removed from the vehicle and clamped in a bench vise. If heat must be applied to brake parts while on the vehicle, remove any adjacent brake hoses or cover them with wet cloths to protect them from heat.

If a frozen bleeder screw is heated with the caliper or cylinder clamped in a vise, position the bleeder screw downward. Any residual fluid in the bore will then run into the screw threads as they are heated and help to free the screw.

Wheel Brake Bleeding Sequences

All vehicle manufacturers recommend a specific sequence in which to bleed the wheel brakes. These recommendations can be found in vehicle service manuals and in aftermarket brake service manuals. Before bleeding the wheel brakes on any vehicle, refer to these recommendations and follow them during the bleeding procedure.

If the manufacturer's recommendations are not available, the following sequence will work on most vehicles:

- ❏ Master cylinder
- ❏ Combination valve or proportioning valve (if fitted with bleeder screws)
- ❏ Right rear
- ❏ Left rear
- ❏ Right front
- ❏ Left front
- ❏ Height-sensing proportioning valve (if fitted with bleeder screws)

This sequence is based on the principle of starting at the highest point in the system and working downward, then starting at the wheel farthest from the master cylinder and working to the closest. A few more general rules also are worth remembering.

If the brake system is split between the front and rear wheels, the rear wheels (which are farthest from the master cylinder) usually are bled first. If the brake system is split diagonally, the most common sequence is: RR-LF-LR-RF (Figure 4-38). This sequence also applies to most systems with a quick take-up master cylinder. If you bleed a quick take-up system in any other sequence, you may chase air throughout the system.

Exceptions to the general rules exist, however. DaimlerChrysler, for example, recommends bleeding both rear brakes before the front brakes, regardless of how the hydraulic system is split.

If a caliper has two bleeder screws, one higher than the other, bleed the lower one first. Similarly, you may run into an older imported car with two single-acting wheel cylinders for a drum brake installation. In this case, also bleed the lower cylinder before the upper one. These instructions may seem contrary to the general rule of bleeding the system from the highest points downward. In the case of calipers and dual wheel cylinders with upper and lower bleeder screws, starting at the lower bleeders will remove air trapped in the lower parts of the assembly. Any air remaining will rise to the top where it can be removed by a final bleeding step at the upper bleeder for an individual wheel brake assembly.

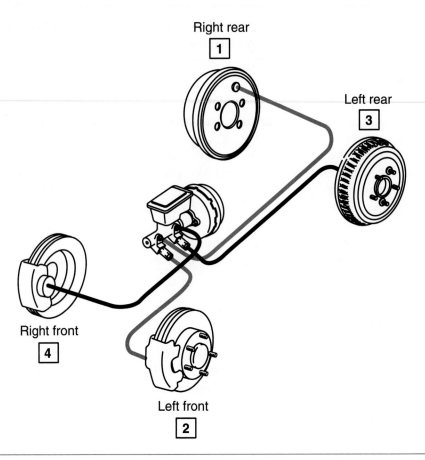

Figure 4-38 Recommended bleeding sequence for a diagonally split brake system.

Some older Japanese imports have rear drum brakes with a bleeder screw on only one wheel cylinder. The manufacturers felt that one bleeder was satisfactory to remove air from both rear brakes, but bleeding may take longer with such an installation.

Remember to press or pull the stem of a metering valve to lock it open when using a pressure bleeder. Many valves require a special clip to hold the metering valve stem open while bleeding the front brakes (Figure 4-39). Also, the pistons of many pressure differential valves must be recentered after bleeding the system to be sure that the instrument panel warning lamp is off. Refer to later sections of this chapter for instructions.

Six methods are commonly used for brake bleeding:

1. Manual bleeding
2. Pressure bleeding
3. Vacuum bleeding
4. Gravity bleeding
5. Surge bleeding
6. Reverse fluid injection bleeding

The following sections explain the advantages and disadvantages of these six methods and contain procedures for their use.

CAUTION: Vehicles with ABSs often require special bleeding procedures or additional steps for the following general procedures. Failure to follow these special instructions may result in damage to ABS components or incomplete bleeding of the system. Refer to Chapter 10.

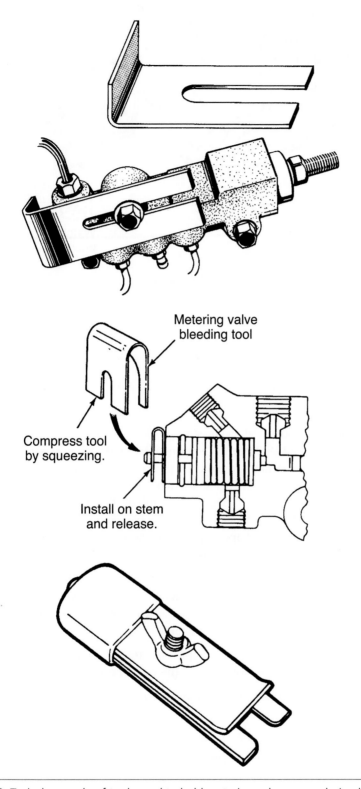

Metering valve
bleeding tool

Compress tool
by squeezing.

Install on stem
and release.

Figure 4-39 Typical example of tools used to hold metering valves open during bleeding.

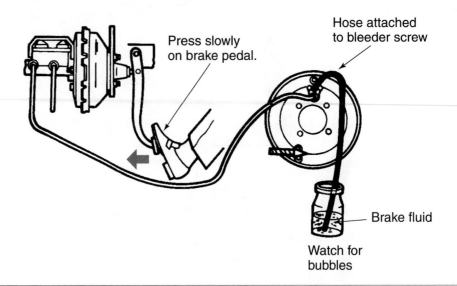

Press slowly
on brake pedal.

Hose attached
to bleeder screw

Brake fluid

Watch for
bubbles

Figure 4-40 Basic setup for manual brake bleeding.

✓ **SERVICE TIP:** Remember to turn on the ignition when bleeding a Teves Mark II ABS. You cannot get fluid to the rear wheels unless the ignition is on and the ABS pump is running.

Manual Bleeding

Manual bleeding uses the brake pedal and master cylinder as a hydraulic pump to expel air and brake fluid from the system when a bleeder screw is opened. **Manual bleeding** is a two-person operation: one person pumps the brake pedal, and the other opens and closes the bleeder screws.

Manual bleeding requires a bleeder screw wrench of the correct size, a container partially full of fresh brake fluid, a length of clean plastic tubing that fits over the top of the bleeder screws, and several clean shop cloths. During manual bleeding, the brake pedal must be applied slowly and steadily (Figure 4-40). Rapidly pumping the pedal will churn air in the system and make it harder to expel. Bleed the system as follows:

1. Check the fluid level in the master cylinder reservoir and be sure that both sections are full. Recheck the fluid level after bleeding each wheel brake and refill as necessary. If the level drops below the ports, air will enter the system. Be sure that the reservoir cover or cap is installed securely during bleeding.

2. Discharge the vacuum or hydraulic pressure reserve in the booster by pumping the brake pedal with the ignition off until the pedal becomes hard.

3. Using a clean shop cloth, wipe dirt away from the bleeder screw on the first wheel in the recommended sequence.

4. Fit the plastic hose over the top of the bleeder screw and submerge the other end in the container of fresh brake fluid.

5. Loosen the bleeder one-half to one turn and have your assistant press the brake pedal slowly and steadily and hold it to the floor. Observe air bubbles flowing from the hose into the fluid container.

6. Tighten the bleeder screw and have your assistant slowly release the brake pedal.

7. Repeat step 5 and step 6 until no more air flows from the tubing into the brake fluid container.

Manual bleeding requires no special tools or equipment.

Manual bleeding is the process of using the brake pedal and master cylinder as a hydraulic pump to expel air and brake fluid from the system.

Special Tools

Brake fluid

Cloths

Catch basin

Brake bleeder kit or Tubing and transparent container

Hand tools

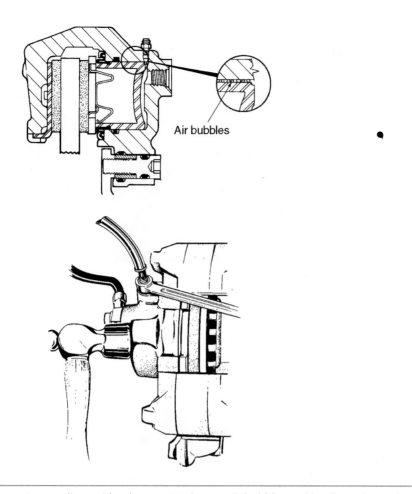

Figure 4-41 You can tap a caliper with a hammer to loosen air bubbles and let them rise to the bleeder screw.

8. Check the fluid level in the master cylinder and add fluid if necessary. Then proceed to the next wheel in the bleeding sequence. Repeat the bleeding sequence as necessary until the brake pedal is consistently firm. Check the fluid level a final time and install the reservoir cover or cap.

If you have trouble getting all of the air out of a disc brake caliper. Tap the caliper lightly but firmly with a hammer to loosen trapped air bubbles and let them rise to the bleeder screw (Figure 4-41).

Photo Sequence 7 shows the proper way to bleed a disc brake caliper manually. The steps shown apply to one bleeder screw and should be repeated at all other bleed points.

Manual Bleeding with Check Valve Bleeder Hose. Bleeding hoses are available with a one-way check valve that only lets fluid flow out of the bleeder screw, not in the reverse direction. This bleeder hose keeps air from being drawn back into the system when the brake pedal is released.

Pressure Bleeding

Pressure bleeding requires special equipment to force brake fluid through the system. Pressure bleeding has two advantages over manual bleeding. It is faster because the master cylinder does not have to be refilled several times, and the job can be done by one person. Therefore, pressure bleeding is the method most often used in the brake service profession. Figure 4-42 shows the basic equipment and setup for pressure bleeding.

Special Tools

Pressure bleeder
Tubing and
    transparent
    container
Cloths

Pressure bleeding is the process of using a tank filled with brake fluid and pressurized with compressed air to expel air and brake fluid from the system.

Photo Sequence 7
Typical Procedure for Manually Bleeding a Disc Brake Caliper

P7-1 Be sure the master cylinder reservoir is filled with clean brake fluid. Recheck it often to replace fluid lost during the bleeding process.

P7-2 Attach a bleeder hose to the bleeder screw.

P7-3 Place the other end of the hose in a capture container partially filled with brake fluid. Be sure that the free end of the hose is submerged in brake fluid. This helps to show air bubbles as they come out of the system and prevents air from being accidentally sucked into the system through the bleeder screw.

P7-4 Have an assistant apply moderate (40 to 50 pounds), steady pressure on the brake pedal and hold it down.

P7-5 Open the bleeder screw.

P7-6 Observe the fluid coming from the submerged end of the hose. At the start, you should see air bubbles.

P7-7 When the fluid is clear and free of air bubbles, close the bleeder screw.

P7-8 Have your assistant release the brake pedal. Wait 15 seconds and repeat step 2 and step 3 until no bubbles are seen when the bleeder screw is opened. Close the bleeder screw at that wheel and move to the next wheel in the bleeding sequence. Bleed all four wheels in the same manner.

P7-9 When the entire system has been bled, turn on the ignition switch.

Photo Sequence 7
Typical Procedure for Manually Bleeding a Disc Brake Caliper (continued)

P7-10 Check the pedal for sponginess.

P7-11 Check the brake warning lamp for an indication of unbalanced pressure. Repeat the bleeding procedure to correct the problem.

P7-12 Top off the master cylinder reservoir to the proper fill level.

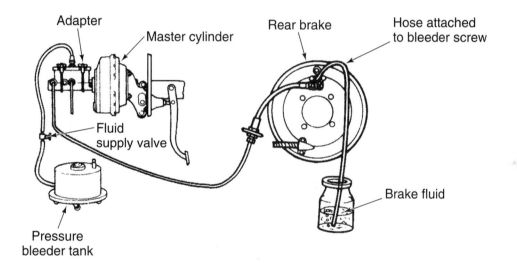

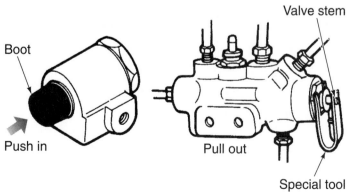

Holding Metering Valve Open

Figure 4-42 Basic setup for pressure brake bleeding.

WARNING: Do not depress the brake pedal when pressure brake bleeding equipment is being used. This could damage seals and piston cups within the brake system.

A pressure bleeder is a small tank that contains brake fluid, which is pressurized by compressed air (Figure 4-43). A hose from the pressure bleeder is connected to the master cylinder by an adapter fitting that fits over the reservoir in place of the reservoir cap. Adapters exist in different configurations for different types of reservoirs. One-piece cylinders with integral reservoirs generally use a flat plate adapter. Some plastic reservoirs require adapters that seal around the ports in the bottom of the reservoir (Figure 4-44).

The hydraulic pressure generated by manual bleeding is enough to open the metering or combination valve and let fluid flow to the front disc brakes. When pressure bleeding, the valve must be held open manually because the pressure bleeder works with pressure in the range where a metering valve is normally closed.

To hold the metering valve open, either push the valve stem in or pull it out, depending on the valve type. Figures 4-39 and 4-42, shown earlier, illustrate several kinds of metering valve tools and ways to open the valve for bleeding. On some combination valves, such as those used by GM, loosen a valve mounting bolt and slide the valve tool under the bolt head. Move the end of the tool toward the valve body until it depresses the metering valve stem. Then tighten the mounting bolt to hold the tool in place. Other valves have a stem that must be held outward by a spring clip tool. The stem on still other valves must be held in by hand while the front brakes are bled.

CAUTION: Prevent brake fluid from coming in contact with the vehicle's finish. Brake fluid damages paint and finish immediately on contact. If fluid contacts finish, wash area thoroughly with running water using soap if possible.

CAUTION: Always clean around any lines or covers before removing or loosening them. Dirt and other contaminants will void the warranty and may damage system components.

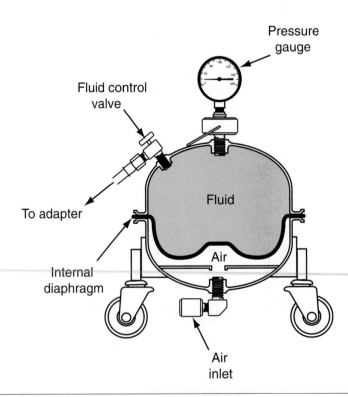

Figure 4-43 The lower chamber is compressed air; the upper chamber holds brake fluid. The chambers are separated by a flexible, air-tight diaphragm.

Figure 4-44 Many different adapters are needed to properly pressure bleed the various master cylinder configurations.

⚠️ **WARNING:** Wear safety glasses and/or face protection when using brake fluid. Injuries to the face and/or eyes could occur from spilled or splashed brake fluid.

In addition to a pressure bleeder with the right adapter and a metering valve tool (if required), pressure bleeding requires a bleeder screw wrench of the correct size, a container partially full of fresh brake fluid, a length of clean plastic tubing that fits over the top of the bleeder screws, and several clean shop cloths. Bleed the system as follows:

1. Fill the pressure bleeder with the type of fluid specified by the vehicle manufacturer. Then charge the bleeder with 25 psi to 30 psi of compressed air according to the equipment instructions.

2. Clean the top of the master cylinder, remove the reservoir cover, and fill the reservoir about half full of fresh brake fluid.

3. Install the adapter on the reservoir and connect the fluid supply hose from the pressure bleeder to the adapter.

4. If required, install the correct override tool on the metering valve.

5. Open the fluid supply valve on the pressure bleeder to let pressurized fluid flow to the reservoir. Check the adapter and all hose connections for leaks and tighten if necessary.

6. Using a clean shop cloth, wipe dirt away from the bleeder screw on the first wheel in the recommended sequence.

7. Fit the plastic hose over the top of the bleeder screw and submerge the other end in the container of fresh brake fluid.

8. Loosen the bleeder one-half to one turn and observe air bubbles flowing from the hose into the fluid container.

9. Tighten the bleeder screw when clean fluid without any air bubbles flows into the container.

10. Repeat step 6 through step 9 at the next wheel in the bleeding sequence. Continue until the last brake in the sequence is bled. Repeat the bleeding sequence as necessary until the brake pedal is consistently firm.

11. Remove the metering valve override tool.

12. Close the fluid supply valve on the pressure bleeder.

13. Wrap the end of the fluid hose at the master cylinder adapter with a clean cloth and disconnect the hose from the adapter.

14. Remove the adapter from the master cylinder and be sure the reservoir is filled to the correct level. Install the reservoir cover or cap.

Vacuum Bleeding

Special Tools

Vacuum bleeder kit

Cloths

Vacuum bleeding is an alternative to pressure bleeding and is preferred by some technicians. As is pressure bleeding, vacuum bleeding is a one-person operation. Depending on the type of equipment, however, the master cylinder may require refilling during the bleeding operation. Two basic types of vacuum bleeding equipment are available: the hand-operated vacuum pump and the system operated by compressed air.

A hand-operated vacuum pump used for brake bleeding is the same kind of vacuum pump used to test and service fuel system and emission control devices. The pump holds a small cup that contains fresh brake fluid, and a length of plastic tubing connects the pump to the bleeder screw (Figure 4-45). Vacuum bleed a brake system as follows:

1. Check the fluid level in the master cylinder reservoir and be sure that both sections are full. Recheck the fluid level after bleeding each wheel brake and refill as necessary. If the level drops below the ports, air will enter the system. Be sure that the reservoir cover or cap is installed securely during bleeding.

2. Install the small fluid container on the vacuum pump according to the equipment instructions. Be sure all connections are tight so that the pump cannot draw air past the fluid container.

3. Fill the small container about half full of fresh brake fluid. Be sure the short hose inside the container is submerged in fluid so that air cannot flow back into the brake system.

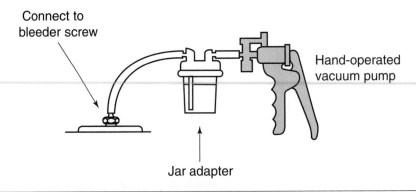

Figure 4-45 The same type of hand-operated vacuum pump used for engine service can be used to bleed brakes with suitable adapters. The bleeder screw can be temporarily sealed by adding silicone dielectric grease to the threads.

4. Using a clean shop cloth, wipe dirt away from the bleeder screw on the first wheel in the recommended sequence.

5. Fit the plastic hose from the vacuum pump over the top of the bleeder screw and operate the pump handle ten to fifteen times to create a vacuum in the container.

6. Using the correct bleeder screw wrench, loosen the bleeder one-half to three-quarters turn. Observe fluid with air bubbles flowing into the fluid container on the pump.

7. After evacuating about 1 inch of fluid into the container, tighten the bleeder screw.

8. Repeat step 5 through step 7 until no more air flows into the brake fluid container. Remove old fluid from the pump container as necessary during the bleeding procedure.

9. Check the fluid level in the master cylinder and add fluid if necessary. Then proceed to the next wheel in the bleeding sequence. Repeat the bleeding sequence as necessary until the brake pedal is consistently firm. Check the fluid level a final time and install the reservoir cover or cap.

Some vacuum bleeding equipment that uses compressed air may include a fresh fluid container that attaches to the master cylinder reservoir with an adapter similar to the type used with a pressure bleeder. This eliminates the need to refill the reservoir repeatedly during bleeding.

A compressed air vacuum pump uses airflow through a venturi to create a vacuum in a chamber connected to the throat of the venturi. This process is called air-aspirated vacuum. The vacuum pump may be connected to the compressed air supply during bleeding, or it may be charged with reserve vacuum and disconnected from the air supply for use.

To use vacuum bleeding equipment (Figure 4-46), fill the master cylinder or connect the fresh fluid container to the master cylinder according to the equipment instructions. Then connect the vacuum pump to the first wheel in the recommended bleeding sequence. Open the wheel bleeder screw and the vacuum valve and let fluid flow from the wheel brake into the pump container until it is free of air bubbles. Close the bleeder screw and move the vacuum pump to the next wheel in sequence. Repeat the bleeding sequence as necessary until the brake pedal is consistently firm and no more air flows from any wheel brake. Remove old fluid from the pump container as necessary during the bleeding procedure. Check the fluid level in the master cylinder a final time and install the reservoir cover or cap.

When you use this type of vacuum bleeding equipment, it is common to see bubbles or foam in the fluid drawn from the brake system. Air is drawn into the evacuated fluid, past the threads of the bleeder screw. The air mixes with the fluid being drawn out of the system and flows to the vacuum bleeder (Figure 4-47). This does not affect the bleeding operation and does not indicate continuous air in the system.

Vacuum bleeding with either a hand-operated pump or a compressed-air system generally does not require overriding the metering valve. Check the equipment operating instructions, however, to verify the exact requirements.

Gravity Bleeding

Gravity bleeding is letting nature take its course in brake bleeding. Atmospheric pressure on the surface of the fluid in the reservoir eventually forces the fluid through the brake system and out the open bleeder screws. Gravity bleeding is the simplest and slowest way to bleed a brake system. It also can be the most effective on some systems. Gravity bleeding works best on a system that does not have a combination valve or a proportioning valve that requires bleeding. Gravity bleeding cannot be used on a system that contains a residual pressure check valve or any other valve that isolates any part of the system at low pressure. Gravity bleeding also may be difficult on a system with high points in the lines that can trap air. Before using the gravity bleeding method, the master cylinder should be thoroughly bench bled and then bled again after it is installed on the vehicle.

Gravity bleeding is the process of letting old brake fluid and air drain from the brake hydraulic system through a wheel bleeder screw.

Special Tools

Four sets of tubing and transparent containers

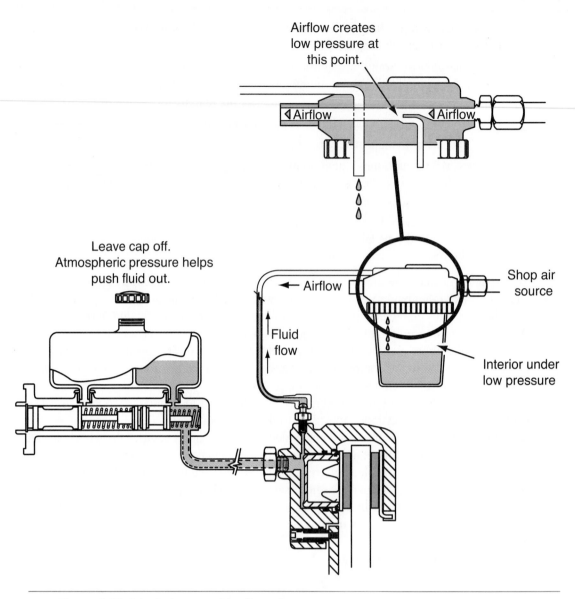

Figure 4-46 A typical vacuum bleeding setup.

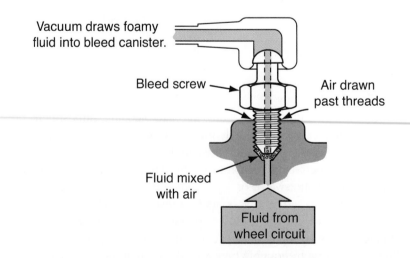

Figure 4-47 Air entering via the bleeder screw threads will not affect the bleeding process.

Gravity bleeding requires that all bleeder screws on the wheel brakes be open at the same time. Install lengths of clean plastic tubing on all bleeder screws and immerse the other end of each hose in a container of clean fluid before opening the bleeder screws. If all bleeder screws were opened to the air, air might be drawn back into the system because the total area of the bleeder openings may be greater than the area of the compensating ports through which fluid must flow to the master cylinder.

Fill the master cylinder reservoir with fresh fluid before starting the bleeding process and check it periodically during gravity bleeding. Open each bleeder screw approximately one turn and verify that fluid starts to flow from each tube. Check the fluid level in the master cylinder periodically during bleeding. If fluid falls below the master cylinder ports, you will have to start the bleeding process all over again. Gravity bleeding can take an hour or more to completely purge the system of air.

When fluid flowing from the bleeder screws is clear and free of air bubbles, close the bleeder screws, disconnect the tubing, and check the master cylinder fluid level. Add fluid as required.

Surge Bleeding

Surge bleeding is a supplementary procedure that can be used to remove air pockets that resist other bleeding methods. **Surge bleeding** is a variation of manual bleeding in which an assistant pumps the brake pedal rapidly to create turbulence in the system.

Surge bleeding should not be used as the only bleeding procedure for a brake system. The agitation will often dislodge air trapped in pockets in the system.

The steps for surge bleeding are basically the same as for manual bleeding with one exception. After connecting plastic tubing to a wheel bleeder screw and immersing the tubing in a container of fresh fluid, open the bleeder screw about one full turn. Then, with the bleeder open, have an assistant pump the brake pedal rapidly several times. Watch for surges of air bubbles to flow from the tubing into the fluid container. Finally, have your assistant hold the pedal to the floor while you close the bleeder screw.

Repeat the surge bleeding procedure at each wheel several times and then let the system stabilize for 5 to 10 minutes. Follow the surge bleeding with any of the other bleeding procedures explained previously.

Other Bleeding Equipment

As mentioned in Chapter 4 of the *Classroom Manual,* one style of brake bleeding equipment can pressure or vacuum the system. Figure 4-48 shows a typical hookup to pressure bleed the brake system with a Phoenix Injector. The injector fluid source (which may be a regular container of brake fluid) has been filled with the correct brake fluid and the injector system bled. The master cylinder has sufficient fluid to seal the port(s) when the injector is removed. Note the capture container attached to the wheel cylinder/caliper bleeder screw. Open the bleeder screw enough to allow fluid to flow from the cylinder/caliper. Support the container in some way to prevent tipping or spillage or have a second technician hold it and open and close the bleeder screw. Insert the nozzle of the injector into the appropriate replenishment port of the master cylinder firmly enough to seal the connection. Pump the handle to force fluid from the source through the injector and throughout the brake system. Repeat until clear, air-free brake fluid is forced from the bleeder screw. Tighten the bleeder screw before moving the injector or capture container. Repeat with the other wheels until the complete system is bled. Remove the injector completely after the last wheel is bled and fill the master cylinder to the proper level. A modified version of this setup could be used to bench bleed the master cylinder on or off the vehicle.

Figure 4-49 shows a typical installation of the Phoenix Injector set up to suction (vacuum) bleed a brake system. In this setup, the injector uses the brake system as the source fluid and pumps the fluid from the open bleeder screw into a capture container. The master cylinder has to be monitored and kept full at all times in this procedure. Ensure that the bleeder screw is closed

Figure 4-48 A pressure bleeding system. (Courtesy of Phoenix Systems LLC)

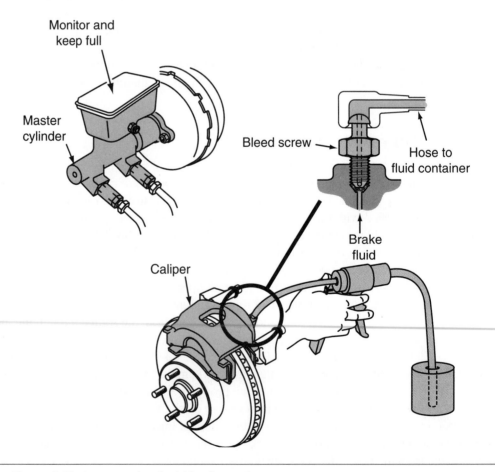

Figure 4-49 A vacuum (suction) bleeding system.

before disconnecting the injector. Repeat with each wheel brake until all show clear, air-free fluid from the cylinder/caliper. Fill the master cylinder to the proper level if needed. This setup would not work well in bench bleeding the master cylinder.

Brake Fluid Replacement: Flushing and Refilling the Hydraulic System

Manufacturers are about evenly divided on whether or not the brake system should be flushed periodically and refilled with fresh fluid. DOT 3 and DOT 4 fluids absorb moisture from the atmosphere. Most vehicles will contain about 2 percent water in their brake fluid after just one year of service. As little as 6 percent moisture in the brake fluid can cut the fluid boiling point in half. These facts may be very good arguments for periodically flushing and refilling the brake hydraulic system.

Currently, more than a dozen carmakers specify periodic brake fluid changes for some, or all, of their models built since 1985. Change intervals vary from as often as every 12 months or 15,000 miles to as infrequently as every 60,000 miles.

If this service is offered, however, there are a few general points to remember. All brake systems accumulate sludge over some period of time. Flushing the system can remove this sludge; but once it has been disturbed, make sure to get *all* of it out of the system. Stirring up sludge from the master cylinder reservoir may cause it to get into ABS valves and pumps if the sludge is not out of the system. The control valves for some rear-wheel ABS installations on some trucks may be particularly susceptible to sludge and dirt contamination.

Brake hoses for disc brakes usually enter the caliper near the top of the caliper body. The bleeder valve also is located at the top of the caliper bore. (It has to be because air rises.) If sludge accumulates in the caliper bore, it collects at the bottom. A quick, superficial bleeding of the caliper just passes a few ounces of fluid across the top of the bore. It does not flush out the sludge and all the old fluid. To flush a caliper thoroughly, pump several ounces of fluid through it. On some vehicles, it may be advisable to remove the caliper from its mounts and retract the piston to force out all the old fluid. Then reinstall it and thoroughly flush it with fresh fluid.

Flushing is done at each bleeder screw in the same manner as bleeding. Open the bleeder screw approximately one and a half turns and force fluid through the system until the fluid emerges clear and uncontaminated. Do this at each bleeder screw in the system. After all lines have been flushed, bleed the system using one of the bleeding procedures explained previously.

✓ **SERVICE TIP:** Flush a brake system by draining the old fluid and adding denatured or isopropyl alcohol to the system. Continue to add alcohol until the system is clean. Flush out the alcohol with new brake fluid until all of the alcohol is removed.

● **CUSTOMER CARE:** A little customer education can go a long way toward safer driving. Explain to your customers the importance of following their vehicle maintenance schedules for brake hydraulic system flushing and refilling. Refilling the system with fresh fluid is cheap insurance against hydraulic failure due to sludge, dirt, and moisture in the system.

Classroom Manual
pages 62–68, 84–86

Special Tools

Flushing requires the same special tools as bleeding.

The tools depend on the type of bleeding method used.

C A S E   S T U D Y

A customer complained of excessive pedal travel on a 10-year-old domestic sedan. The master cylinder reservoir was full. A test drive confirmed the condition. The car was placed on the lift. Lines and hoses were in good condition and leak free at all connections. The wheels were pulled and the caliper and wheel cylinders were inspected for leakage. All checked out okay.

The technician then suspected problems with the master cylinder or an improperly adjusted pushrod. The first problem was more likely, and a fluid bypass test confirmed the problem. The technician watched the fluid levels in the reservoirs while a helper pressed the brake pedal slowly and then released it quickly. The level in one reservoir rose, while the level in the other reservoir dropped. But the total level remained the same.

A leaking primary piston cup seal was allowing the fluid to bypass the seal and move between reservoirs. The technician replaced the master cylinder, and the problem was solved.

Terms to Know

Bench bleeding	Manual bleeding	Specific gravity
Brake bleeding	Pressure bleeding	Surge bleeding
Bleeder screw	Refractometer	Vacuum bleeding
Gravity bleeding		

ASE-Style Review Questions

1. As part of the brake system inspection and diagnosis:
   Technician A checks for correct wheel alignment, inspects the tires, and notes any unbalanced loading of the vehicle.
   Technician B performs a test drive on a smooth, level road, testing the brakes at various speeds.
   Who is correct?
   A. A only
   B. B only
   C. Both A and B
   D. Neither A nor B

2. *Technician A* says that unequal fluid levels in the master cylinder reservoir chambers may be caused by normal lining wear.
   Technician B says that a slight squirt of brake fluid from one or both master cylinder reservoir chambers when the brake pedal is applied is normal. It is caused by the fluid displacement through the reservoir replenishing ports.
   Who is correct?
   A. A only
   B. B only
   C. Both A and B
   D. Neither A nor B

3. *Technician A* says that if the brake pedal drops under steady foot pressure, the master cylinder may have an internal leak or there may be an external leak in a brake line or hose.
   Technician B says that a slight trace of brake fluid on the booster shell below the master cylinder mounting flange indicates a master cylinder leak, and the unit should be replaced.
   Who is correct?
   A. A only
   B. B only
   C. Both A and B
   D. Neither A nor B

4. *Technician A* says that checking brake pedal travel is best learned through experience, and good technicians develop a "feel" over time for a good travel range.
   Technician B says that brake pedal travel is a set specification that is found in the service manual and measured using a gauge or a tape measure.
   Who is correct?
   A. A only
   B. B only
   C. Both A and B
   D. Neither A nor B

5. Testing for open replenishing ports on the master cylinder is being discussed:
   Technician A says that ripples or a small geyser should be visible in the master cylinder reservoir when the brake pedal is pumped.
   Technician B says that these ripples may only be visible in the front reservoir of a quick take-up master cylinder.
   Who is correct?
   A. A only
   B. B only
   C. Both A and B
   D. Neither A nor B

6. *Technician A* automatically replaces all rubber O-rings and seals when rebuilding a master cylinder. *Technician B* inspects all rubber O-rings and seals when rebuilding a master cylinder to determine if replacement is needed.

Who is correct?

A. A only   **C.** Both A and B

B. B only   **D.** Neither A nor B

7. *Technician A* recommends cleaning a master cylinder body in the shop solvent tank cleaning system when rebuilding a master cylinder. *Technician B* says that the bore of aluminum master cylinder bodies should be honed when rebuilding a master cylinder to remove any small pits or burrs that have developed.

Who is correct?

A. A only   **C.** Both A and B

B. B only   **D.** Neither A nor B

8. *Technician A* says that if the master cylinder pushrod is too long, it causes the master cylinder piston to close off the vent port, resulting in brake drag. *Technician B* says that if the pushrod is too short, the brake pedal will be low and piston stroke length will be reduced with a loss of braking power.

Who is correct?

A. A only   **C.** Both A and B

B. B only   **D.** Neither A nor B

9. *Technician A* says that the master cylinder should be bench bled before being installed on the vehicle. *Technician B* says that after installing the master cylinder on the vehicle, the entire system should be bled at each wheel.

Who is correct?

A. A only   **C.** Both A and B

B. B only   **D.** Neither A nor B

10. *Technician A* says that the reservoir of a composite master cylinder can be removed and should be replaced if cracked or warped. *Technician B* says that on many ABSs, the master cylinder is part of the ABS hydraulic modulator and master cylinder assembly that is removed as a unit and then separated.

Who is correct?

A. A only   **C.** Both A and B

B. B only   **D.** Neither A nor B

ASE Challenge Questions

1. Brake fluids are being discussed:
   Technician A says that a vehicle that shows signs of rubber deterioration at the wheels' brake component but not at the master cylinder indicates the wrong fluid has been added to the systems.
   Technician B says that DOT 4 fluid in a DOT 5 system may cause the system rubber components to deteriorate.
   Who is correct?
   A. A only   **C.** Both A and B
   B. B only   **D.** Neither A nor B

2. The master cylinder is being discussed:
   Technician A says that a sinking pedal during initial brake application indicates a possible internal leak.
   Technician B says that leaking secondary piston cups will cause a sinking pedal after the vehicle is stopped and brake moderately applied.
   Who is correct?
   A. A only   **C.** Both A and B
   B. B only   **D.** Neither A nor B

3. *Technician A* says that a leak in the rear system will cause the pedal to travel much farther but be firm once the brakes are applied.
   Technician B says that a too long pushrod will cause turbulence within the master cylinder reservoir during brake application.
   Who is correct?
   A. A only   **C.** Both A and B
   B. B only   **D.** Neither A nor B

4. The results of a road test are being discussed:
   Technician A says that the vehicle "dived" to the left during braking indicating the master cylinder may be supplying pressure to the right wheel only.
   Technician B says that the same problem could be caused by improper wheel alignment.
   Who is correct?
   A. A only   **C.** Both A and B
   B. B only   **D.** Neither A nor B

5. A vehicle has a spongy feeling to the pedal during normal braking:
   Technician A says that the brake fluid probably has water in it.
   Technician B says that the brake fluid has air in it.
   Who is correct?
   A. A only   **C.** Both A and B
   B. B only   **D.** Neither A nor B

Job Sheet 8

8

Name: _____ Date: _____

Brake Fluid

Upon completion of this job sheet, you should be able to demonstrate an understanding of brake fluid.

NATEF Correlation

This job sheet is related to NATEF Brake Task: Select, handle, store, and fill brake fluids to proper level.

Tools and Materials

MSDS
Brake fluid containers

Procedure

Task Completed

1. Select containers of DOT 3, DOT 4, DOT 5, DOT 3/4, and DOT 5.1 (if possible). ☐

2. Locate MSDS for DOT 3, DOT 4, DOT 5, DOT 3/4, and DOT 5.1 (if possible). ☐

3. Determine and list the boiling point and hygroscopic properties of each type of fluid listed.

   DOT 3 _____

   DOT 4 _____

   DOT 5 _____

   DOT 3/4 _____

   DOT 5.1 _____

4. Determine and list the general and specific hazards of each fluid.

   DOT 3 _____

   DOT 4 _____

   DOT 5 _____

   DOT 3/4 _____

   DOT 5.1 _____

5. Describe the first aid measures for each fluid.

   DOT 3 _____

   DOT 4 _____

   DOT 5 _____

   DOT 3/4 _____

   DOT 5.1 _____

6. List which fluids can be mixed or used as a replacement. List all that apply.

DOT 3 can be replaced with DOT _____

DOT 4 can be replaced with DOT _____

DOT 5 can be replaced with DOT _____

DOT 3/4 can be replaced with DOT _____

DOT 5.1 can be replaced with DOT _____

7. Explain the general storage limitations and procedures for each fluid.

DOT 3 _____

DOT 4 _____

DOT 5 _____

DOT 3/4 _____

DOT 5.1 _____

Problems Encountered _____

_____

Instructor's Response _____

_____

174

Job Sheet 9

9

Name: _____ Date: _____

Replace Non-ABS Master Cylinder

Upon completion of this job sheet, you should be able to replace a master cylinder.

NATEF Correlation

This job sheet is related to NATEF Brake Task: Remove, bench bleed, and reinstall master cylinder.

Tools and Materials

Safety glasses
Hand tools
Cloths
Fender cover

Describe the Vehicle Being Worked On

Year _____ Make _____ Model _____

VIN _____ Engine type and size _____

Procedure

WARNING: Always clean around any lines or covers before removing or loosening them. Dirt and other contaminants will void the warranty and may damage the system components.

WARNING: Do not remove protective shipping seals, covers, or plugs before installing the device. Dirt and other contaminants may enter and damage the system components or void warranty.

CAUTION: Wear safety glasses or face protection when using brake fluid. Injuries to the face or eyes could occur from spilled or splashed brake fluid.

WARNING: Prevent brake fluid from coming in contact with the vehicle's finish. Brake fluid damages paint and finish immediately on contact. If fluid contacts finish, wash area thoroughly with running water and soap if possible.

1. List the service data.

   Power assisted?_____ Type _____

   Master cylinder mounting fasteners torque _____
   Does the service manual specify procedures for checking or adjusting the brake pushrod? If yes, outline the procedures and tools. _____

   _____

2. Explain why this master cylinder is being replaced. _____

   _____

   _____

☐   **3.** Place a fender cover over the vehicle. Place cloths under the master cylinder.

☐   **4.** Remove the brake lines from the master cylinder.

☐   **5.** Remove the master cylinder fasteners and the master cylinder.

☐   **6.** Remove the reservoir from the master cylinder and install it on the new one if necessary.

☐   **7.** Refer to Job Sheet 10 for bench bleeding the master cylinder.

> **CAUTION:** Ensure the fitting is not cross-threaded when reconnecting. This could damage the fitting, the component, or both.

> **CAUTION:** Do not use paints, lubricants, or corrosion inhibitors or fasteners(s) un less specified by the manufacturer. Add-on coating of any type may affect clamping or damage the fastener(s).

> **CAUTION:** Use the correct fastener torque and tightening sequence when installing components. Incorrect torque or sequencing could damage the fastener(s) or component(s).

☐   **8.** Install the master cylinder to the cowl or power booster.

☐   **9.** Connect the brake lines to the master cylinder. Do not completely tighten the lines at this time.

☐   **10.** Have a coworker slowly depress the brake pedal while step 10 through step 12 are being accomplished.

☐   **11.** Loosen the rear brake line slightly. Check the fluid for air. Tighten the brake line before the pedal achieves the halfway point.

☐   **12.** Loosen the front brake line and repeat step 10.

☐   **13.** Have a coworker release the pedal.

☐   **14.** Repeat step 9 through step 13 once or twice more to ensure the master cylinder is clear of air.

☐   **15.** Operate the brake pedal to full brake with the engine off and on (if power assisted).

☐   **16.** Does the brake pedal have the correct feel? _____
Are the brakes operative? _____

☐   **17.** If either answer to step 16 is no, determine the needed repairs or actions. Consult the instructor.

☐   **18.** When the repair is complete, clean the area, store the tools, and complete the repair order.

Problems Encountered _____

_____

Instructor's Response _____

_____

_____

Job Sheet 10

10

Name: _____ Date: _____

Bench Bleed a Master Cylinder

Upon completion of this job sheet, you should be able to bench bleed a master cylinder.

NATEF Correlation

This job sheet is related to NAETEF Brake Task: Remove, bench bleed, and reinstall master cylinder.

Tools and Materials

Safety glassess

Chemical-resistant gloves

Hand tools

Vise

Cloths

Small rod to push master cylinder primary piston

Describe the Vehicle Being Worked On

Year _____ Make _____ Model _____

VIN _____ Engine type and size _____

Procedure

Task Completed

CAUTION: Do not remove protective shipping seals, covers, or plugs before installing the device. Dirt and other contaminants may enter and damage the system components and/or void warranty.

WARNING: Wear safety glasses and/or face protection and gloves when using brake fluid. Injuries to the face and/or eyes and skin could occur from spilled or splashed brake fluid.

1. Remove master cylinder from vehicle if necessary. See Job Sheet 9. ☐
2. Place the new master cylinder in the vise with the front end slightly elevated. ☐

WARNING: Do not reuse brake fluid. Discard old brake fluid according to EPA and local regulations.

3. Fill the reservoir about one-third full with the proper brake fluid. ☐
4. Place cloths and/or a catch basin under the master cylinder. ☐
   ☐
   ☐

☐   **5.** Remove the shipping plugs and install the bleeder kit.

☐   **6.** Use the rod to slowly push the primary piston inward.

☐   **7.** Continue until fluid is forced from the front line (secondary).

      8. Observe the fluid action around the end of the bleeder hoses. Describe.

_____

_____

☐   **9.** Repeat step 6 through step 8 until there is no air coming from the bleeder hoses.

☐   **10.** Remove the bleeder hoses and lightly install the port shipping plugs.

☐   **11.** Install master cylinder on vehicle. See Job Sheet 9.

☐   **12.** When the repair is complete, clean the area, store the tools, and complete the repair order.

Problems Encountered _____

_____

Instructor's Response _____

_____

_____

Job Sheet 11

Name: _____ Date: _____

Manual Bleed a Brake System

NATEF Correlation

This job sheet addresses the following NATEF tasks: Select, handle, store, and fill brake fluids to proper level; bleed (manual, pressure, vacuum, or surge) brake system.

Objective

Upon completion of this job sheet, you will be able to bleed/flush a brake system using manual procedures.

Tools and Materials

Service manual, paper or computerized

Appropriate type of brake fluid

Appropriate line wrenches

Flexible hose about 18 inches long

Clear capture container

Lift or jack and safety stands

Assistant

Procedure

1. Determine the following information from the service manual.

Vehicle make _____ Model _____ Year _____

Type of split system _____

Bleeder screw torque _____

Brake fluid type _____

Bleeding sequence _____

ABS cautions, if equipped _____

CAUTION: If the vehicle is raised on a lift with the assistant inside, ensure the vehicle is well balanced, pads are placed correctly, and the manual locks are engaged when the vehicle is at working height. If a jack and jack stands are used, ensure both are placed correctly. Failure to follow safety procedures and proper use of lifting equipment could result in damage or injury.

Task Completed

2. Check the brake fluid level and top off as needed. ☐

3. Lift the vehicle until the wheels are free of the floor. Install jack stands if needed. ☐

4. Remove the rubber dust cap from the first wheel to be bled. For this job sheet, it is assumed the sequence is RR, LR, RF, LF. ☐

☐

5. Ensure the capture container is about $^1/_4$ full of clean brake fluid appropriate to the vehicle being serviced.

☐

6. Connect one end of the flexible hose onto the bleeder screw. Insert the other end into the fluid in the container.

☐

7. Place the line wrench on the bleeder screw and ask the assistant to pump the pedal several times and then hold the brake pedal down.

☐

8. After the assistant has the brake pedal depressed, open the bleeder screw one complete turn. Observe the container for air bubbles.

☐

9. When alerted by the assistant that the pedal is on the floor, close the bleeder screw and ask the assistant to pump up the pedal again.

☐

10. Repeat step 8.

☐

11. Are air bubbles still visible in the container as the fluid is expelled from the system? If yes, repeat steps 8 and 9 until the expelled fluid is clear of air. Once air is cleared, continue with steps 8 and 9 until clear, new fluid is expelled.

☐

12. Does expelled brake fluid appear to be new fluid? If yes, go to step 13. If not, go back to steps 8 and 9.

☐

13. Disconnect the hose from the bleeder screw and install the dust cap.

☐

14. Check the fluid level in the master cylinder reservoir. Is the reservoir dry? If so, top off and go back to step 4 and repeat the process with the wheel(s) that have been bled. If not, top off and go to step 15.

☐

15. Connect the hose and capture container to the next wheel in the sequence.

☐

16. Follow steps 4–14 to bleed this wheel.

☐

17. Continue to bleed each wheel in sequence. Remember to check and top off the fluid level after each wheel.

Problems Encountered _____

_____

Instructor's Response _____

_____

Job Sheet 12

12

Name: _____ Date: _____

Pressure Bleed a Brake System

NATEF Correlation

This job sheet addresses the following NATEF tasks: Select, handle, store, and fill brake fluids to proper level; bleed (manual, pressure, vacuum, or surge) brake system.

Objective

Upon completion of this job sheet, you will be able to bleed/flush a brake system using an air-operated pressure bleeder.

Tools and Materials

Service manual, paper or computerized

Appropriate type of brake fluid

Appropriate line wrenches

Flexible hose about 18 inches long

Clear capture container

Lift or jack and safety stands

Air-operated brake bleeder

Procedure

1. Determine the following information from the service manual.

   Vehicle make _____ Model _____ Year _____

   Type of split system _____

   Bleeder screw torque _____

   Brake fluid type _____

   Bleeding sequence _____

   Brake bleeder operating air pressure _____

   ABS cautions, if equipped _____

Task Completed

_____

   ◼ **CAUTION:** If the vehicle is raised on a lift, ensure the vehicle is well balanced, pads are placed correctly, and the manual locks are engaged when the vehicle is at working height. If a jack and jack stands are used, ensure both are placed correctly. Failure to follow safety procedures and proper use of lifting equipment could result in damage or injury.

   ◼ **CAUTION:** Before adding fluid to the brake bleeder, ensure the air pressure in the lower chamber has been exhausted down to zero. Eye and skin injuries could result if the upper chamber is opened with air pressure present in the lower chamber.

2. Ensure the lower chamber is at zero air pressure. Remove the top and fill the upper chamber with the appropriate brake fluid.

☐

☐   **3.** Pressurize the bleeder's lower chamber with the proper air pressure.

☐   **4.** Select the master cylinder/bleeder adapter kit and install the kit onto the master cylinder.

☐   **5.** Connect the bleeder's outlet hose to the adapter.

☐   **6.** Turn on the bleeder hydraulic outlet valve. Check the adapter kit for leaks at the master cylinder.

☐   **7.** Lift the vehicle until the wheels are free of the floor. Install jack stands if needed.

☐   **8.** Remove the rubber dust cap from the first wheel to be bled. For this job sheet, it is assumed the sequence is RR, LR, RF, LF.

☐   **9.** Ensure the capture container is about $\frac{1}{4}$ full of clean brake fluid appropriate to the vehicle being serviced.

☐   **10.** Connect one end of the flexible hose onto the bleeder screw. Insert the other end into the fluid in the container.

☐   **11.** Place the line wrench on the bleeder screw.

☐   **12.** Open the bleeder screw one complete turn. Observe the container for air bubbles.

☐   **13.** Continue to observe the container for air bubbles and new fluid. When the air is cleared and new fluid is being expelled, close the bleeder screw.

☐   **14.** Disconnect the hose from the bleeder screw and install the dust cap.

☐   **15.** Connect the hose and capture container to the next wheel in the sequence.

☐   **16.** Follow steps 10–15 to bleed this wheel.

☐   **17.** Continue to bleed each wheel in sequence.

☐   **18.** After the last wheel is bled, lower the vehicle to the floor.

☐   **19.** Turn off the bleeder valve.

☐   **20.** Exhaust the air pressure in the lower chamber until the gauge reads zero pressure.

☐   **21.** Place some rags under the master cylinder. Disconnect the bleeder's hose and remove the adapter kit from the master cylinder.

☐   **22.** Remove the rags and, if necessary, clean any spilled brake fluid from the vehicle.

☐   **23.** Clean and store the brake bleeder.

Problems Encountered _____

_____

Instructor's Response _____

_____

Job Sheet 13

13

Name: _____ Date: _____

Suction Bleed a Brake System

NATEF Correlation

This job sheet addresses the following NATEF tasks: Select, handle, store, and fill brake fluids to proper level; bleed (manual, pressure, vacuum, or surge) brake system.

Objective

Upon completion of this job sheet, you will be able to bleed/flush a brake system using a suction/vacuum bleeder.

Tools and Materials

Service manual, paper or computerized

Appropriate type of brake fluid

Appropriate line wrenches

Flexible hose about 18 inches long

Clear capture container

Lift or jack and safety stands

Brake suction equipment

Note: For this job sheet, a typical hand-operated vacuum pump with capture container is used. Other suction equipment will be used in a similar manner.

Procedure

1. Determine the following information from the service manual.

Vehicle make _____ Model _____ Year _____

Type of split system _____

Bleeder screw torque _____

Brake fluid type _____

Bleeding sequence _____

Brake bleeder operating air pressure _____

ABS cautions, if equipped _____

CAUTION: If the vehicle is raised on a lift, ensure the vehicle is well balanced, pads are placed correctly, and the manual locks are engaged when the vehicle is at working height. If a jack and jack stands are used, ensure both are placed correctly. Failure to follow safety procedures and proper use of lifting equipment could result in damage or injury.

CAUTION: Before adding fluid to the brake bleeder, ensure the air pressure in the lower chamber has been exhausted down to zero. Eye and skin injuries could result if the upper chamber is opened with air pressure present in the lower chamber.

☐ 2. Set up the suction equipment. Check its operation by placing a finger over the hose that goes to the bleeder screw and pump the handle. A vacuum should be felt.

☐ 3. Lift the vehicle until the wheels are free of the floor. Install jack stands if needed.

☐ 4. Remove the rubber dust cap from the first wheel to be bled. For this job sheet, it is assumed the sequence is RR, LR, RF, LF.

☐ 5. Connect the input (vacuum) end of the vacuum pump onto the bleeder screw.

☐ 6. Place the line wrench on the bleeder screw.

☐ 7. Open the bleeder screw one complete turn.

☐ 8. Operate the vacuum pump until fluid flows into the capture container. Observe the container for air bubbles and new fluid.

☐ 9. Continue to operate the pump until the expelled fluid is clear of air and new fluid is being expelled.

☐ 10. Close the bleeder screw and disconnect the hose. Install the dust cap.

☐ 11. Lower the vehicle if necessary and check the fluid level in the master cylinder reservoir. If the reservoir is dry, then top off and return to the wheel(s) already bled and repeat steps 4–8 on each wheel. If the reservoir is not dry, top off as needed.

☐ 12. Connect the hose and capture container to the next wheel in the sequence.

☐ 13. Follow steps 4–9 to bleed this wheel and check the brake fluid.

☐ 14. Continue to bleed each wheel in sequence.

☐ 15. After the last wheel is bled, lower the vehicle to the floor.

Problems Encountered _____

_____

Instructor's Response _____

_____

Hydraulic Line, Valve, and Switch Service

Upon completion and review of this chapter, you should be able to:

❏ Bleed and flush the hydraulic system.

❏ Pressure test the brake hydraulic system.

❏ Reset a pressure differential valve (warning lamp switch).

❏ Inspect brake lines and fittings for leaks, dents, kinks, rust, cracks, or wear. Tighten loose fittings and supports.

❏ Inspect brake hose for leaks, kinks, cracks, bulging, or wear. Tighten or replace hoses as necessary.

❏ Remove and replace double flare and ISO brake lines, hoses, fittings, and supports.

❏ Fabricate replacement brake tubing, including the forming of double inverted flare or ISO flare ends and correct bends to fit the vehicle chassis.

❏ Diagnose poor stopping and brake pull or grab conditions caused by problems in the brake lines or brake hoses and perform needed repairs.

❏ Diagnose poor stopping and brake pull or grab conditions caused by problems in the hydraulic system valves and perform needed repairs.

❏ Inspect, test, and replace metering valves, proportioning valves, pressure differential valves, and combination valves.

❏ Inspect, test, adjust, and replace a height-sensing proportioning valve.

❏ Diagnose electrical problems in circuits for brake system switches and sensors.

❏ Test, adjust, repair, or replace the brake stoplamp switch and wiring.

❏ Inspect, test, and replace the brake warning lamp, switch, and wiring.

❏ Test and repair the parking brake indicator lamps, switches, and wiring.

❏ Inspect, test, and replace the master cylinder fluid level sensor or switch.

❏ Repair electrical wiring and connectors.

Introduction

All hydraulic systems need lines and hoses to contain the fluid. All hydraulic systems require valves to control and direct the fluid if work is to be done in an efficient manner. Take the dam across a river, for instance. The dam is, in fact, a valve that controls water flow, reducing flooding below the dam and directing flow through causeways to drive electric turbines. Releasing the water completely at one time would only destroy the electric plant and the surrounding areas. The hydraulic brake system shares some of the same concerns with regard to fluid containment and direction. Without the proper lines and hoses, the brake fluid would just "flood" and there would be no braking effect. If control valves were present, there would be uneven or no braking effect at all, which brings up the problems associated with the old mechanical brake system. Over the last 90 years, hydraulic brake systems have been improved to be almost fail-safe. But those lines, hoses, and valves require some services during the life of the vehicle, even if the only reason is old age. This chapter outlines some of those common services. Some valves have been computerized, and, where appropriate, those valves are discussed in Chapter 10, Electronic Braking Systems.

Recentering a Pressure Differential Valve (Failure Warning Lamp Switch)

Almost every techni-
cian refers to the
pressure differential
valve as a combina-
tion valve.

After bleeding or flushing and refilling some brake systems, the pressure differential valve (or warning lamp switch) may be actuated and the warning lamp may be lit. Opening a bleeder screw creates a pressure differential between the two halves of the hydraulic system. This pressure differential has the same effect as a leak, and the valve piston moves toward the low-pressure side to close the lamp circuit.

If the warning lamp stays lit after bleeding, the valve piston may need to be recentered. First, however, verify that the parking brake is not applied and that its linkage is properly adjusted. Also check the fluid level in the master cylinder reservoir. Both the parking brake warning circuit and the low-fluid warning circuit usually use the same warning lamp as the pressure differential valve. It is not uncommon to find the parking brake out of adjustment after relining the rear brakes or to find the fluid level low after bleeding the system.

Three different types of pressure differential valves have been used in domestic and imported vehicle brake systems. On older vehicles, the valve may be a separate part in the system. On newer vehicles, the valve is often part of the combination valve (Figure 5-1), or it may be built into the master cylinder (Figure 5-2). Whether a separate component or combined with other valve functions, a pressure differential valve of a given design operates in the same way. The following paragraphs explain how to recenter that valve piston when required.

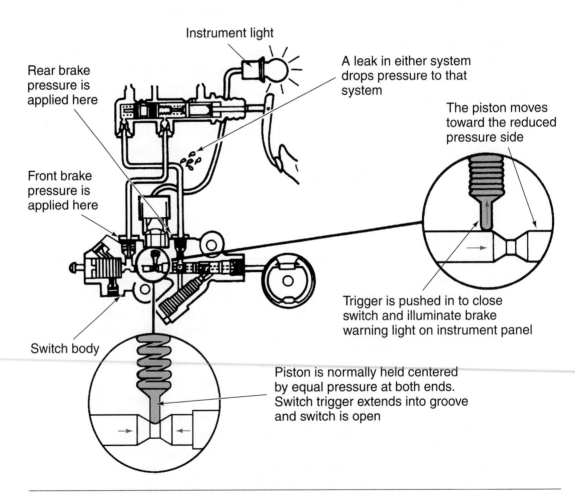

Instrument light

Rear brake pressure is applied here

A leak in either system drops pressure to that system

The piston moves toward the reduced pressure side

Front brake pressure is applied here

Trigger is pushed in to close switch and illuminate brake warning light on instrument panel

Switch body

Piston is normally held centered by equal pressure at both ends. Switch trigger extends into groove and switch is open

Figure 5-1 Many late-model pressure differential valves (warning lamp switches) are part of a combination valve.

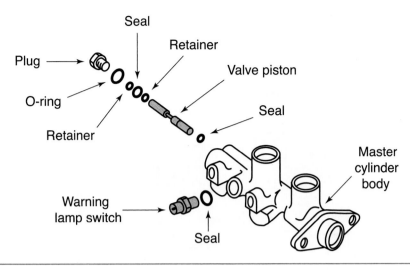

Figure 5-2 The pressure differential valve and the warning lamp switch are built into this master cylinder body.

Single-Piston Valve with Centering Springs

The most common pressure differential valve has a single piston and centering springs (Figure 5-3). The warning lamp lights only when the brakes are applied and a pressure difference exists between the two halves of the hydraulic system. When the brakes are released, pressure is low in both halves of the system. The springs then recenter the valve piston, and the lamp should turn off.

The piston in this type of valve (or switch) should recenter automatically with no special action required. Occasionally, however, the piston may stick at one side of its bore or the other and leave the lamp lit. If the lamp is lit with the ignition on, apply the brakes rapidly with moderate to heavy force two or three times. Hydraulic pressure usually frees a stuck piston, and the springs will recenter it.

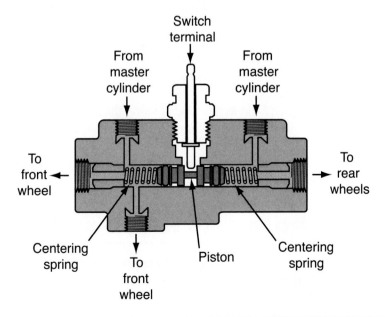

Figure 5-3 Pressure differential valve (warning lamp switch) with single piston and centering springs.

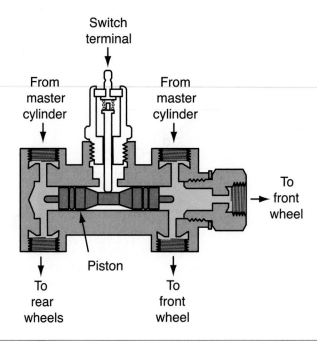

Figure 5-4 Pressure differential valve (warning lamp switch) with single piston and no centering springs.

If the lamp stays lit and the parking brake or fluid level switch is not closed (lamp lit), try to recenter the piston, following the instructions in the next paragraph. If the lamp is still lit, test the circuit for an accidental ground. If the circuit is not otherwise grounded, replace the pressure differential valve.

Single-Piston Valve Without Centering Springs

Some other vehicles, particularly older Ford products and some imports, have a single-piston pressure differential valve without centering springs (Figure 5-4). This type of valve often leaves the warning lamp lit after system bleeding. Recentering the piston is a two-person job.

After ensuring that the parking brake is off and the fluid level is correct, turn on the ignition and verify that the lamp is lit. Open a bleeder screw on the side opposite from the side last bled. Have your assistant slowly press the brake pedal by hand until the warning lamp turns off. Tighten the bleeder screw.

When trying to recenter the piston in this type of valve, the piston often goes over center in the opposite direction. This causes the lamp to turn off momentarily and then relight. If this happens, open a bleeder screw on the opposite side of the hydraulic system and repeat the procedure. Two or three tries may be needed to get the piston properly centered. A solution to this problem is the use of a two-piston valve (Figure 5-5).

Brake Line, Fitting, and Hose Service

Classroom Manual
pages 95–106

Hydraulic system lines are made of steel tubing and rubber hoses. Rigid hydraulic lines are made of double-wall, welded steel tubing that is coated to resist corrosion. Brake hoses must be free to flex and move as the wheel moves up and down or turns. Brake hoses also must be able to withstand the high pressures within the system. Exposure to the elements, road salts in winter, salt air, water, and contaminants in the system, all contribute to rusting and corrosion of brake fittings, lines, and hardware.

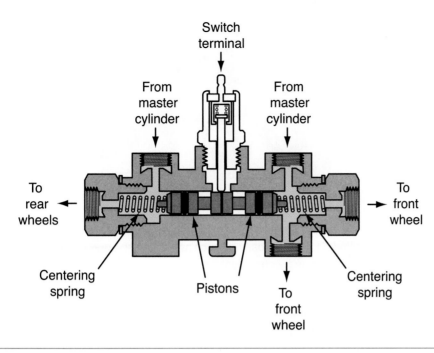

Figure 5-5 Pressure differential valve (warning lamp switch) with two pistons.

> ✓ **SERVICE TIP:** As mentioned in Chapter 5 of the *Classroom Manual*, there is a copper-nickel alloy brake tubing that meets SAE Standard J1047 and ISO 4038. It can be serviced in the same manner as the more common steel tubing.

Brake Line Inspection

All carmakers include brake line inspection on their vehicle maintenance schedules. Most manufacturers recommend inspecting brake hoses twice a year, but it is good practice to check them whenever the vehicle is getting lubrication service. Steel brake lines and fittings should be checked for damage and leakage once a year or whenever the vehicle gets brake service.

Brake line inspection is more than a quick glance to see if all the parts are in place. Physical damage may be apparent from the outside, but wear and deterioration also can occur inside tubing and hoses. To inspect brake lines thoroughly, be sure to cover the points described in the following paragraphs.

> ✓ **SERVICE TIP:** Very small leaks at brake line fittings that appear only under pressure can be hard to find. Often, pressure must be applied for a long time before seepage appears. Lift the vehicle and clean each valve and line connection. This will make it easier to spot a leak. To help pinpoint such a problem, apply the brakes with a brake pedal depressor used for wheel alignment. A brake pedal depressor is a long bar with a twist stop that rests against the driver's seat. Place the end of the bar on the brake pedal and apply as much hand and arm force as possible. With the force applied, slide the twist lock against the forward edge of the seat and slowly release the force. The lock will hold the pedal applied. A brake pedal depressor is usually found in the area of the wheel alignment bay. Leave the depressor applied for several hours and then check for small leaks. When using this test method, however, remove the stoplamp fuse to keep from discharging the battery.

Tubing Inspection. Steel tubing is more durable than rubber hoses, but it can suffer rust, corrosion, impact damage, and cracking. Water trapped around brake tubes, fittings, and mounting

Special Tools

Cloths
Brake pedal
   depressor

Special Tool

Cloths

Figure 5-6 Inspect brake tubing for damage such as kinks, rust, abrasion, and looseness. All bends should be smooth and without kinks.

clips can rust and corrode steel tubing. Corrosion can be particularly severe in areas that use a lot of salt to melt ice on the roads during the winter. Mounting clips are necessary to hold brake lines to the body or frame, but they can trap salt water and hide severe corrosion. Therefore, inspect all mounting points closely.

Missing mounting clips can cause other problems. If brake tubing is not mounted securely, vibration can cause the tubing to fracture and leak. Brake tubing that hangs below the body or frame can be snagged and torn loose. Inspect all brake tubing for damage and looseness (Figure 5-6). Also look for empty screw holes or scuff marks on body and frame parts that indicate missing clips.

Brake tubing can be damaged by objects thrown up by the tires, particularly if the vehicle is used off-road. Road impact damage is far less common, however, than impact damage caused by improperly installed towing chains or improper placement of a floor jack or lift. Inspect brake tubing for dents, kinks, and cracks caused by careless service practices.

Hose Inspection. Damage to the outside of a brake hose is easier to spot than internal damage or deterioration. Inspect brake hoses for abrasion caused by rubbing against chassis parts and for cracks at stress points, particularly near fittings. Look for fluid seepage indicated by softness in the hose accompanied by a dark stain on the outer surface. Look at each hose closely for general damage and deterioration such as cracks, soft or spongy feel or appearance, stains, blisters, and abrasions. Extreme rust between the clamp and hose or a crimped clamp could pinch the hose.

To check for internal hose damage, have an assistant pump the brakes while you feel the hose for swelling or bulging as pressure is applied internally (Figure 5-7). Have your assistant apply and release the brakes; then quickly spin the wheel. If the brake at any wheel seems to drag after pressure is released, the brake hose may be restricted internally. No conclusive way exists to inspect or test for internal restriction, but replacing a hose if a brake has these symptoms can be good insurance.

Finally, inspect the mounting clips, brackets, and fittings where hoses are connected to rigid brake lines. Replace any missing clips and be sure that the hoses are not kinked or twisted. At the front brakes, rotate the steering from lock to lock and verify that the brake hoses do not rub on chassis parts or twist and kink when the wheels turn.

SERVICE TIP: Make sure the hose is not twisted between the frame or body anchor and the point of attachment to the disc brake caliper. Sometimes during a rush job or do-it-yourself job, the hose gets twisted and is not noticed. This damages the hose, particularly during turns when it gets stretched because the twist makes the hose a little too short. Over a fairly short period of time, this breaks the inner lining and causes a "check-valved" condition.

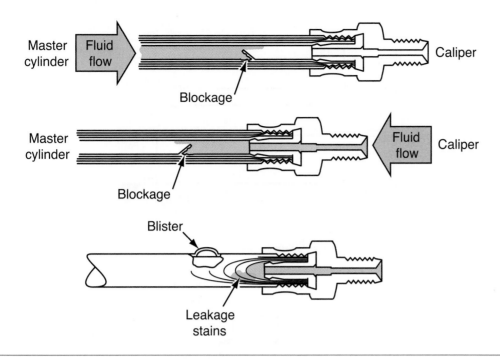

Figure 5-7 Possible internal defects in a brake hose.

✓ **SERVICE TIP:** A piece of rubber can break partly loose inside a brake hose and act as a check valve to trap pressure at the brake. If you suspect that a dragging brake is caused by a "check-valved" hose, open the bleeder screw and turn the wheel by hand. If the brake changes from locked up to free, the hose is probably the culprit.

To confirm the diagnosis, install a pressure gauge in place of the bleeder screw, pump up the brakes and release the pedal. If the brake locks again and a pressure reading stays on the gauge after the pedal is released, loosen the fitting that connects the brake pipe to the hose. If the brake does not loosen and the gauge pressure does not drop to zero, the hose is retaining the pressure. Replace it.

Brake Hose Removal and Replacement

■ **CAUTION:** Always clean around any lines or covers before removing or loosening them. Dirt and other contaminants will void the warranty and may damage system components.

✎ **AUTHOR'S NOTE:** A replacement brake hose must be the same length as the original one. A hose that is too long may rub on the chassis. A hose that is too short may break when the movable component reaches the limits of its travel. Some hoses may appear to be too short if the vehicle is lifted and the suspension extends to its maximum. Check the hose length closely in this instance.

Hose Removal. Some brake hoses have a swivel fitting at one end and a fixed fitting that cannot be rotated at the other (Figure 5-8). Disconnect the swivel fitting first on this type of hose. Other hoses have a fixed male fitting on one end and a fixed female fitting on the other. The female end of such a hose is connected to a flare nut on a rigid brake tube; disconnect this end first. If the hose has a banjo fitting on one end—usually for connection to a caliper—disconnect the banjo fitting first (Figure 5-9), then the other end of the hose. Occasionally, some caliper

Special Tools

Hand tools
Cloths

When removing a banjo fitting, check it for sealing washers. If equipped with washers, it is recommended to replace them.

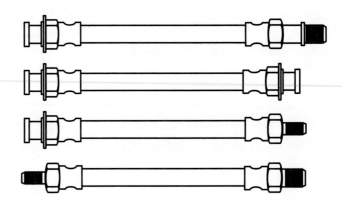

Figure 5-8 Typical brake hose end fittings.

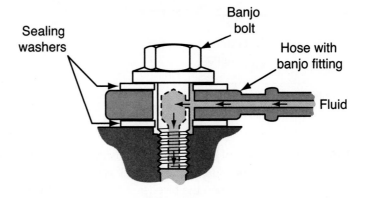

Figure 5-9 Note the sealing washers on this banjo fitting.

fittings and hoses have left-hand threads. Sometimes the left-hand fasteners are noted with a slash through the flat surfaces of the nut or bolt.

Follow these guidelines to remove a brake hose:

1. Clean dirt away from the fittings at each end of the hose to keep it from entering the system.

2. Use a flare-nut wrench to disconnect the flare nut from the female end of the hose (Figure 5-10) or loosen and disconnect the swivel end of the hose. When loosening

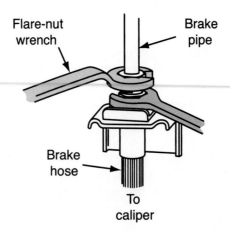

Figure 5-10 Use a flare-nut (line) wrench to disconnect the brake line (tubing) from the hose.

one fitting at the end of a hose, hold the mating half of the fitting with another flare-nut wrench, which will make fitting removal easier and prevent damage to mounting brackets and clips.

3. Remove the hose retaining clip from the mounting bracket with a pair of pliers.

4. Separate the hose from the mounting bracket and any other clips used to hold it in place.

5. Use a flare-nut wrench to disconnect the other end of the hose from the caliper or wheel cylinder.

6. If a replacement hose is not going to be installed immediately, cap or plug open fittings on the vehicle to keep dirt out of the system.

Hose Installation. When installing a brake hose, be sure the new hose is the correct length and determine whether or not the right-hand and left-hand hoses are the same or different. Route the new hose in the same location as the original and provide ¾ inch to 1 inch of clearance between the hose and suspension and wheel parts in all positions. If the original hose had special mounting clips or brackets, the replacement should have the same.

WARNING: Ensure that the fitting is not cross-threaded when reconnecting. This could damage the fitting or the component, or both.

CAUTION: Always replace brake hoses in axle sets. This will eliminate brake malfunction(s) caused by bad hoses and help in diagnosing ongoing brake problems.

SERVICE TIP: Some brake hoses are not DOT approved. Check with your supervisor or parts vendor as to the hose warranty or desirability of using non-approved hose.

CAUTION: Never use low-pressure hydraulic hoses or oil hoses as replacements for brake hoses. These components cannot withstand the high pressure of the brake system. Fluid leakage, line rupture, and system failure can result.

Follow these guidelines to install a brake hose:

1. If the hose has a fixed male end, install it into the wheel cylinder or caliper first. If the connection requires a copper gasket, install a new one.

2. If one end of the hose has a banjo fitting for attachment to a caliper, install the banjo bolt and a new copper gasket on each side of the fitting shown earlier in Figure 5-9. Leave the banjo bolt loose at this time; tighten it after connecting and securing the other end of the hose.

3. Route the hose through any support devices and install any required locating clips.

4. Insert the free end of the hose through the mounting bracket.

5. Depending on hose design, connect the flare nut on the steel brake line to the female end of the hose or connect the swivel end of the hose to the mating fitting.

6. Use a flare-nut wrench to tighten the fitting and hold the hose with another flare-nut wrench to keep it from twisting (see Figure 5-10). Check the colored stripe or the raised rib on the outside of the hose to verify that the hose has not twisted during installation.

7. Install the retaining clip to hold the hose to its mounting bracket (Figure 5-11). Install any other clips as required.

8. If the banjo bolt was left loose in step 2, position the banjo fitting to provide the best hose position and tighten the bolt.

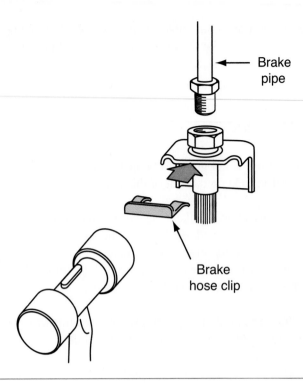

Figure 5-11 The locking clip is tapped into place with a small hammer.

After installing the new brake hose, check the hose and line connections for leaks and tighten if needed. Check for clearance during suspension rebound and while turning the wheels. If any contact occurs, reposition the hose, adjusting only the female end or the swivel end.

Brake Tubing Removal and Replacement

■ **CAUTION:** Always clean around any lines or covers before removing or loosening them. Dirt and other contaminants will void the warranty and may damage system components.

 SERVICE TIP: Install the fittings onto the tubing before flaring the ends. Many feet of tubing have been wasted over the years because the fitting will not fit over a finished flare.

Special Tools

Hand tools

Cloths

Removing and replacing a length of brake tubing looks like a straightforward job, but the following guidelines will make the task easier. Start by cleaning dirt away from the fittings at each end of the tubing. Do not remove the tubing mounting clamps yet. Leaving them in place will keep the tubing from moving around and make it easier to disconnect the fittings.

Use a flare-nut wrench to disconnect the fittings at each end of the tubing. If the tubing is attached to a hose, use another flare-nut wrench to hold the hose fitting. If the tubing is attached to a rigidly mounted junction block or cylinder, a second wrench is not needed. If replacement tubing is not going to be installed immediately, cap or plug open fittings on the vehicle to keep dirt out of the system.

Remove the mounting clips from the chassis and remove the brake tubing. Inspect the clips and their screws to determine if they are reusable. If they are not, install new ones. If the brake tubing has any protective shields installed around it, save them also for installation with the new tubing. If you must fabricate a new section of tubing, save the old section for a bending guide.

To install a length of brake tubing, position it on the chassis and install the mounting clips loosely. Leaving the new tubing slightly loose will help to align the tube fittings. Next, use the appropriate flare-nut wrenches to connect the fittings at both ends of the tubing. Tighten the fittings securely and then tighten the mounting clips.

Fabricating Brake Tubing

Brake pipes or tubes are normally $\frac{3}{16}$-inch double-wall steel tubing. Stock brake tubing is available in various lengths with ends preflared and flare nuts installed (Figure 5-12). Always use prefabricated tubing whenever possible. If original equipment tubing that is formed to the required bends for a specific vehicle is available, it may be your best choice from a time and overall cost basis.

Special Tool

Tubing cutter

> ■ **CAUTION:** Double-wall steel brake tubing is the only type of tubing approved for brake lines. Never use copper tubing or any other tubing material as a replacement; it cannot withstand the high pressure or the vibrations to which a brake line is exposed. Fluid leakage and system failure can result.

Lacking a preformed original equipment manufacturer (OEM) tube, a straight section of brake tubing of the proper length is your next best choice. It may be necessary in many cases, however, to fabricate a replacement brake line from bulk tubing. The following sections explain how to cut and bend tubing and how to form the required flared ends. Photo Sequence 8 shows the basic steps of cutting and bending tubing and forming an SAE double flare on the tubing ends.

Cutting Tubing. Bulk tubing comes in large rolls. To form a replacement length of tubing, you must cut it from a roll as follows:

1. Determine the exact length of replacement tubing needed; add $\frac{1}{8}$ inch for each flare that is to be made. Measure the old tubing as accurately as possible. Use a string, if necessary, as shown in Photo Sequence 8.

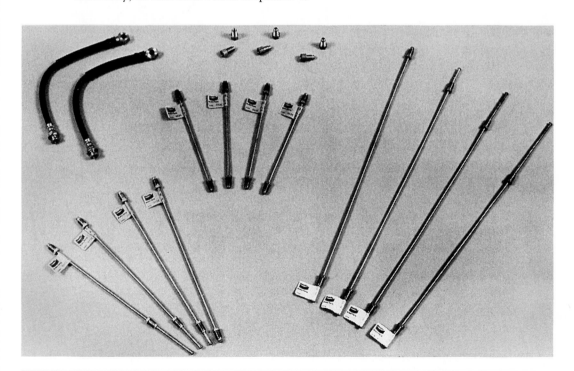

Figure 5-12 Prefabricated brake lines come in various lengths with the fittings installed and the ends already flared.

Photo Sequence 8
Typical Procedure for Fabricating and Replacing a Brake Line

P8-1 Be sure to use the recommended bulk ³⁄₁₆-inch double-wall steel brake tubing and the correct size and type of tube nuts.

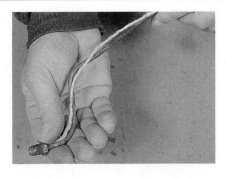

P8-2 To determine the correct length, measure the removed tube with a string and add about ⅛ inch for each flare.

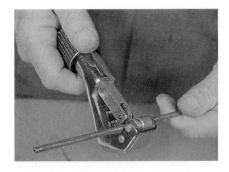

P8-3 Cut the tubing to the required length with a tubing cutter.

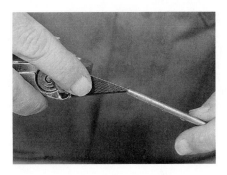

P8-4 Clean any burrs after cutting.

P8-5 Place the tube nut in the correct direction.

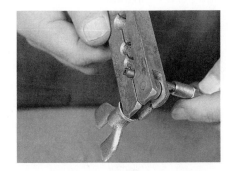

P8-6 Place the tubing in the flaring bar with the end protruding slightly above the face of the bar. Use the adapter shown in P5-8 to set the correct length of the protruding portion.

P8-7 Firmly clamp the tube in the bar so the force exerted during flaring does not push the tubing down through the bar.

P8-8 Place the adapter (anvil) in place over the tube opening to form the first stage of the flare.

P8-9 Tighten down the flaring clamp.

Photo Sequence 8
Typical Procedure for Fabricating and Replacing a Brake Line (continued)

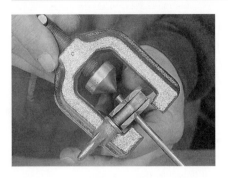

P8-10 Loosen the flaring clamp, remove the adapter (anvil), and check to see that the end of the tubing is properly belled.

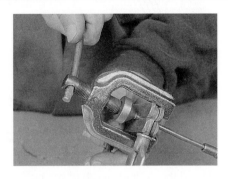

P8-11 Install the cone onto the tube opening and retighten the flaring clamp.

P8-12 The cone completes the double flare by folding the tubing back on itself. This doubles its thickness and creates two sealing surfaces.

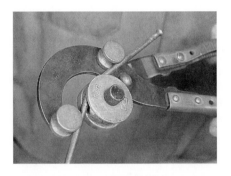

P8-13 Bend the replacement tube to match the original tube using a tubing bender.

P8-14 Clean the brake tubing by flushing it with clean brake fluid.

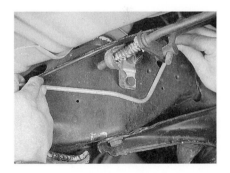

P8-15 Install the replacement brake tube, maintaining adequate clearance to metal edges and moving or vibrating parts.

P8-16 Install the brake tube and tighten the tube nuts to shop manual specifications with an inch-pound torque wrench.

P8-17 Bleed the lines.

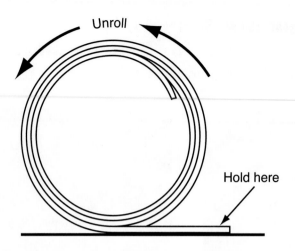

Unroll

Hold here

Figure 5-13 Unroll the desired length of tubing from the bulk roll in this manner.

2. Hold the free outer end of the tubing against a flat surface with one hand and unroll the roll in a straight line with the other hand (Figure 5-13). Do not lay the roll flat and pull one end toward you; this will twist and kink the tubing.

CAUTION: Do not use a hacksaw to cut tubing. The uneven pressure of the blade will distort the tubing end, and the teeth will leave a jagged edge that cannot be flared properly.

3. Mark the tubing at the point to be cut and place a tubing cutter on the tubing. Tighten the cutter until the cutting wheel contacts the tubing at the marked point.

4. Turn the cutter around the tubing toward the open side of the cutter jaws. After each revolution, tighten the cutter slightly until the cut is made.

5. Ream the cut end of the tubing with a reaming tool (usually attached to the cutter) to remove burrs and sharp edges. Hold the end downward so that metal chips fall out. Ream only enough to remove burrs; then blow compressed air through the tubing to be sure all chips are removed.

Special Tool

Tubing bender

Bending Tubing. Whether a replacement brake tube is a straight length of preflared tubing or made from bulk material, it is usually necessary to bend the new line to match the old one. Steel tubing can be bent by hand to form gentle curves.

CAUTION: Do not try to bend tubing into a tight curve by hand; you will usually kink the tubing. Because a kink in a brake tube weakens the line, never use a kinked tube. To avoid kinking, use a bending tool. Several types are available.

For large-radius bends on small diameter tubing, you can use a bending spring. Slip the coil spring over the tubing and bend it slowly by hand (Figure 5-14). Bend the tubing slightly further than required and back off to the desired angle. This releases spring tension in the bender so it can be easily removed. A bending spring can be used more easily before the tubing ends are flared. To use a bending spring on flared tubing, you must use an oversize spring that will slip over the flares. Bending may be more difficult because of the looseness of the spring.

On larger diameter tubing, or where tighter bends are needed, use a lever-type or gear-type bender (Figure 5-15). Slip the bender over the tubing at the exact point the bend is required.

Spring-type bender

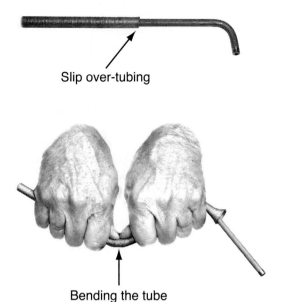

Slip over-tubing

Bending the tube

Figure 5-14 Large-radius bends on small diameter brake tubing can be made with a bending spring.

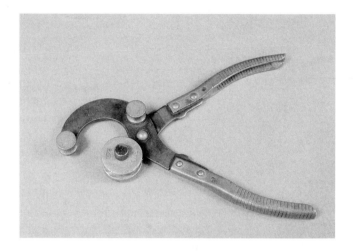

Figure 5-15 One type of tube bending tool that will form tight radius bends in steel tubing.

If you are bending the tubing near an end that is to be flared, leave about 1½ inches of straight tubing at the end. After the tubing is bent to the proper shape, assemble the flare nuts on the tubing before flaring the tube ends. Once the ends are flared, the flare nuts will not fit over the end of the tubing.

Observe these additional guidelines when bending and fabricating tubing:

1. Avoid straight lengths, or runs, of tubing from fitting to fitting. They are hard to install and subject to vibration damage.
2. Ensure that the required clips and brackets will fit a replacement length of tubing, especially on long sections.
3. Bend tubing to provide necessary clearance around exhaust components and suspension parts.
4. Be sure that tubing ends align with the fittings on mating components before mounting the tubing securely.
5. When installing tubing, connect the longest straight section first.

Flare Fittings

After tubing is cut and bent to fit, flares must be formed on unfinished ends. Both the SAE and the International Standards Organization (ISO) were formed to research and establish automotive standards and improvements, and both organizations have established brake tubing flare standards: SAE 45-degree double flares and ISO flares or bubble flares (Figure 5-16). These flares and their fittings are not interchangeable. Be sure to form the proper type of flare required by the

Special Tool

Tubing flare kit

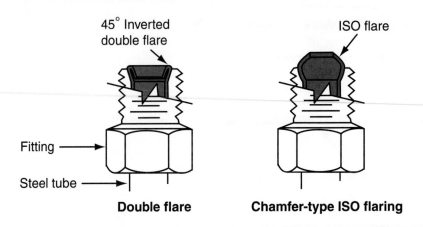

45° Inverted double flare

ISO flare

Fitting

Steel tube

Double flare

Chamfer-type ISO flaring

Figure 5-16 The two common flare types used on brake tubing. The 45-degree inverted double flare uses fractional inch-size fittings. The ISO flare uses metric fittings. See Figure 5-18 and Figure 5-22 for more detailed drawings of the angles and fitting surfaces.

vehicle system. Always place the flare nut on the tubing with the threads facing the end of the tube *before* forming the flare.

Flare nuts do not usually corrode or rust, but the tubing that passes through them may. If the line corrodes and freezes to the nut, the line will twist if you try to loosen the nut with a flare-nut wrench. To free a flare nut frozen to the line, apply penetrating oil to the connection. You also can heat the connection with a torch if all plastic and rubber parts are removed from the immediate area. Using a pencil-thin flame, apply heat to all sides of the flare nut, never to the line. When the steel nut begins to glow from the heat, try to loosen the nut with the flare wrench. If the nut cannot be freed, cut the line. The component to which the line is connected may be reusable. If heat is used to free a frozen flare nut, the entire length of tubing should be replaced.

Always use the correct size of flare-nut wrench when tightening or loosening hydraulic fittings. When loosening or tightening a nut on a fitting or a union, use two wrenches: one to turn the nut and one to hold the fitting or union (Figure 5-17). When connecting a tube to a hose, a tube connector, or a brake cylinder, use an inch-pound torque wrench to tighten the tube fitting nut to specifications. On fittings requiring gaskets, always install new copper gaskets. Used gaskets have taken a set and will not seal properly if reinstalled.

Special tools are required to flare tubing, and the tools for SAE and ISO flares are different. The following sections provide more instructions for forming SAE and ISO flares. Observe these precautions about flaring and installing brake line fittings.

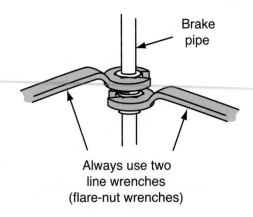

Brake pipe

Always use two line wrenches (flare-nut wrenches)

Figure 5-17 Use flare-nut (line) wrenches to loosen line fittings.

CAUTION: Never use copper fuel line fittings as a replacement for steel brake line fittings. Copper cannot withstand the high pressure or the vibrations to which a brake line is exposed. Fluid leakage and system failure can result.

CAUTION: Never use spherical-sleeve compression fittings in brake lines. Spherical compression fittings are low-pressure fittings for applications such as fuel lines. They will fail and leak under the high pressures and vibrations of a brake hydraulic system.

CAUTION: Do not interchange metric and inch-size fittings. They have different threads and cannot be mixed. Do not interchange ISO flare nuts with SAE inch-size flare nuts. Either of these conditions can cause fluid leakage and system failure.

Forming an SAE 45-Degree Double Flare. A double flare is made in two stages using a special flaring tool. A typical flaring tool consists of a flaring bar and a flaring clamp. The double flaring process is shown in Figure 5-18 and demonstrated in Photo Sequence 8.

Special Tool

Tubing flare kit

The angle of the flare and the nut is 45 degrees, while the angle of the seat is 42 degrees. When the nut is tightened into the fitting, the difference in angles—called an interference angle—causes both the seat and the flared end of the tubing to wedge together. When correctly assembled, brake lines connected with flare fittings provide joints that can withstand high hydraulic pressure.

Follow these guidelines to form an SAE 45-degree double flare:

1. Select the forming die from the flaring kit that matches the inside diameter of the tubing.
2. Be sure the flare nut is installed on the tubing; then clamp the tubing in the correct opening in the flaring bar with the end of the tube extending from the tapered side of the bar the same distance as the thickness of the ring on the forming die.
3. Place the pin of the forming die into the tube and place the flaring clamp over the die and around the flaring bar.

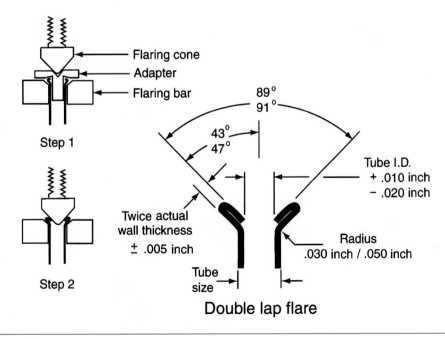

Figure 5-18 The two steps for forming a double flare, the flaring tools, and the dimensions of a double 45-degree flare are shown.

4. Tighten the flaring clamp until the cone-shaped anvil contacts the die. Continue to tighten the clamp until the forming die contacts the flaring bar.

5. Loosen the clamp and remove the forming die. The end of the tubing should be mushroomed as shown in Figure 5-18.

6. Place the cone-shaped anvil of the clamp into the mushroomed end of the tubing. Be careful to center the tip of the cone and verify that it is touching the inside diameter of the tubing evenly. If the cone is not centered properly before tightening the clamp, the flare will be distorted.

7. Tighten the clamp steadily until the lip formed in the first step completely contacts the inner surface of the tubing.

8. Loosen and remove the clamp and remove the tubing from the flaring bar. Inspect the flare to be sure it has the correct shape as shown in Figure 5-18. If it is formed unevenly or cracked, you must cut off the end of the tubing and start over again.

Forming an ISO Flare. An ISO flare or bubble flare has several advantages. When tightened, the shoulder of the nut bottoms in the body of the part to create uniform pressure on the tube flare. In addition, the design is not subject to overtightening. Simply tighten the nut firmly on the seat to produce the correct sealing pressure.

✓ **SERVICE TIP:** NAPA and other part and tool suppliers sell an ISO flaring tool that looks very similar to the one used for SAE flares. Do not use the wrong forming die or clamp. Use only the die and clamp included with the respective tools. The angle of the die and the diameter of the clamp are not the same between the SAE and ISO tools. Using the wrong die or clamp results in a waste of time and wasted tubing.

Figure 5-19 shows the parts of the special tool used to make an ISO flare. Form an ISO flare as follows:

1. Cut the tubing to length and install the fittings before forming the flare.

2. Clamp the ISO flaring tool in a bench vise. Select the proper size collet and forming mandrel for the diameter of steel tubing being used. Insert the mandrel into the body of the flaring tool. Hold the mandrel in place with your finger and thread in the forcing screw until it contacts and begins moving the mandrel (Figure 5-20). After contact is felt, turn the forcing screw back one full turn.

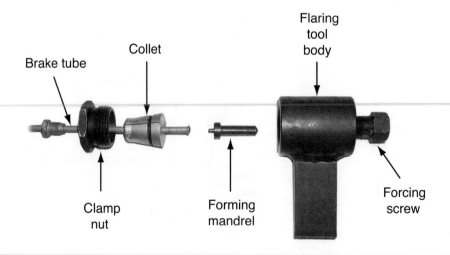

Figure 5-19 The special flaring tool used to form an ISO flare.

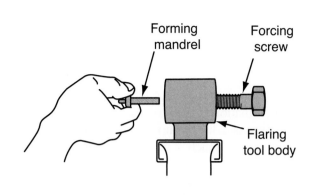

Figure 5-20 Insert the correct size mandrel against the forcing screw in the flaring tool body.

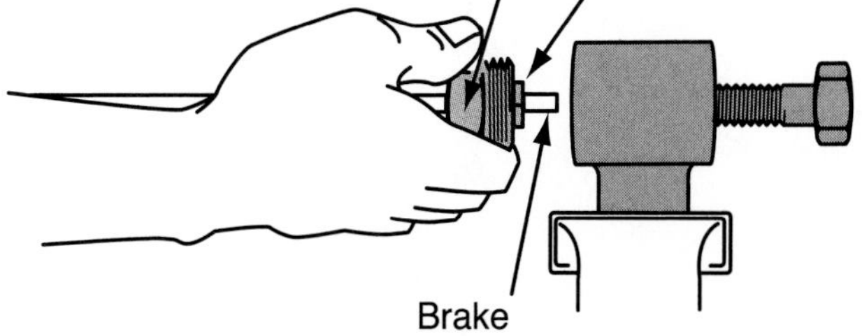

Figure 5-21 Slide the brake tube (pipe) through the clamping nut and collet.

3. Slide the clamping nut over the tubing and insert the tubing into the correct collet. Leave about ¾ inch of tubing extending out of the collet (Figure 5-21).

4. Insert the assembly into the tool body so that the end of the tubing contacts the forming mandrel. Tighten the clamping nut into the tool body very tightly to prevent the tubing from being pushed out during the forming process.

5. Using a wrench, turn in the forcing screw until it bottoms out. Do not overtighten the screw or the flare may be oversized.

6. Back the clamping nut out of the flaring tool body and disassemble the clamping nut and collet assembly.

7. Inspect the flare to be sure it has the correct shape (Figure 5-22). If it is formed unevenly or cracked, you must cut off the end of the tubing and start over again.

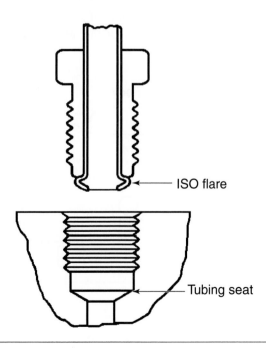

Figure 5-22 The completed ISO flare.

Servicing Hydraulic System Valves

Classroom Manual
pages 106–120

CAUTION: Always clean around any lines or covers before removing or loosening them. Dirt or other contaminants will void the warranty and may damage system components.

Metering valves, proportioning valves, and combination valves are used by auto manufacturers to regulate pressures within the system. Valves usually are mounted on or near the master cylinder, except a height-sensing proportioning valve, which is mounted under the rear of the vehicle.

Not all types of valves are used on all vehicles, so check the vehicle service manual for the exact type and location of hydraulic valves. All valves in the hydraulic system should be inspected whenever brake work is performed or a problem exists in the system. Uneven braking or premature wear of pad or shoe lining may indicate a faulty metering, proportioning, or combination valve.

SERVICE TIP: Most vehicles equipped with ABS or other electronic braking systems no longer use mechanical or hydraulic valves. Instead they use the hydraulic modulator operating on computer commands to control the brake pressures to the different wheels. This includes the metering and height-sensing proportioning pressures and the activation of the warning lamp. Control is more precise and quicker and some parts that could fail from damage, corrosion, or lack of maintenance are eliminated. Testing and service of the hydraulic modulator is covered in Chapter 10.

Special Tools

Brake pressure
  bleeder
Cloths
Tubing with
  transparent
  container
Service manual

Metering Valves

Inspect the metering valve (Figure 5-23) whenever the brakes are serviced. Fluid leakage inside the boot on the end of the valve means the valve is defective and should be replaced. A small amount of moisture inside the boot does not necessarily indicate a bad valve.

Metering valves are not adjustable or repairable. If a valve is defective, replace it. Always be sure to mount the new valve in the same position as the old valve.

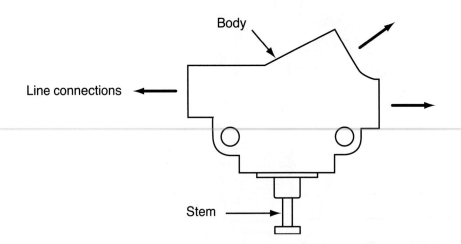

Figure 5-23 Whether it is an individual part or part of a combination valve, inspect the metering valve during every brake job.

A faulty metering valve can allow the front brakes to apply prematurely and possibly lock, especially on wet pavement. Premature front pad wear or a tendency for front brakes to lock may indicate a bad metering valve.

If you suspect a metering valve problem, have an assistant apply the brakes gradually while you watch and feel the valve stem. It should move as pressure increases in the system. If it does not, replace the valve.

You also can check metering valve operation with a pressure bleeder. Charge the bleeder tank with compressed air to about 40 psi and connect it to the master cylinder as explained earlier in this chapter. Do not override the metering valve manually or with a special tool. Pressurize the hydraulic system with the pressure bleeder and open a front bleeder screw. If fluid flows at pressure from the bleeder, the metering valve is not closing at the right pressure and must be replaced. If you need to test a metering valve more precisely, you can do it with two pressure gauges connected to the system as explained in the following section.

✓ **SERVICE TIP:** The master cylinder pressure test set shown in Figure 2-20 in Chapter 2 of this text can be used to perform the following tests and other brake pressure tests.

Special Tools

Two 500-psi gauges
Service manual
Coworker

Metering Valve Pressure Test. This test requires two pressure gauges that measure from 0 psi to at least 500 psi, plus the help of an assistant:

1. Use a T-fitting to connect one gauge to the line from the master cylinder to the metering valve.

2. Use another T-fitting to connect the other gauge to the line from the metering valve to the front brakes.

3. Have an assistant apply the brakes gradually but firmly while you watch the gauges. The gauge readings should be as follows:

   A. As pressure is first applied, the readings of both gauges should increase together until the closing pressure of the valve is reached. This should be from 3 psi to 30 psi, depending on valve design.

   B. Above the closing pressure of the valve, the inlet pressure should continue to increase while the outlet pressure stays constant.

   C. As inlet pressure continues to increase, the valve will reopen. This should be from 75 psi to 300 psi, depending on valve design. At that point, the reading on the outlet gauge should rise to match the reading on the inlet gauge. Both gauges should then read the same as pressure continues to rise.

4. If the gauge readings do not follow the patterns described in step 3, replace the metering valve.

Proportioning Valves

Proportioning valve designs and locations vary more than those of metering valves. Older proportioning valves may be a single valve installed in a line to the rear drum brakes. Diagonally split hydraulic systems often have two small proportioning valves installed in the master cylinder outlet ports to the rear brakes (Figure 5-24). Some master cylinders have two proportioning valves built into the master cylinder body. Some dual proportioning valves are separate assemblies, installed in the brake lines near the master cylinder (Figure 5-25). Height-sensing proportioning valves are installed under the rear of the vehicles and are connected to the rear axle or suspension (Figure 5-26). Finally, many proportioning valves are part of a combination valve. Regardless of the design and location, all proportioning valves work in basically the same way.

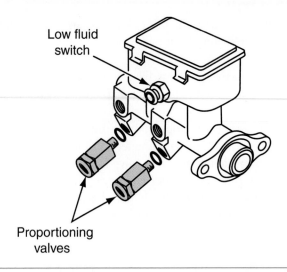

Figure 5-24 Proportioning valves installed between the master cylinder and brake lines.

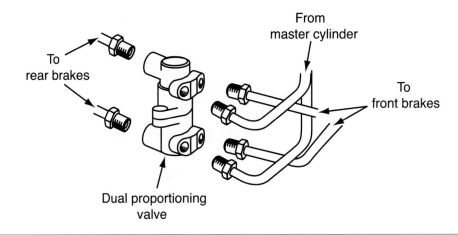

Figure 5-25 A separate dual proportioning valve.

Split point is the pressure at which a proportioning valve closes during brake application and reduces the rate at which further pressure is applied to rear drum brakes.

If rear drum brakes lock up during moderate-to-hard braking and all other possible causes of lockup have been eliminated, the proportioning valve is the likely problem. If the valve is leaking, replace or service it. In many cases, the valve cannot be disassembled and serviced; it must be replaced.

Proportioning Valve Pressure Test. If a proportioning valve has fittings on the inlet and outlet lines that allow you to connect pressure gauges with T-fittings, you can test valve operation similarly to the text explained earlier for a metering valve. If a proportioning valve is built into the master cylinder with no access for a gauge, it can still be tested by connecting pressure gauges to the bleeder ports on a front caliper and a rear wheel cylinder as explained later. If the hydraulic system is split front to rear, only a single test is needed. If the hydraulic system is split diagonally, both gauges must be connected twice to test the left and right rear brakes individually. This test requires two pressure gauges that measure from 0 psi to 1,000 psi, plus the help of an assistant. For accurate testing, you also should know the **split point** of the proportioning valve. The split point can be found in some service manuals.

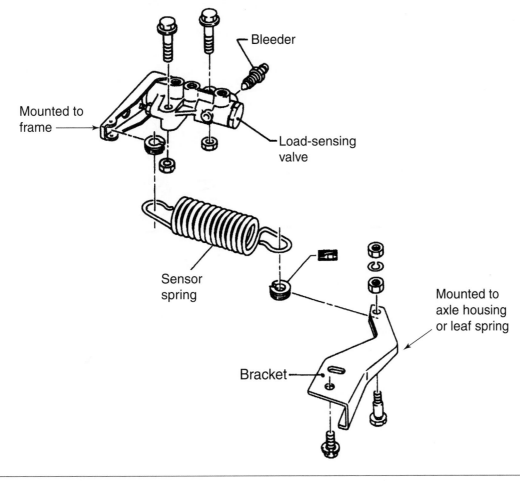

Figure 5-26 A height-sensing proportioning valve is installed under the rear of a vehicle and connected to the rear suspension or axle housing.

To test proportioning valve pressure:

1. Connect one gauge to the proportioning valve inlet pressure port by one of the following methods:

   A. Use a T-fitting to connect one gauge directly to the line from the master cylinder to the proportioning valve inlet port.

   B. If you cannot connect a T-fitting to the valve inlet line, remove a bleeder screw and connect the gauge to the bleeder port of one front caliper (Figure 5-27). Pressure to the front brakes should be the same as pressure at the proportioning valve inlet.

2. Connect the other gauge to the proportioning valve outlet pressure port by one of the following methods:

   A. Use another T-fitting to connect the other gauge directly to the line from the proportioning valve to the rear brakes.

   B. If you cannot connect a T-fitting to the valve outlet line, remove a bleeder screw and connect the gauge to the bleeder port of one rear wheel cylinder. Pressure at the rear wheel cylinder is the same as outlet pressure at the proportioning valve. If a diagonally split hydraulic system has two proportioning valves, the gauge must be connected twice: once to each wheel cylinder.

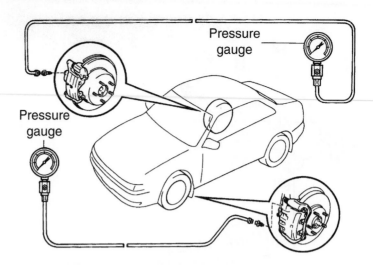

Figure 5-27 Install pressure gauges at diagonally opposite front and rear brakes to test proportioning valve operation.

3. Have an assistant apply the brakes gradually but firmly while you watch the gauges. The gauge readings should be as follows:

   A. The readings of both gauges should increase together until the split point pressure of the valve is reached.

   B. Above the split point, the outlet pressure rises more slowly than the inlet pressure.

4. If the gauge readings do not follow the patterns described above, replace the proportioning valve.

Many carmakers do not provide pressure test specifications for proportioning valves. As a general rule, however, the split point should occur at 300 psi to 500 psi. Maximum outlet pressure should be one-half to two-thirds of the maximum inlet pressure, or a slope of 50 percent to 67 percent.

> ✔ **SERVICE TIP:** If the ABS activates during moderate braking on just one rear wheel of a diagonally split brake system, check the proportioning valve for that rear wheel. If it does not modulate brake pressure correctly for the rear wheel, premature lockup can occur.

Special Tools

Service manual

Hand tools

Denatured alcohol

Proportioning Valve Servicing. On some vehicles, proportioning valves are built into the master cylinder body (Figure 5-28). These valves often can be serviced with reconditioning kits as follows:

1. Remove the master cylinder reservoir, the proportioning valve caps, and the cap O-rings. Discard the O-rings.

2. Using a needle nose pliers, remove the proportioning valve piston springs and valve pistons. Take care not to damage or scratch the piston stems. Remove the valve seals from the pistons.

3. Wash all parts with clean denatured alcohol and dry them with unlubricated compressed air. Inspect the pistons for corrosion and damage. Replace them as needed.

4. Lightly lubricate the new proportioning valve cap O-rings, valve seals, and piston stems with the special lubricant supplied in the repair kit.

5. Install the valve seals on the valve pistons so the seal lips are facing upward toward the caps.

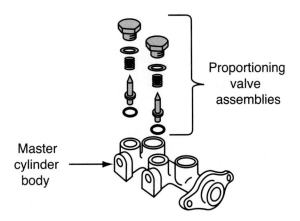

Proportioning valve assemblies

Master cylinder body

Figure 5-28 Repair kits often are available for these proportioning valves built into the master cylinder.

6. Install the pistons and seals into the master cylinder body, followed by the valve springs.

7. Place the new cap O-rings into the grooves in the proportioning valve caps and install the valve caps, torquing them to specifications.

8. Reinstall the master cylinder reservoir.

Height-Sensing Proportioning Valves

A height-sensing proportioning valve regulates the amount of brake pressure according to vehicle load. More weight allows more braking from the rear brakes, while less weight reduces rear brake pressure.

A height-sensing proportioning valve must be adjusted correctly if it is to balance rear braking force to the load. Height-sensing valves also are calibrated to work with the stock suspension. Any modification that improves load-carrying capability, such as helper springs or air-assist shocks, can adversely affect valve operation.

Modifications that make the suspension stiffer can prevent the suspension from deflecting the normal amount during hard braking or heavy load conditions. As a result, the proportioning valve may not increase rear brake effort enough, and stopping distance may increase dangerously. Because of this potential problem, modifications to the rear suspensions of these vehicles should be avoided.

Adjustment of Height-Sensing Proportioning Valves. There are as many different ways to adjust a height-sensing proportioning valve as there are valve installations. All adjustments, however, involve setting the operating rod to a specified stroke to ensure that the proportioning action takes place at the right pressure in relation to vehicle load and height.

The following example illustrates the adjustment principles for one Ford installation (Figure 5-29). The details will be different for other valve designs. Because it is difficult to accurately measure the operating rod in this example, a short length of ¼-inch plastic tubing is cut to size and used as a gauge to accurately set the rod length.

To adjust the height-sensing proportioning valve:

1. Raise the vehicle on a lift or an alignment rack so that its wheels are on a flat surface and the vehicle is at normal ride height.

2. Back off the valve adjuster setscrew but do not change the position of the upper nut.

Special Tools

Service manual
Hand tools
Short ¼-inch tubing

209

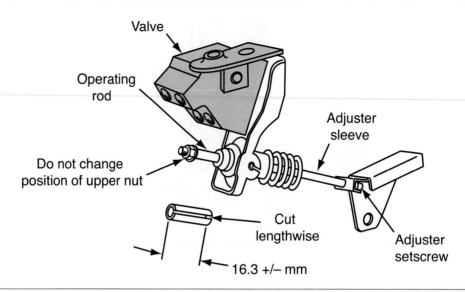

Figure 5-29 This adjustment drawing for a Ford height-sensing proportioning valve represents typical adjustments for most such valves.

3. Cut the length of ¼-inch tubing to the adjustment specification length (in this example 16.3 mm) and slit the tubing lengthwise so you can install it on the operating rod.

4. Slip the tubing onto the operating rod to set the proper operating length between the valve body and the upper nut.

5. With the adjuster sleeve resting on the lower mounting bracket, tighten the adjusting setscrew to lock the setting.

6. Remove the tubing from the operating rod. The operating rod adjustment is now set for normal driving conditions.

Test drive the vehicle. If you find that rear braking pressures are too little or too great, slight adjustments can be made. With the suspension at normal ride height, loosen the adjuster setscrew and move the adjuster sleeve toward or away from the brake pressure control valve. Each 1 mm the adjuster sleeve is moved changes braking pressure by 60 psi. Move the adjuster sleeve down away from the valve body on the operating rod to increase braking pressure. Move the adjuster sleeve up toward the valve body to decrease braking pressure. When the setting is properly adjusted, tighten the setscrew.

Removal and Replacement of Height-Sensing Proportioning Valves. Remove and replace a typical height-sensing proportioning valve (Figure 5-30) as follows:

CAUTION: Always clean around any lines or covers before removing or loosening them. Dirt and other contaminants will void the warranty and may damage system components.

CAUTION: Ensure the fitting is not cross-threaded when reconnecting. This could damage the fitting, the component, or both.

CAUTION: Use the correct fastener torque and tightening sequence when installing components. Incorrect torque or sequencing could damage the fastener(s) or component(s).

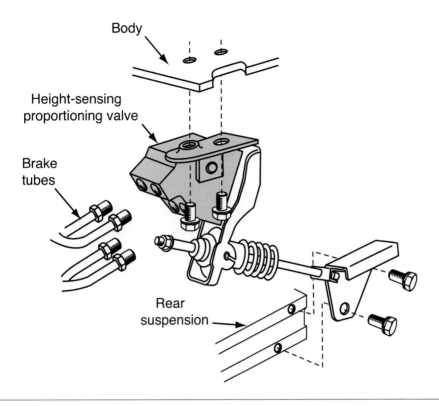

Body

Height-sensing
proportioning valve

Brake
tubes

Rear
suspension

Figure 5-30 Typical height-sensing proportioning valve installation.

1. Raise the vehicle on a lift and disconnect the brake lines from the valve body. Tag the line positions to be certain they are reinstalled correctly.

2. Remove the fastener securing the height-sensing valve bracket to the rear suspension arm and bushing.

3. Remove the screws that secure the valve bracket to the underbody and remove the assembly.

4. Before installing the new valve be certain the rear suspension is in full rebound. If the valve has a red plastic gauge clip, make sure it is in position on the proportioning valve and the operating rod lower adjustment screw is loose.

5. Position the valve and install the bolts that hold it to the underbody.

6. Secure the lower mounting bracket to the rear suspension arm and bushing using the retaining screw. Tighten all fasteners to specifications.

7. Ensure that the valve adjuster sleeve rests on the lower bracket and then tighten the lower adjuster setscrew.

8. Reconnect the brake lines to their original ports on the valve body and bleed the rear brakes. Remove the plastic gauge clip and lower the vehicle to the ground.

 SERVICE TIP: The rear wheels must be supporting the vehicle weight during bleeding or the valve will not open.

The new height-sensing proportioning valve automatically becomes operational when the suspension is at the normal ride height. Remember that the addition of extra leaf springs to increase load capacity, spacers to raise the vehicle height, and air shocks to allow heavier loads without sagging should not be used on vehicles with height-sensing proportioning valves.

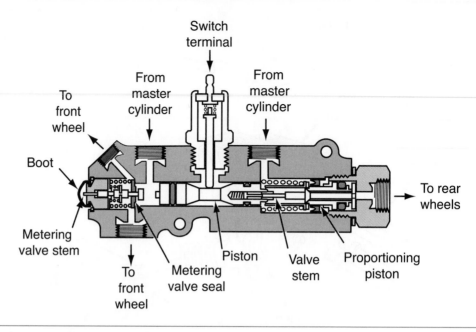

Figure 5-31 Inspect a combination valve whenever the brakes are serviced.

● **CUSTOMER CARE:** When you are troubleshooting problems on a sport utility vehicle (SUV) or light truck be sure to explain to the owner that modifications such as lift kits, oversize wheels and tires, and extra shock absorbers can affect brake performance.

Combination Valves

Inspect a combination valve whenever the brakes are serviced. If there is leakage around the large nut on the proportioning end, the valve is defective and must be replaced. A small amount of moisture inside the boot or a slight dampness around the large nut does not indicate a defective valve. Combination valves are nonadjustable and nonrepairable. If a valve is defective in any way, it must be replaced (Figure 5-31).

Brake Electrical and Electronic Component Service

Classroom Manual
pages 120–126

Most electrical components in the brake system are switches and sensors that warn of hydraulic system problems. Because of their relationship with the hydraulic system, it is appropriate to include them in this chapter with hydraulic control valves.

Stoplamps are the most basic part of the brake electrical system, so the sections on brake electrical components begin with stoplamp diagnosis and service. The following sections also cover:

❏ Brake system warning lamps (failure indicators)
❏ Parking brake indicator lamps
❏ Master cylinder fluid level switches (sensors)
❏ Basic wiring and connector repairs

Logical Electrical Troubleshooting

To isolate and repair an electrical problem, you must follow a logical troubleshooting procedure. Service manual circuit diagrams or schematics make it easy to identify common circuit problems, which will help narrow the problem to a specific area. If several circuits fail at the same time, check for a common power or ground connection. If part of a circuit fails, check the connections between the functioning areas of the circuit and the failed areas. Also remember that a problem in one system could result in a symptom in another system.

Proceed as follows to troubleshoot an electrical problem:

1. Verify the problem. Review the work order, operate the system, and list symptoms in order to check the accuracy and completeness of the owner's complaint. If the problem is intermittent, try to re-create the problem.

2. Determine the possible causes of failure. Refer to the circuit diagram for clues to the problem. Locating and identifying the circuit components may help determine where the problem is.

3. Determine if the problem is located in a parallel circuit or the parallel portion of a series-parallel. If it is, check the other matching components wired parallel to the failed component; for example, one head lamp is inoperative. If one head lamp works, then the switch, fuse, relay, and most of the wiring are good. The same diagnostic observation can be applied to almost all exterior light circuits and many of the other parallel and series-parallel circuits on the vehicle.

4. Identify the faulty circuit by studying the circuit diagram to determine circuit operation for the problem circuit. You should have enough information to narrow the failure to one component or one portion of the circuit.

5. Locate the failed component or element. Service manual procedures or diagnostic charts give a step-by-step approach to diagnosing a symptom. The test procedures are given in numerical sequence and should be followed in that order.

6. Make the repair and verify that the repair is complete by operating the system.

Refer to Chapter 2 in this *Shop Manual* for an explanation of common electrical troubleshooting equipment. Many of the procedures in the following sections require the use of a digital multimeter (DMM) or a DVOM, a probe light, and some jumper wires.

Stoplamp Testing and Switch Adjustment

AUTHOR'S NOTE: On most older vehicles (1998 and earlier) and a few newer ones, the stoplamp switch is located in the "hot," or positive, side of the stoplamp circuit. Closing the switch connects the lights to battery power. Modern electrical circuits may have the same switch acting as a grounding sensor in that when closed it completes a circuit in a lighting module, which then completes the stoplamp circuit. Diagnosing these computerized lighting circuits can be confusing to older technicians (and some new ones). The following section deals with stoplamp switch testing in general. The testing tools and diagnostic procedures may be different for the computerized version and even between different models from the same manufacturer, however. Many newer SUVs and light trucks are equipped with computer-controlled trailer light circuits. Always consult the service manual and study the wiring diagrams and system description carefully. That is the only way to properly perform diagnostics on the new lighting systems.

If one stoplamp lights but the other does not, you know you have electric power through the fuses and switches to the rear of the car. In this case, the problem is usually a burned-out bulb. Occasionally, an open circuit may exist in the branch of the circuit to one of the stoplamps, but this is much less common. If the bulb is not burned out, but will not light when installed, use a DMM or probe light to pinpoint the open circuit.

SERVICE TIP: The CHMSL on many vehicles is wired directly from the brake lamp switch to the lamp itself, whereas the two lower brake lamps are wired through the turn signal switch. If the CHMSL is lit when the brake pedal is depressed but the two lower lamps do not light, then the problem is most commonly found in the turn signal switch. Even if the turn signals work correctly, the switch cannot be ruled out as the brake lamp fault.

If both stoplamps do not light, check the circuit fuse. If the fuse is okay, check the bulbs. Occasionally, both lamp bulbs may fail at the same time. If the fuse and bulbs are okay, continue with the tests outlined next.

Two other common problems with stoplamp circuits are that the lamps do not light at all or they are lit continuously. Lamps that do not light indicate an open-circuit problem. Lamps that are lit continuously indicate a short-circuit problem. Begin troubleshooting either condition by observing the stoplamps with the brakes applied and released.

If the stoplamps *do not light at all,* check the following points:

1. Locate the switch under the instrument panel or under the hood.

2. Disconnect the harness connector and connect a jumper wire between the two terminals of the connector.

CAUTION: Do not use a jumper wire until it has been absolutely determined that the stoplight switch is not a sensor for a lighting module. Consult the wiring diagram and system description carefully. If the switch is acting as a sensor, use the service manual directions and a DMM to test the switch and circuit. Use of improper test tools or jumpers can do serious damage to electronic circuits.

3. Check the stoplamps; they should light. If they light, replace or adjust the switch. If the lamps do not light, continue testing for an open circuit condition between the switch and the lamps.

If the stoplamps *are lit continuously,* check the following points:

1. Locate the switch under the instrument panel or under the hood.

2. Disconnect the harness connector and check the stoplamps. If the lamps are still lit, locate and repair the short circuit to battery voltage in the wiring harness. If the lamps turn off, adjust or replace the switch.

Since the introduction of split hydraulic systems in 1967, almost all cars and trucks have had mechanical stoplamp switches operated by the brake pedal lever. Three basic adjustment methods exist for mechanical stoplamp switches:

1. If the switch has a threaded shank and locknut, disconnect the electrical connector and loosen the locknut. Then connect an ohmmeter or a self-powered test light to the switch. Screw the switch in or out of its mounting bracket until the ohmmeter or the test light indicates continuity with the brake pedal pressed about $\frac{1}{2}$ inch. Tighten the locknut and reconnect the electrical harness.

2. If the switch is adjusted with a spacer or feeler gauge, loosen the switch mounting screw. Press the brake pedal and let it return freely. Then place the spacer gauge between the pedal arm and the switch plunger. Slide the switch toward the pedal arm until the plunger bottoms on the gauge. Tighten the mounting screw, remove the spacer, and check stoplamp operation.

3. If the switch has an automatic adjustment mechanism, insert the switch body in its mounting clip on the brake pedal bracket and press it in until it is fully seated. Then pull back on the brake pedal to adjust the switch position. The switch click as it ratchets back in its mount. Repeat this step until the switch no longer clicks in its mounting clip. Finally, check stoplamp operation to be sure they go on and off correctly.

AUTHOR'S NOTE: A procedure for adjusting the type of stop switch noted in step 3 can be found in Chapter 4 of this text in discussing the adjustment of the brake pedal free play section.

Replacing Stoplamp Bulbs. Some taillamp and stoplamp bulbs can be replaced without removing the lens assembly. Remove the bulb and socket by twisting the socket slightly and pulling it out of the lens assembly (Figure 5-32). Push in on a brass base bulb and turn it counterclockwise. Plastic base bulbs are pulled straight out. When the lugs align with the channels of the socket, pull the bulb out to remove it.

On other vehicles, the complete taillamp lens assembly must be removed for access to the bulbs. Figure 5-33 shows how a typical lens assembly mounts to the vehicle body. It is normally held in position by several nuts or special screws.

Brake Warning Lamp Circuit Troubleshooting

The brake warning lamp on most vehicles performs multiple warning functions. Several circuits are connected to the same lamp on the instrument panel (Figure 5-34). Typically, this lamp lights when:

❏ Pressure is lost in half of the split hydraulic system and the pressure differential valve closes a switch. This warns the driver of a hydraulic system problem.
❏ If equipped, low pad sensors are grounded.
❏ The parking brake is applied.
❏ When the brake fluid level is low.
❏ As a circuit test while the engine is cranking.

Special Tools

Multimeter
Vehicle wiring
 diagram

Special Tools

Multimeter
Vehicle wiring
 diagram

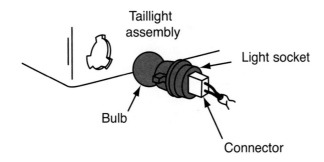

Figure 5-32 Taillamp with removable socket.

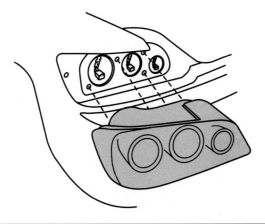

Figure 5-33 Many vehicles require the complete rear lamp lens be removed to replace a lamp.

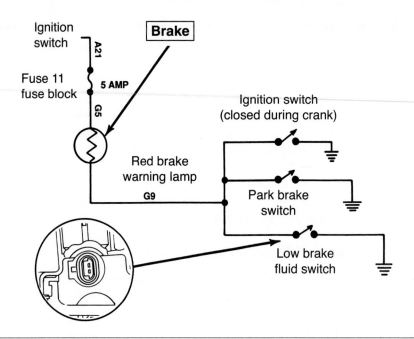

Figure 5-34 The brake warning lamp can be controlled by three different switches.

✓ **SERVICE TIP:** If only the amber ABS warning light is lit, then the problem is in the ABS system. If both brake warning lights (red and amber) are lit, the problem is probably in some shared component. If only the red is lit, then the trouble is usually within the service or parking brake system.

Additionally, on some vehicles with ABSs, the control module turns on the red brake warning lamp if it finds a problem in the amber warning lamp circuit.

Battery voltage reaches the brake warning lamp when the ignition switch is in the run, bulb test, and start positions. When any of the other switches connected to the brake warning lamp close, the circuit is grounded and the lamp lights. The ignition switch completes the circuit to ground when it is in the bulb test and start positions. The parking brake switch provides a ground when the parking brake is applied.

On vehicles with daytime running lamps, the parking brake switch completes ground to the stoplamp through a diode in the daytime running lamp module. On some vehicles, the parking brake switch is an input to the body control module that controls the daylight running lamps.

The brake fluid level switch closes to light the brake warning lamp when the brake fluid in one of the two reservoirs falls below switch level. This can be caused by a leak in one of the brake lines or by simply neglecting the fluid level.

Operational Check. Check the basic operation of the brake warning lamp as follows:

1. Turn the ignition switch to the start position or a point halfway between on and start. The warning lamp should light.

2. Release the ignition switch to the run position. With the parking brake off, the warning lamp should turn off. On some vehicles, the lamp may light for a few seconds with the ignition on and then go off.

3. With the ignition on, apply the parking brake. The warning lamp should light.

4. Release the parking brake. The warning lamp should turn off.

If the warning lamp operates as described in these four steps, the system is working properly. If the warning lamp stays lit with the ignition on and the parking brake off, the following problems may be present:

Special Tools

Multimeter

Vehicle wiring
diagram

Special Tools

Solderless connectors

Splicing/crimping
pliers

❏ Hydraulic leak or failure in one-half of the hydraulic system
❏ Low fluid level in the master cylinder reservoir
❏ Low pad sensor(s) grounded
❏ Parking brake not fully released or parking brake switch shorted or grounded
❏ ABS problem

Circuit Troubleshooting. As always, the vehicle service manual should be your first source of information for electrical troubleshooting. Figure 5-34 is a typical brake warning lamp circuit. Besides checking the lamp bulb and the switches, check the power fuse in the fuse block and the individual grounds for each circuit branch to be sure that battery voltage and a good ground are available.

When checking the warning lamp bulb, it is often hard to tell if the filament is good. If you have any doubt about the bulb condition, replace it.

It is highly unlikely that all of the indicator lamps on the instrument panel would fail at the same time. If other indicator lamps are not operating properly, check the fuses first. Next, check for voltage at the last common connection. If no voltage is present here, trace the circuit back to the battery. If voltage is found at the common connection, test each branch of the circuit in the same manner.

> **CAUTION:** Do not use a jumper wire until it has been absolutely determined that the warning switch is not a sensor for a lighting module. Consult the wiring diagram and system description carefully. If the switch is acting as a sensor, use the service manual directions and a DMM to test the switch and circuit. Use of improper test tools or jumpers can do serious damage to electronic circuits.

To test the electrical circuit between the switch in a pressure differential valve and the warning lamp, disconnect the wire from the pressure switch terminal. Connect a jumper wire from the wire to a ground on the engine or chassis. Turn the ignition key on. The brake warning lamp should light. If the lamp does not light, inspect and service the bulb, wiring, and connectors as required. If the warning lamp lights with the jumper connected, turn the ignition switch off and reconnect the wire to the pressure differential switch terminal.

Low pad sensors activate the brake warning light when the pad(s) become worn enough to be replaced. Before testing this circuit, ensure that the parking brake is off and the brake fluid is at the proper level. Determine if there is a sensor at each wheel or only at the two front wheels. It may be possible to disconnect the wiring without removing the wheel assembly. If not, lift and support the front axle of the vehicle and remove all affected wheels. Disconnect the sensor wiring at the wheel and switch the ignition on, engine off. If the light remains on, switch the ignition off and connect the wiring. Repeat with the other wheels until either the warning light goes off with the ignition on, engine off or it is determined that the circuit is bad or the pads need replacing. Many times the act of removing the wheels and inspecting the brake pads will determine that the system is working correctly.

Refer to the procedures earlier in this chapter to check the operation of the pressure differential valve.

Parking Brake Switch Test

If the brake warning lamp does not light with the ignition on and the parking brake applied but otherwise works properly, a problem exists with the parking brake switch or circuit. To check the parking brake switch and its wiring, locate the switch on the pedal, lever, or handle and disconnect it with the ignition off.

If the switch connector has a single wire, connect a jumper from that wire to ground. If the switch connector has two wires, connect a jumper between the two wires in the connector. Turn the ignition on and check the brake warning lamp. If the lamp is lit, replace the parking brake switch. If the lamp is still off, find and repair the open circuit in the wiring harness between the lamp and the switch.

Brake Fluid Level Switch Test

With the ignition on and the brake fluid level switch closed, the brake warning lamp lights to alert the driver of a low-fluid condition in the master cylinder. Some switches are built into the reservoir body; others are attached to the reservoir cap. Test principles are similar for both types.

Begin by ensuring that the fluid level is at or near the full mark on the reservoir. Turn the ignition on and observe the warning lamp. If it is lit, disconnect the wiring connector at the switch. If the lamp then goes out, replace the switch. If the lamp does not go out, find and repair the short circuit between the switch and the lamp.

To verify that the warning lamp will light when the fluid level is low, manually depress the switch float or remove the cap with an integral switch and let the float drop. If the lamp does not light with the switch closed, check for an open circuit between the switch and the lamp. If circuit continuity is good, replace the switch.

As a final check, disconnect the wiring harness from the switch and connect a jumper wire between the two terminals in the harness connector. The warning lamp should light. If it does not, find and repair the open circuit between the switch and the lamp.

Electrical Wiring Repair

An **American wire gauge (AWG)** is a system for specifying wire size (conductor cross-sectional area) by a series of gauge numbers; the lower the number, the larger the wire cross section.

A **fusible link** is wire of smaller gauge that is connected into a circuit to act as a fuse.

Rosin flux solder is solder used for electrical repairs.

Heat-shrink tubing is plastic tubing that shrinks in diameter when exposed to heat.

Wire size is determined by the amount of current, the length of the circuit, and the voltage drop allowed. Wire size is specified in either the **American Wire Gauge (AWG)** system or in metric cross-sectional area. The higher the number in AWG the smaller the conductor. A 20 gauge is much smaller than a 12 gauge.

CAUTION: Never replace a wire with one of a smaller size or replace a **fusible link** with a wire of larger size. Using the incorrect size could cause repeated failure and damage to the vehicle electrical system.

When replacing a wire, the correct size wire must be used as shown on applicable wiring diagrams or in parts books. Each harness or wire must be held securely in place to prevent chafing or damage to the insulation due to vibration. Always use **rosin flux solder** to splice a wire and use insulating tape or **heat-shrink tubing** to cover all splices or bare wires. Rosin flux cleans the connection during soldering without eroding the material as does acid-based flux. Applying heat to shrink tubing causes the tubing to contract and completely seal the wiring and connections. Utility companies used heat-shrink tubing to seal underground electrical supply cables.

Many electrical system repairs require replacing damaged wires. It is important to make these repairs in a way that does not increase the resistance in the circuit or lead to shorts or grounds in the repaired area. Several methods are used to repair damaged wire, with many factors influencing the choice. These factors include the type of repair required, accessibility of the wiring, the type of conductor and size of wire needed, and the circuit requirements. The three most common repair methods are:

1. Wrapping the damaged insulation with electrical tape
2. Crimping the connections with a solderless connector
3. Soldering splices

When deciding where to cut a damaged wire, avoid points close to other splices or connections. As a rule, do not have two splices or connections within 1.5 inches (40 mm) of each other. Use a wire of the same size or larger than the wire being replaced.

Crimping. A solderless connection uses a compressed junction to connect two conductors. Some manufacturers require the use of solderless connections on all repairs, whereas others require soldered protected repairs. Crimping solderless connections is an acceptable way to splice wire that is not exposed to weather, dirt, corrosion, or excessive movement. To make a splice using a **solderless connector:**

1. Strip enough insulation from the end of the wire to allow it to completely penetrate the solderless connector.

2. Wrap the connection in electrical tape or install a length of heat-shrink tubing over the connection. The heat-shrink tubing may have to be installed over the wire before the connection is made.

3. Position the wire in the connector and crimp the connector (Figure 5-35). To ensure a good crimp, place the open area of the connector facing toward the anvil. Make certain you compress the wire under the crimp.

4. Insert the stripped end of the other wire into the connector and crimp it in the same way.

5. Use a heat gun to shrink the tubing.

The tap splice connector is another style of crimping connector. As shown (Figure 5-36), this connector allows for adding an additional circuit to an existing wire without stripping the wires. This type of splice connection should be considered a temporary fix. As the figure shows, there are a lot of places in the splice for contaminants to get into and add resistance to the current flow. However, many DIYs use this type of splice to connect a trailer electrical wiring to the vehicle harness in probably the worst place to ever use this splice: under the rear of the vehicle, exposed to every road hazard and to shorting when the boat trailer is backed into the water. The good part about this DIY work is that sooner or later the vehicle will show up at the shop to get the lights and other electrical problems repaired. A good starting point when a vehicle shows up for lighting or electrical problems is to check for a trailer hitch and then to check for the connection under the vehicle where the trailer's harness is connected. As an aside, if a vehicle is presented for electrical problems and there are $2,000 worth of electrical add-ons in a $200 car, the installation of those items should be the first thing to check.

Soldering. Soldering is the best way to splice wires. Photo Sequence 9 shows a typical soldering procedure.

> **CAUTION:** Do not use acid-core solder for electrical repairs. The acid will corrode the wire and increase its resistance.

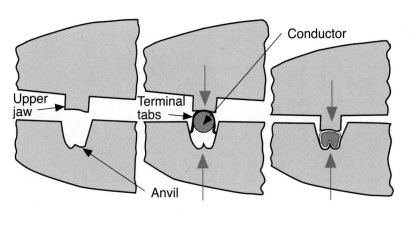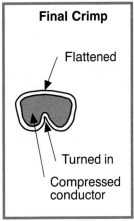

Figure 5-35 Crimping a solderless connector.

Photo Sequence 9
Soldering Two Copper Wires Together

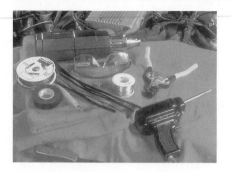

P9-1 Tools required to solder copper wire: 100-watt soldering iron, 60/40 rosin core solder, crimping tool, splice clip, heat-shrink tube, heating gun, and safety glasses.

P9-2 Disconnect the fuse that powers the circuit being repaired. Note: If the circuit is not protected by a fuse, disconnect the ground lead of the battery.

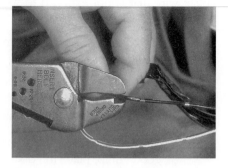

P9-3 Cut out the damaged wire.

P9-4 Using the correct size stripper, remove about ½ inch of insulation from both wires.

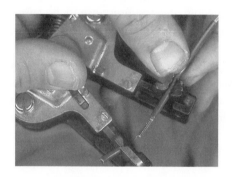

P9-5 Now remove about ½ inch of the insulation from both ends of the replacement wire. The length of the replacement wire should be slightly longer than the length of the wire removed.

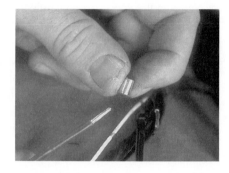

P9-6 Select the proper size splice clip to hold the splice.

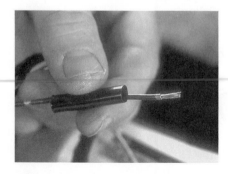

P9-7 Slide the correct size and length of heat-shrink tube over the open ends of the wire.

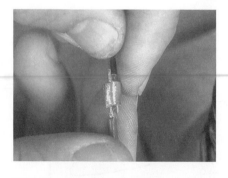

P9-8 Overlap the two splice ends and center the splice clip around the wires, making sure the wires extend beyond the splice clip in both directions.

P9-9 Crimp the splice clip place.

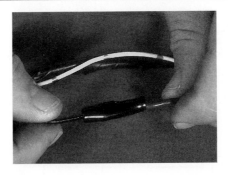

P9-10 Heat the splice clip with the soldering iron while applying solder to the opening of the clip. Do not apply solder to the iron. The iron should be 180 degrees away from the opening of the clip.

P9-11 After the solder cools, slide the heat-shrink tube over the splice.

P9-12 Heat the tube with the hot air gun until it shrinks around the splice. Do not overheat the heat-shrink tube.

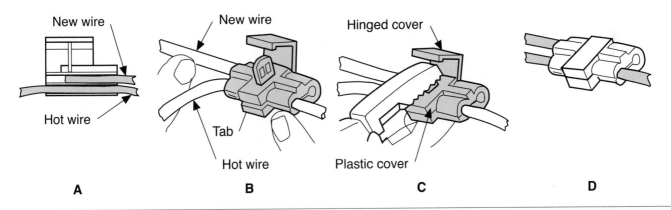

Figure 5-36 Using a tap connector to splice one wire to another. (A) Place wires in position in the connector, (B) close the connector around the wires, (C) use pliers to force the tab into the conductors, and (D) close the hinged covers.

A **splice clip** is a special connector used with solder to create a good connection. The splice clip differs from a solderless connector in that it does not have insulation. The hole is used for applying the solder (Figure 5-37).

A second way to solder wire uses a wire joint instead of a splice clip. Begin by removing about 1 inch of insulation from the wire ends. Join the wires using one of the techniques shown in Figure 5-38. Heat the twisted connection with the soldering iron and apply solder to the wires.

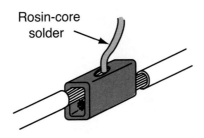

Figure 5-37 Using a splice clip to join wires.

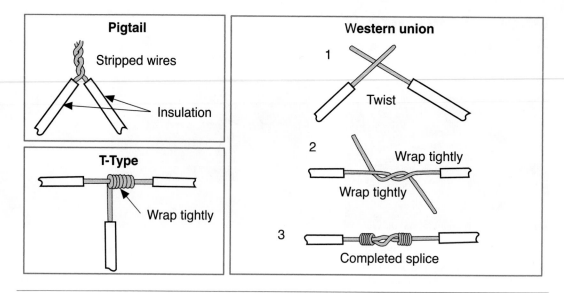

Figure 5-38 Twist wires together in one of these ways before soldering them.

Do not apply the solder directly to the iron. Heat the wire and allow the solder to flow onto it (Figure 5-39). Insulate the soldered connection with heat-shrink tubing.

Electrical Connector Repair

It is important to inspect the condition of the connectors when diagnosing electrical problems. Check connectors for cracks or signs of overheating and for contacts that are bent, scorched, corroded, or missing. If any of these problems are present, the connector should be replaced.

Molded Connectors. Molded connectors (Figure 5-40) are one-piece connectors that cannot be separated. If the connector is damaged, it must be cut out and a new connector spliced in wire by wire. This can be a time-consuming job, so work carefully around these connectors to avoid damaging them.

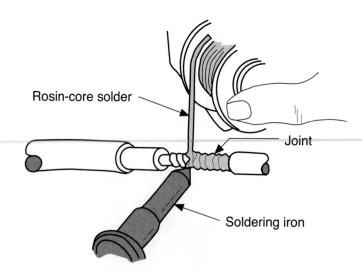

Figure 5-39 Heat the rolled joint with the soldering iron and let the heat draw solder into the joint. Do not apply solder directly to the iron.

Figure 5-40 Typical molded connector.

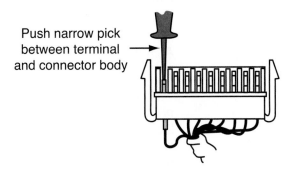

Push narrow pick
between terminal
and connector body

Figure 5-41 Depress the locking tang to remove the terminal from a hard-shell connector.

Hard-Shell Connectors. Hard-shell connectors allow removing the terminals for repair. Use a pick or special tool to depress the locking tang of the connector (Figure 5-41). Pull the lead back to release the locking tang from the connector. Remove the pick and pull the lead completely out of the connector. Repair the terminal using the same procedure as for repairing copper wire.

CASE STUDY

The owner of a 1994 small SUV with antilock brakes complained that the left front brake would intermittently lock and not release. The little 4×4 had rear-wheel antilock (RWAL) brakes, and their operation seemed normal.

Raising the vehicle on a hoist and closely inspecting the entire brake system revealed that the hose to the left front brake was not secured in its mounting bracket as it was supposed to be. The front struts had been replaced recently, and apparently the hose had not been properly secured or had come loose from its mounting shortly afterward. Certain driving maneuvers—but not others—twisted and kinked the hose, causing it to trap pressure in the brake caliper. Repositioning the hose and mounting it securely fixed the problem.

Terms to Know

American wire gauge (AWG)	Rosin flux solder	Splice clip
Fusible link	Solderless connector	Split point
Heat-shrink tubing		

ASE-Style Review Questions

1. All of the following brake hydraulic valves are used to control an individual brake or a pair of brakes EXCEPT
   - **A.** a metering valve.
   - **B.** a proportioning valve.
   - **C.** a combination valve.
   - **D.** a pressure differential valve.

2. *Technician A* says that double-wall steel tubing with single flare fittings is commonly used for brake lines. *Technician B* says that copper tubing with double-flare or ISO-flare fittings is acceptable for brake system use.
   Who is correct?
   - **A.** A only
   - **B.** B only
   - **C.** Both A and B
   - **D.** Neither A nor B

3. The red brake warning light is lit:
   Technician A says that the pressure differential piston may be off-center.
   Technician B says that this could be caused by an improper fluid type in the master cylinder reservoir.
   Who is correct?
   - **A.** A only
   - **B.** B only
   - **C.** Both A and B
   - **D.** Neither A nor B

4. *Technician A* says that copper-nickel alloy tubing requires special tools and processes for flaring. *Technician B* says that some aftermarket SAE brake tubing may have ISO flare nuts installed. Who is correct?
   - **A.** A only
   - **B.** B only
   - **C.** Both A and B
   - **D.** Neither A nor B

5. *Technician A* says that metering valves must be held open using a special tool during the bleeding operation to ensure good results.
   Technician B says that a pressure bleeder requires special adapters to connect it to the master cylinder reservoir.
   Who is correct?
   - **A.** A only
   - **B.** B only
   - **C.** Both A and B
   - **D.** Neither A nor B

6. *Technician A* says that changing the suspension or load-carrying capacities of the vehicle with a height-sensing (load-sensing) proportioning valve can adversely affect valve operation.
   Technician B says that height-sensing (load-sensing) proportioning valves are factory set and are nonadjustable.
   Who is correct?
   - **A.** A only
   - **B.** B only
   - **C.** Both A and B
   - **D.** Neither A nor B

7. *Technician A* says that a broken or frayed wire can cause an unintentional grounded circuit. *Technician B* says that dirt and grease buildup at terminals and connections can cause the same problem.
   Who is correct?
   - **A.** A only
   - **B.** B only
   - **C.** Both A and B
   - **D.** Neither A nor B

8. *Technician A* says that dual-filament lamp bulbs provide a backup filament in the bulb to double the life expectancy of the bulb.
   Technician B says that a dual-filament bulb serves two distinct functions, such as stoplamps and turn signals.
   Who is correct?
   - **A.** A only
   - **B.** B only
   - **C.** Both A and B
   - **D.** Neither A nor B

9. *Technician A* says that the brake system warning lamp can be activated by different switches such as the parking brake switch or the low fluid level switch in the master cylinder reservoir.
   Technician B says that the brake system warning lamp should light when the engine is cranking.
   Who is correct?
   - **A.** A only
   - **B.** B only
   - **C.** Both A and B
   - **D.** Neither A nor B

10. *Technician A* says that stoplamps receive fused voltage directly from the battery and will light with the ignition off if the brake pedal is pressed.
    Technician B says that the stoplamps are controlled by a normally open switch on the brake pedal.
    Who is correct?
    - **A.** A only
    - **B.** B only
    - **C.** Both A and B
    - **D.** Neither A nor B

ASE Challenge Questions

1. Brake hydraulic valves are being discussed:
   Technician A says that premature wear on the front disc could indicate a bad proportioning valve.
   Technician B says that two gauges are needed to test a metering valve.
   Who is correct?
   - **A.** A only
   - **B.** B only
   - **C.** Both A and B
   - **D.** Neither A nor B

2. A customer complains that the vehicle "nose dives" during braking:
   Technician A says that the metering valve is probably inoperative.
   Technician B says that the proportioning valve piston is stuck in the forward position. Who is correct?
   - **A.** A only
   - **B.** B only
   - **C.** Both A and B
   - **D.** Neither A nor B

3. A vehicle's rear brakes lock almost every time the brakes are heavily (not panic) applied:
   Technician A says that the metering valve may be at fault.
   Technician B says that the frame-mounted proportioning valve is set too low (as if vehicle is loaded).
   Who is correct?
   - **A.** A only
   - **B.** B only
   - **C.** Both A and B
   - **D.** Neither A nor B

4. The proportioning valve is being discussed:
   Technician A says that if the pedal feels spongy the valve's pin must be locked open so pressure can be directed to the rear wheels.
   Technician B says that a bad valve could cause rear wheel lockup.
   Who is correct?
   - **A.** A only
   - **B.** B only
   - **C.** Both A and B
   - **D.** Neither A nor B

5. Brake warning lights are being discussed:
   Technician A says that all electrical connections should be soldered during repairs.
   Technician B says that low brake fluid should turn on the amber light.
   Who is correct?
   - **A.** A only
   - **B.** B only
   - **C.** Both A and B
   - **D.** Neither A nor B

Job Sheet 14

14

Name: _____ Date: _____

Resetting Brake Warning Light Switches

NATEF Correlation

This job sheet is related to NATEF brake tasks: Inspect, test, and/or replace components of brake warning light system.

Objective

Upon completion of this job sheet, you should be able to manually reset the brake warning light switch.

Tools and Materials

Service manual

Vehicle

Hand tools

Tubing and transparent container

Describe the Vehicle Being Worked On

Year _____ Make _____ Model _____

VIN _____ Engine type and size _____

ABS _____ yes _____ If yes, type _____

Split front/rear or diagonally? _____

Low-fluid level switch? _____

Type of failure warning light switch _____

Procedure

Task Completed

1. Before lifting the vehicle, perform the following test to isolate the light and possible fault. ☐

2. Turn the ignition key on, engine off. ☐

3. Did red and amber warning lights come on? _____

4. If the red light is on, are the parking brakes on? _____

5. If yes, apply the service brakes and release the parking brakes. Did the brake light go out? _____

6. If not, check the brake fluid level, if monitored. Is the level low? _____

7. If the level is not low or not monitored, repair the parking brake warning light circuit before proceeding. ☐

8. If the parking brake circuit is correct or brake fluid is correct, proceed with the next series of tests. ☐

9. Start engine. Observe the brake warning lights. Did each light go out? If not, which stayed on? _____

☐ 10. If amber light is lit, do not continue on this job sheet. Consult the instructor.

☐ 11. If the red light stays on, ensure that the parking brakes are fully released.

☐ 12. Assuming that the parking brakes were off, turn the engine off.

☐ 13. If not done before, check and top off brake fluid level.

☐ 14. Lift the vehicle to a good working height. A coworker needs to be in the vehicle.

NOTE: If equipped with low pad sensors, go to step 15. If not, go to step 22.

☐ 15. Remove any wheel assembly equipped with a low pad sensor at the caliper. It may be possible to do step 16 without removing the wheel assemblies.

☐ 16. Make sure the parking brake is off and the fluid level is correct, then disconnect the sensor wire at one wheel.

☐ 17. Turn the ignition switch to on, engine running.

☐ 18. If light is out, the problem is either the sensor or that the pads are worn.

☐ 19. If the light remains on, switch off the engine and reconnect the wiring.

☐ 20. Move the next wheel and disconnect the sensor wire.

☐ 21. Repeat step 17 through step 20 until all sensor circuits have been checked using this method.

☐ 22. Locate and access the brake failure warning lamp switch.

☐ 23. Have coworker switch the ignition key on, engine running, and apply the service brakes gently.

☐ 24. Instruct coworker to observe the red light. If it goes out, he should hold the pedal in place and notify the technician at the wheels.

☐ 25. Open a front bleeder screw with the brake applied. Keep the bleeder open until the light goes out or the pedal goes completely down.

☐ 26. Close the bleeder screw before releasing the pedal.

☐ 27. Assuming that the previous steps did not clear the warning light, move to a rear wheel that is on the opposite side of the first wheel bled.

☐ 28. Repeat step 23 through step 26 to the rear wheel.

☐ 29. If the light goes off at either wheel, lower the vehicle and top the fluid if necessary. Go to step 31.

☐ 30. If the light did not go out under any condition covered here, consult the instructor.

☐ 31. If the repair is complete, clean the area, store the tools, and complete the repair order.

Problems Encountered _____

_____

Instructor's Response _____

_____

Job Sheet 15

15

Name: _____ Date: _____

Replace a Brake Combination Valve

NATEF Correlation

This job sheet is related to NATEF brake tasks: Inspect, test, and/or replace metering (hold-off), proportioning (balance), pressure differential, and combination valves; inspect, test, and adjust height (load) sensing proportioning valve.

Objective

Upon completion of this job sheet, you should be able to replace a combination valve.

Tools and Materials

Service manual

Vehicle

Lift or jack with stands

Hand tools

Tubing or transparent container

Catch basin

Describe the Vehicle Being Worked On

Year _____ Make _____ Model _____

VIN _____ Engine type and size _____

ABS _____ yes _____ If yes, type _____

NOTE: If equipped with ABS do not proceed with this job sheet.

Are there specific procedures for bleeding this type of valve?

If yes, explain. _____

_____

_____

Procedure

Task Completed

1. Lift the vehicle to a good working height. ☐

2. Place a catch basin under the valve. ☐

■ **CAUTION:** Always clean around any lines or covers before loosening or removing them. Dirt and other contaminants will void the warranty and may damage system components.

▲ **WARNING:** Wear safety glasses or face protection when using brake fluid. Injuries to the face or eyes could occur from spilled or splashed brake fluid.

▲ **WARNING:** Brake fluid may be irritating to the skin and eyes. In case of contact wash skin with soap and water or rinse eyes thoroughly with water.

■ **CAUTION:** Prevent brake fluid from coming in contact with the vehicle's finish. Brake fluid damages paint and finish immediately on contact. If fluid contacts finish, wash area thoroughly with running water using soap if possible.

☐ **3.** Use a line wrench to loosen all brake lines at the valve.

☐ **4.** Remove all mounting fasteners on the valve. Remove the valve.

☐ **5.** Install the new valve. Do not tighten the fasteners at this time.

■ **CAUTION:** Ensure the fitting is not cross-threaded when reconnecting. This could damage the fitting or the component, or both.

■ **CAUTION:** Use the correct fastener torque and tightening sequence when installing components. Incorrect torque or sequencing could damage the fastener(s) or component(s).

☐ **6.** Screw each line fitting loosely into the valve.

☐ **7.** When all line fittings are started, torque the valve's mounting fasteners.

☐ **8.** Tighten each line.

☐ **9.** Bleed the valve following the instructions in the service manual.

☐ **10.** Top off the brake fluid.

▲ **WARNING:** Road test a vehicle under safe conditions and while obeying all traffic laws. Do not attempt any maneuver that could jeopardize vehicle control. Failure to adhere to this precaution could lead to serious personal injury and vehicle damage.

☐ **11.** Test drive if necessary.

☐ **12.** When the repair is complete, clean the area, store the tools, and complete the repair order.

Problems Encountered _____

_____

Instructor's Response _____

_____

_____

Job Sheet 16

16

Name: _____ Date: _____

Inspecting and Diagnosing Brake Lines and Hoses

NATEF Correlation

This job sheet addresses the following NATEF tasks: Inspect brake lines, flexible hoses, and fittings for leaks, dents, kinks, rust, cracks, bulging or wear; tighten loose fittings and supports; determine necessary action.

Objective

Upon completion of this job sheet, you should be able to inspect and diagnose brake lines, hoses, and connecting fittings.

Tools and Materials

Basic hand tools

Describe the Vehicle Being Worked On

Year _____ Make _____ Model _____

VIN _____ Engine type and size _____

ABS _____ yes _____ no _____

Engine Compartment

Procedure

Task Completed

1. Inspect all lines to the vacuum (hydro-boost) booster unit, if equipped, for crimps, damage, or leaks. Record your findings. ☐

   _____

   _____

2. Inspect the master cylinder for leaks or damage to the cap, the outlet fittings, and the mounting to the booster or fire wall. Record your findings and suggested repairs. ☐

   _____

   _____

3. If any brake hydraulic valves are visible, inspect them for leaks or damage. Record your findings and suggested repairs. ☐

   _____

   _____

4. Lift the vehicle to gain access to the undercarriage.

5. Inspect the valve(s) located in the area of the left rear engine compartment for leaks or damage. Valves are normally mounted below the master cylinder on the frame. Record your findings and suggested repairs. ☐

   _____

   _____

6. Inspect the steel lines from the valve to each of the front wheels for leaks or damage. Record your findings and suggested repairs.

_____

_____

7. It may be necessary to remove the wheel assembly to inspect the front hoses. If so, perform step 8 through step 12 before inspecting the front brake hoses.

8. Inspect the hoses from the steel lines to the caliper or wheel cylinder and their fittings for leaks or damage. Record your findings and suggested repairs.

_____

_____

9. Inspect the steel lines from the valves to the rear wheels for damage or leaks. Is there more than one steel line?

_____

_____

10. At which point on the vehicle does the steel line(s) connect to a brake hose(s)?

_____

_____

11. Inspect the hose(s) and the fittings from the steel line to the caliper or wheel cylinder or differential for leaks or damage. Record your findings and suggested repairs.

_____

_____

12. If the vehicle has a solid rear axle, inspect the steel lines from the hose connection to each rear caliper or wheel cylinder for leaks or damage. Record your findings and suggested repairs.

_____

_____

13. If the wheels must be removed to inspect the front brake hoses, remove the wheels at this point and perform step 8. Record your findings and suggested repairs.

_____

_____

14. Was any damage noted on the undercarriage that may affect the brake lines or hoses at a future time? If yes, note the damage and suggested repairs to possibly prevent future line or hose damage.

_____

_____

Problems Encountered   _____

_____

Instructor's Response   _____

_____

Job Sheet 17

17

Name: _____ Date: _____

Constructing an ISO Fitting

NATEF Correlation

This job sheet addresses the following NATEF tasks: Fabricate and/or install brake lines (double flare and ISO types); replace hoses, fittings, and supports as needed.

Objective

Upon completion of this job sheet, you should be able to construct an ISO flare fitting.

Tools and Materials

Basic hand tools
ISO flaring tool including tubing cutter

Describe the Vehicle Being Worked On

Year _____ Make _____ Model _____

VIN _____ Engine type and size _____

Procedure

Task Completed

1. Secure the tubing (steel line) and mark the exact length plus $\frac{1}{8}$ inch. What is the total length to be used? What is the inside diameter of the tubing? ☐

   _____

   _____

2. Install the tubing cutter square and tighten the blade to the tubing. Do not overtighten. ☐
3. Rotate the cutter completely around the tube. Adjust the cutter blade to a deeper depth during each rotation. ☐
4. Repeat step 4 until the tubing is cut through. ☐
5. Using the reamer on the tubing cutter, clean and square the end of the tubing to be flared. ☐
6. Slide the flare nut over the tubing, ensuring it is facing in the right direction. ☐
7. Clamp the ISO flaring tool in a bench vise. ☐
8. Select the collet and mandrel for the size of tubing to be flared. ☐
9. Install the mandrel into the body of the flaring tool and tighten the forcing nut until the mandrel begins to move. ☐
10. Back off the forcing screw one full turn. ☐
11. Slide the clamping nut over the tubing followed by the correct collet. About $\frac{3}{4}$ inch should be exposed after the collet. ☐
12. Insert the tubing and collet into the flaring tool body until the tubing bottoms against the mandrel. ☐

☐

☐

☐

☐

☐

13. Hold the tubing in place as the collet is slid into place and the clamping nut is threaded.

14. Tighten the clamping nut very tightly to hold the tubing in place during the process.

15. Using a wrench, tighten the forcing screw until the mandrel bottoms. Do not overtighten.

16. Loosen the clamping nut and remove the collet and tubing from the flaring tool.

17. Inspect the ISO flare for any cracks or deformities. Record your findings and any suggested repairs.

_____

_____

Problems Encountered _____

_____

Instructor's Response _____

_____

Job Sheet 18

18

Name: _____ Date: _____

Replace a Brake Hose

NATEF Correlation

This job sheet addresses the following NATEF tasks: Fabricate and/or install brake lines (double flare and ISO types); replace hoses, fittings, and supports as needed.

Objective

Upon completion of this job sheet, you should be able to replace a brake hose.

Tools and Materials

Basic hand tools

Describe the Vehicle Being Worked On

Year _____ Make _____ Model _____

VIN _____ Engine type and size _____

ABS _____ yes _____ no _____

Procedure

Task Completed

1. Lift the vehicle, if necessary, and remove the left front tire and wheel assembly. Different wheels may be selected. ☐

2. Explain why (or what circumstances would cause) this brake hose needs to be replaced. ☐

_____

_____

NOTE: This job sheet uses a front disc brake hose as an example. Other brake hoses are replaced in a similar manner.

3. Does this brake hose use banjo-type fittings at either end? If so, adjust your tool choice to perform the following steps. ☐

_____

_____

4. Use flare-nut (line) wrenches to disconnect the hose from the steel line on or near the vehicle frame. Plug the steel line. What size wrench was used? ☐

_____

_____

5. Use a small prybar or pliers to remove the hose retainer at the frame. What tool was used and how was the procedure performed? ☐

_____

_____

6. Use a flare-nut (line) wrench to disconnect the hose from the caliper. ☐

☐   **7.** Remove any washer present and plug the caliper port.

☐   **8.** Select the proper length and diameter hose. Ensure that the hose meets the manufacturer's specifications.

☐   **9.** Slide the washer, if any, over the caliper end of the new hose.

☐   **10.** Remove the caliper plug and thread the hose into the port. Tighten to torque specifications. What is the torque specification on this fitting?

_____

_____

☐   **11.** Install the frame end of the hose through its mount and install the hose retainer.

☐   **12.** Align and thread the flare nut on the steel line into the hose end. Tighten to specification. What is the torque specification on this fitting and which wrenches were used to tighten the fitting?

☐   **13.** Use job sheet 11, 12, or 13 to bleed the brake system. Which job sheet was utilized and why was it selected?

_____

_____

☐   **14.** Install the tire and wheel assembly. What is the torque specification on the lug nuts?

_____

_____

☐   **15.** Lower the vehicle and road test.

Problems Encountered _____

_____

Instructor's Response _____

_____

Power Brake Service

Upon completion and review of this chapter, you should be able to:

❏ Diagnose vacuum power booster problems.

❏ Test pedal free play with and without the engine running to check power booster operation.

❏ Check vacuum supply to a vacuum booster using a vacuum gauge.

❏ Inspect a vacuum power booster for vacuum leaks.

❏ Inspect the vacuum check valve for proper operation.

❏ Remove and install a vacuum booster and properly adjust the pedal linkage and pushrod.

❏ Disassemble, repair, and adjust a vacuum booster as required to restore proper operation.

❏ Troubleshoot an auxiliary vacuum pump. Remove and replace the pump as required.

❏ Inspect and test a hydro-boost and accumulator for leaks and proper operation.

❏ Adjust or replace components on a hydro-boost system as needed.

❏ Flush and bleed a hydro-boost system.

❏ Test an electro-hydraulic brake system.

Types of Power Brake Systems

Two types of power brake systems are used on most late-model vehicles: vacuum-boost systems and hydraulic-boost systems. Both systems multiply the force exerted on the master cylinder piston by the driver. This increases the hydraulic pressure delivered to the wheel cylinder and caliper pistons, resulting in increased stopping performance.

Vacuum boosters (Figure 6-1) use engine vacuum and, in some cases, vacuum from an external vacuum pump to help apply the brakes. Hydro-boost power systems (Figure 6-2) use hydraulic pressure from the vehicle power steering pump. Other hydraulic-boost power brake systems use an independent fluid supply for booster pressure.

Power brakes are simply conventional brakes with an added power booster. When troubleshooting and servicing power brake systems, however, keep the two systems separate. Check for faults in the master cylinder and hydraulic system first. As with conventional brakes, a spongy pedal in a power brake system is caused by air in the hydraulic lines. Brake grab may be caused by grease on the brake linings. Check out all basic brake components before moving on to the power-assist system.

Except for the master cylinder pushrod adjustment, vacuum and hydraulic power-assist units are not adjusted in normal service. If the booster is suspect, it is removed and replaced with a new or rebuilt unit or it can be rebuilt in the shop. Overhaul kits are available.

Vacuum-Boost Systems

Vacuum boosters (Figure 6-3) generate their application energy through the **pressure differential** between engine vacuum and atmospheric pressure. A flexible diaphragm and a power piston use this energy to provide brake assistance. Modern vacuum boosters are **vacuum-suspended** units. This means the booster diaphragm is suspended in a vacuum on both sides when the brakes are not applied. When the brake pedal is pressed, an air control valve attached to the brake pedal pushrod opens. This valve admits atmospheric pressure to the back of the diaphragm. Atmospheric

Classroom Manual
pages 132–134

Basic Tools

Basic technician's tool set

Flare-nut wrench

Vacuum gauge

Regardless of type or operation, technicians refer to power brake boosters as the brake booster.

Figure 6-1 Typical power brake vacuum booster.

Figure 6-2 Hydro-boost power brake booster.

pressure forces the diaphragm forward where it increases the amount of force applied to the pushrod and master cylinder piston.

Vacuum boosters can have one or two diaphragms, but most are single-diaphragm units. Single-diaphragm boosters are larger in diameter than dual-diaphragm vacuum boosters or tandem boosters.

All vacuum boosters have vacuum check valves. The check valve is located between the engine manifold and the booster. Vacuum can reach the booster through the one-way check valve, but it cannot leak back past the valve. As a result, vacuum is maintained inside the booster even after the engine is turned off.

Hydraulic Power Brakes

Diesel engines and some gasoline engine installations do not produce enough intake manifold vacuum to operate a power brake booster. One way to handle a lack of vacuum or low-vacuum conditions is to eliminate vacuum as a power source and use hydraulic power instead. The two kinds of hydraulic boosters are:

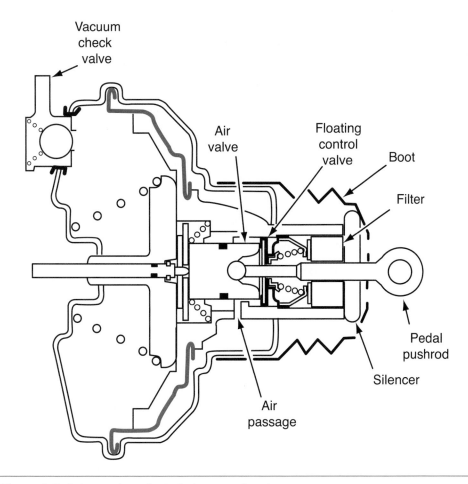

Figure 6-3 Cutaway view of a typical vacuum booster.

1. A mechanical hydraulic power-assist system operated with pressure from the power steering pump. This unit is a Bendix design called hydro-boost.

2. An electro-hydraulic power-assist system with an independent hydraulic power source driven by an electric motor. This unit is a General Motors design called PowerMaster.

Vacuum Booster Testing and Diagnosis

Vacuum boosters are usually trouble free, and many last the lifetime of a vehicle. An operational check and inspection of the vacuum booster are simple operations, however, and should be part of every brake service job.

Any condition that reduces the amount of vacuum the engine generates will affect power brake performance. These conditions include vacuum leaks, faulty valves, improper valve timing, and incorrect ignition timing. An engine rebuilt with a high-performance camshaft may also produce lower vacuum. When investigating poor brake performance, check engine vacuum before inspecting the vacuum booster system. In most systems, at least 14 in. Hg of vacuum is required for proper power brake operation.

Insufficient manifold vacuum, leaking or collapsed vacuum lines, punctured diaphragms, or leaky piston seals can cause weak booster operation. A steady hiss when the brake is held down indicates a leak that also can cause poor booster operation. Hard brake pedal is usually the first signal that the booster is failing.

Classroom Manual
pages 132–137

Special Tools

Vacuum
   gauge/pump

At times a leak in the vacuum booster may affect engine operation.

If the brakes do not release completely, they may have a tight or misaligned connection between the booster pushrod and the brake pedal linkage. If the pedal-to-booster linkage appears in good condition, loosen the connection between the master cylinder and the brake booster. If the brakes release, the trouble is in the power unit. A piston, diaphragm, or bellows return spring may be broken. If the brakes do not release when the master cylinder is loosened from the booster, one or more brake lines may be restricted, or a problem may exist in the brake hydraulic circuit.

If the brakes grab, look for common causes such as greasy linings or scored drums before checking out the booster. If the problem is in the booster, it may be a damaged reaction control. The reaction control assembly is a diaphragm, spring, and valving that tends to resist pedal action. This feature gives brake pedal feel to the driver.

On a vehicle with an engine-driven or electric vacuum pump, test it as the manufacturer recommends.

Basic Vacuum-Boost Operational Test

Conduct a basic operating test of the vacuum booster as follows:

1. Turn off the engine.

2. Repeatedly pump the brake pedal to remove all residual vacuum from the booster.

3. Hold the brake pedal down firmly and start the engine.

4. If the system is working correctly, the pedal should move downward slightly and then stop. Only a small amount of pressure should be needed to hold down the pedal.

5. If you do not get the results described in step 4, do the following tests.

Vacuum Supply Tests

If the booster is giving weak braking assistance or no assistance at all, a problem may exist with the vacuum supply to the unit. Vacuum boost efficiency is affected by loose or kinked vacuum lines and clogged air intake filters. Another cause may be the check valve, which retains vacuum in the booster when the engine is off. You can check this valve with a vacuum gauge to determine if it is restricted or stuck open or closed. Photo Sequence 10 shows the details.

A vacuum supply hose that is restricted but not completely blocked will allow a normal reading on a vacuum gauge but will delay the buildup of full vacuum in the booster; that is, it can reduce the volume of vacuum for rapid, repeated brake applications. Check for a restricted vacuum hose by disconnecting it from the booster with the engine running. If the engine does not stumble suddenly and almost stall, the hose is probably restricted. Install a new hose and recheck booster operation. Also, if the vacuum hose contains a vacuum filter and the filter is clogged, the same kind of delayed vacuum application symptoms may occurs. Photo Sequence 10 shows a typical procedure for vacuum booster testing.

If the vacuum booster seems to have a normal vacuum supply but the brakes take more effort than normal to apply, the booster may have an internal vacuum leak. Check for this by starting the engine and letting it idle to develop normal vacuum at the booster. Then roll up the windows to reduce outside noise and slowly but firmly apply the brakes. Listen to the engine. If it stumbles or runs roughly, and a hissing sound increases around the pedal pushrod, the booster has an internal leak in the diaphragm.

Do not mistake normal booster breathing for a vacuum leak. When the pedal is pressed, air rushing through the filter in the rubber boot of the booster input pushrod causes a slight breathing sound, which is normal. A diaphragm vacuum leak will cause a louder, continuing hiss.

Photo Sequence 10
Typical Procedure for Vacuum Booster Testing

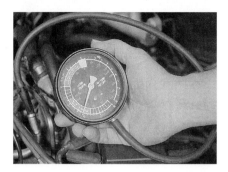

P10-1 With the engine idling, attach a vacuum gauge to an intake manifold port. Any reading below 14 in. Hg of vacuum may indicate an engine problem.

P10-2 Disconnect the vacuum hose that runs from the intake manifold to the booster and quickly place your thumb over it before the engine stalls. You should feel strong vacuum.

P10-3 If you do not feel a strong vacuum in step 2, shut off the engine, remove the hose and see if it is collapsed, crimped, or clogged. Replace it if needed.

P10-4 To test the operation of the vacuum check valve, shut off the engine and wait 5 minutes. Apply the brakes. There should be power assist on at least one pedal stroke. If there is no power assist on the first application, the check valve is leaking.

P10-5 Remove the check valve from the booster.

P10-6 Test the check valve by blowing into the intake manifold end of the valve. There should be a complete blockage of airflow.

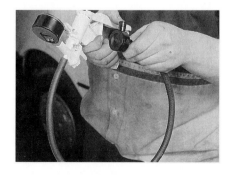

P10-7 Apply vacuum to the booster end of the valve. Vacuum should be blocked. If you do not get the stated results in step 6 and step 7, replace the check valve.

P10-8 Check the booster air control valve by performing a brake drag test. With the wheels of the vehicle raised off the floor, pump the brake pedal to exhaust residual vacuum from the booster.

P10-9 Turn the front wheels by hand and note the amount of drag that is present.

P10-10 Start the engine and allow it to run for 1 minute, then shut it off.

P10-11 Turn the front wheels by hand again. If drag has increased, this indicates that the booster control valve is faulty and is allowing air to enter the unit with the brakes unapplied. Replace or rebuild the booster.

Fluid Loss Test

If the fluid level in the master cylinder reservoir level is low but there is no sign of an external leak, remove the vacuum hose from the intake manifold to the booster. Inspect it carefully for signs of brake fluid. If evidence of brake fluid is found, the master cylinder secondary seal may be leaking. If so, the master cylinder requires rebuilding or replacement. See Chapter 4, Master Cylinder and Brake Fluid Service, for details.

✓ **SERVICE TIP:** Do not overlook the brake vacuum booster as the cause of drivability and comfort complaints. A leak in the booster vacuum hose or at the point that the hose connects to the manifold or the booster can cause a rough idle, misfire, hesitation, or surge. Depending on the size of the hole and the point where the hose connects to the manifold, the leak can affect the air-fuel mixture to one cylinder or several cylinders. Complaints about the air-conditioning system may also be a clue. If the air-conditioning plenum uses vacuum diaphragms to move the air delivery doors, the owner may complain that the air conditioner changes to full heat on the windshield when he or she drives up a long hill or accelerates.

Brake Pedal Checks

Special Tools

Rule

The brake pedal must be adjusted properly for correct power-assist operation. In addition to the mechanical checks and brake travel check presented in Chapter 4, Master Cylinder and Brake Fluid Service, you should also check pedal free play and pedal height settings. Excessive play or low pedal height may limit the amount of power assist generated by the vacuum booster.

Pedal Free Play Inspection

Pedal free play is the first easy movement of the brake pedal before the braking action begins to engage. Pedal free play should be only about 1/16 inch to 3/16 inch in most cases. To determine the amount of free play present, gently press the pedal down by hand until you feel an increase in effort (Figure 6-4). Hold a ruler alongside the pedal to measure the amount of free play.

Figure 6-4 Checking pedal free play by hand rather than with your foot will give you a more precise measurement.

Pedal Height Adjustment

Brake pedal height specifications are listed in most vehicle service manuals. Typical heights range from 6 inches to 7 inches. Most pedal height measurements are taken from the floor mat to the base of the pedal, but in some cases the measurement is taken from a special point below the floor mat (Figure 6-5).

Before adjusting pedal height, loosen the brake switch locknut and back off the brake switch until it no longer touches the brake pedal. Then use pliers to screw the pushrod in or out as needed (Figure 6-6).

When the proper pedal height is set, adjust the stoplamp switch as required. Figure 6-7 is an example of one stoplamp switch adjustment method. After adjusting pedal height, check for proper stoplamp and cruise control operation. Be sure that the stoplamps are off with the brakes released and on with the brakes applied. Test drive the vehicle and verify that the cruise control disengages when the brake pedal is pressed. Refer to Chapter 5 of this *Shop Manual* for more information on stoplamp and other brake pedal switch adjustments.

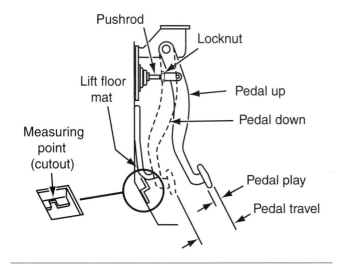

Figure 6-5 Pedal height is measured on some cars from a specific spot on the floor.

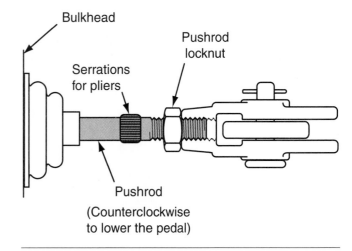

Figure 6-6 Pedal height is adjusted on most vehicles by shortening or lengthening the pushrod.

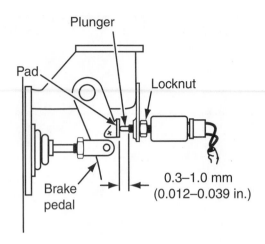

Figure 6-7 Typical stoplamp switch adjustment.

Vacuum Booster Removal and Installation

Classroom Manual
pages 137–145

Special Tools

Cloths

Follow these general steps to remove and replace a typical vacuum booster (Figure 6-8). Detailed steps will vary from one vehicle to another, but most removal and installation procedures follow this basic sequence. Photo Sequence 11 illustrates the basic steps.

CAUTION: Always clean around any lines or covers before removing or loosening them. Dirt and other contaminants will void the warranty and may damage the system components.

Figure 6-8 Many times there are wiring harnesses and other components that must be moved or shifted to gain access to the booster fasteners and make room to extract the booster from the vehicle.

Photo Sequence 11
Typical Procedure for Replacing a Vacuum Booster

P11-1 The master cylinder and booster may be easy or difficult to access depending on the components in the surrounding area.

P11-2 With the surrounding components removed or shifted to the side, the master cylinder fasteners can be accessed.

P11-3 Once the fasteners are removed, pull the master cylinder back from the booster. Take care to prevent damage to the brake lines.

P11-4 Many times some lower portion of the dash must be removed to gain access to the brake pedal. If the technician is tall or heavily built, it may be easier to work in this area with the front seat removed.

P11-5 On many vehicles, it is necessary to remove the stoplamp switch to prevent damage to it.

P11-6 The spring clip retaining the pushrod to the pedal can usually be unlocked and removed with a small pocket flat-tip screwdriver.

P11-7 The mounting nuts for the booster are sometimes far up on the bulkhead. Usually a ratchet, socket, and universal and a short extension will be needed.

P11-8 Once the fasteners are removed, the booster can be pulled away from the bulkhead and from the engine compartment.

P11-9 Slide the new booster into place on the bulkhead. If available, have an assistant hold the booster in place until the four fasteners under the dash have been started onto the studs.

Photo Sequence 11
Typical Procedure for Replacing a Vacuum Booster (continued)

P11-10 Tighten the four booster fasteners to specifications.

P11-11 Connect the pushrod to the pedal and install the retaining clip. Replace the clip if so directed by the service manual.

P11-12 Install the stoplamp switch and reconnect its electrical harness.

P11-13 Install and clean the underdash panels and any other components that were removed for access to the brake pedal.

P11-14 Position the master cylinder and tighten the fasteners to specification. Check the brake lines for possible bends or kinks.

P11-15 Install all of the components that were removed for access to the master cylinder and booster.

> ⚠ **WARNING:** Wear safety glasses or face protection when using brake fluid. Injuries to the face or eyes could occur from spilled or splashed brake fluid.

Booster Removal

To remove a vacuum booster:

1. Set the parking brake and disconnect the battery ground (negative) cable.
2. Disconnect the vacuum hose at the booster check valve.
3. Remove all fasteners securing the master cylinder to the brake booster and carefully lift the master cylinder out of the engine compartment.
4. Pull the master cylinder back from the booster, taking care not to crimp a brake line. There is usually enough movement in the lines to allow this. If not, the lines must be disconnected and master cylinder outlets plugged.
5. With the master cylinder removed, move inside the vehicle and disconnect the stoplamp switch wiring connector from the switch (Figure 6-9).
6. Remove the switch from its mounting pin.

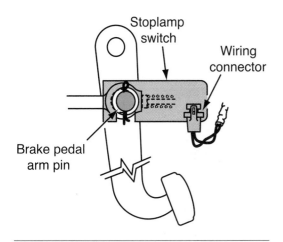

Figure 6-9 Disconnecting the stoplamp switch connector.

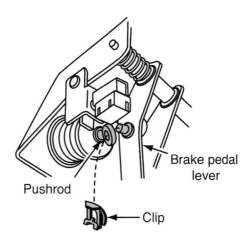

Figure 6-10 The retaining clip holding the pushrod to the brake pedal has to be replaced if it is ever removed. The removal damages the spring action of the clip.

7. Remove the nuts fastening the vacuum booster to the passenger compartment side of the bulkhead.

8. Slide the booster pushrod and bushing off the brake pedal pin (Figure 6-10).

9. Return to the engine compartment and clear the area around the booster. This may require removing a vacuum reservoir or manifold (Figure 6-11), moving a wiring harness, or removing the transmission shift cable and bracket.

10. When the area is cleared of obstacles, move the booster forward so that its studs clear the engine compartment bulkhead.

11. Lift the booster out of the engine compartment.

Figure 6-11 During the removal or shifting of the components, take care not to damage any of the electrical devices or braces.

● **CUSTOMER CARE:** The placement of most vacuum boosters requires the technician to work in a close space in the hood hinge area. This places the technician's belt buckle and clothing in closer contact with the vehicle's panels than with other repairs. Slide the belt buckle around to the side. Use a fender cover during this repair to protect the finish and other items found in and around the booster location.

Booster Installation

■ **CAUTION:** Ensure that the fitting is not cross-threaded when reconnecting the vacuum booster. Cross-threading could damage the fitting or the component, or both.

Check the booster pushrod length as described later in this chapter under "Vacuum Booster Pushrod Length Check." Install the vacuum booster following these eleven general steps:

1. Line up the brake support bracket from inside the vehicle.
2. Have a helper set the booster in position on the engine compartment bulkhead.
3. Thread the nuts onto the studs from inside the vehicle.
4. Reinstall the pushrod onto the brake pedal pin using a new pushrod bushing.
5. Tighten the booster retaining nuts to the specified torque.
6. Reinstall the stoplamp switch on the pedal pin. Reconnect the electrical connector to the switch body.
7. Return to the engine compartment and reposition and install the wiring harness, transmission shift cable, and vacuum manifold.
8. Connect the manifold vacuum hose to the booster check valve.
9. Reinstall the master cylinder, connect all brake lines, and bleed the system of all air.
10. On a vehicle with a manual transmission or cruise control, adjust the manual shift linkage and cruise control dump valve.
11. Reconnect the battery and test drive the vehicle to be sure the booster is operating properly.

☑ **SERVICE TIP:** You can make some quick tests of a vacuum booster with a hand-held vacuum pump. Start by disconnecting the booster vacuum hose from the intake manifold and connect your vacuum pump to the hose. Apply 17 in. Hg to 20 in. Hg of vacuum. The gauge reading on the pump should hold steady. If it drops, the booster or the hose is leaking.

Then hold the 17 in. Hg to 20 in. Hg of vacuum and have an assistant apply and hold the brake pedal for 30 seconds. Vacuum should drop to no less than 6 in. Hg when the brakes are first applied and should leak down no more than another 2 in. Hg during 30 seconds.

Booster Overhaul
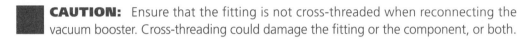

Classroom Manual
pages 137–145

Special Tools

Booster holding
  fixture

With a proper overhaul kit, a typical tandem booster can be overhauled (Figure 6–12) by following manufacturer's instructions. Detailed steps will vary from one vehicle to another, but most overhaul procedures follow a basic sequence.

☑ **SERVICE TIP:** The vacuum boosters are usually replaced rather than rebuilt in the shop. This saves time and money for the shop and the customer. However, replacement costs for larger vehicles may be expensive, and, therefore, it may be more feasible economically to rebuild.

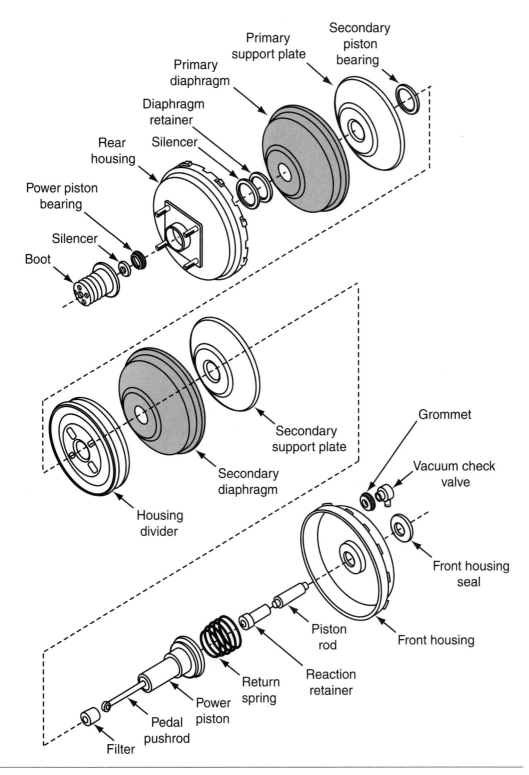

Figure 6-12 Typical tandem-diaphragm vacuum booster.

CAUTION: The return spring in the booster is strong, and the booster should not be separated unless it is mounted in a clamp-type holding device. Damage and injury could result if the spring is released suddenly and without control.

On passenger cars and light trucks, vacuum boosters are usually not overhauled but instead are replaced. Like the master cylinder, however, there may be a reason to actually overhaul a vacuum booster. This type of repair will be few and far between in most shops. Should it be necessary to overhaul a vacuum booster, ensure that a service manual is present and that the instructions are followed exactly.

There are some repairs that can be performed on a hydraulic booster like the one used in hydro-boost systems, but these repairs usually consist of just replacing some seals instead of an overhaul.

Vacuum Booster Pushrod Length Check

Classroom Manual
page 131–132,
143–145

Proper adjustment of the master cylinder pushrod is essential for the safe operation of vacuum power brake systems. If the pushrod is too long, the master cylinder piston will block the compensating port, preventing the hydraulic pressure from being released and resulting in brake drag.

If the pushrod is too short, the brake pedal will be low and the pedal stroke length will be reduced, resulting in a loss of braking power. When the brakes are applied with a short pushrod, groaning noises may be heard from the booster.

During assembly, the pushrod is matched to the booster. It is normally adjusted only when the vacuum booster or the master cylinder is serviced.

The pushrod length is checked by observing fluid action at the master cylinder compensating ports when the brakes are applied. Remove the master cylinder cover and have an assistant apply the brake pedal. Observe the fluid reservoirs. A small ripple or geyser in the reservoirs should be visible as the brakes are applied. If there is no turbulence, loosen the bolts securing the master cylinder to the booster about ⅛ inch to ¼ inch and pull the cylinder forward, away from the booster. Hold it in this position and have the assistant apply the brakes again. If turbulence (indicating compensation) now occurs, the brake pedal pushrod or the booster pushrod needs adjustment.

Booster pushrod adjustment is usually checked with a gauge, which measures from the end of the pushrod to the booster shell. Two basic gauge designs are used: Bendix and Delco-Moraine (Delphi Chassis). Be certain the pushrod is properly seated in the booster when making the gauge check.

Bendix Pushrod Gauge Check

Bendix pushrod
gauge

Bendix vacuum boosters are used on most Ford brake systems, as well as on vehicles from other manufacturers. Check and adjust pushrod length as follows with the booster installed on the vehicle and vacuum applied:

1. Disconnect the master cylinder from the vacuum booster housing, leaving the brake lines connected. Secure or tie up the master cylinder to prevent the lines from being damaged.

2. Start the engine and let it run at idle.

3. Place the gauge over the pushrod and apply a force of about 5 pounds to the pushrod (Figure 6-13). The gauge should bottom against the booster housing.

4. If the required force is more or less than 5 pounds, hold the pushrod with a pair of pliers and turn the self-locking adjusting nut with a wrench until the proper 5 pounds of preload exists when the gauge contacts the pushrod.

5. Reinstall the master cylinder.

6. Remove the reservoir cover and observe the fluid while your assistant applies and releases the brake pedal. If the fluid level does not change, the pushrod is too long. Disassemble and readjust the rod length.

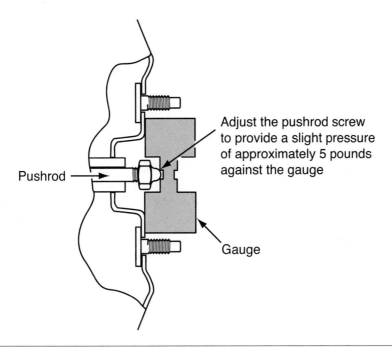

Adjust the pushrod screw
to provide a slight pressure
of approximately 5 pounds
against the gauge

Pushrod

Gauge

Figure 6-13 Bendix pushrod adjustment gauge.

Delco-Moraine Pushrod Gauge Check

On most Delco-Moraine (or Delphi Chassis) brake systems, used primarily by GM, the master cylinder pushrod length is fixed. If the pushrod length needs to be adjusted after master cylinder or booster service, an adjustable pushrod must be installed. Check the pushrod length with the booster on or off the vehicle as follows:

Special Tools

Delco-Moraine
    pushrod gauge

1. With the pushrod fully seated in the booster, place the go/no-go gauge over the pushrod (Figure 6-14).

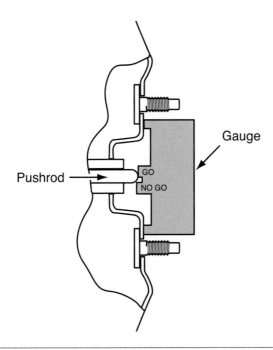

Gauge

Pushrod

GO
NO GO

Figure 6-14 Delco-Moraine pushrod adjustment gauge.

2. Slide the gauge from side to side to check the pushrod length. The pushrod should just touch the longer, no-go, area and just miss the shorter, go, area.

3. If the pushrod is not within the limits of the gauge, replace the original pushrod with an adjustable one. Adjust the new pushrod to the correct height.

4. Install the vacuum booster and check the adjustment. The master cylinder compensating port should be open with the engine running and the brake pedal released.

✔️ **SERVICE TIP:** You can check master cylinder pushrod clearance by putting a small ball of putty or modeling clay on the end of the pushrod from the vacuum booster and then bolting the master cylinder to the booster. Then remove the master cylinder and measure the putty or clay thickness and compare to manufacturer's specifications if available. It is usually about 0.015 inch of clearance.

Without proper pushrod clearance, the master cylinder can be held partially applied and cause the brakes to drag. Adjust the pushrod or shim the master cylinder away from the booster with cork gasket material.

Adjusting the Booster Pushrod on a Honda

If necessary, remove the master cylinder from the vehicle. Install the special tool onto the master cylinder (Figure 6-15). Use the adjusting nut to move the tool's center shaft until it contacts the primary piston of the master cylinder. Install the special tool onto the booster without moving the center shaft. Use the master cylinder attaching nuts to hold the tool in place. Tee in a vacuum gauge into the engine or booster vacuum line. Operate the engine to a continuous 20 in. Hg (66 kPa) during the adjustment phase (Figure 6-16).

✔️ **SERVICE TIP:** If the engine cannot achieve 20 in. Hg (66 kPa) of vacuum, disconnect the vacuum line from the engine, and with an appropriate adapter(s), attach a hand-operated vacuum pump. Operate the hand pump until 20 in. Hg (66kPa) of vacuum is achieved. Observe the gauge to ensure that the vacuum is holding. If vacuum cannot be maintained, then there is a problem with the pump connection or the diaphragm in the booster.

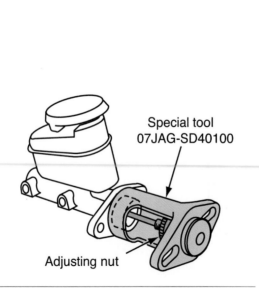

Figure 6-15 Install Honda's special tool, 07JAG-SD40100, into the master cylinder.

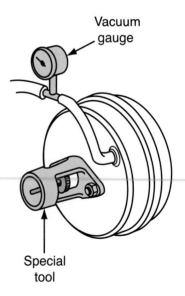

Figure 6-16 Tee a vacuum gauge into the engine or booster hose and run the engine at a speed that will provide 20 in. Hg (66 kPa) of vacuum.

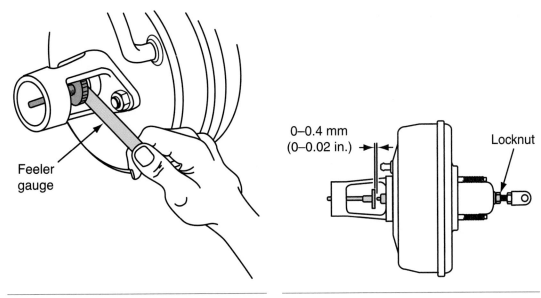

Figure 6-17 Measure the gap between the adjusting nut and the tool body.

Figure 6-18 Loosen the locknut and turn the adjuster to gain the proper clearance.

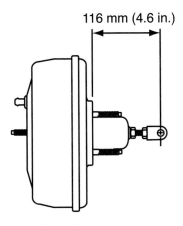

Figure 6-19 If the booster has not been installed on the vehicle yet, measure the pushrod length and adjust as needed. This procedure will not work with the booster installed on the vehicle.

With the tool in place and sufficient vacuum supplied, use a feeler gauge to measure the gap between the adjusting nut and the gauge body (Figure 6-17). If the gap is not between 0 mm and 0.4 mm (0 inch and 0.02 inch), then the pushrod must be adjusted. Locate the locknut on the pedal pushrod (opposite side of the booster) and turn the adjusting nut until the correct clearance is obtained (Figure 6-18). If the booster is off the vehicle, the length of the pushrod can be checked and adjusted as needed (Figure 6-19).

Auxiliary Vacuum Pumps

Diesel-fueled vehicles have an auxiliary pump to provide vacuum to operate the vacuum brake booster. These auxiliary vacuum pumps are usually the diaphragm or vane type. Auxiliary vacuum pumps for passenger cars are driven by an electric motor or by a belt, gear, or cam from the engine.

Classroom Manual
page 147–149

Electric Motor-Driven Pumps

A typical electric auxiliary vacuum pump consists of a small electric motor, a pump, and an electronic controller combined in a single assembly (Figure 6-20). The controller has an integral on–off switch, a timer relay, and a piston and umbrella valve assembly. Vacuum hoses connect the vacuum pump to the vacuum-operated components. The vacuum pump outlet hose is usually connected to the intake manifold. A charcoal filter located in the outlet line prevents fuel vapors from entering the vacuum pump.

The electric auxiliary pump is supplied with current whenever the ignition switch is on. Under low-vacuum conditions, the internal vacuum switch closes, completing the timer-relay circuit. This activates the vacuum pump motor, which generates auxiliary vacuum to help maintain 14.0 in. Hg of vacuum the brake booster requires. When 14.0 in. Hg of vacuum is reached, the internal vacuum switch opens. The timer relay continues to supply current to the pump motor for several seconds before turning off current to the pump motor. A low-vacuum warning switch in the inlet hose line activates an instrument panel warning lamp when a low-vacuum condition exists.

In addition to providing increased vacuum power for the brake system, an auxiliary vacuum pump provides extra vacuum for the heating and air-conditioning system, various emission controls, and cruise control.

Electric Pump Troubleshooting

CAUTION: Do not use test lights on electronic circuits unless instructed to do so by the manufacturer. A digital multimeter (DMM) should be used to check electronic circuits. Damage to the circuits could result from using improper test equipment.

Some larger utility trailers use vacuum over hydraulics for brakes. The vehicle towing this type of trailer may have a small electric vacuum pump to supply the vacuum necessary for controlling this brake system.

Special Tools

Digital multimeter

Service manual or Vehicle wiring diagram

Jumper wires

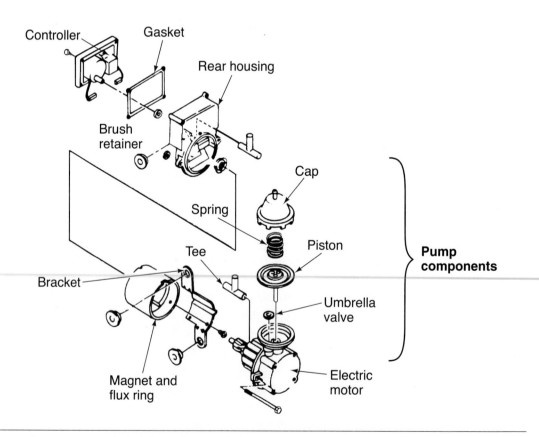

Figure 6-20 Exploded view of a typical electric vacuum pump.

Before troubleshooting the auxiliary vacuum pump, check the hydraulic system, engine vacuum system, and the brake booster. If these systems appear to be in order, check the condition of the vacuum hoses and the connections at the vacuum pump. Also check for loose mounting bolts, an improperly tensioned drive belt, or improper engine idle speed on mechanically operated vacuum pumps. On inoperative electrically operated vacuum pumps, check for power at the pump motor electrical connector to ensure the problem is in the pump motor, not the circuit to it.

The ports on a vacuum pump typically include the inlet port, a vacuum switch port, a pump housing port, and the pump outlet port. You can use a vacuum gauge (or the gauge on a hand-operated vacuum pump) for most diagnostic procedures. Connect the gauge to various ports on the auxiliary vacuum pump. Reading vacuum at these connections can pinpoint problems in the auxiliary pump.

Refer to the service manual electrical wiring diagrams to locate the power and ground terminals on the electric vacuum pump connector.

CAUTION: When testing a vacuum pump, do not block the outlet nozzle. Permanent damage to the diaphragm can result.

Diagnostic Step 1. If excessive brake pedal effort is needed, if the brake warning lamp is on, or if the vacuum pump is inoperative, troubleshoot the pump as follows:

1. Turn on the ignition. Disconnect the electrical connector from the vacuum pump. Use a digital voltmeter to check for battery voltage across the appropriate connector pins. Also check for proper ground. If battery voltage is present at the connector, proceed to step 2. If there is no power, check the fusible link, fuse, and wiring circuits and repair as needed.

2. Connect a 12-volt battery directly to the appropriate connector terminals and ground the circuit via the correct terminal pin. If the pump is still inoperative, proceed to step 3. If the pump now operates, go on to diagnostic step 2.

3. Remove the vacuum pump. Disassemble and check for sticking or shorted brushes. Check the controller for damaged wires and repair or replace as needed.

Diagnostic Step 2. If the vacuum pump operates but the brake warning lamp still lights and excessive brake pedal effort is needed, troubleshoot as follows:

1. Remove the pump inlet hose and connect a vacuum gauge (or pump) to the fitting (Figure 6-21). Turn on the ignition.

2. If the gauge indicates a vacuum of 10 in. Hg to 15 in. Hg and the electric pump runs for 5 to 10 seconds, check for vacuum leaks in other parts of the system. If the pump runs intermittently and fails to maintain steady vacuum, proceed to step 3. If the pump runs continously, go to diagnostic step 3.

3. Remove the auxiliary vacuum pump from the vehicle. Remove the controller and vacuum connector from the pump. Connect a hand-operated vacuum pump to the vacuum switch inlet port and apply approximately 20 in. Hg of vacuum. If the switch leaks more than 2 in. Hg per minute, replace the controller. If the switch holds vacuum, go to step 4.

4. Connect the hand-operated vacuum pump to the pump inlet. Apply 20 in. Hg of vacuum. If the pump leaks more than 2 in. Hg per minute, the pump umbrella valve is leaking and must be replaced. If the pump holds vacuum, proceed to step 5.

5. Connect the hand-operated vacuum pump to the pump outlet. Plug the pump housing inlet and apply 20 in. Hg of vacuum. If the pump leaks more than 2 in. Hg per minute, check for a loose bonnet. Repeat the test. If the pump continues to leak, replace the piston. If the pump holds vacuum, the system is operating properly.

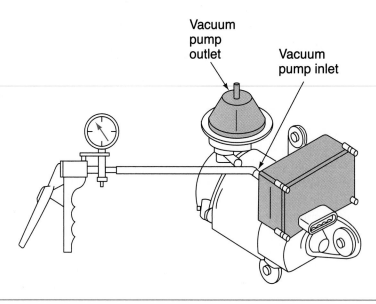

Vacuum
pump
outlet

Vacuum
pump inlet

Figure 6-21 Testing an electric vacuum pump.

Diagnostic Step 3. If the vacuum pump runs continuously with normal or excessive brake pedal effort and the brake warning light is on or off, troubleshoot as follows:

1. Check all vacuum hoses for leaks and repair or replace as needed. If no leaks are found, go to step 2.

2. With the ignition off, disconnect the inlet hose from the vacuum connection. Attach the hand-operated vacuum pump and turn on the ignition. If the pump runs then stops when the vacuum reaches 10 in. Hg to 15 in. Hg, the system is normal. Bleed off the vacuum to below 10 in. Hg. At levels below 10 in. Hg, the pump should start, run until the vacuum exceeds 10 in. Hg to 15 in. Hg, and automatically shut off. If the pump runs continuously and draws a vacuum of 10 in. Hg or greater, the controller is defective and should be replaced.

Electric Pump Component Service

Special Tools

Lift or jack with
stands

Remove and replace various electric pump components as follows.

Vacuum Pump. If necessary for access to the pump, raise the vehicle on a lift or safety stands. Remove any protective shields for access to the pump. Carefully disconnect the electrical connector and vacuum hoses from the pump. Remove all retaining nuts and carefully remove the pump.

 CAUTION: When using a jack always block the wheels that remain on the floor. The vehicle and jack could roll causing damage to the vehicle and/or jack.

WARNING: Do not work under or on any vehicle supported only by a jack. Always use a jack and safety stands to support the vehicle during repairs. Injuries could occur if the vehicle slips from the jack or the jack rolls.

Controller. To remove the controller, first remove the vacuum pump. Next, remove all protective shields from around the controller and remove the fasteners that hold the rear housing and pump housing together. Carefully separate the rear housing from the pump housing and remove the fasteners securing the controller to the rear housing. Remove the brushes from the brush holders; then remove the controller, gasket, and washer.

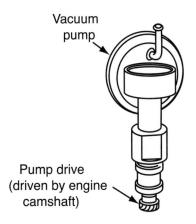

Figure 6-22 A cam-driven vacuum pump.

Install the controller using a new gasket and washer. Place both brushes in the rear brush holder cavity. Preload the brush springs by locking them in the slots provided above the spring access slots. Position the brushes in channels. Install the brush retainer in the proper position and return the springs to the load position. Install the rear housing on the pump housing, replace all controller shields, and reinstall the vacuum pump.

Piston Assembly. To remove and replace the pump piston assembly, remove the pump upper and lower shields. Carefully release the tabs of the cap; then remove the cap and spring.

CAUTION: The bonnet is spring loaded. The bonnet and pump housing must be held together during disassembly or component damage may occur.

Remove the piston from the housing and the umbrella valve. Replace the umbrella valve, piston assembly, and spring. Place the bonnet over the spring and compress the spring and bonnet onto the pump housing. Bend the bonnet tabs around the pump housing, checking to be sure that the bonnet is secure. Reinstall the shields and install the pump.

Belt-Driven Vacuum Pumps

If you suspect that a belt-driven pump is performing poorly, check for worn or misadjusted belts or pulleys. Connect a vacuum gauge to the pump inlet and plug the outlet port. The gauge should read approximately 15 in. Hg after running the engine at idle for about 30 seconds. If the pump is overly noisy or does not operate to specifications, replace it.

Special Tools

Vacuum gauge
Belt tension gauge

Gear-Driven Vacuum Pumps

Gear-driven vacuum pumps are usually driven by a camshaft gear in the engine (Figure 6-22). They are diaphragm pumps that require no maintenance. Replace the entire pump if a vacuum reading of approximately 20 in. Hg cannot be attained at the pump with the engine running at idle.

Special Tools

Vacuum gauge

Hydro-Boost Power Brakes

A hydro-boost power brake system consists of the booster assembly, the accumulator, and the power steering fluid circuit (including the pump and reservoir). The hydro-boost is mounted in the engine compartment the same as a vacuum booster (Figure 6-23). The pedal pushrod is connected at the booster input rod end. The master cylinder is bolted to the other end of the hydro-boost and is operated by a pushrod projecting from the booster cylinder bore.

Classroom Manual
pages 149–155

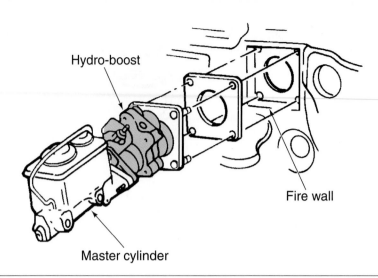

Hydro-boost

Fire wall

Master cylinder

Figure 6-23 Typical hydro-boost installation.

Power steering fluid is delivered to the booster from the power steering pump at 100 psi to 150 psi (Figure 6-24). The flow of fluid inside the booster is controlled by a hollow-center **spool valve.** The fluid presses against the rear of the power piston, increasing the application force. When the brake pedal is released, pressure in the power cavity is released. A return spring pushes the power piston back to the unapplied position.

The hydro-boost accumulator holds a charge of pressurized fluid as a reserve if fluid flow from the steering pump is lost or reduced. Inside the accumulator, the fluid compresses an internal piston and spring or a gas charge. If steering fluid pressure is lost, the force of the compressed

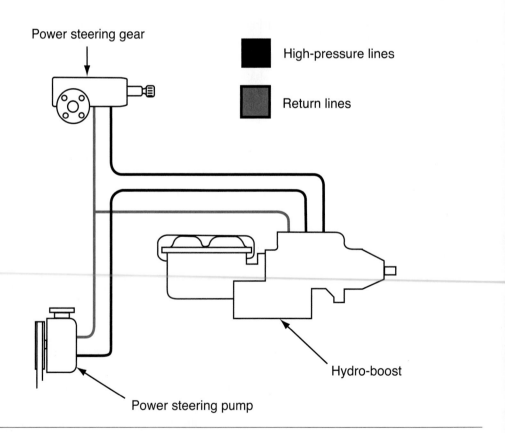

Power steering gear

High-pressure lines

Return lines

Hydro-boost

Power steering pump

Figure 6-24 The hydro-boost system shares hydraulic power with the power steering system.

spring or gas charge forces the pressurized fluid out of the accumulator to assist in applying the brakes. Depending on the system, one to three power applications are possible after power steering fluid pressure is lost.

> **CAUTION:** Use the recommended power steering fluid in the boost system circuit of a hydro-boost system. The proper fluid is essential to system operation. Do not mix brake fluid with power steering or other hydraulic fluids. The use of improper fluids will damage the gaskets, O-rings, and other seals. Keep dirt out of the hydro-boost system.

Hydro-Boost System Inspection

Before any detailed testing of hydro-boost components, check basic engine conditions and power steering operation that could affect hydro-boost performance. If both brake application and steering require more than normal effort, the cause is probably related to fluid pressure and delivery.

Special Tools

Belt tension gauge

> **CAUTION:** Always clean around any lines or covers before removing or loosening them. Dirt or other contaminants will void the warranty and may damage the system components.

Follow these steps to inspect the hydro-boost system:

1. Inspect fluid level in the power steering pump. Some pump reservoirs have dipsticks marked to check the fluid only when at normal operating temperature. Others have dipsticks with fluid level markings for both warm and cool fluid. Follow the carmaker's instructions to check fluid level accurately and use only the type of fluid specified for the vehicle.

2. Also inspect the condition of the power steering fluid. If it is dirty or smells burned or appears to be contaminated in any way, flush and refill the power steering system before proceeding further.

3. Inspect the condition of the power steering pump drive belt and replace it if it is cracked, glazed, grease soaked, or otherwise badly worn. Check the tension of a V-belt with a strand tension gauge or be sure that a serpentine belt is installed properly and correctly positioned on the belt tensioner.

4. Inspect all hoses and steel lines in both the hydro-boost system and the power steering system for leakage.

> **CAUTION:** Do not hold the steering at full lock for more than 5 seconds while testing system pressure. System damage may result from prolonged high-pressure operation.

5. To verify a leak, have an assistant run the engine at fast idle and alternately apply the brakes and turn the steering from lock to lock. These actions develop high pressure and will force fluid from small leaks. Tighten connections or replace lines and hoses as required.

6. If you find leakage around the pump, clean and tighten all fittings and bolts. If the leak continues, rebuild or replace the pump. Figure 6-25 shows common leak locations in a power steering pump and in a steering gear.

7. Check brake fluid level in the master cylinder and add fluid if necessary. If air has entered the brake system and given the pedal a spongy feel, it will be difficult or impossible to troubleshoot hydro-boost operation accurately. Bleed the brake system if you suspect air has entered the lines and cylinders.

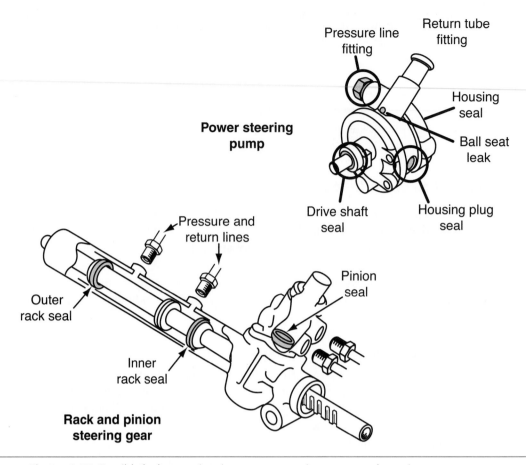

Figure 6-25 Possible leakage points in a power steering pump and steering gear.

8. Check engine idle speed and adjust it if necessary. Also check for engine speed control by the powertrain control module (PCM) as you turn the steering to full lock. If the engine does not drive the power steering pump at the required speed, the pump will not develop full pressure.

✓ **SERVICE TIP:** Many vehicles have a power steering switch mounted either on the pump's outlet or in the high-pressure line. This switch sends a signal to the vehicle's engine controller under full-lock steering or when the power steering pressure reaches a preset limit. Failure of this switch could cause engine stalling or low idle during maneuvers like parking. If there is a problem with steering pump speed or output during high-pressure tests, check the power steering switch.

Power Steering Operation and Hydro-Boost Service

Special Tools

Power steering
   pressure gauge

The hydro-boost system requires a continuous supply of power steering fluid at the proper pressure and volume from the power steering pump. In addition to the inspection of power steering components listed previously, you may need to test the pressure developed by the pump. If so, follow the instructions of the pressure-test equipment manufacturers. Power steering pressures are listed in vehicle service manuals.

The power steering pump should not be operated without fluid in the reservoir. The pump bearings and seals can be damaged. If the pump has failed, check the fluid carefully for abrasive dirt and metal particles that can damage the hydro-boost. If you find such particles in the fluid from any source, flush the system thoroughly. Depending on the severity of the contamination, you may need to remove the booster and disassemble it to clean it completely. Also, flush the power steering and hydro-boost lines thoroughly to remove particulate contamination.

Booster Fluid Leakage

Figure 6-26 shows possible leakage points at seals in the hydro-boost. If the booster is leaking at any of these points, if is often most practical to replace the booster with a new or rebuilt unit. All of these seals can be replaced individually, however, with seal repair kits. Most require that the booster be removed from the vehicle and disassembled. The spool valve plug seal and the accumulator seal can be replaced with the booster installed in the vehicle if component access permits. If leakage appears around the return port fitting, torque the fitting to 7 foot-pounds (10 Nm). If leakage continues, replace the O-ring under the fitting.

Basic Operational Test

Check hydro-boost basic operation in two steps as follows:

1. With the engine off, pump the brake pedal repeatedly to bleed off the residual hydraulic pressure stored in the accumulator.

2. Hold firm pressure on the brake pedal and start the engine. The brake pedal should move downward, then push up against your foot.

Accumulator Test

 WARNING: Block the wheels when power testing the brakes. If the foot slips from the pedal or the brakes fail, injury could occur.

Test accumulator operation as follows:

1. With the engine running, rotate the steering wheel until it stops and hold it in that position for no more than 5 seconds.

2. Return the steering wheel to the center position and shut off the engine.

3. Pump the brake pedal. You should feel two to three power-assisted strokes.

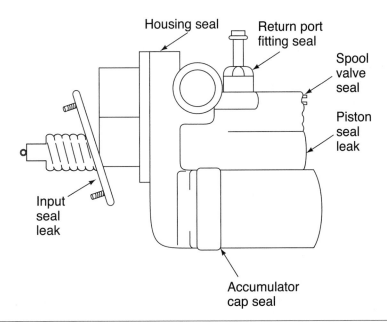

Figure 6-26 Possible leakage points in a hydro-boost booster.

4. Now repeat step 1 and step 2. This will pressurize the accumulator.

5. Turn the engine off and wait 1 hour; then pump the brake pedal. There should be one or two power-assisted strokes.

Bad valves are the most common accumulator problem. If the valves are leaking, the accumulator may hold a charge for only a short period or may fail to hold a charge at all. In either case, the booster must be disassembled and the valves replaced.

Noise Troubleshooting

Because hydro-boost is part of another major system in the vehicle—the power steering system—problems in the steering system can affect the booster, and a problem in the booster can affect the steering.

Certain noises often occur with the hydro-boost system and may be cause for customer complaint. Some noises are normal and usually occur for a short time. Other noises are a sign of wear in the system or the presence of air in either the hydro-boost or the steering system.

Servicing the Hydro-Boost

Classroom Manual
pages 149–155

Special Tools

C-clamp

Vise

Torque wrench

The hydro-boost can be overhauled or rebuilt in the field. Repair kits and service parts are available. In many cases, however, it is more practical to replace the booster with a new or rebuilt unit.

Booster Removal

Remove a hydro-boost by following these general steps. Details will vary for different vehicle installations.

⚠️ **WARNING:** When fully charged, the accumulator holds more than 1,000 psi of hydraulic pressure. Before removing the hydro-boost, discharge the accumulator by firmly applying the brake pedal at least six times with the engine off. Pedal application force should increase as the accumulator discharges. Failure to discharge the accumulator could cause serious injury.

1. Before removing a hydro-boost from a vehicle, turn the engine off and pump the brake pedal several times to exhaust accumulator pressure.

2. Disconnect the master cylinder from the booster, but leave the service brake hydraulic lines connected to the master cylinder.

3. Carefully lay the master cylinder aside, being careful not to kink or bend the steel tubing. Support the master cylinder from a secure point on the vehicle with wire or rope. Do not support the master cylinder on the brake lines.

4. Disconnect the hydraulic hoses from the booster ports. Plug all tubes and the booster ports to prevent fluid loss and system contamination.

5. Detach the pedal pushrod from the brake pedal. Remove the nuts and bolts from the booster support bracket and remove the booster from the vehicle.

Hydro-Boost Air Bleeding

Classroom Manual
pages 149–155

Whenever the hydro-boost or power steering components are removed and reinstalled, the hydraulic system must be bled of all air. Air also can enter the hydro-boost system if the power steering fluid level drops below the minimum safe level in the pump reservoir, and air can be drawn into the fluid through loose fittings in hydraulic lines and hoses.

Check the steering fluid for signs of air. Aerated fluid looks milky. The level in the steering fluid reservoir will also rise when the engine is turned off if air has been compressed in the fluid. If the fluid has air in it that cannot be purged using the following procedure, the problem may lie in the steering system pump. Refer to the vehicle service manual for further troubleshooting instructions on the power steering system.

1. Fill the power steering pump reservoir to the full mark and allow it to sit undisturbed for several minutes.

2. Start the engine and run it for approximately 1 minute.

3. Stop the engine and recheck the fluid level. Repeat step 1 and step 2 until the level stabilizes at the full mark.

4. Raise the front of the vehicle on a hoist or safety stands.

5. Turn the wheels from lock to lock. Check and add fluid if needed.

6. Lower the vehicle and start the engine.

7. Apply the brake pedal several times while turning the steering wheel from lock to lock.

8. Turn off the engine and pump the brake pedal five or six times.

9. Recheck the fluid level. If the steering fluid is extremely foamy, allow the vehicle to stand for at least 10 minutes with the engine off. Then repeat step 7 through step 9 until fluid in the pump reservoir is clear and free of air bubbles.

Special Tools

Lift or jack with stands

Large catch basin

Coworker

Sufficient power steering fluid

Servicing Vacuum Boosters on Vehicles with EHB or Vehicle Stability Assist (VSA)

More vehicles are being equipped with either EHB or VSA or some version in between. Also, EHB and VSA usually utilize an ABS. In most cases, diagnosing and/or replacing the boosters follows the same procedures except for some additional steps. Most of those additional steps concern the electronics associated with EHB and VSA. This section highlights some of those steps. As always the best guide is the service manual.

 AUTHOR'S NOTE: There are several names used for stability systems. Honda has the VSA and DaimlerChrysler has the Electronic Stability Program (ESP).

The Honda S2000 vacuum booster is shown in Figure 6-27. This booster is common to the standard, ABS, and VSA systems, and it is not particularly difficult to replace. The DaimlerChrysler booster shown in Figure 6-28 is more representative of the booster configuration used with newer

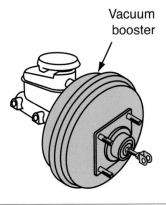

Vacuum booster

Figure 6-27 The booster on this Honda is part of a VSA system.

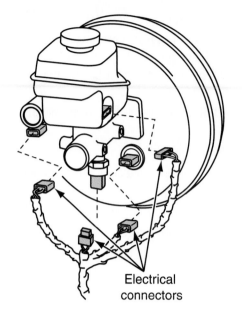

Electrical
connectors

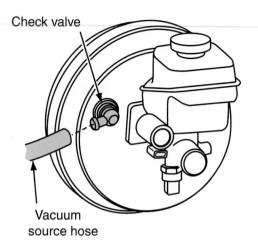

Check valve

Vacuum
source hose

Figure 6-28 Before removing the booster fasteners, disconnect all of the ESP electrical connectors on the booster and master cylinder.

Figure 6-29 Do not remove the check valve from this type of booster. Remove the hose from the check valve instead.

ABS and ESP/VSA systems. As mentioned before, some ESP/VSA systems have no vacuum booster at all. Instead they use the computer-controlled hydraulic modulator as a booster. This section discusses the DaimlerChrysler vacuum booster with ESP.

The testing of the booster is done in the same manner as testing a typical vacuum booster. With the engine off, pump the brake pedal several times until it is firm and high. Hold the pedal down and start the engine. The pedal should fall away, stopping about 2 inches from the floorboard. If it does not respond accordingly, refer back to the vacuum booster testing and follow those guidelines.

Replacing the vacuum booster in an ESP requires some additional steps. Disconnect and isolate the negative cable from the battery. Remove the windshield wiper module and components to gain access to the booster.

Disconnect the electrical connections at the booster and remove the master cylinder. Move the master cylinder back from the booster. Do not bend or damage the brake lines. Disconnect the vacuum hose from the check valve, but do not remove the check valve from the booster (item 2, Figure 6-29).

> ⚠ **WARNING:** Before working in or around the steering column, ensure that the air bag system has had time to discharge. Failure to properly disarm the air bag system could result in serious injury.

Move inside the passenger compartment, and, if sufficient time has elapsed for the air bags to disarm, disconnect and remove the stoplamp switch (item 2, Figure 6-30). The switch will be replaced with a new one upon installation of the booster. Use a screwdriver to remove the retaining clip from the booster pushrod, and slide the pushrod from the pedal pin (items 4 and 5, Figure 6-30). Remove the booster's four mounting nuts, and remove the booster from the engine compartment.

Before installing the new booster, ensure that a new booster seal is present on the bulkhead side of the booster (Figure 6-31). Slide the booster into place through the bulkhead and tighten the four mounting nuts to specifications. Position the booster pushrod over the pedal pin and install

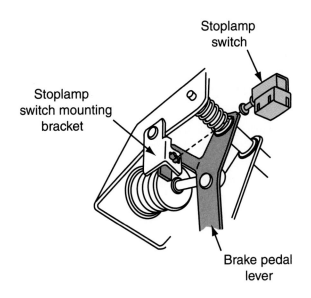

Stoplamp
switch

Stoplamp
switch mounting
bracket

Brake pedal
lever

Figure 6-30 Remove the stoplamp switch, the pushrod retaining clip, and the pushrod from the pedal.

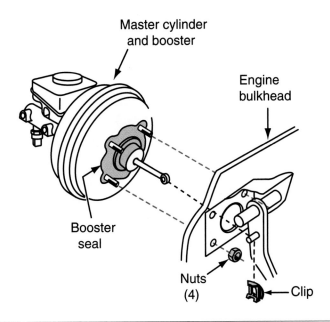

Master cylinder
and booster

Engine
bulkhead

Booster
seal

Nuts
(4)

Clip

Figure 6-31 Always install a new booster seal before positioning the new booster in place on the bulkhead. Reverse the removal procedures to complete the installation.

a new retaining clip. Install and adjust the new stoplamp switch. Under the hood, install the master cylinder onto the booster and reconnect all electrical connections. Install the wiper module and other removed components. Connect the battery and road test the vehicle.

Servicing a PowerMaster System

One quirk associated with a PowerMaster system is the method to check the fluid level in the booster's reservoir. This reservoir is under the master cylinder reservoir cover, but it is partitioned off from the two master cylinder reservoirs. If the accumulator is fully charged, then the reservoir appears to be nearly empty. This is the correct level in this condition. To properly check the level, pump the brake pedal at least ten times with the engine off. This will discharge the accumulator, and the fluid will return to the reservoir. With the accumulator completely discharged, check and top off the fluid level as needed. The operational test of the system is done in a similar manner to a booster supplied by an electric auxiliary vacuum pump. Depress the pedal several times until the accumulator is exhausted, and switch the ignition switch to RUN. The pump should switch on and the pedal should fall away as the pressure is increased and stabilized. If it fails to function correctly, check the power supply to the pump and the pressure switch for operation. If the fluid level is correct, power is reaching the pump, and the switch is functioning properly, then the unit must be rebuilt or replaced. Like many similar units, it is usually more economically feasible to replace the unit, although some components can be replaced individually (Figure 6-32).

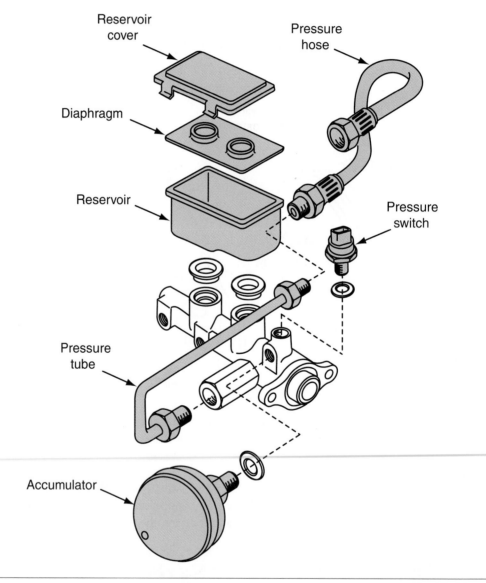

Figure 6-32 The highlighted parts can be replaced without removing the PowerMaster assembly from the vehicle.

The problem with a 1995 imported SUV was that the brakes would barely work first thing in the morning. After the vehicle was warmed up for a few minutes and driven about a half mile, the brakes started working normally. The owner suspected a stuck master cylinder, an ABS problem, or some similar "high-tech" cause. The tech working on the car took a simpler approach.

After the vehicle sat outside overnight, the technician started the engine and removed the vacuum line from the power brake booster. Engine speed jumped up several hundred rpm. The tech then shook the hose and blew through it several times. The engine speed then dropped back to normal.

On many imported vehicles, the vacuum check valve for the power booster is in line in the hose, not in the fitting on the booster. On this Isuzu, the valve was sticking open when cold. A few minutes of driving and some engine heat would get it working again. Replacing the hose with a new OEM unit that contained a new check valve solved the problem.

Terms to Know

Pressure differential   Spool valve   Vacuum suspended

ASE-Style Review Questions

1. *Technician A* says that if a vacuum booster is in good condition, starting the engine after the vacuum is exhausted should cause the brake pedal to drop slightly under foot pressure.
   Technician B says that the pedal should pulsate lightly after it drops.
   Who is correct?
   A. A only
   B. B only
   C. Both A and B
   D. Neither A nor B

2. The brake pedal of a high-mileage car is sluggish and moderately hard to apply:
   Technician A says that vacuum leaks or restrictions may exist between the engine and the booster.
   Technician B says that mechanical wear or damage to the engine may have reduced the available vacuum.
   Who is correct?
   A. A only
   B. B only
   C. Both A and B
   D. Neither A nor B

3. *Technician A* says that a diesel-powered vehicle may have an auxiliary vacuum pump for the power brake booster.

Technician B says that a diesel-powered vehicle may have a hydraulic power brake booster.
Who is correct?
A. A only
B. B only
C. Both A and B
D. Neither A nor B

4. *Technician A* says that a defective vacuum check valve in the power brake booster may cause a hard-pedal problem.
   Technician B says that a vacuum check valve may not be installed in a vacuum booster.
   Who is correct?
   A. A only
   B. B only
   C. Both A and B
   D. Neither A nor B

5. *Technician A* says that the check valve in the booster is used to keep atmospheric air pressure from leaking from the booster when the brakes are applied.
   Technician B says that the check valve is used to keep vacuum from being lost from the booster when the engine is turned off.
   Who is correct?
   A. A only
   B. B only
   C. Both A and B
   D. Neither A nor B

6. *Technician A* says that any condition that reduces engine vacuum may also reduce vacuum booster performance.
   Technician B says that a blocked passage in the power piston, a sticking air valve, and a broken piston return spring are all causes of hard pedal.
   Who is correct?
   A. A only
   B. B only
   C. Both A and B
   D. Neither A nor B

7. *Technician A* says that with the engine off and residual vacuum in the vacuum booster, pressing and holding the brake pedal should result in a slight pedal drop after a short time.
   Technician B says that there should never be even the slightest drop in pedal when this test is performed.
   Who is correct?
   A. A only
   B. B only
   C. Both A and B
   D. Neither A nor B

8. *Technician A* says that whenever a hydro-boost is serviced the power steering system must be bled.
   Technician B says that power steering fluid and brake fluid are interchangeable in the hydro-boost system in small amounts.

Who is correct?
   A. A only
   B. B only
   C. Both A and B
   D. Neither A nor B

9. A customer complains of a hard brake pedal feel during parking or low speed operation:
   Technician A says that the power steering fluid may be low.
   Technician B says that the power steering pump may be weak.
   Who is correct?
   A. A only
   B. B only
   C. Both A and B
   D. Neither A nor B

10. *Technician A* says that the accumulator must be discharged prior to removing a PowerMaster brake booster.
    Technician B says that the master cylinder can be serviced in a PowerMaster brake system.
    Who is correct?
    A. A only
    B. B only
    C. Both A and B
    D. Neither A nor B

ASE Challenge Questions

1. Vacuum boosters are being discussed:
   Technician A says that a reading of 16 in. Hg on an auxiliary pump may cause the brake pedal to have a soft feel.
   Technician B says that a poorly performing engine could indicate the booster's air valve is bad.
   Who is correct?
   A. A only
   B. B only
   C. Both A and B
   D. Neither A nor B

2. *Technician A* says that excessive braking effort could indicate air is leaking into the rear chamber of the booster on a vacuum-suspended booster.
   Technician B says that a continuous air leaking sound heard in the passenger compartment is a sign that the vacuum-suspended booster diaphragm is holed.
   Who is correct?
   A. A only
   B. B only
   C. Both A and B
   D. Neither A nor B

3. A vehicle has no brake boost until the engine has run for a few seconds:
   Technician A says that the vacuum check valve is leaking.

   Technician B says that the engine is just worn out and cannot hold a vacuum during shutdown.
   Who is correct?
   A. A only
   B. B only
   C. Both A and B
   D. Neither A nor B

4. *Technician A* says that the tandem-diaphragm booster will have no loss of boost even if one diaphragm is holed.
   Technician B says that a hard feel to the pedal is common when driving in traffic and a series of quick braking applications is executed.
   Who is correct?
   A. A only
   B. B only
   C. Both A and B
   D. Neither A nor B

5. *Technician A* says that a vehicle with hydro-boost may have a hard pedal if the pressure switch is open.
   Technician B says that the PowerMaster system may lose boost if the drive belt is slick or loose.
   Who is correct?
   A. A only
   B. B only
   C. Both A and B
   D. Neither A nor B

Job Sheet 19

19

Name: _____ Date: _____

Vacuum Booster Testing and Diagnosis

NATEF Correlation

This job sheet addresses the following NATEF tasks:

Inspect the vacuum-type power booster unit for vacuum leaks; inspect the check valve for proper operation; determine necessary action.

Objective

Upon completion of this job sheet, you should be able to properly test and diagnose a vacuum booster system.

Tools and Materials

Basic hand tools

Describe the Vehicle Being Worked On

Year _____ Make _____ Model _____

VIN _____ Engine type and size _____

ABS equipped? _____ EHB/VSA/ESP _____

Procedure

Task Completed

1. Inspect the vacuum booster for obvious damage. Describe any damage to the booster, master cylinder, and lines or hoses.

   _____

   _____ ☐

2. With the engine off, pump the pedal several times to exhaust the power booster reserve. Does the pedal feel become firm and high?

   _____

   _____ ☐

3. With the pedal depressed, start the engine. What pedal action was observed? How much clearance is now available between the pedal and the floorboard?

   _____

   _____ ☐

4. Release the pedal, pause for a few seconds, and then reapply the brakes. What is the pedal feel at this point? How much clearance is between the pedal and the floorboard? Was there any air (hissing) noise as the pedal was applied?

   _____

   _____ ☐

☐

5. Is the vacuum booster operating properly? If not, go to step 6.

_____

_____

☐

The student may make a diagnostic decision on the vacuum at steps 7, 8, or 9.

6. Turn off the engine.

7. Disconnect the hose or valve directly from the vacuum booster body. Did a hissing sound occur as the hose or valve was disconnected? If yes, what repairs are indicated? If not, go to step 8.

_____

☐

_____

8. Connect a vacuum gauge to the outlet side (booster end) of the valve and start the engine. Record the gauge reading, if any. Is a full-engine vacuum (15–20 in. Hg) available at this connection? If yes, what repairs are indicated? If not, go to step 9.

_____

☐

_____

9. Remove the booster hose from the engine intake nipple and connect the vacuum gauge in its place. Is a full-engine vacuum (15–20 in. Hg) available at this connection? If yes, what repairs are indicated?

_____

☐

_____

10. Based on your observations in steps 7, 8, and 9, what is the condition of the booster system? Are repairs indicated? What are they?

_____

☐

_____

Problems Encountered _____

_____

Instructor's Response _____

_____

_____

Job Sheet 20

20

Name: _____ Date: _____

Replace a Vacuum Booster

NATEF Correlation

This job sheet addresses the following NATEF tasks: Inspect the vacuum-type power booster unit for vacuum leaks; inspect the check valve for proper operation; determine necessary action; measure and adjust master cylinder pushrod length.

Objective

Upon completion and review of this job sheet, you should be able to replace a vacuum brake booster.

Tools and Materials

Service manual

Hand tools

Measuring device as specified by service manual

Describe the Vehicle Being Worked On

Year _____ Make _____ Model _____

VIN _____ Engine type and size _____

ABS _____ yes _____ If yes, type _____

Procedure

Task Completed

1. Pushrod adjustment? _____ Explain procedures _____

_____

Booster fastener(s) torque _____

Master cylinder fastener(s) torque _____ ☐

⚠ **WARNING:** Disable the supplemental inflatable restraint system (SIRS) or air bag system when working on or near the system wiring. Accidental discharge of the air bag could cause injury.

■ **CAUTION:** Before working on the brakes of a vehicle with an ABS, consult the service manual for precautions and procedures. Failure to follow procedures to protect ABS components during routine brake work could damage the components and cause expensive repairs.

2. Disarm SIRS (air bag) if equipped. ☐

3. Disconnect stoplight switch. ☐

4. Disconnect brake pedal pushrod from pedal. ☐

Task Completed

_____

> ■ **CAUTION:** Always clean around any lines or covers before removing or loosening them. Dirt or other contaminants will void the warranty and may damage the system components.

☐   **5.** Remove booster fasteners from under the dash if necessary.

☐   **6.** Disconnect vacuum hose and electrical connections, as needed, from booster.

☐   **7.** Remove the master cylinder from the booster fasteners.

> ■ **CAUTION:** Do not twist or bend the hydraulic lines leading to the master cylinder. Most master cylinder lines can flex enough for the master cylinder to clear the booster without disconnecting the lines. Violent jerking motions could kink or twist a brake line and prevent full brake application.

☐   **8.** Move the master cylinder away from the booster.

☐   **9.** Remove booster fasteners as needed.

☐   **10.** Remove the booster from the vehicle.

> ■ **CAUTION:** Do not remove protective shipping seals, covers, or plugs before installing the device. Dirt and other contaminants may enter and damage the system components or void warranty.

☐   **11.** Install the new booster and torque fasteners.

☐   **12.** Reinstall the master cylinder and torque fasteners.

☐   **13.** Connect electrical connections and the vacuum hose.

☐   **14.** Connect the pushrod to the brake pedal.

☐   **15.** Measure and adjust the pushrod to manufacturer's specifications. Explain the procedures.

☐   **16.** Connect the stoplight switch.

17. Start engine and test brake booster. Explain the results. _____

_____

_____

☐   **18.** Stop the engine and pump the brake pedal to exhaust the booster.

19. Hold the brake pedal down and start the en gine. The pedal should fall away but retain firmness without being spongy. Results _____

_____

20. Test the stoplights. Results _____

_____

NOTE: A test drive may be necessary to test the operation of the brake switch on vehicles equipped with antilock brakes or cruise control. If necessary, do so with caution.

> ▲ **WARNING:** Before moving the vehicle after a brake repair, pump the pedal several times to test the brake. Failure to do so may cause an accident with damage to vehicles or the facility, or personal injury.

WARNING: Road test a vehicle under safe conditions and while obeying all traffic laws. Do not attempt any maneuvers that could jeopardize vehicle control. Failure to adhere to this precaution could lead to serious personal injury and vehicle damage.

21. Arm the air bag system as needed. ☐

22. When the repair is complete, clean the area, store the tools, and complete the repair order. ☐

Problems Encountered _____

_____

Instructor's Response _____

_____

_____

Job Sheet 21

Name: _____ Date: _____

Testing Electric Motor Vacuum Pumps

NATEF Correlation

This job sheet addresses the following NATEF task: Check vacuum supply (manifold or auxiliary pump) to vacuum-type power booster.

Objective

Upon completion of this job sheet, you should be able to perform operational and electrical tests on an electric motor vacuum pump.

Tools and Materials

Service manual

Wiring diagrams

Component location

Hand tools

Digital multimeter

Describe the Vehicle Being Worked On

Year _____ Make _____ Model _____

VIN _____ Engine type and size _____

ABS _____ yes _____ If yes, type _____

Procedure

Task Completed

1. Location of pump _____

Fuse location and size _____ Pump's wiring colors _____

_____

_____

Does the system have an external vacuum sensor or switch? _____

If yes, give location and test procedures. _____

_____ ☐

2. Describe the test procedures for determining if boost is being applied. _____

_____

_____ ☐

NOTE: KOEO is key on, engine off.

3. Perform test(s) from step 2 and record results. _____

_____

_____ ☐

☐

4. Determine if the pump is operating.

Results _____

_____

If the pump is operating, do not complete this job sheet. Perform vacuum tests instead.

☐

5. Use the multimeter to test the pump's fuse. There should be 12 volts on each side of the fuse in KOEO or the removed fuse should show continuity.

Results _____

_____

☐

6. If equipped with an external vacuum switch, test the switch using the following routine.

Unplug harness from switch. In KOEO, use the multimeter to test for 12 volts on the conductor from the ignition switch.

Results _____

_____

☐

If voltage is present, use the multimeter to check for continuity through the switch.

Results _____

_____

☐

7. Use the multimeter to test the pump's supply voltage (12 volts) at the pump.

Results _____

_____

☐

8. Use the multimeter to test the pump's ground (continuity with key off).

Results _____

_____

☐

9. Based on the above results, determine and record the needed repairs. _____

_____

_____

_____

☐

10. Consult the instructor for additional instructions.

☐

11. When the repair is complete, clean the area, store the tools, and complete the repair order.

Problems Encountered _____

_____

Instructor's Response _____

_____

Disc Brake Service

Upon completion and review of this chapter, you should be able to:

- ❑ Diagnose disc brake problems, including poor stopping, pulling, or dragging caused by problems in the caliper, the caliper installation, the hydraulic system, or the rotor.

- ❑ Remove, inspect, and replace brake pads.

- ❑ Remove and replace a caliper.

- ❑ Overhaul a caliper, including disassembly, inspection, adjustment, replacement, and reassembly of all parts.

- ❑ Clean a disc brake installation using vacuum or aqueous cleaning equipment.

- ❑ Remove and replace brake rotors.

- ❑ Inspect and measure rotors for wear.

- ❑ Machine a rotor to correct dimensions and finish on a brake lathe.

- ❑ Reinstall wheels, torque lug nuts, and make final brake system checks and adjustments.

Service Precautions

Specific CAUTIONS and WARNINGS are given throughout this chapter where needed to emphasize safety. The following general precautions apply to many different disc brake service operations and are presented at the beginning of this chapter to highlight their overall importance.

- ❑ When servicing disc brakes, never use an air hose or a dry brush to clean the brake assemblies. Use OSHA-approved cleaning equipment to avoid breathing brake dust. See Chapter 1 in this *Shop Manual* for details on working safely around airborne asbestos fibers and brake dust.
- ❑ Do not spill brake fluid on the vehicle; it may damage the paint. If brake fluid does contact the paint, wash it off immediately. To keep fluid from spraying or running out of lines and hoses, wrap the fittings with shop cloths when disconnecting them.
- ❑ Always use the DOT type of brake fluid specified by the vehicle manufacturer. Do not mix different brands of brake fluid.
- ❑ During servicing, keep grease, oil, brake fluid, or any other foreign material off the brake linings, calipers, surfaces of the rotors, and external surfaces of the hubs. Handle brake rotors and calipers carefully to avoid damaging the rotors or nicking or scratching brake linings.

Basic Tools

Basic technician's tool set

Rotor micrometer

Dial indicator

Diagnosing Disc Brake Problems

Diagnosing has always been a key to good first-time automotive repair, but it has become an increasingly important function as the vehicle systems become more interlocking and electronically controlled or actuated. Poorly performing disc brakes usually result from worn brake pads or other parts, poorly fitted or incorrectly assembled parts, or rotor problems, such as grooving, distortion, or grease and dirt on the rotor surface. Worn pads increase the braking effort needed to stop the vehicle, but the same problem can be caused by a sticking or sluggish caliper piston. Installing the wrong type of brake pad can result in brake fade.

A lot of technical effort, knowledge, and skill are covered by the single word "inspect" on a vehicle maintenance schedule as it relates to the brake system. "Inspect" means different things to different people; but in any case, it means more than a quick glance at the calipers,

Classroom Manual
pages 163–169

hoses, lines, and master cylinder fluid level. If problems are suspected in the disc brakes, road test the vehicle. Instructions for a safe, complete road test are given in Chapter 7 of this manual. Even driving across the driveway and into the service bay can reveal a lot about brake system condition.

As the pedal is applied, check for excessive travel and sponginess. Listen for noises: not just the obvious sounds of grinding pads or pad linings, but mechanical clanks, clunks, and rattles. A vehicle that pulls to one side when the brakes are applied may have a bad caliper or loose caliper at one wheel. Grease or brake fluid may have contaminated the pads and rotor, or the pad and lining may be bent or damaged. Grabbing brakes also may be caused by grease or brake fluid contamination or by a malfunctioning or loose caliper. Worn rotors or pads also may cause roughness or pedal pulsation when the brakes are applied.

✅ **SERVICE TIP:** A customer may complain of a clicking noise that occurs every time the brakes are applied or released. Sometimes it happens in forward braking or in reverse braking, but it is noticeable only at slow speeds. This noise can occur when the pad hardware is either missing or damaged and the pads are moving back and forth in the caliper. Check the history of the vehicle for recent brake repairs and then check the brake caliper for the correct hardware.

For a complete inspection, the wheels must be removed for a clear view of the brake pads and caliper mounting hardware. Some manufacturers, such as Ford, say to lubricate the caliper slides. Now is a good time to do it; and on any vehicle, it is a good idea to verify smooth operation of the calipers and lubricate if necessary.

Inspect the wheel and brake assembly for obvious damage that could affect brake system performance. Check the following:

1. Tires for excessive or unusual wear or improper inflation
2. Wheels for bent or warped rims
3. Wheel bearings for looseness or wear
4. Suspension system for worn or broken components
5. Improper mounting of caliper and rotor
6. Brake fluid level in the master cylinder
7. Signs of leakage at the master cylinder, in brake lines or hoses, at all connections, and at each wheel

If the vehicle is driven in an area where salt is used on the road during the winter, it is very important to check brake assemblies frequently. Salt water and rust damage caliper slides and pad-mounting hardware. Cars with rear disc brakes often have complicated parking brake mechanisms, so inspect them closely; lubricate where necessary.

✅ **SERVICE TIP:** Most brake squeal occurs under light (not hard) braking at low to moderate (not high) speeds. Sometimes brakes will squeal when stopping in reverse but not when going forward. When troubleshooting brake noise complaints, check brake operation carefully under all moderate driving conditions.

✅ **SERVICE TIP:** When you are troubleshooting a brake-pull problem with disc brakes, remember to check the caliper slides (ways) or pins for burrs, gouges, rust, or any other problem that could keep the caliper from moving properly. If damaged or defective caliper mountings delay or interfere with caliper movement, the vehicle will pull to the opposite side as that brake applies first. Loose mounting bolts on a caliper support can cause similar brake-pull problems.

Inspecting Brake Pads

Disc brake pad linings should be inspected regularly at the time or mileage intervals recommended by the carmaker. Manufacturers' recommendations vary from as often as every 7,500 miles to as seldom as every 30,000 miles. Most pad inspection recommendations, however, seem to be at 12,000-, 15,000-, or 30,000-mile intervals or every 12 to 24 months. A convenient time to check pad wear is when the wheels are removed for rotation, which is typically every 5,000 to 7,000 miles. Because of the wide range of carmakers' inspection recommendations, always check the preventive maintenance schedule for the vehicle being serviced.

> ■ **CAUTION:** Older brake linings and some aftermarket linings, discs, and drums may still contain asbestos fibers. The fibers may cause serious eye and breathing injuries.

Follow these basic steps for any disc brake pad inspection:

1. Raise the vehicle on a hoist or safety stands. Be sure it is properly centered and secured on the stands or hoist.

2. Mark the relationship of the wheel to the hub to ensure proper wheel balance on reassembly. An easy way to do this is to make a chalk mark on one wheel stud and on the wheel next to that stud. If the rotor and the hub are a two-piece floating assembly, chalk mark the rotor and the hub after removing the wheel. The chalk marks ensure that all rotating parts are indexed to each other and help to avoid wheel balance problems.

3. Remove the wheel and tire from the brake rotor. Be careful not to hit the brake caliper, the rotor splash shield, and the steering knuckle or suspension parts.

4. If the rotor and hub are a two-piece assembly, reinstall two wheel nuts to hold the rotor on the hub.

5. On most disc brakes, the pads can be inspected without removing the calipers. Check both ends of the outboard pad by looking in at each end of the caliper. As shown in Figure 7-1, these are the areas where the highest rate of pad wear occurs. Also check the lining thickness on the inboard pad to be certain it has not worn prematurely. If the caliper has an opening in the top, look through it to view the inboard pad and lining (Figure 7-2). Some calipers do not have such openings, and the center areas of the pads cannot be inspected without removing the caliper.

Classroom Manual
pages 174–182

Special Tools

Lift or jacks with
  stands
Impact tools
Chalk

As the disc brakes are inspected, observe the bushings and linkages of the steering and suspension mechanism. Worn steering or suspension may cause a noise or pull as the brakes are applied or released.

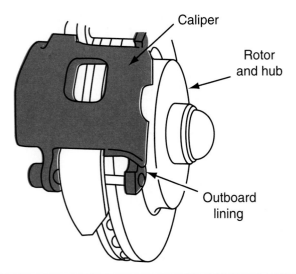

Caliper

Rotor
and hub

Outboard
lining

Brake pads

Figure 7-1 Inspect both ends of the outboard pad by looking in at each end of the caliper. If the caliper has an opening at the top, check the inboard and outboard pads.

Figure 7-2 Most calipers have an opening in the top of the casting that provides a good view of both pads.

SERVICE TIP: Many pads have a vertical groove about midway the length of the pad (see Figure 7-5). When checking the pads through the caliper opening, the groove is visible. If pad wear is down to or near the bottom of the groove, then it is time for new pads.

6. On vehicles with floating or sliding calipers, check for uneven wear on the inboard and outboard linings. If the inboard pad shows more wear than the outboard pad, the caliper should be overhauled or replaced. If the outboard pad shows more wear, the sliding components of the assembly may be sticking, bent, or damaged. In any case, uneven brake wear is a sign the pads or calipers need service.

7. To inspect the lining surfaces, remove the pads from the calipers. This is not usually necessary as part of routine brake inspection; but if the customer complains of poor brake performance, inspect the lining surfaces regardless of thickness. Replace the pads if the linings are glazed (shiny and smooth), heat damaged, cracked, or contaminated with dirt or brake fluid.

If the customer complains that the brakes are making a high-pitched squeal, immediately suspect an audible brake pad wear indicator, signaling that the system needs service (Figure 7-3). Instrument panel warning lamps also are indicators that the pads have worn past specifications.

✓ **SERVICE TIP:** When the pads wear to the point that the sensor contacts the rotor, the brake warning lamp on the instrument panel lights (Figure 7-4). After the pads are replaced, the lamp may stay lit until the car has been driven about a quarter mile. The electronic control unit simply has to "learn" that the worn pads have been replaced before it will turn off the lamp.

CAUTION: Use extreme care when repairing brakes equipped with electronic wear indicators. Use only the specified tools and procedures to prevent damage to the components.

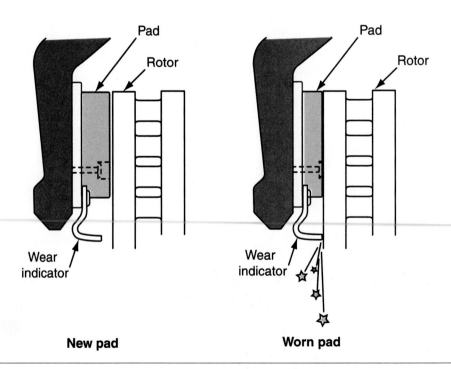

Figure 7-3 As the pad linings wear, the wear indicator eventually hits the rotor and makes noise to warn the driver that new pads are needed.

Figure 7-4 This lamp indicates that the brake pads should be checked for wear and replaced as needed.

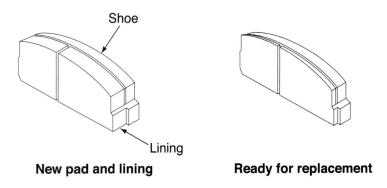

New pad and lining   Ready for replacement

Figure 7-5 Any lining worn to the thickness of the metal backing pad needs replacement.

All service manuals specify a minimum lining thickness, but it can be measured accurately only if the pads are removed from the calipers. Most carmakers specify a minimum lining thickness of $\frac{1}{32}$ inch. Ford, however, wants you to change brake pads when the lining is $\frac{1}{8}$ inch thick or less. By any measure, $\frac{1}{32}$ inch may be the minimum thickness to prevent rotor scoring, but braking efficiency is dramatically reduced with worn linings. Worn linings cannot dissipate heat adequately, and the last $\frac{1}{32}$ inch of lining will wear and peel off much faster than the first $\frac{1}{32}$ inch of fresh linings. When visually evaluating pad lining wear, consider that any lining worn to the thickness of the metal backing pad needs replacement (Figure 7-5).

✓ **SERVICE TIP:** If a customer complains about noisy disc brakes, compare the pads installed on the car to the OEM pads. The grooves and chamfers on OEM pads are often there to reduce squeaks and squeals. "Economical" aftermarket pads often do not include these features. OEM pads may cost a bit more, but they can be the cure for a customer complaint.

Disc Brake Service Operations

Disc brake service consists of these three general operations:

1. Pad replacement
2. Caliper overhaul or replacement
3. Rotor resurfacing when necessary

Classroom Manual
pages 169–182

Pad replacement is the most frequent disc brake service because pads are intended to wear as they operate. Calipers may require overhaul to replace worn parts or fix leaks and corrosion. Calipers need not be overhauled at every pad replacement, however, if they are in good shape. Years ago, rotors were routinely resurfaced at every brake job. Lighter rotors on late-model vehicles do not have the mass of iron found in rotors of the 1960s and 1970s. General practice today is to resurface rotors if they are grooved, scored, pulsating, or otherwise badly worn or if runout and parallelism measurements are out of limits. Rotors cannot be resurfaced beyond minimum thickness specifications, which are discussed later in this chapter.

Figure 7-6 Remove some brake fluid from the master cylinder reservoir. This reduces the chance of an overflow if fluid is forced back into the reservoir when the piston is forced back into its bore.

Vehicle Preparation

Special Tools

Lift or jack with
  stands

Syringe

Impact tools

Brake service
  cleaning equipment

Tubing with catch
  basin

For pad, caliper, or rotor service, raise the vehicle on a hoist or safety stands, remove the wheels, and disassemble the brakes to the extent necessary for the planned service. Follow these four general preparation steps for all disc brake service:

1. Use a brake fluid siphon, or syringe, to remove approximately two-thirds of the brake fluid from the front or disc brake reservoir on a front-rear split system (Figure 7-6). On a diagonally split system, remove about half the fluid from both master cylinder reservoirs.

2. Raise the vehicle on a hoist or safety stands and support it safely.

3. Remove the wheels from the brakes to be serviced. Brakes are always serviced in axle sets, so remove both front wheels, both rear wheels, or all four.

4. Vacuum or wet-clean the brake assembly to remove all dirt, dust, and fibers.

Removing brake fluid from the master cylinder is an important preliminary step for all disc brake service. If fluid is not removed, it may overflow the reservoir and spill as the caliper pistons are moved back into the caliper bores for caliper or pad removal. After siphoning fluid from the reservoir, replace the reservoir cover and safely discard the removed brake fluid.

The best method to prevent reservoir overflow is to open the caliper bleeder screw and run a bleeder hose into a container to catch the fluid expelled when the piston is forced back into its bore. Opening the bleeder screw also makes it easier to move the piston. This is the recommended way to push pistons back on any brake system but is especially important on ABS, to prevent damaging the hydraulic valve body. It is still good practice, however, to remove some fluid from the reservoir even when the bleeder screw is opened at the caliper.

Loaded Calipers

Loaded calipers are rebuilt calipers that come with brake pads and mounting hardware fully installed (Figure 7-7). They eliminate the need to overhaul calipers and prevent many of the errors commonly committed when performing caliper service. These mistakes include forgetting

Figure 7-7 A loaded caliper is a rebuilt caliper with new hardware, seals, and pads.

to bend brake pad locating tabs that reduce vibration and noise, leaving off antirattle clips and pad insulators, or reusing worn or corroded mounting hardware that limits caliper movement and reduces pad life. Loaded calipers provide an attractive alternative to pad replacement and caliper overhaul for disc brake service on many vehicles.

Use of loaded calipers reduces labor time and ensures that all components that should be replaced are replaced. However, be sure the loaded calipers have quality pad materials. Calipers should also be matched side to side with the same type of friction materials. When one caliper is bad, both should be replaced with the same type of loaded caliper.

Whether or not to use loaded calipers for a particular brake job is as much a business decision as a technical decision. If a shop's labor rate is high, loaded calipers may actually be more economical for the customer than the labor costs of overhauling the original calipers. In addition, loaded calipers usually have a manufacturer's warranty for all materials and labor that went into the rebuilding.

However, not all calipers for all vehicles are available as rebuilt, loaded assemblies. The job of a skilled brake technician still includes the ability to overhaul a caliper thoroughly and properly. Many shop owners also prefer to overhaul all calipers for complete brake jobs so that they know the condition of all calipers and the quality of the parts and materials used in their service.

If you use loaded calipers for brake service on a particular vehicle, do not overlook the necessary inspection, cleaning, and lubrication of the caliper supports on the vehicle suspension. Also do not overlook brake rotor inspection and service. Rotors may require resurfacing if grooved or warped or replacement if worn past their discard dimensions. Even the best loaded calipers cannot operate properly if installed on damaged or rusted mounting hardware or if they are forced to operate with defective rotors.

✓ **SERVICE TIP:** Incorrectly installed clips, springs, and other hardware used to hold a disc brake pad to the caliper or the piston often cause a low brake pedal and the inability to bleed the hydraulic system properly. If the hardware is installed incorrectly, the pads may appear to fit and may stay in place but be out of position. Out-of-position pads often cause the caliper and piston to travel farther than normal to apply the brakes. This requires a greater-than-normal fluid volume in the caliper and more pedal travel to supply the extra fluid. All the air may be removed from the system by normal bleeding, but no amount of bleeding will cure the low pedal until the pads are installed correctly.

Brake Pad Replacement for Floating or Sliding Calipers

Classroom Manual
pages 174–186

Special Tools

Lift or jack with
  stands

Hand tools

C-clamp or small
  prybar

If one pad is worn
much more than
its mate, suspect a
sticking piston or
slide mechanism
on the caliper.

Replace pads and linings in axle sets only. An axle set contains four pads: the inner and outer pads for the caliper at each wheel. Do not try to interchange pads between right and left calipers to equalize wear. Even if only one pad in a set of four is badly worn, replace all four pads after fixing the problem that caused the uneven wear.

To replace brake pads, raise the vehicle on a hoist or safety stands and prepare it as explained previously in this chapter. The details of pad replacement vary with the design of a particular caliper, but some general steps are common to all calipers. Photo Sequence 12 shows a typical replacement procedure for floating caliper disc brake pads. Figure 7-8 is an exploded view of a typical floating caliper disc brake, showing the pad installation. If unfamiliar with a particular caliper design, work on one side of the car at a time. The assembled caliper on the opposite side will be a guide for correct installation of pads and other parts.

✔ SERVICE TIP: Disc brake rotor and pad diagnoses complement each other. The pads cannot be evaluated separately from the rotor condition and vice versa. For example, if used pads are worn pretty thin and seem to be harder than most, check the rotors carefully for hot spots and checking. Hard, thin pads can act like cutting tools on the rotors.

Pad Removal

On most calipers, it is necessary (or at least convenient) to push the piston into the caliper bore to provide clearance between the linings and the rotor and ease caliper removal. Before retracting the caliper piston, install a hose on the caliper bleeder screw and place the other end into a container partially filled with clean brake fluid. Then loosen the bleeder screw to relieve hydraulic pressure and prevent the return of dirty fluid back to the master cylinder during piston retraction. Verify that some brake fluid has been removed from the master cylinder to make room for any fluid that may be forced back from the retracting piston. If fluid has not been removed, do so now.

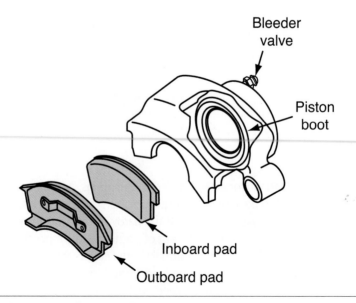

Figure 7-8 Exploded view of a typical floating caliper and pads.

Photo Sequence 12
Typical Procedure for Replacing Brake Pads

P12-1 Begin front brake pad replacement by removing brake fluid from the master cylinder reservoir, or reservoirs, for the disc brakes. Use a syphon to remove about one-half of the fluid.

P12-2 Raise the vehicle on the hoist, making certain it is positioned correctly. Mark and remove the wheels.

P12-3 Inspect the brake assembly, including the caliper, brake lines and hoses, and rotors. Look for fluid leaks, broken or cracked lines or hoses, and a damaged brake rotor. Fix any problems found before replacing the pads.

P12-4 Loosen and remove the caliper mounting pins or bolts.

P12-5 Lift and rotate the caliper assembly up and off of the rotor.

P12-6 Remove the old pads from the caliper.

P12-7 To avoid damaging the caliper, suspend it from the underbody with a strong piece of wire.

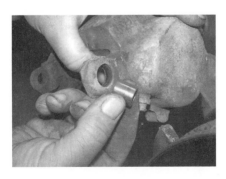

P12-8 Check the condition of the caliper mounting pin insulators and sleeves. Clean and lubricate with brake lubricant.

P12-9 Install a hose from the bleeder screw to the container. Open the bleeder screw.

P12-10 Use a C-clamp and wood to force the piston back into its bore. Just before the piston bottoms, close the bleeder screw.

P12-11 Close the bleeder screw just before the piston bottoms in the bore. Remove the C-clamp and check the piston boot. Install the new pads. If needed, install new mounting pins and sleeves.

P12-12 Set the caliper with its new pads onto the rotor and install the mounting pins. Check the assembly for proper position. Torque the pins to specifications.

P12-13 Install the tire and wheel and torque to specifications. Then press slowly on the brake pedal several times to set the brakes.

⚠ **WARNING:** Brake fluid may be irritating to the skin and eyes. In case of contact, wash skin with soap and water or rinse eyes thoroughly with water.

■ **CAUTION:** Prevent brake fluid from coming in contact with the vehicle's finish. Brake fluid damages paint and finish immediately on contact. If fluid contacts the finish, wash area thoroughly with running water using soap if possible.

⚠ **WARNING:** Wear safety glasses or face protection when using brake fluid. Injuries to the face and/or eyes could occur from spilled or splashed brake fluid.

■ **CAUTION:** Do not depress the brake pedal when the pads are being removed, installed, or have been removed from the caliper. Without the pads positioned properly, the piston could be forced from the bore when the pedal is depressed.

Several methods may be used to retract caliper pistons. One method is to install a large C-clamp over the top of the caliper and against the back of the outboard pad (Figure 7-9). Slowly tighten the clamp to push the piston into the caliper bore far enough so that the caliper can be lifted off the rotor easily.

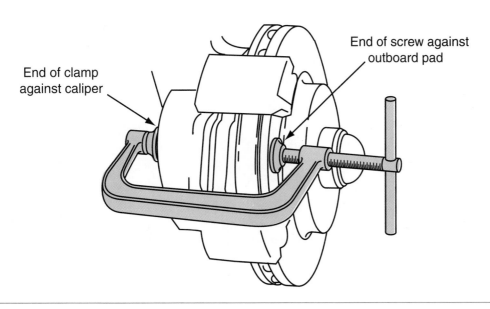

Figure 7-9 Use a C-clamp to retract the piston into the caliper bore.

On some calipers, the piston can be forced back in its bore with a prybar or a large screwdriver placed between the inboard pad lining and the rotor prior to removal (Figure 7-10). On other calipers, large slip-joint pliers placed over the inboard side of the caliper and a tab on the inboard pad can be used to squeeze the piston into its bore (Figure 7-11). Care must be taken to prevent cocking the piston. On older, fixed-caliper brakes, prying the pistons back into their bores

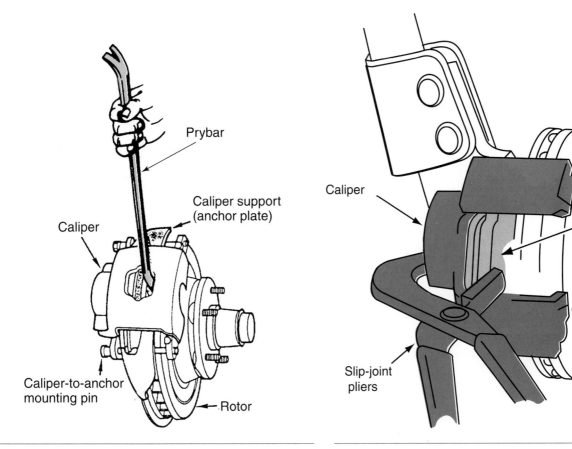

Figure 7-10 You can retract some caliper pistons with a prybar or large screwdriver.

Figure 7-11 You can retract other calipers by squeezing them with large slip-joint pliers.

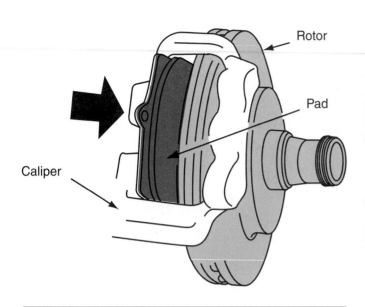

Figure 7-12 Pads in the fixed calipers in older Corvettes can be removed with the caliper in place on the car. The arrows are pointing to clips that hold the pistons retracted for pad replacement.

Figure 7-13 Some calipers can be rotated up on the upper mounting pin to replace the pads.

is the usual way to provide clearance for pad removal. Then the pads can be removed without demounting the caliper from the caliper support (Figure 7-12).

Some calipers can be rotated off the rotor on one guide pin, usually the top one (Figure 7-13). Remove the bottom pin and rotate the lower end of the caliper up. The pads can then be removed without completely removing the caliper from the vehicle. It is best, however, to either remove the upper pin or slide the caliper from the upper pin so the pin can be cleaned and lubricated.

Do not use a C-clamp, slip-joint pliers, or a screwdriver to retract the piston of a rear caliper with a parking brake mechanism that moves the piston to apply the parking brakes. Trying to force this kind of piston back into its bore against the parking brake mechanism usually damages the parking brake. Ford rear disc brakes of this type, for example, require a special tool to rotate the piston back into its bore on the automatic adjuster screw of the brake mechanism (Figure 7-14). If unfamiliar with a rear caliper parking brake, refer to the vehicle service manual before trying to retract the caliper piston. Another tool for retracting some rear caliper pistons is shown in Figure 7-15. The same tool can be used on front calipers.

Remove the caliper bolts and sleeves from a floating caliper or the support keys from a sliding caliper. Then remove the caliper from the caliper support. You do not need to disconnect the brake hose from the caliper if the caliper is not going to be removed from the vehicle. Suspend the caliper from the vehicle underbody or suspension with a heavy length of wire or rope. Do not let the caliper hang from the brake hose.

If the outboard pad is clipped to the caliper, pry it off with a screwdriver (see Figure 7-15) or tap it off with a small hammer. Inspect the pad hardware before removing the pads. The hardware must fit in a certain way and can be broken if forced. Also note the amount and type of hardware used. Many times the hardware consists of small pieces that can easily get lost if care is not exercised. Then remove the inner pad from the piston and the inner part of the caliper housing. Note the position of all springs, clips, shims, and other hardware used to attach the pads and to prevent noise. Inspect the mounting bolts and sleeves and other miscellaneous hardware for cor-

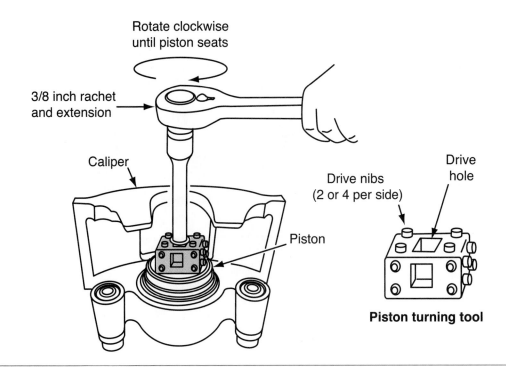

Figure 7-14 One type of piston retracting tool.

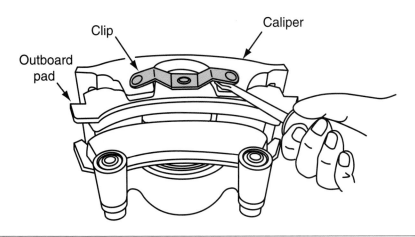

Figure 7-15 Some outboard pads can be pried off with a screwdriver.

rosion and damage. Inspect all rubber or plastic bushings for cuts and nicks. If any part is damaged, install new parts when the caliper is reinstalled. Do not try to polish away corrosion.

 SERVICE TIP: Tapered pad wear up to ⅛ inch is normal for some floating caliper disc brakes. Before installing new pads, compare wear on all four pads in an axle set.

Pad Installation

A complete set of springs, clips, shims, and other miscellaneous pieces may be available as a pad hardware kit for popular brake assemblies. Installing such a kit is often a practical and economical choice when replacing pads or doing other caliper service.

Special Tools

Vernier caliper or
precision scale

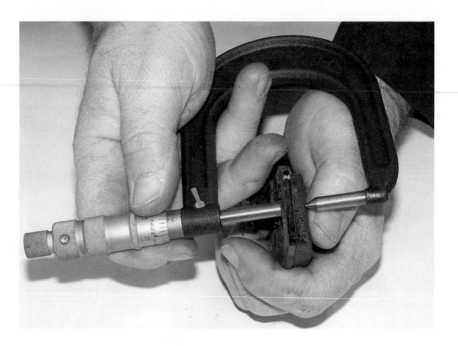

Figure 7-16 If necessary, measure the pad to determine if it is serviceable or to compare wear between different pads on the same axle.

If the pads appear serviceable, use a **vernier caliper** precision scale, outside micrometer, or other precision measuring instrument to measure the thickness of each brake pad lining (Figure 7-16). Compare the pad thickness against service manual specifications. For example, the standard brake pad thickness may be 0.50 inch (12.5 mm). The service limit may be 0.060 inch (1.6 mm). This measurement does not include the pad backing plate thickness. If the lining thickness is close to or less than the service limit, replace all pads on both calipers as a set. From a practical standpoint, if the old pads are removed, it is unlikely that they will be reinstalled unless they are almost brand-new and undamaged. If the pads are serviceable, however, they can be reinstalled in their original positions. Switching pad positions may reduce braking power. The pads must also be free of grease or brake fluid if they are to be reused. Replace contaminated brake pads and wipe any excess grease off the parts.

Measuring pad thickness precisely as shown in Figure 7-16 can sometimes be helpful when troubleshooting brake pulling problems or abnormal pad wear.

Before installing the new pads, wipe the outside of the piston dust boot with denatured alcohol. Use a C-clamp to bottom the piston in the caliper bore with the bleeder screw open, taking care not to damage the piston or the dust boot. Do not rush the piston. Apply slow steady force and the piston(s) will slide in easily without cocking or building excessive pressure. Excessive pressure will blow brake fluid out of the reservoir or capture container and onto the finish. When the piston is bottomed in the bore, lift the inner edge of the boot next to the piston and press out any trapped air. The boot must lie flat.

When specified in the vehicle service manual, apply any recommended noise suppression compound or shims to the backs of pads during reassembly to minimize noise (Figure 7-17). Some technicians prefer to use shims or compound on all pad installations. Others use shims or compound only when recommended by the pad manufacturer or carmaker. In all cases, shims and compound are antinoise devices and do not affect pad attachment to the caliper. Do not use both noise suppression compound and a shim on the same pad, unless directed by the manufacturer's service manual or bulletin.

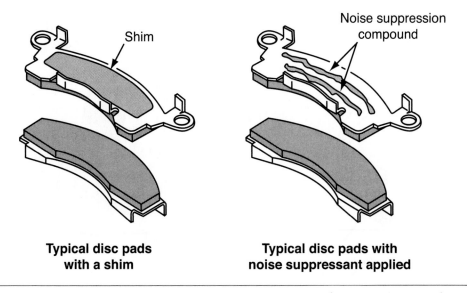

Typical disc pads with a shim

Typical disc pads with noise suppressant applied

Figure 7-17 You can use noise suppression compound or a soft shim on the back of a brake pad to help reduce noise, but do not use both on the same pad.

If the replacement pads have audible wear sensors, the pad must be installed so that the sensor is at the leading edge of the pad in relation to wheel rotation. This does not mean at the top or front of the caliper. The rotor must contact the edge of the pad that holds the sensor first as it rotates through the caliper; this is the pad leading edge. If the vehicle has electronic lining wear sensors that light a warning lamp on the instrument panel, install the pads and sensors according to the manufacturer's service manual procedures.

Install the inboard pad and lining by snapping the pad retaining spring into the piston's inside opening (Figure 7-18). The pad retainer spring is usually staked to the inboard pad. The pad must lie flat against the piston. After the pad is installed, verify that the boot is not touching the pad. If it is, remove the pad and reposition the boot.

Install the outboard pad in the caliper and secure any locking tabs as necessary. The exact methods and parts used for pad installation vary with brake design (Figure 7-19 and

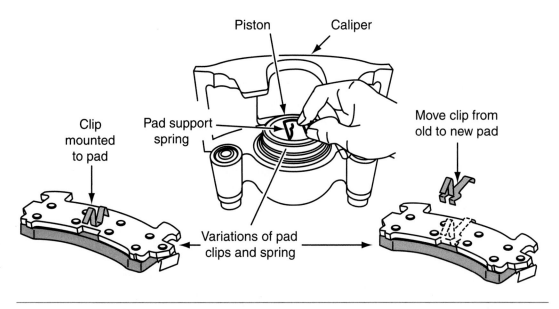

Figure 7-18 Many inboard pads have retainer springs that snap into the inner opening of the piston.

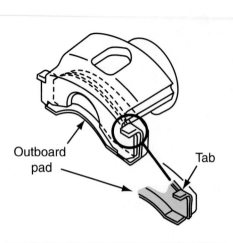

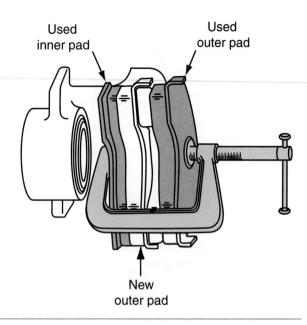

Figure 7-19 Many outboard pads have retaining flanges or tabs that snap around the outboard part of the caliper.

Figure 7-20 Some outboard pads can be snapped into place with a C-clamp and a couple of old pads or pieces of wood.

Figure 7-20). You may have to bend the retaining flanges or tabs of some pads to ensure secure attachment to the caliper (Figure 7-21). If you do not bend the retaining flanges, a rattle during braking may occur. If you service one side of the car at a time, the opposite caliper is available as a guide for pad installation.

On floating calipers, liberally coat the inside diameter of the bushings with brake lubricant before installing the mounting bolts and sleeves (Figure 7-22). For sliding calipers, lubricate the caliper ways on the caliper support and the mating parts of the caliper housing with the recommended lubricant (Figure 7-23). Caliper ways may be known as caliper slides. This is where a floating caliper slides on its support.

CAUTION: Before trying to move the vehicle, press the brake pedal several times to make sure the brakes work. Bleed the brakes as necessary and then road test the vehicle.

After the new pads are installed and the calipers remounted on the caliper supports, add fresh brake fluid to the master cylinder reservoir to bring it to the correct level. Then start the engine without moving the vehicle and apply the brake pedal. The pedal probably will go to the floor as the caliper pistons move out to take up the clearance between the new pads and the rotors. Do not pump the pedal fast when seating the pads. Slow, smooth pumps fill the caliper bores without agitation and tend to extend the piston more smoothly. Recheck the brake fluid level and add more fluid as needed. Apply the pedal several more times until it becomes firm and verify that the fluid is at the correct level. A four-wheel brake repair is an ideal time to flush the system. It is best to check for air anyway and a few minutes more will replace all of the old fluid.

Theoretically, if the brake lines were not disconnected and the pistons were not removed from the calipers, air will not have entered the system and bleeding should not be necessary. It is good practice, however, to bleed the brakes to ensure that the hydraulic system is free of air. If the pedal seems at all spongy or low after pad replacement, brake bleeding is required. If brake bleeding does not restore proper pedal action and brake performance, inspect the system thoroughly for leaks. The calipers may need to be removed and disassembled to check for corroded caliper bores and leaking seals that let air into the system.

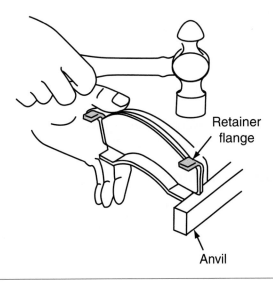

Figure 7-21 Bend the outboard pad retaining flange when required for secure attachment.

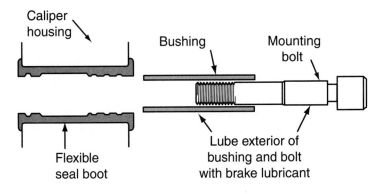

Figure 7-22 Lubricate the inside of the caliper bushing as highlighted here.

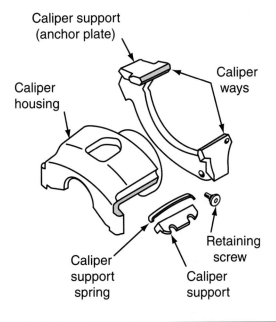

Figure 7-23 For a sliding caliper, lubricate the caliper ways and the mating areas of the caliper housing as highlighted here.

After proper pedal action is restored, recheck fluid level, turn on the ignition, and release the parking brake. Verify that the brake warning lamp on the instrument panel is not lit. If it is, adjust the pressure differential valve and switch as explained in Chapter 4 of this *Shop Manual.*

✓ **SERVICE TIP:** Avoid the temptation to improve pad cooling by cutting a groove in new pad linings that were made without a groove. The manufacturer knows best. It did not groove the pad lining for a reason. Cutting a groove in the field can weaken the lining material and actually cause overheating because of the reduced contact area.

Road Test and Pad Burnishing

Burnishing or **bedding in** is the process of applying friction materials to each other to create a desired wear pattern.

Whenever new brake pads are installed, they need a short period of controlled operation that is called a **burnishing** or **bedding in** period. Burnishing polishes the pads and mates them to the rotor finish. Road testing the vehicle performs this burnishing procedure and verifies that the brakes work properly.

⚠ **WARNING:** Road test a vehicle under safe conditions and while obeying all traffic laws. Do not attempt any maneuvers that could jeopardize vehicle control. Failure to adhere to this precaution could lead to serious personal injury.

New pads require burnishing to establish full contact with the rotor and to heat and cure any resin left uncured in the friction material. Whether the rotors were refinished or not, new pads do not initially make full contact with the rotor surfaces but require a period of light wear to establish this contact. Also, when brake linings are manufactured, some of the resin materials may remain uncured until the pads are put into service. If fresh pads are subjected to hard braking, the resins can boil to the surface of the pads and cause glazing when they cool. The pads then may never operate properly.

Burnish the brake pads during the initial road test by driving at 30 mph to 35 mph (50 kph to 60 kph) and firmly but moderately applying the brakes to fully stop the car. Do this five or six times with 20 to 30 seconds of driving time between brake applications to let the pads cool. Then drive at highway speeds of 55 mph to 60 mph (85 kph to 90 kph) and apply the brakes another five or six times to slow the car to 20 mph (30 kph). Again allow about 30 seconds of driving time between brake applications to let the brake pads cool. Finally, advise the customer to avoid hard braking for the first 100 miles of city driving or the first 300 miles of highway driving.

✓ **SERVICE TIP:** Excessive brake temperatures are usually associated with brake fade problems, but overheated brakes also can cause pulling or grabbing problems. For example, an overheated caliper piston can stick in its bore and cause the brake at one wheel to grab. Similarly, cheap brake linings of uncertain frictional coefficient can grab and pull when they get hot.

Disc Brake Cleaning

Classroom Manual
pages 188–193

Disc brakes stay much cleaner than drum brakes. A disc brake is partly shrouded by the wheel, but it is largely exposed to circulating air. Dust and dirt created by pad wear and accumulated through normal driving are thrown off by the centrifugal force of the spinning rotor. Some brake dust accumulates inside the wheel and partially on the caliper. Proper and safe cleaning of brake assemblies and components is as important for disc brakes as it is for drum brakes.

⚠ **WARNING:** Do not blow dust and dirt off brake assemblies with compressed air outside of a brake-cleaning enclosure. Airborne dust and possible asbestos fibers are an extreme respiratory hazard.

Vacuum enclosures and **aqueous** or water-based cleaning equipment are among the most popular for brake service and should always be used to clean brake assemblies and components.

General-purpose parts washers also can be used to clean brake parts after they are removed from the car.

Chapter 2 of this *Shop Manual* describes various kinds of brake-cleaning equipment and tools. Always ensure that cleaning equipment is in proper working condition and follow the manufacturer's instructions for the specific equipment in your shop.

Asbestos waste must be collected, recycled, and disposed of in sealed impermeable bags or other closed, impermeable containers. Any spills or release of asbestos-containing waste material from inside of the enclosure or vacuum hose or vacuum filter should be immediately cleaned up using vacuuming or wet-cleaning methods. Review the asbestos safety instructions in Chapter 1 of this *Shop Manual*. The following paragraphs contain general information about brake-cleaning equipment that can be used to clean the caliper and rotor while they are installed on the vehicle. A general-purpose parts washer can be used to clean parts removed from the vehicle.

Cleaning with Vacuum-Enclosure Cleaning Systems

Vacuum-enclosure cleaning and containment systems consist of a tightly sealed protective enclosure that covers and contains the brake assembly (Figure 7-24). The enclosure has built-in, impermeable sleeves and gloves that let you inspect and clean the brake parts while preventing the release of dust and potential asbestos fibers into the air. Examine the condition of the enclosure and its sleeves

Figure 7-24 This full-enclosure asbestos vacuum system traps brake dust and helps keep the shop's air free of dust.

Brake-cleaning tools/
   equipment

One of the simplest
brake-cleaning tools
is a hand-operated
spray bottle filled
with soapy water.
Place a catch basin
under the brake
assembly and spray
liberally. The waste
water can be poured
into a hazardous
waste container for
later disposal.

before beginning work. Inspect the enclosure for leaks and a tight seal. See Photo Sequence 2 in Chapter 2 for details.

A high-efficiency particulate air (HEPA) filter vacuum is used to keep the enclosure under negative pressure as work is done. Because particles cannot escape the enclosure, compressed air can be used to remove dust and dirt from brake parts. Once the dirt is loose, draw it out of the enclosure with the vacuum port. The dust is then trapped in the vacuum cleaner filter. When the vacuum cleaner filter is full, spray it with a fine mist of water, then remove it and immediately place it in an impermeable container. Label the container as follows:

 WARNING: CONTAINS ASBESTOS FIBERS. AVOID CREATING DUST. CANCER AND LUNG DISEASE HAZARD.

Cleaning with Wet-Cleaning Systems

Low-pressure wet-cleaning systems wash dirt from the brake assembly and catch the cleaning solution in a basin (Figure 7-25). The cleaner reservoir contains water with a nonpetroleum solvent or wetting agent. To prevent any asbestos-containing brake dust from becoming airborne, control the flow of liquid so that the brake assembly is gently flooded. The solvent also is effective for removing brake fluid and oil or grease from brake parts.

Figure 7-25 Typical low-pressure, wet-cleaning equipment.

Some wet-cleaning equipment uses a filter. When the filter is full, first spray it with a fine mist of water, then remove the filter and place it in an impermeable container. Label and dispose of the container as described earlier.

Cleaning with Vacuum-Cleaning Equipment

Several types of vacuum-cleaning systems are available to control brake dust in the shop. The vacuum system must have a HEPA filter to handle asbestos dust (Figure 7-26). A general-purpose shop vacuum is not an acceptable substitute for a special brake vacuum cleaner with a HEPA filter. After vacuum cleaning, wipe any remaining dust from components with a damp cloth.

Because they contain asbestos fibers, the vacuum cleaner bags and any cloth used in asbestos cleanup are classified as hazardous material. Such hazardous material must be disposed of in accordance with OSHA regulations. Always wear your respirator when removing vacuum cleaner bags or handling asbestos-contaminated waste. Seal the cleaner bags and cloths in heavy plastic bags. Label and dispose of the container as described previously.

Special Tools

Vacuum brake cleaner

Cleaning with Parts Washers

Almost all shops have a parts washer that is used to clean miscellaneous small and medium-size parts that are removed from vehicles (Figure 7-27). The equipment reservoir holds solvent that is recirculated by an electric pump. A nozzle is used to apply solvent to parts, and heavy dirt can be loosened with a cleaning brush. Filters in the reservoir trap dirt and grease.

Traditionally, parts washers have used petroleum solvent, but most current models use water-based solvents and detergents. Do not use petroleum-based solvents on rotors and drums. Cast iron is porous and will absorb the solvent. Solvent comes out during braking and will ruin new pads and shoes. A parts washer that uses only water-based solvents and detergents is usually safe for brake cleaning with no extra steps.

Special Tools

Parts washer

Figure 7-26 A HEPA-equipped vacuum cleaner is a good tool for cleaning brake components.

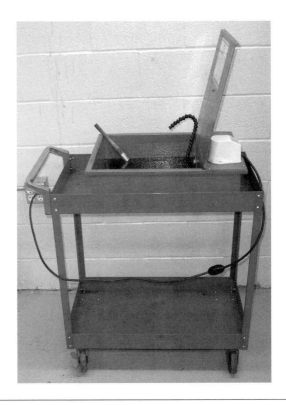

Figure 7-27 Use a parts washer to clean the outside of calipers and similar dirty parts.

Brake Caliper Service

Classroom Manual
pages 182–188

Special Tools

Lift or jack with
  stands

Impact tools

C-clamp or small
  prybar

Cloths

Catch basin

Syringe

Do not forget to
remove some of the
brake fluid from the
reservoir as needed.

In the 1960s when disc brakes started to become standard equipment, brake calipers were overhauled routinely whenever pads were replaced. This is no longer a universal practice. When changing brake pads, however, the calipers should be inspected for damage, leakage, and general wear. Remember that a set of brake pads usually lasts at least 30,000 miles (48,000 km). If the technician does not overhaul or replace the calipers, he must be confident that the calipers will operate properly and safely until the next pad replacement.

Generally, calipers are more likely to need overhaul or replacement if the car is 5 years old or older or has more than 50,000 miles (80,000 km) and the calipers have never been serviced. If the vehicle is driven hard or operated in very hot or cold temperatures, the calipers will need service sooner than calipers on a car used conservatively in a moderate climate.

The details of caliper overhaul vary with caliper design, but all overhaul procedures share common principles. Photo Sequence 13 shows the major overhaul steps for a typical floating caliper. The following sections explain general guidelines and common steps that apply to most calipers.

Caliper Removal

Much of the caliper can be inspected while it is mounted on the car, but it must be removed to closely examine the dust boot, the pistons and seals, and the mounting hardware, including bushings and sleeves. Removal procedures are different for different calipers, but the following are general guidelines.

A brake hose can be attached to a caliper with a banjo fitting, a swivel fitting, or a rigid (non-swivel) fitting. If the hose is attached to the caliper with a banjo fitting or a swivel fitting, disconnect it from the caliper and be careful not to twist the hose as you loosen the fitting. Plug or cap the open end of the hose to keep dirt out of the brake lines. If the brake hose is attached to the caliper with a rigid fitting, disconnect the brake pipe from the hose at the hose mounting bracket and cap the end of the pipe to keep out dirt (Figure 7-28).

AUTHOR'S NOTE: Sometimes the right size plug may not be readily available. Try using the cap from a ballpoint pen. "Screw" the cap onto the flared or threaded fitting. The plastic cap will not damage the flare or threads and will work sufficiently as a plug.

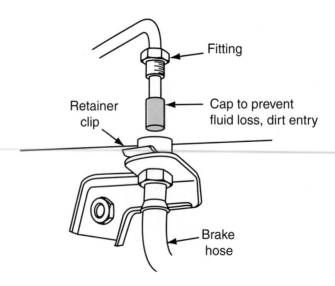

Figure 7-28 Cap or plug disconnected brake lines on the vehicle to keep out dirt.

Photo Sequence 13
Typical Procedure for Rebuilding a Disc Brake Caliper

P13-1 Disconnect the brake hose from the caliper. Remove the caliper bolts, then remove the caliper completely from the vehicle and move to the workbench.

P13-2 Inspect the bushings for cuts and nicks. Replace them if damage is found.

P13-3 Place a wooden block or shop rag in the caliper opposite the piston, then carefully remove the piston from the caliper by applying air pressure through the brake line hole.
WARNING: Do not place your fingers in front of the piston. Do not use high air pressure; use an OSHA-approved 30-psi nozzle instead. Cover the caliper with a shop towel to catch fluid spray.

P13-4 Remove the piston boot and seal, taking care not to damage the cylinder bore. Use a small wooden or plastic tool to remove the seal.

P13-5 Inspect for wear, nicks, corrosion, or damage.

P13-6 Use crocus cloth to polish out light corrosion. Replace the caliper if light polishing does not remove corrosion from around the seal groove.

P13-7 Clean the piston, caliper bore, and all parts with clean, denatured alcohol.

P13-8 Dry all parts with unlubricated compressed air. Blow out all passages in the caliper and the bleeder valve.

Photo Sequence 13
Typical Procedure for Rebuilding a Disc Brake Caliper (continued)

P13-9 Screw the bleeder valve and bleeder valve cap into the caliper housing. Tighten to specifications.

P13-10 Apply brake fluid to a new piston seal, then install the piston seal in the cylinder groove. Make sure the seal is not twisted.

P13-11 Apply brake lubricant or brake fluid to a new piston boot, then install the boot onto the piston.

P13-12 Lubricate the caliper bore and piston with brake fluid. Install the piston into the caliper bore and push it to the bottom of the bore.

P13-13 Seat the boot in the caliper housing counter bore using the proper seating tool.

P13-14 Lubricate the beveled end of the bushings with brake lubricant. Pinch the bushing and install it bevel end first into the caliper housing. Push the bushing through the housing bore.

P13-15 If desired, the caliper can be bench bled by using a hand-operated pressure bleeder. Pump new brake fluid through the hose opening until the fluid flows from the bleeder opening. Install the bleeder screw.

P13-16 Install new pads as shown in Photo Sequence 12. Then reinstall the caliper on the vehicle. Fill the master cylinder reservoir and bleed the system.

Depending on the caliper design, clearance around the caliper installation, and the amount of pad wear, it may be easier to remove the pads before removing the caliper or to leave them installed and remove the complete assembly. If you remove the pads first, inspect the mounting hardware and set it aside for reinstallation or for closer examination and replacement.

Remove clips and keys from sliding calipers and slide each caliper off its ways on the caliper support (Figure 7-29). Remove mounting pins or bolts from floating calipers and similarly slide each caliper off its caliper support (Figure 7-30). Inspect the mounting hardware for excessive wear, corrosion, and other damage. Remove all bolts holding a fixed (stationary) caliper to its support.

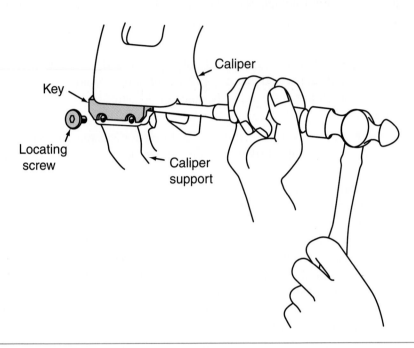

Figure 7-29 On this sliding caliper, remove the locating screw and then drive the key out with a drift punch.

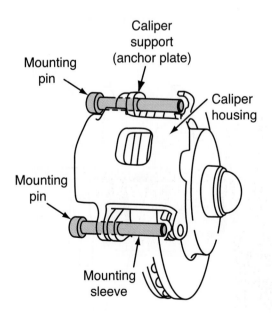

Figure 7-30 Remove the mounting pins or bolts from a floating caliper.

Lift the caliper off the rotor and take it to a bench for further service. If you have not serviced a particular caliper design previously, work on the caliper for one side of the car at a time and use the caliper for the other side for assembly and installation reference.

Caliper Inspection

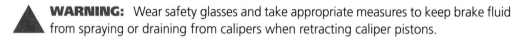
Special Tools

C-clamp or other tool to compress piston

Note that any damage to the boot or any part of the caliper requires a rebuild at minimum.

When replacing brake pads as explained previously in this chapter, inspect the calipers closely. Routine, preventive maintenance brake inspection is another important time to check the calipers. Inspection can be more thorough and complete if the calipers are off the car and the pads removed. Look at the following five areas and check for these conditions during complete caliper inspection:

1. Inspect the entire outside of the caliper body for cracks and other major damage. Replace any damaged caliper.

2. Inspect the piston dust boot closely for holes and tears. Be sure it is correctly installed in the piston and caliper and that no openings exist that could let dirt or water into the caliper bore. If the dust boot is damaged or defective in any way, replace or overhaul the caliper.

3. If the dust boot is okay, inspect the caliper closely for leakage. Any sign of leakage means that a piston seal is leaking and the caliper must be replaced or overhauled.

⚠️ **WARNING:** Wear safety glasses and take appropriate measures to keep brake fluid from spraying or draining from calipers when retracting caliper pistons.

4. With the caliper off the car, use a C-clamp (Figure 7-31), large slip-joint pliers, or a hammer handle to slowly force the piston to the bottom of its bore. Be careful not to damage the dust boot or seal. Note how the piston feels as it moves. If it sticks or moves unevenly, remove the piston to check for rust and scoring. Refer to specific service manual procedures to retract the pistons on rear calipers with parking brake linkage attached to the pistons.

5. Inspect all caliper mounting parts for rust and damage.

If a caliper passes these general checks and the vehicle is relatively new or has low mileage, it is reasonable to replace the pads without overhauling or replacing the calipers. If a caliper shows any signs of damage, excessive wear, or age, it should be replaced or overhauled. Just as pads are replaced in axle sets, calipers must always be serviced in pairs for each end of the vehicle.

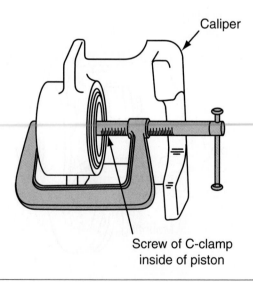

Caliper

Screw of C-clamp inside of piston

Figure 7-31 On some calipers it is helpful to compress the piston in its bore when inspecting it. This lets you check piston movement and inspect the dust boot closely.

Bleeder Screw Removal

Before starting to overhaul any caliper, try to loosen the bleeder screw (Figure 7-32) with a bleeder screw wrench or a six-point socket. Bleeder screws often get stuck or frozen into the caliper, particularly in aluminum caliper bodies or on cars driven on salted roads in the winter. A frozen bleeder screw is often a sign of extreme wear or age that immediately makes the caliper a candidate for replacement. If replacement calipers are not readily available, however, the frozen screw must be replaced to restore the caliper to proper operation.

If the bleeder screw does not loosen easily with a wrench, do not continue to force it until it breaks. Many frozen screws can be loosened by applying penetrating oil to the outer threads and letting it work its way into the screw hole. Wait 10 minutes after applying penetrating oil and then place a short piece of pipe over the screw so that the pipe does not contact the screw but rests on the caliper body. Hit the socket several times with a hammer to break the surface tension between the screw and the hole and to help the oil work into the threads. Repeat the penetrating oil and hammer steps several times and try to loosen the screw.

> ⚠ **WARNING:** Do not apply heat to a sealed or semisealed hydraulic system. If heat must be applied to a caliper, remove the piston and fluid first. The port for the hydraulic line could act like a jet, spurting hot brake fluid out. Heat could cause a rupture and cause injury.

> ■ **CAUTION:** Do not apply heat to a caliper made of aluminum or other lightweight material. The caliper will be weakened at a minimum and destroyed in most cases.

If the bleeder screw still will not turn, heat may help to loosen it. These steps should be completed with the caliper removed from the car and clamped in a bench vise. Heat the area of the caliper body around the screw with a torch and apply penetrating oil to the outer screw threads. Repeat this several times and try to turn the screw with a wrench. For a particularly stubborn bleeder screw, it may help to invert the caliper in the vise so that the screw is at the lowest point of the caliper. Ensure that some brake fluid is in the caliper bore and covers the inner end of the bleeder screw. Add an ounce or two through the hose connection port if necessary. Make sure the hose connection port is opened during heating. This will allow the pressure to escape. Once again, heat the caliper body. Brake fluid can work its way past the tip of the hot bleeder and into the inner threads to help loosen the screw.

Special Tools

Torch

Penetrating oil

Figure 7-32 Be sure the bleeder screw can be removed before proceeding further with caliper overhaul.

If the bleeder screw breaks or cannot be loosened by these methods, replace the caliper. If a replacement is not available, the broken screw can be drilled out and the caliper rethreaded. Because bleeder screws are made of hard steel and because of the awkwardness of the caliper, drilling and retaping should be done by a machine shop.

Caliper Piston Removal

Caliper pistons fit in their bores with only a few thousandths of an inch of clearance so even a small amount of rust, corrosion, or dirt can make them hard to remove. Caliper pistons are rarely loose enough in their bores to remove them by hand so three common methods are used for removal: compressed air, mechanical, and hydraulic. The following sections explain piston removal for single-piston calipers.

> ⚠️ **WARNING:** Wear safety glasses whenever removing caliper pistons. Keep your hands away from the piston when using compressed air or hydraulic pressure for removal. Cover the open hydraulic ports and the perimeter of the piston bore with clean shop cloths and take other necessary precautions to avoid spraying brake fluid. Clean up any spilled fluid immediately.

Special Tools

Air pressure regulator

OSHA-approved blowgun

Compressed Air Piston Removal. Using compressed air to remove a caliper piston is probably the oldest and most common method. It requires several precautions, however. Compressed air removal works best on pistons that are somewhat free or only moderately stuck in their bores. Compressed air contains a lot of energy that is released suddenly when the piston pops free, and it is hard to control the amount of air pressure applied to the piston. If the piston is not restrained when it is loosened, component damage or personal injury can occur. Observe all precautions in the preceding and following warnings when using compressed air to remove a piston.

> ⚠️ **WARNING:** Never use full shop air to remove a piston from a caliper unless that is the only option available. A regulator should be used to apply air gradually. Do not exceed about 30 psi of air pressure without taking additional precautions for containing the piston. Serious injuries could occur as the piston exits the bore.

Follow these seven steps for compressed air piston removal:

1. Remove the caliper from the car and clamp it securely in a bench vise by one of its mounting points.
2. Place a wooden block wrapped in shop cloths in the outer pad area. The block should be thin enough to let the piston move outward in its bore, but thick enough to keep the piston from leaving the bore completely.
3. Cover the caliper with cloths as necessary to keep brake fluid from spraying during removal. Old brake pads could be used to contain the piston's travel, but check to see if they are thick enough.
4. If an OSHA-approved safety nozzle that limits nozzle air pressure to 30 psi is available, use it as first choice to remove the piston.
5. If an OSHA-approved nozzle is not available but the air pressure can be regulated, start with about 30 psi and gradually increase line pressure. Use the lowest possible air pressure to loosen the piston. If regulated line pressure is not possible, apply air in short increments. Do not exceed normal maximum shop pressure of 90 psi to 100 psi.
6. Insert an air nozzle with a rubber tip into the brake hose port and gradually apply air pressure (Figure 7-33). Keep your fingers away from the piston.
7. When the piston pops free, remove the air nozzle and the wooden block and extract the piston the rest of the way by hand.

Mechanical Piston Removal. Mechanical piston removal is a method that uses a special tool to grip the inner bore or opening of the piston and then provides leverage to twist the piston out of

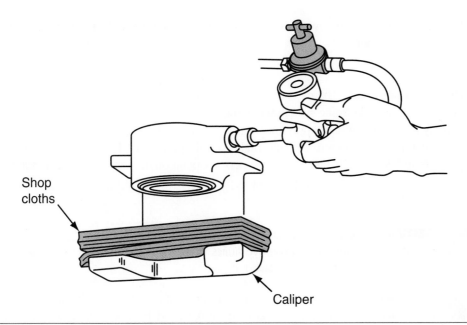

Shop cloths

Caliper

Figure 7-33 Carefully pop the piston out of the caliper bore with compressed air. Cushion the piston with shop cloths and use the lowest air pressure possible.

the caliper. This method works best with pistons that are only mildly stuck in their bores. It is often the fastest way to remove multiple pistons from a caliper because when one of two pistons in one side of a caliper is removed, the fluid passages to that piston must be blocked to apply air pressure to the other piston.

Follow these four steps for mechanical piston removal:

1. Remove the caliper from the car and clamp it securely in a bench vise by one of its mounting points.

2. Insert the removal tool into the piston opening and grip the inner surface of the piston securely (Figure 7-34). With a pliers-type tool, squeeze the handles tightly. With an expanding tool, turn the locking bolt until the tool grips the piston tightly.

Special Tools

Piston removal tool

Using a tool to remove the piston is the safest and should be tried first.

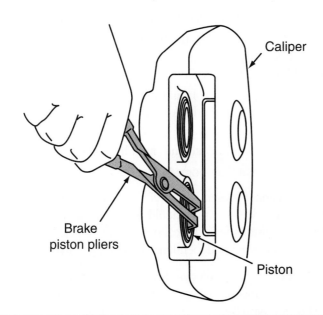

Caliper

Brake piston pliers

Piston

Figure 7-34 Use piston removal pliers to remove multiple pistons from a caliper.

3. Rotate the piston while working it back and forth in its bore until it loosens and slides out of the caliper.

4. If the piston cannot be removed manually, use compressed air or hydraulic pressure.

Hydraulic Piston Removal. The most common way to use hydraulic pressure to remove caliper pistons is to use the vehicle brake system while the calipers are still connected to the brake lines. The vehicle hydraulic system can apply much more pressure than can shop compressed air, and it is easier to control that pressure through brake pedal application. Hydraulic removal requires no special tools and is relatively fast. It is a two-person operation, however. Vehicle hydraulic pressure cannot be used for piston removal if air is present in the lines or if the master cylinder is bypassing internally.

 SERVICE TIP: If the piston is stuck so badly that other methods did not work, replace the caliper. Do not use hydraulics to remove completely a near-frozen piston. The danger is not worth the money saved, if any.

WARNING: Because hydraulic piston removal uses the vehicle brake system, you must take extra precautions to keep brake fluid from spraying or draining onto the vehicle finish or onto yourself. If a piston pops out of its bore with several hundred pounds of hydraulic pressure behind it, almost a pint of fluid will spray from the caliper and lines.

SERVICE TIP: Use the hydraulic method to remove a piston with the caliper on the vehicle only if the piston seems to be free.

Follow these nine steps for hydraulic piston removal:

1. Check the fluid level in the master cylinder and add fluid if necessary. Ensure that reserve vacuum is available in the vacuum booster.

2. Raise the vehicle on a hoist or safety stands and support it securely.

3. Remove both front or both rear calipers from the caliper supports and hang them from the chassis or suspension with heavy wire.

4. Place a wooden block in the caliper opening that is thick enough to keep the piston from being ejected from its bore while still allowing piston movement.

5. Place large pans under the calipers to catch any fluid that may drain from the calipers.

6. Have an assistant gradually apply the brake pedal to force the pistons from their bores. If several pedal applications are needed to loosen the pistons, it may be necessary to add fluid to the master cylinder.

7. When the pistons are loosened from their bores and pressed against the wooden blocks, disconnect and cap the brake hoses and remove the calipers from the vehicle.

8. Take the calipers to a work bench and drain brake fluid into a suitable container.

9. Remove the pistons from their bores by hand.

If hydraulic pressure cannot loosen a severely stuck caliper piston, it is probably so badly rusted or corroded that the caliper assembly should be replaced. Special hydraulic service equipment is available to remove pistons with the caliper off the vehicle. Such equipment makes piston removal a one-person operation, independent of the vehicle hydraulic system.

SERVICE TIP: Piston removal from fixed (stationary) calipers with two or four pistons requires some special techniques. If compressed air or hydraulic pressure is used, the loosest piston will come out first and the caliper will no longer hold pressure.

Mechanical removal tools can be used effectively on multiple-piston calipers, but they may not provide enough force to remove a badly stuck piston. If a set of fixed calipers must be overhauled, remember these tricks for piston removal:

❑ If the caliper can be separated into two halves, the pistons may be removed more easily from each side with air or hydraulic pressure or mechanical removal tools.

❑ Install a C-clamp onto each piston, or install blocks so that the pistons can move about half their lengths. This will allow the loosest piston to move and then stop so the other piston can be moved. Once both pistons have reached the blocks (C-clamps), move or loosen the blocks (C-clamps) and reapply pressure. This time when the pistons are stopped, remove the blocks (C-clamps). The pistons should now come out the remaining distance with hand force.

❑ If the pads can be removed while the calipers are installed on the car, more space will exist between the pistons and the rotor. Cover the rotor with shop cloths to protect its surface and use compressed air to pop the pistons out of the caliper. They will move far enough to loosen them, but the rotor will keep them from moving out of the bores completely so that they retain enough air pressure to loosen all pistons.

❑ With the caliper removed from the car, place a block of wood in the caliper that is thin enough to let the pistons loosen in their bores, but thick enough to hold them in the bores. Then apply air pressure to loosen the pistons.

Dust Boot and Seal Removal—Caliper Cleaning

After the piston is removed, remove the dust boot from the caliper. If the boot stays attached to the caliper, use a small screwdriver to pry it from its groove (Figure 7-35). If the dust boot is attached to the piston, remove and discard it.

Use a wooden or plastic pick to remove the piston seal from the caliper bore (Figure 7-36). To avoid scratching the caliper bore, do not use a screwdriver or metal pick. Before discarding the old piston seal, examine its shape. If it is anything but a square-cut seal, remember the direction in which it was installed so that the new seal can be installed correctly. Sketch the seal shape if it helps to remember the correct installation direction.

Special Tools

Catch basin

Crocus cloth

Denatured alcohol

Brush

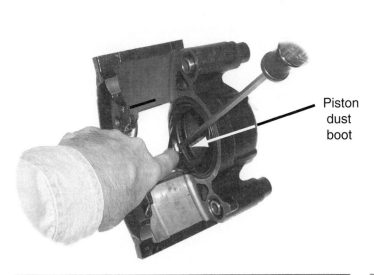

Figure 7-35 Remove the piston dust boot from its groove in the caliper.

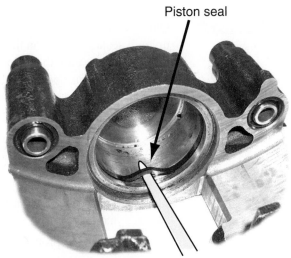

Figure 7-36 Use a wooden or plastic (not metal) pick to remove the piston seal from its groove.

After the caliper is disassembled, clean it with brake-cleaning solvent or alcohol and a soft-bristled brush. Use a plastic Scotchbrite scouring pad to remove dirt and varnish from caliper bores. Do not use a wire brush on the caliper bore, the piston, or dust boot attaching points. A wire brush will scratch and nick caliper bore and piston surfaces, which will cause leaks after reassembly.

Use crocus cloth lubricated with brake fluid to remove light rust, dirt, corrosion, or scratches from the caliper bore and the piston. Wash any areas cleaned with crocus cloth using brake-cleaning solvent or alcohol to remove all abrasive materials. If light corrosion cannot be removed from the caliper bore with crocus cloth, the bore can be honed as explained later in this chapter. Heavy corrosion or rust or severe pitting and scratching on the piston or its bore indicate that the complete caliper should be replaced.

Caliper Internal Inspection

If the caliper appears serviceable after it is disassembled and cleaned, inspect the piston and the caliper bore closely. If there is any doubt about the serviceability of the caliper or piston, replace with a new caliper. Scratched or rusted calipers may be honed if necessary, but excessive honing may create too much clearance between the piston and the bore. In such a case, the caliper must be replaced.

> ✓ **SERVICE TIP:** Standard-use calipers are almost never honed. If honing is required, the best repair is caliper replacement. Hone only those calipers that cannot be replaced or are very expensive. Generally, the labor and warranty cost of honing are not worth the trouble to the shop or the customer.

Inspect metal pistons for corrosion and scratching. Remember that the outside surface of the piston provides the sealing surface, so even minor scratches can create leakage. If the piston is chrome plated, make sure that the plating is not scuffed or flaked away. Similarly, the outer anodized surface of an aluminum piston must not be worn away. Replace any piston with a damaged surface finish.

Inspect phenolic plastic pistons for cracks and chips. Phenolic is a thermosetting resin normally used to mold or cast plastics. This type of piston had problems during its first years of use. It has since been refined and works well in most instances. Generally speaking, a phenolic piston will not damage the caliper bore. Small surface chips or cracks away from sealing surfaces are acceptable, but the best choice is to replace any piston that has surface damage.

Caliper Honing and Piston Clearance

Special Tools

Denatured alcohol

Hone

Vise

Older fixed calipers that had pistons with stroking seals were routinely honed to provide a fresh, uniform sealing surface on the caliper bore. Single-piston, floating, or sliding calipers are not routinely honed because the caliper bore is not the sealing surface. After cleaning as explained above, a caliper bore can operate satisfactorily with light scratches or pits. Heavier corrosion, scratches, or pits often can be removed with a hone as follows:

> **CAUTION:** Do not use machining oil as a lubricant for honing. This oil promotes metal removal and the caliper bore may become oversized quickly. Use only brake fluid as the honing lubricant.

1. Remove the caliper from the car and clamp it securely in a bench vise by one of its mounting points.
2. Install the proper size hone with fine-grit stones in a variable-speed electric drill.
3. Lubricate the bore and the hone with clean brake fluid.
4. Insert the hone in the bore and operate the drill at about 500 rpm.
5. Move the hone moderately back and forth in the bore but do not let the hone come completely out of the bore while the drill is running.

6. After about 10 seconds, stop the drill and remove the hone.

7. Wipe the bore with a clean cloth and inspect it to see if it has been cleaned satisfactorily.

8. Repeat the honing operation until corrosion, scratches, and pits are satisfactorily removed.

9. After honing the bore, clean it thoroughly with brake-cleaning solvent, alcohol, or fresh brake fluid. If alcohol or brake-cleaning solvent is used, make a final cleaning pass with brake fluid. Be sure to remove all abrasive materials from the seal groove and fluid passages.

Honing the bore of a floating or sliding caliper is a cleaning procedure, not a machining procedure for a sealing or bearing surface. Hone the bore only enough to remove the worst corrosion and surface wear. Piston clearance is more important than the final surface finish of the bore. Generally, the maximum piston-to-bore clearance is 0.002 inch to 0.005 inch (0.06 mm to 0.13 mm) for metal pistons and 0.005 inch to 0.010 inch (0.13 mm to 0.26 mm) for phenolic pistons. Refer to the specific vehicle service manual for exact specifications.

After cleaning the honed bore, lubricate the bore and piston with fresh brake fluid. Then insert a feeler gauge with a thickness of the maximum allowable clearance into the bore. Try to insert the piston into the bore next to the feeler gauge. If the piston slides into the bore, the clearance is too great. Double-check the fit with a new piston or measure the piston and bore diameters with micrometers to be sure. If clearance exceeds specifications, replace the entire caliper assembly.

Caliper Assembly

After cleaning and inspecting the caliper and ensuring that it is suitable for overhaul, obtain any replacement parts that you need, along with new seals, dust boots, and miscellaneous small parts. Assemble the caliper on a clean workbench and be sure that your hands and tools are clean and free of grease and oil.

Follow these six steps to assemble most floating or sliding calipers:

1. Lubricate the piston seal and the seal groove in the caliper bore with fresh brake fluid. If the seal is any shape other than square cut, be sure it is installed in the proper direction. Square-cut seals can be installed with either side facing in either direction.

2. Insert one edge of the seal in its groove and guide it into place by hand (Figure 7-37). Be careful not to twist, roll, or nick the seal.

Press seal into place with finger

Figure 7-37 Lubricate a new piston seal and install it by hand in the caliper.

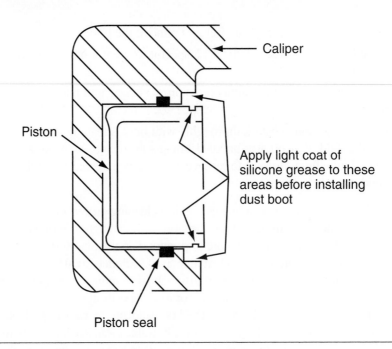

Caliper

Piston

Apply light coat of
silicone grease to these
areas before installing
dust boot

Piston seal

Figure 7-38 Lubricate the dust boot groove in the caliper.

3. Lubricate the piston and the caliper groove for the dust boot with fresh brake fluid (Figure 7-38). Ensure that the piston enters the caliper bore smoothly. Keep the piston square in the bore as it is being inserted into the bore.

4. If the dust boot attaches to the piston, install the boot as shown in Figure 7-39 and Figure 7-40.

5. Attach the dust boot to the caliper by one of the following three general methods, depending on dust boot design:

   A. If the dust boot is a press fit into the caliper, slide the piston into the dust boot and caliper bore until the boot snaps into its groove in the piston (Figure 7-41). Then drive the boot into place with the proper driver (Figure 7-42).

   B. If the dust boot is held in the caliper by a retaining ring, slide the piston into the dust boot and caliper bore until the boot snaps into its groove in the piston. Then push the outer edge of the boot into its groove in the caliper body. Ensure that the boot is seated correctly and install the metal lock ring to hold it in the groove.

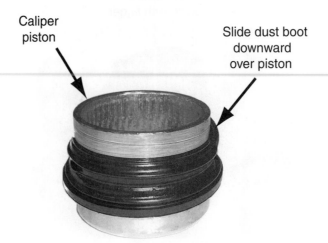

Caliper
piston

Slide dust boot
downward
over piston

Figure 7-39 Install a new dust boot on the piston and pull it downward, inside out.

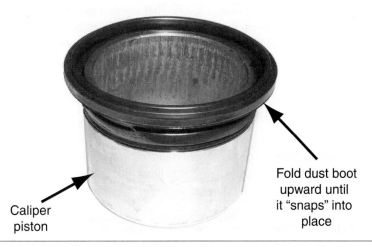

Caliper
piston

Fold dust boot
upward until
it "snaps" into
place

Figure 7-40 Then fold the boot back upward until it is in this position.

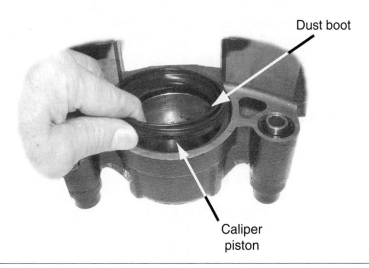

Dust boot

Caliper
piston

Figure 7-41 Slide the piston and boot into the caliper bore.

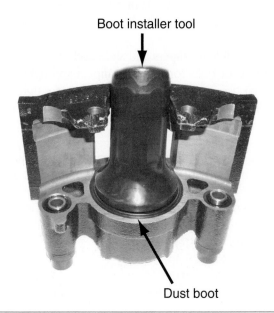

Boot installer tool

Dust boot

Figure 7-42 Using the proper size driver, tap the dust boot into its retaining groove in the caliper.

C. If the dust boot is retained by a lip that fits in a groove in the caliper bore, obtain special installation rings to make the job easier. Slide the boot over the correctly sized ring. Install the lip of the boot in the caliper groove; then slide the piston through the ring and boot into its bore. Remove the ring from the piston so that the inner lip of the dust boot snaps into the piston groove.

6. After the piston and dust boot are installed, finish assembling the caliper by following the steps for pad installation given earlier in this chapter.

☑ **SERVICE TIP:** Some caliper rebuilding kits contain silicone lubricant for the seals and dust boots. If supplied, use this to lubricate the seal and boot during installation. The silicone lubricant improves corrosion protection in these areas.

☑ **SERVICE TIP:** Some manufacturers and some technicians recommend bench bleeding the rebuilt or new caliper before installation. A hand-operated pressure bleeder can be used for this task. Remove the bleeder screw and hold the caliper over a capture container with the bleeder opening on top. Pump brake fluid through the hose fitting until it comes out of the bleeder opening. Plug the hose fitting and then install the bleeder screw. This does not remove the requirement to bleed the system on-vehicle, but it reduces overall time. On some calipers it will save a lot of time because of the close area around the mounted caliper.

Caliper Installation

Refer to the caliper removal guidelines earlier in this chapter before installing a new or overhauled caliper. Installation is basically the opposite of removal, but vehicle service is never quite that simple. A few different steps always exist. Installation procedures are different for different calipers, but they should follow these guidelines.

Depending on the caliper design and clearance around the caliper installation, it may be easier to install the pads before or after caliper installation. If installing the pads first, be sure that all antirattle springs and clips are in place and that any necessary antinoise compound is applied.

One of the most important points of caliper installation is lubrication of all moving parts on sliding or floating calipers. Apply the carmaker's recommended brake lubricant to contact points on caliper surfaces and on the caliper ways. Lubricate sleeves and bushings of floating calipers according to the manufacturer's directions. On all caliper installations, be sure to keep lubricant off rotor and pad surfaces.

Install clips and keys on sliding calipers and slide each caliper onto its ways on the caliper support. Install mounting pins or bolts for floating calipers and tighten all bolts to the specified torque with a torque wrench. Do not guess about brake fastener tightness. Install all bolts holding a fixed (stationary) caliper to its support. When installing a banjo fitting, check it for sealing washers. If equipped with washers, replace them with new washers.

■ **CAUTION:** Ensure the fitting is not cross-threaded when reconnecting. Cross-threading could damage the fitting or the component, or both.

■ **CAUTION:** Do not use paint, lubricants, or corrosion inhibitors on fastener(s) unless specified by the manufacturer. Add-on coating of any type may affect torquing, clamping, or may damage the fastener(s).

■ **CAUTION:** Use the correct fastener torque and tightening sequence when installing components. Incorrect torque or sequencing could damage the fastener(s) or component(s).

SERVICE TIP: If the brake hose was disconnected from the vehicle's brake line, it is easiest to connect the hose to the line before connecting it to the caliper. This allows the hose to be "wiggled" so the fitting threads are more easily aligned. The typical brake fitting can be easily damaged by not aligning the matching threads properly and attempting to tighten the fitting with a wrench. This method does not apply to a hose with a male fitting that screws into the caliper opening. There are two methods that can be used in that case. The first is to connect the hose to the line as noted above and then rotate the caliper like a nut so it screws onto the hose. This may not properly position the hose without some twisting once the caliper is installed on the vehicle, however. The best method with a male-fitted hose is to screw it into the caliper opening and torque the hose to specification. Then use care in aligning and connecting the hose to the line on the vehicle.

Connect the brake hose to the caliper, being careful not to twist the hose as you tighten the fitting.

SERVICE TIP: Give your customers a short course in brake noise that is easy to understand and that will make them more responsible car owners. Explain that a low-pitched grinding or rumble during braking is often the metal-to-metal sound of worn pads or shoes grinding away the rotors or drums. A shrill screech that sounds like fingers being dragged across a chalkboard is usually a pad wear indicator telling the driver that it is time for brake service. Some motorists have the idea that light squeaks and squeals are normal. If squeaky brakes develop suddenly where no noise had been present before, however, it can mean that the pads are wearing down toward the replacement point. A thick new pad may not vibrate at an audible frequency, but the vibration frequency can change and produce noise when the pad wears.

Rotor Service

Brake rotors are not routinely resurfaced as part of every brake job as they were many years ago. Rotor service is still an important part of disc brake service, however, because maximum braking performance cannot be obtained with a rotor that is damaged or defective. Rotor service consists of these general operations:

❑ Rotor inspection
❑ Rotor measurement
❑ Rotor resurfacing or turning
❑ Rotor finishing

Classroom Manual
pages 169–174

Rotor Inspection

Inspect disc brake rotors (Figure 7-43) when the pads or calipers are serviced or when the wheels are rotated or removed for other work. Many rotor problems may not be apparent on casual inspection. Rotor thickness, parallelism, runout, flatness, and depth of scoring can be measured only with precision gauges and micrometers. Other rotor conditions should be checked with equal precision and thoroughness. Inspect a rotor as follows after removal of the caliper:

1. If the rotor is dirty enough to interfere with inspection, clean it with a shop cloth dampened in brake-cleaning solvent or alcohol. If the friction surface is rusted, remove the rust with medium-grit sandpaper or emery cloth and then clean with brake cleaner or alcohol.

2. Turn the rotor through a full revolution to view the inboard surface at an accessible point, usually in the area where the caliper was mounted.

Figure 7-43 Rotor inspection begins before you remove the rotor from the vehicle.

3. Inspect both rotor surfaces for scoring and grooving. Scoring or small grooves up to 0.010 inch (0.25 mm) deep are usually acceptable for proper braking performance (Figure 7-44). To determine the depth of grooves and whether or not the rotor can be resurfaced within its thickness limits, use a brake (rotor) micrometer with a pointed anvil and spindle designed for this use. Rotor measurements are described in the next section.

4. Inspect the rotor thoroughly for cracks or broken edges. Replace any rotor that is cracked or chipped, but do not mistake small surface checks in the rotor for structural cracks. Surface checks will normally disappear when a rotor is surfaced just a few thousandths of an inch. Structural cracks, however, can become more visible when surrounded by a freshly turned rotor surface.

5. Inspect the rotor surfaces for **heat checking,** which appears as many small interlaced cracks on the surface (Figure 7-45). Heat checking lowers the heat dissipation ability and friction coefficient of the rotor surface. Heat checking is more noticeable than minor surface checking and does not disappear with resurfacing.

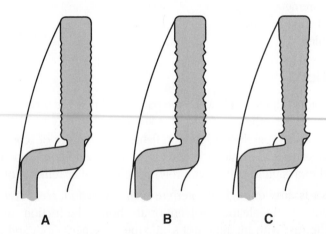

A   B   C

Figure 7-44 The small grooves on rotor A are less than 0.010 inch (0.25 mm) and acceptable for continued use. Rotor B has deeper scoring and requires refinishing if the thickness allows. Rotor C has extreme taper and probably will not stand refinishing.

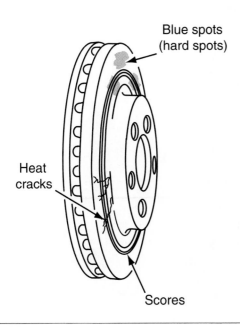

Figure 7-45 Some typical rotor defects to check for.

Figure 7-46 The groove in this rotor surface is machined to aid in cooling and noise reduction.

6. Also inspect the surfaces for hard spots, which appear as round, shiny, bluish areas on the friction surface as shown in Figure 7-45. It may be possible to machine hard spots out of a rotor if the rotor has not been turned previously and has enough thickness for extended machining. Hard spots are difficult to machine, however, and may require special cutting bits for the brake lathe. Because of the difficulty and uncertainty involved in machining a rotor with hard spots, most carmakers recommend replacing a rotor with this kind of damage.

7. Inspect the fin areas of vented rotors for cracks and rust. Rust in cracks near the fins can cause the rotor to expand and lead to rotor parallelism, or thickness, variations and excessive runout, or even exploding the rotor. Machining the rotor may remove runout and thickness variations, but rotor expansion due to rust may cause these problems to reappear soon. Replacing rotors with rust damage of this kind can avoid service comeback problems.

Some rotors have a single deep groove manufactured into each surface (Figure 7-46). This groove helps to keep the pads from moving radially outward and reduces operating noise.

If a brake rotor passes this preliminary inspection, proceed to measure it as explained in the following sections. A rotor cannot be returned to service or resurfaced if it does not meet all thickness specifications.

Rotor Measurement

Various micrometers and dial indicators are needed to measure a rotor. A micrometer is used to measure rotor thickness and **parallelism,** as well as taper. Parallelism is achieved when both sides of the rotor are exactly parallel to each other. Loss of parallelism happens during normal wear and tear. A dial indicator is used to measure rotor runout. Some rotors require an additional surface depth measurement.

Special Tools

Outside micrometer or vernier caliper

Service manual

Parallelism is the thickness uniformity of a disc brake rotor; both rotor surfaces must be parallel with each other within 0.0001 inch or less.

Minimum thickness
specification

Figure 7-47 The discard dimension, or minimum thickness specification, is cast or stamped into all disc brake rotors.

Rotor Thickness and Parallelism. All brake rotors except the earliest examples from the mid-1960s should have a discard-thickness dimension cast into them (Figure 7-47). If a discard dimension cannot be found on a rotor or if it is hard to read, check a service manual for thickness specifications. The **discard dimension** seems like a simple specification, but you must understand its complete meaning and how to apply it to rotor service.

A third measurement or specification that is of more importance to the technician than the discard dimension is the maximum refinishing limit. If the rotor measures less than the refinishing limit, then the rotor is discarded. On newer vehicles, this limit is not very much smaller than the standard thickness. See Table 7-1, which was quoted from the 2000–2005 Honda S2000 service manual for rear disc rotors. Most vehicle manufacturers are listing rotor refinishing limits in their service manuals.

✔ **SERVICE TIP:** There is a special tool available to check parallelism at the rotor. The tool is U-shaped with a dial indicator plunger forming one side of the U. Placing the tool on the rotor and rotating the rotor through the tool causes the dial indicator needle to deflect back and forth from zero. The amount of deflection is the amount of warpage.

Table 7-1   Rotor Specifications

Brake disc thickness:

Standard: 11.9–12.1 mm (0.459–0.476 in.)

Maximum refinishing limit: 10.0 mm (0.39 in.)

Brake disc parallelism: 0.015 mm (0.0006 in.)

The specifications for a 2005 Honda S2000 rear brake rotor. Note the very small allowance for parallelism, only 6/10,000th of an inch.

Rotor discard-thickness dimensions or refinishing limits are given in two or three decimal points (hundredths, thousandths, or ten-thousandths of an inch or hundredths of a millimeter), such as 1.25 inch, 1.375 inch, 0.750 inch, or 24.75 mm. When measuring rotor thickness, subtract 0.015 inch to 0.030 inch (0.40 mm to 0.75 mm) to allow for wear after the rotor is returned to service. If resurfacing the rotor, it must similarly be 0.015-inch to 0.030-inch thicker than the discard dimension after surfacing to allow for wear. A rotor should not be returned to service, with or without resurfacing, if its thickness is at or near the refinishing limit.

If unsure about the resurfacing limits or replacement requirements for any rotor, check the vehicle service manual.

Rotor parallelism refers to thickness variations in the rotor from one measurement point to another around the rotor surface. Thickness variations and excessive rotor runout are the major causes of brake pedal pulsation. Thickness variation, or parallelism, is measured at the same time as basic rotor thickness.

Use a micrometer graduated in ten-thousandths of an inch or hundredths of a millimeter to measure rotor thickness as follows:

1. Raise the vehicle on a hoist or safety stands and remove the wheels. If the rotor is removable from the hub (a floating, two-piece rotor), place flat washers on all wheel studs and reinstall the wheel nuts. Torque the nuts to specifications in the specified tightening pattern to minimize any possible runout (Figure 7-48).

2. If the caliper must be removed to measure the rotor, hang it from the underbody or suspension with heavy wire so it will not drop. Do not let the caliper hang from the brake hose.

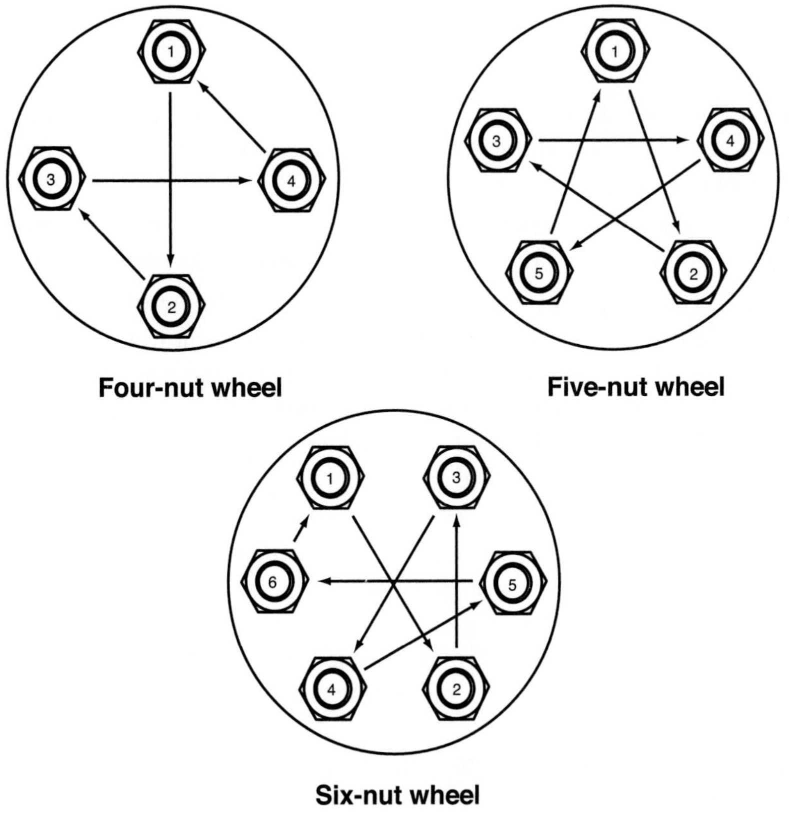

If disk or drum wear exceeds the refinishing limits, the disc or drum must be replaced. A disc or drum should never be machined to the discard dimension.

Four-nut wheel

Five-nut wheel

Six-nut wheel

Figure 7-48 Reinstall the wheel nuts in the specified pattern to hold a hubless rotor on the hub or axle flange for measurement.

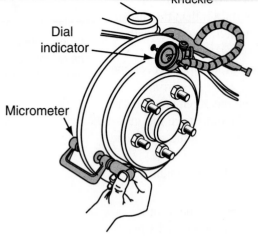

Attach dial
indicator clamp
to steering
knuckle

Dial
indicator

Micrometer

Figure 7-49 The rotor and hub runout is measured with a dial indicator mounted to some fixed point on the vehicle (left). The rotor thickness and parallelism can be measured with a standard outside micrometer, as shown on the right, but is more easily done with an electronic brake micrometer.

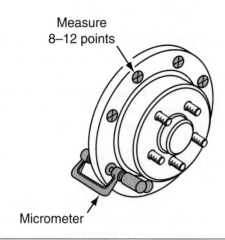

Measure
8–12 points

Micrometer

Figure 7-50 Check for thickness variations by measuring at 8 to 12 points around the rotor.

3. Place an outside micrometer or a brake micrometer about $\frac{3}{8}$ inch (10 mm) in from the outer edge of the rotor and measure the thickness (Figure 7-49). Compare the measurement to specifications.

4. Also check the vehicle service manual for an allowable thickness variation. Many carmakers hold tolerances on thickness variations as close as 0.0005 inch (0.013 mm).

5. Repeat the measurement at about eight points, equidistant (45 degrees) around the surface of the rotor, and compare each measurement to specifications (Figure 7-50). Take all measurements at the same distance from each edge so that rotor taper does not affect measurement comparisons. If the rotor is thinner than the minimum thickness at any point or if thickness variations exceed limits, the rotor must be replaced.

6. If the rotor is deeply grooved, it must be thick enough to allow the grooves to be completely removed without turning the rotor to less than its minimum thickness. To measure the depth of the grooves on both sides of the rotor, obtain a micrometer with a pointed anvil and spindle. Measure rotor thickness to the bottom of the deepest grooves in about eight places. If rotor thickness at the bottom of the deepest grooves is at or near the discard dimension, replace the rotor.

✓ **SERVICE TIP:** You can use a dime to make a simple check. Place the dime into the groove with Roosevelt's head toward the scored groove. If the dime goes into the groove far enough that the top of his head disappears, the groove exceeds 0.060 inch and may need to be resurfaced or replaced.

Rotor Lateral Runout. Excessive **rotor lateral runout** causes the rotor to wobble from side to side as it rotates (Figure 7-51). This wobble knocks the pads farther back than normal, which causes the pedal to pulse or vibrate as it is applied. There may be an increase in brake pedal travel because the pistons must move a greater distance to contact the rotor surface. For best brake performance, lateral runout should be less than 0.003 inch (0.08 mm) for most vehicles. Some manufacturers, however, specify runout limits as small as 0.002 inch (0.05 mm) or as great as 0.008 inch (0.20 mm).

Runout measurements are taken only on the outboard surface of the rotor. Using a dial indicator and suitable mounting adapters, measure runout as follows:

1. Raise the vehicle on a hoist or safety stands and remove the wheels. If the rotor is removable from the hub (a floating, two-piece rotor), place flat washers on all wheel studs and reinstall the wheel nuts. Torque the nuts to specifications to remove looseness from runout measurements.

2. If the caliper must be removed to measure runout, hang it from the underbody or suspension with heavy wire so it will not drop.

3. If the caliper is not removed but the rotor drags heavily on the pads as it is rotated, remove fluid from the master cylinder and push the pistons back in the calipers as explained previously in this chapter.

4. If the rotor is mounted on adjustable wheel bearings, readjust the bearings to remove **bearing end play** from the runout measurements. Do not overtighten the bearings. Bearing end play is the amount of preload applied to the bearing. Preload is the bearing with the right amount of tension to prevent side-to-side movement and wobble, but allows the bearing and its supported device to spin free. Bearing adjustment instructions are in Chapter 3 in this *Shop Manual*. On rotors bolted solidly to the axles of front-wheel-drive (FWD) cars, bearing end play is not a factor in rotor runout measurement. If excessive bearing end play is noted, the assembly must be replaced. Bearing end play is best checked with a dial indicator set up to measure wheel wobble.

Special Tools

Dial indicator

Lift or jack with stands

Impact tools

Service manual

Rotor lateral runout refers to rotor wobble from side to side as it rotates.

Bearing end play is the designed looseness in a bearing assembly.

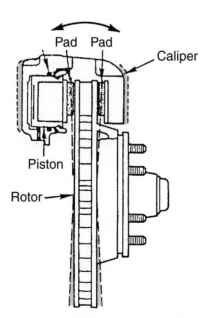

Figure 7-51 Excessive lateral runout of the rotor knocks the piston back into its bore and causes pedal pulsation and increased pedal travel.

Figure 7-52 The dial indicator can be mounted in several ways as long as the face remains stable.

5. Clamp the dial indicator support to the steering knuckle or other suspension part that will hold it securely as you turn the rotor (Figure 7-52).

6. Be sure the rotor surface is free of dirt and rust in the area where you will place the dial indicator tip.

7. Position the dial indicator so that its tip contacts the rotor at 90 degrees. Place the indicator tip on the friction surface about midway between the inner and outer edges of the friction area of the rotor. Do not place the dial indicator in a grooved or scored area.

8. Turn the rotor until the lowest reading appears on the dial indicator; then set the indicator to zero.

9. Turn the rotor through one complete revolution and compare the lowest to the highest reading on the indicator. This is the maximum runout of the rotor. If the dial indicator reading exceeds specifications, resurface or replace the rotor.

If the rotor is a floating type, try repositioning it on the hub one or two bolt positions from its original location and repeat the runout measurement. If repositioning a floating rotor fixes an excessive runout condition, make index marks on the rotor and hub so that the rotor will be installed in the correct position.

If the rotor is mounted on nonadjustable wheel bearings that have any amount of end play, you must account for this end play in rotor runout measurements. To make this measurement, press in on the rotor and turn it to the point of the lowest dial-indicator reading. Set the indicator to zero and pull outward on the rotor. Read the indicator, which now shows the amount of bearing end play. Then turn the rotor to the point of the highest reading and pull outward. Subtract the end play reading from the reading at this point to determine runout of the rotor alone.

Rotor Friction Surface Depth Measurement. Ford Motor Company calls for an additional rotor measurement after the rotor has been removed for resurfacing. This measurement indicates how much metal can be safely removed from the inboard surface of the rotor. It requires a

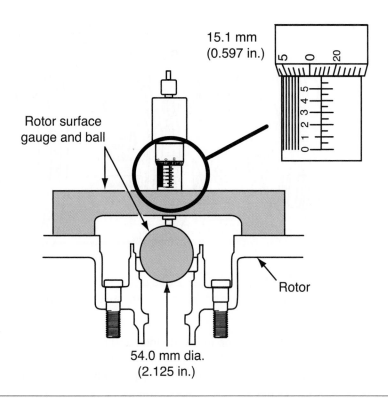

Figure 7-53 A special gauging with a depth micrometer and a gauge ball is needed to ensure the inner surface depth of a rotor.

depth micrometer and a special gauge. One-piece rotor and hub assemblies use a ball-type gauge (Figure 7-53); removable (floating) rotors use a pyramid-shaped template.

With the rotor off the vehicle, remove the grease seal and inner bearing from a one-piece rotor. Place the gauge ball or template inside the rotor. Ensure that the inboard friction surface is clean and place the depth micrometer on the surface with the spindle centered over the gauge. Set the micrometer to the initial setting given in the vehicle service manual. Then turn the micrometer until the spindle contacts the gauge. The difference between the initial reading and the reading with the micrometer touching the gauge is the maximum amount of metal that can be removed during resurfacing.

If the micrometer contacts the gauge at the initial setting or if you must retract the micrometer to place both legs flat on the friction surface, the rotor cannot be safely resurfaced according to Ford.

> ✓ **SERVICE TIP:** Manufacturers try to hold rotor thickness variation to less than 0.0002 inch. A rotor machined to this tolerance will wear evenly with little or no thickness variation during its service life. When initial thickness variation exceeds 0.0008 inch, rotor wear will be rapid and uneven.

Removing a Rotor

To remove a rotor from the vehicle, raise the vehicle on a hoist or safety stands and remove the wheel and tire. Remove the caliper from the rotor and suspend it from the vehicle underbody as explained earlier in this chapter.

Before you remove a rotor, mark it "L" or "R" for left or right so that it gets reinstalled on the same side of the vehicle from which it was removed.

Special Tools

Depth micrometer

Ford's special rotor gauge

Service manual

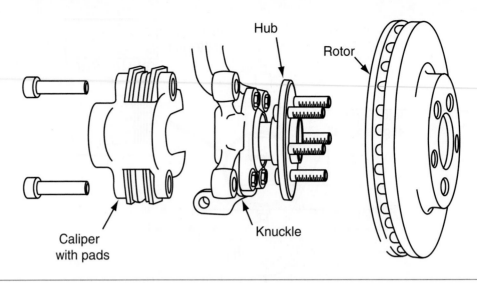

Hub

Rotor

Caliper
with pads

Knuckle

Figure 7-54 Remove a floating (two-piece) rotor from the mounting flange.

If the rotor is a two-piece floating rotor, index the rotor and hub, then remove it from the hub by pulling it off the lug studs (Figure 7-54). If it cannot be pulled off by hand, apply penetrating oil on the front and rear rotor-to-hub mating surfaces. Strike the rotor between the studs with a rubber or plastic hammer. If this does not free the rotor, attach a three-jaw puller to the rotor and pull it off.

When separating a floating rotor from its mounting flange, clean any rust or dirt from the mating surfaces of the hub and rotor. Neglecting to clean rust and dirt from the rotor and hub mounting surfaces before installing the rotor will result in increased rotor lateral runout, leading to premature brake pulsation and other problems.

If the rotor and hub are a one-piece assembly, remove the outer wheel bearing and lift the rotor and hub off the spindle. Be careful not to hit the inner bearing on the spindle when removing the hub.

SERVICE TIP: Hub and rotor mating areas can be cleaned with an electric drill and a special wire brush tool. See Figure 7-61 and Chapter 2 of this *Shop Manual* for the details of this tool.

Installing a Rotor

WARNING: Always check ventilated rotors prior to installing them on the vehicle. Installed on the wrong side of the vehicle, they will cause rapid heat buildup during braking and will damage or destroy the rotor.

New rotors come with a protective coating on the friction surfaces. To remove this coating, use carburetor cleaner, brake cleaner, or solvent recommended by the carmaker.

Whether the original or refinished rotor or a new rotor is being reinstalled, sand the rotor surfaces with 80-grit to 150-grit aluminum oxide sandpaper, as recommended by the carmaker. After sanding the rotor, wash it with denatured alcohol; then do not touch the surface with your hands.

Sanding a rotor establishes a **nondirectional finish** on the surface that will not start a premature wear pattern on the pads. The nondirectional sanded surface also helps pad break in and reduces brake noise.

If the rotor is a two-piece floating rotor, make sure all mounting surfaces are clean. Apply a small amount of silicone dielectric compound to the pilot diameter of the disc brake rotor before installing the rotor on the hub. Reinstall the caliper as explained in the caliper rebuilding procedures. Install the wheel and tire on the rotor and torque the wheel nuts to specifications, following the recommended tightening pattern (Figure 7-55). Failure to tighten in the correct pattern may result in increased lateral runout, brake roughness, or pulsation.

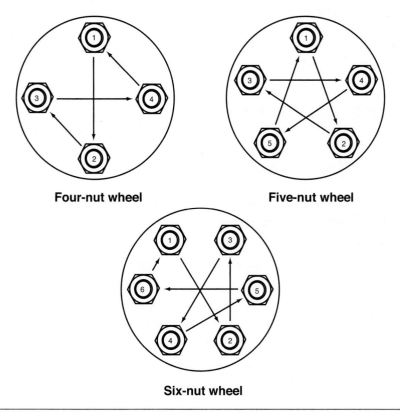

Four-nut wheel   **Five-nut wheel**

Six-nut wheel

Figure 7-55 Tighten wheel nuts in the specified sequence to the specific torque value.

If the rotor is a fixed, one-piece assembly with the hub that contains the wheel bearings, clean and repack the bearings and install the rotor as explained in Chapter 3 of this *Shop Manual*.

After lowering the vehicle to the ground, pump the brake pedal several times before moving the vehicle. This positions the brake linings against the rotor. If so equipped, turn the air suspension service switch back on before starting the engine.

Preventing Runout and Thickness Problems

Runout and thickness, or parallelism, problems can compound each other. For example, excessive runout causes the high spot of the rotor to hit the pad harder each time it rotates with the brakes off, which, in turn, causes the high spot to wear faster, gets much hotter, and leads to a parallelism problem. To avoid—or at least reduce—runout and parallelism problems, observe these guidelines during various steps of rotor service:

❏ Measure rotor runout before removing the rotor from the vehicle and measure it again after mounting the rotor on a bench lathe. Runout should be the same on the vehicle and on a bench lathe. If it is not, machining may actually add runout to the rotor.

❏ Check the runout of a bench lathe arbor frequently. As little as 0.002 inch of runout in a lathe arbor can add runout to a rotor during machining.

❏ Do not overtighten the rotor mounting on a bench lathe arbor. Overtightening the mounting can add runout during machining; tighten by hand.

❏ If bearing races are to be replaced in a hub-type (one-piece) rotor, replace them before machining the rotor, not after.

❏ Reindexing a hubless or floating rotor on a wheel hub may reduce runout.

❏ A lightweight hub used with a hubless or floating rotor can actually warp and acquire runout. If such a problem is suspected, remove the rotor and measure runout of the hub mounting surface while the hub is on the vehicle.

✓ **SERVICE TIP:** Rotor runout and thickness variation, or lack of parallelism, work together to cause pedal pulsation and premature rotor wear. Runout causes the high spot on the rotor to hit the pads harder than the other rotor areas, which causes rotor thickness to wear unevenly.

Refinishing Brake Rotors

Brake rotors should be refinished only under the following conditions:

❏ If the rotor fails lateral runout or thickness variation checks
❏ If there is noticeable brake pulsation
❏ If there are heat spots or excessive scoring that can be removed by resurfacing
❏ If the rotor exceeds refreshing limits

⚠ **WARNING:** Wear safety glasses or face protection when using machining equipment. Injuries to the face or eyes could occur from flying chips of metal.

The minimum thickness specification for all brake rotors is the discard dimension. Do not use a brake rotor that does not meet these specifications. A refinished rotor must be thicker than its discard dimension. Refer to the service manual for exact specifications. A rotor that has been refinished too thin will not have proper heat transfer capabilities and should be replaced.

⚠ **WARNING:** Do not attempt to use a brake lathe without training. The minimum training is studying the lathe's operator manual. Serious injury or damage could occur if the lathe is improperly set up.

When refinishing a rotor, remove the least amount of metal possible to achieve the proper finish, which helps to ensure the longest service life from the rotor.

Never turn the rotor on one side of the vehicle without turning the rotor on the other side. Left- and right-side rotors should be the same thickness, generally within 0.002 inch to 0.003 inch. Similarly, equal amounts of metal should be cut off both surfaces of a rotor. This is particularly critical on rotors used with stationary, fixed calipers. If the rotor is not surfaced equally on both sides, it will not be centered in the caliper opening when installed (Figure 7-56).

⚠ **WARNING:** Do not use rotors that are below or near minimum thickness. A too-thin rotor could shatter during braking, possibly causing an accident or serious injury. Legal action could be taken against the shop and technician.

Ideally, a rotor used with a floating or sliding caliper should be turned equally on both sides. Heat transfer capabilities may be reduced on a ventilated rotor if unequal amounts of metal are removed from the surfaces. From a practical standpoint, one side of a rotor used with a movable caliper can be turned a few thousandths of an inch more than the other as long as total thickness is within limits. Ford rotors described in the section on friction surface depth measurement have limits on the amount of metal that can be removed from the inboard surface.

Turning New Rotors

✓ **SERVICE TIP:** Never assume that new rotors are parallel. Always check them before installing. If they are more than 0.002 inch out of parallel, swap them with the parts vendor.

Most carmakers and parts manufacturers recommend against refinishing new rotors unless measured runout exceeds specifications, which does not happen often. New rotors have the correct surface finish, which may be disturbed by turning on a lathe. Making a light cut on a new rotor may produce excessive lateral runout and result in brake shudder after only a few thousand miles. Clean any oil film off of the rotor with brake-cleaning solvent or alcohol and let the rotor air dry

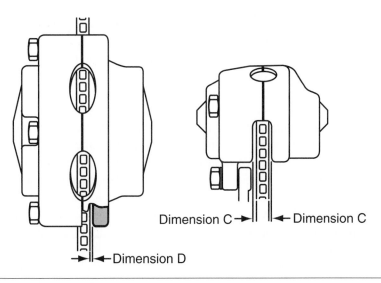

Dimension C → ← Dimension C

→ ← Dimension D

Figure 7-56 A fixed caliper must be centered and parallel with the rotor. In a few cases, you can shim the caliper mounting for proper alignment but not always.

before installing it on the vehicle. Lacquer thinner cleans the shipping compound from a rotor quicker. However, the thinner must be used in an open, ventilated area and must be stored in a tightly capped, labeled container.

Brake Lathes

Both bench-type, off-vehicle lathes (Figure 7-57) and on-vehicle lathes (Figure 7-58) are used to refinish rotors. Regardless of the type of lathe used, the equipment should be serviced regularly according to the manufacturer's maintenance procedures to maintain correct operation. If this is

Figure 7-57 AAMCO brake lathes like the one shown are found in most brake repair shops and are preferred by many older technicians.

Figure 7-58 A typical on-the-car brake lathe.

the first time this lathe is being used, it may be advisable to use a dial indicator to check the trueness of the lathe shaft, stationary and rotating.

On a bench lathe, the rotor is mounted on the lathe arbor and turned at a controlled speed while cutting bits pass across the rotor surfaces to remove a few thousandths of an inch of metal. The lathe turns the rotor perpendicularly to the cutting bits so that the entire rotor surface is refinished. Most lathes can operate at slow, medium, and fast speeds through a series of drive belts.

Different cutting assemblies are used for rotors and for drums. Most rotor cutting assemblies have two cutting bits. The rotor mounts between the bits and is pinched between them. As the cut is made, the same amount of surface material is cut from both sides of the rotor. Some lathes use only one bit for rotor refinishing and separate cuts of equal amounts must be taken from each side.

When using a lathe with only one cutting bit, do not take the rotor off the lathe until both sides are cut. Always cut both sides of the rotor without removing it from the arbor. This ensures that both sides of the rotor will be parallel after refinishing. Check the accuracy of the cuts with a dial indicator and an outside micrometer. Even a 0.001-inch to 0.002-inch variation in positioning the rotor on the lathe can add runout to a rotor. This will cause pedal pulsation and reduce braking efficiency.

The attaching adapters, tool holders, vibration dampers, and cutting bits must be in good condition. Make sure mounting adapters are clean and free of nicks. Always use sharp cutting tools or bits and use only replacement cutting bits recommended by the equipment manufacturer. Dull or worn bits leave a poor surface finish, which will affect braking performance.

The tip of the cutting bit should be slightly rounded, not razor sharp. This rounded tip is more important for turning a drum where a pointed-tip bit can cut a spiral groove into the drum. Even when turning a rotor, however, a pointed bit can leave small grooves that work against the nondirectional finish needed on a rotor.

Mounting a Rotor on a Bench Lathe

The mounting procedure for a rotor depends on whether the rotor has wheel bearings mounted in its hub. A one-piece rotor with bearing races in the hub mounts to the lathe arbor with tapered or spherical cones (Figure 7-59). A two-piece rotor removed from its hub is centered on the lathe arbor with a spring-loaded cone and clamped in place by one or two large cup-shaped adapters (Figure 7-60).

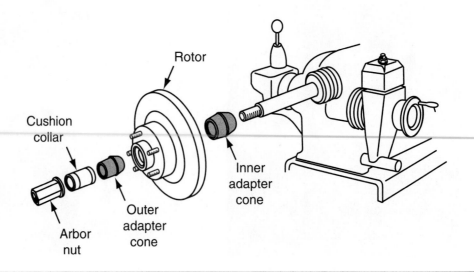

Figure 7-59 Rotors with hubs mount on a bench lathe by using adapter cones in the bearing races.

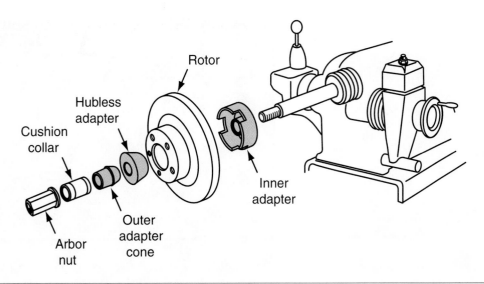

Figure 7-60 Hubless rotors mount on a bench lathe by using hubless adapters that sandwich the rotor between them.

CAUTION: Do not overtighten the rotor cones and adapter onto the lathe spindle. Damage to the rotor or the lathe may occur.

For one-piece rotors and hubs with bearings installed, remove the outer bearing and the inner bearing and grease seal before mounting the rotor on the lathe arbor. When mounting a one-piece rotor and hub, check the inner bearing races (cones) to be sure they are secure in the hub. If either race is loose, replace all bearings and races. Refer to Chapter 3 in this *Shop Manual* for bearing removal and installation instructions.

Remove all grease and dirt from the bearing races before mounting the rotor. It is sometimes necessary to steam clean the grease out of the hub. Index the rotor on the wheel-bearing races to ensure that the machining is accurately indexed to the axis of the rotor. Use the appropriate cones and spacers to lock the rotor firmly to the arbor shaft (Figure 7-61). For hubless rotors, clean all

 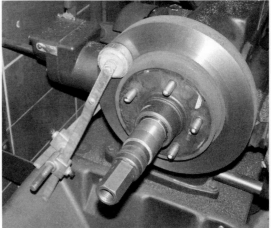

Figure 7-61 On the left is a spring-type vibration damper installed around the edge of the rotor. The right shows a puck-type vibration damper that resembles a disc brake pad. There is one on each side of the rotor.

Figure 7-62 Clean dirt and rust from the hub-mounting area of a hubless rotor.

Figure 7-63 A close-up of a spring-type vibration damper installed around the edge of the rotor. This damper is growing in use.

rust and corrosion from the hub area with coarse sandpaper or a wire brush (Figure 7-62). As with bearing-type rotors, use the proper cones and spacers to mount the rotor to the arbor shaft.

When the rotor is on the lathe, install a rubber, spring-type, or puck-type vibration damper on the outer circumference of the rotor (Figure 7-63) to prevent the cutting bits from chattering during refinishing. Use of the vibration damper results in a smoother finished surface. The damper also helps reduce unwanted noise. Photo Sequence 14 shows the general procedures for mounting a floating rotor on a brake lathe. For a one-piece disc/hub, refer to Photo Sequence 19 for general mounting procedures.

 SERVICE TIP: A disc brake rotor should have a smooth, nondirectional finish for optimum braking and pad life, but how smooth is smooth? Actually, a "smooth" rotor surface is pretty well polished to a 30-, 40-, or 50-microinch finish. A quick way to judge the surface finish is to draw a line with a common ballpoint pen along the radius of the rotor at any point. If the line looks like a solid smooth line and not a series of dots or dashes, the surface is close to the desired finish.

Machining a Rotor on a Bench Lathe

Special Tools

Bench-mounted brake lathe

Sufficient and correct rotor adapters

Service manual

Before removing any metal from the rotor, verify that it is centered on the lathe arbor and that extra runout has not been created by the lathe mounting. If the rotor is not centered and square with the arbor, machining can actually add runout. To check rotor mounting, make a small scratch on one surface of the rotor, as follows:

WARNING: Wear safety glasses or face protection when using machining equipment. Injuries to the face or eyes could occur from flying chips of metal.

1. Begin by backing the cutting assembly away from the rotor and turning the rotor through one complete revolution to be sure there is no interference with rotation.
2. Start the lathe and advance the cutting bit until it just touches the rotor surface near midpoint.

Photo Sequence 14
Mounting a Rotor on a Bench Brake Lathe

P14-1 Clean the inside and outside of the rotor where the lathe's centering cone and clamps will be in contact with the rotor. Select a centering cone that fits into the rotor's center hole without protruding too much in either direction.

P14-2 Slide the inner clamp onto the arbor followed by the centering cone tension spring.

P14-3 Slide the cone onto the arbor followed by the rotor. The rotor is mounted in the same way as on the vehicle.

P14-4 Push on the rotor to compress the spring, and then hold the rotor in place as the outer clamp is fitted onto the arbor. The two clamps, inner and outer, must be the same size.

P14-5 Keep the pressure on the install parts as the spacers and bushing are installed onto the rotor.

P14-6 Hold all parts in place as the arbor nut is installed and tightened to lathe manufacturer specifications.

P14-7 Install the vibration damper. The damper may be a flexible strap or spring that fits the outer circumference of the rotor, or two padlike devices that are forced against each side of the rotors similar to installed disc brake pads.

P14-8 The rotor is now properly mounted and ready for the final check before machining.

> ⚠️ **WARNING:** A brake lathe can produce a lot of torque. Do not wear loose clothing or unrolled long-sleeved shirts while machining or setting up the lathe.

3. Let the cutting bit lightly scratch the rotor, approximately 0.001 inch (0.025 mm) deep.

4. Move the cutting bit away from the rotor and stop the lathe. If the scratch is all the way around the rotor, the rotor is centered and you can proceed with resurfacing.

> ⚠️ **WARNING:** Do not attempt to make adjustments or perform other actions in or near the cutting head. Allow the rotor to come to a complete halt before loosening the nut. The lathe produces sufficient torque to break bones or cause other injuries.

5. If the scratch appears as a crescent (Figure 7-64), either the rotor has a lot of runout or it is not centered on the arbor. In this case, loosen the arbor nut and rotate the rotor 180 degrees on the arbor; then retighten the nut.

6. Repeat step 2 through step 4 to make another scratch about ¼ inch away from the first.

7. If the second scratch appears at the same location on the surface as the first, the rotor has significant runout, but it is properly centered on the lathe and you can proceed with machining.

8. If the second scratch appears opposite the first on the rotor surface, remove the rotor from the lathe arbor and recheck the mounting.

> ⚠️ **WARNING:** Do not allow the lathe to operate without close monitoring. Do not allow other persons to work near the lathe until it stops running. Inattention or lack of monitoring could cause an accident.

In extreme cases, the lathe arbor shaft may be bent. If the second test scratch appears as a crescent in a different area from the first scratch, the rotor is wobbling on the arbor 0.003 inch to 0.004 inch or more. Do not resurface the rotor because additional runout could be machined into

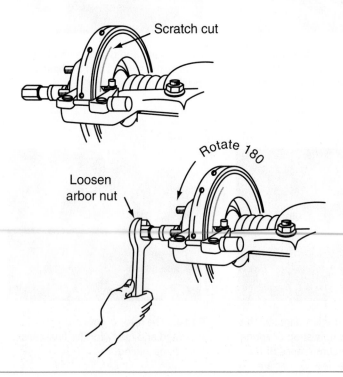

Figure 7-64 Check rotor runout by making a light scratch cut (top). Then rotate the rotor 180 degrees on the lathe arbor (bottom) and make the second cut. The position of the second cut helps to indicate the cause of any runout present.

the rotor, causing a pedal pulsation. One of three conditions is the likely cause of the wobble: a bent lathe arbor, distorted mounting adapters, or excessive rotor runout. The last is the most common.

To determine if the arbor is bent, mount a dial indicator on the lathe and disconnect lathe power. Release the pulley belt tension by moving the controlling lever; then rotate the arbor slowly by turning the drive pulleys. Observe the dial indicator needle. Movement of the needle more than one division (0.001 inch) indicates a bent arbor. Contact the lathe manufacturer for lathe service information.

A distorted mounting adapter sometimes can be corrected by installing it on a precision metal-working lathe and machining it. It is usually more practical to replace a defective adapter, however.

Rotors with excessive runout can be resurfaced if the amount and position of runout is marked when the rotor is on the vehicle and these same conditions can be reproduced on the bench lathe. Extreme runout correction is explained in a later section of this chapter.

Determining Machining Limits. To determine the approximate amount of metal to be removed, turn on the lathe and bring the cutting bit up against the rotating disc until a slight scratch is visible as you did to verify rotor centering. Turn off the lathe and reset the depth-of-cut dial indicator to zero (Figure 7-65). Find the deepest groove on the face of the rotor and move the cutting bit to that point without changing its depth-of-cut setting. Now use the depth-of-cut dial to bottom the tip of the cutter in the deepest groove. The reading on the dial now equals, or is slightly less than, the amount of metal to be removed to eliminate all grooves in the rotor. For example, if the deepest groove is 0.019 inch deep, the total amount to be removed may be 0.020 inch. For best results with cuts that have a total depth greater than 0.015 inch, take two or more shallow cuts rather than one deep cut.

Using a disc rotor micrometer is a sometimes quicker way to determine the amount of material to be removed from a rotor.

Adjusting Lathe Settings. Before starting to machine the rotor, consider and adjust three lathe settings: lathe speed (rpm), cross-feed, and depth of cut. The lathe speed usually stays constant throughout the machining operations, but most lathes have at least two or three speed settings. Select the best speed for the rotor you are machining according to the lathe manufacturer's instructions. Most rotors can be refinished and sanded satisfactorily at 150 rpm to 200 rpm.

The **cross-feed** is the distance the cutting bit moves across the friction surface during each lathe revolution. The cross-feed and the depth of cut work together for a correct finish. A cross-feed that is too fast and a deep depth of cut will cause the rotor to vibrate and leave gouges in the material. The rotor will have to be replaced in this instance. A cross-feed of 0.006 inch to

Figure 7-65 The depth-of-cut dial is graduated in thousandths of an inch to set the cutting depth.

0.010 inch (0.15 mm to 0.25 mm) per revolution is good for rough cuts on most rotors. Make the finish cut at a slower cross-feed of about 0.002 inch (0.05 mm) per revolution.

The depth of cut is the amount of metal removed by the cutting tool in each pass across the rotor. Limit the depth of cut to 0.006 inch or 0.007 inch (about 0.15 mm) for each pass. Make the finish cut at the same depth for organic pads, but semimetallic linings work best with a finish cut on the rotor made at a depth of 0.002 inch (0.05 mm).

Special Tools

Bench-mounted
  brake lathe

Sufficient and correct
  rotor adapters

Micrometer

Power or hand
  sanding disc

Hot soapy water
  with basin

Service manual

Machining the Rotor. To make the series of refinishing cuts, proceed as follows:

1. Reset the cutter position so the cutting bits again just touch the ungrooved surface of the inside of the rotor.

2. Zero the depth-of-cut indicators on the lathe.

3. Turn on the lathe and let the arbor reach full running speed.

4. Turn the depth-of-cut dials for both bits to set the first pass cut. Turning these dials moves the bits inward. The dial is calibrated in thousandths-of-an-inch or millimeter increments. The first cut should only be a portion of the total anticipated depth of cut.

5. When the cutting depth is set for the first cut, activate the lathe to move the cutting bits across the surface of the rotor. After the first cut is completed, turn off the lathe and examine the rotor surface. Areas that have not yet been touched by the bits will be darker than those that have been touched (Figure 7-66).

6. If there are large patches of unfinished surface, make another cut of the same depth. When most of the surface has been refinished, make a shallow finishing cut at lower arbor speed. Repeat the slow finishing cut until the entire rotor surface has been refinished.

7. Do not remove any more metal than necessary for a uniform surface finish, free of grooves. Make sure the refinished rotor thickness is above or beyond its service limit. To ensure this, remeasure the refinished rotor with a micrometer to determine its minimum thickness and compare this measurement to the manufacturer's minimum refinished thickness specification.

It is very important that the rotor surface be made nondirectional after machining (Figure 7-67). Dress the rotor surfaces with a sanding disc power tool with 120-grit to 150-grit aluminum oxide sandpaper. Sand each rotor surface with moderate pressure for at least 60 seconds. You

Figure 7-66 The first cut should remove most of the worn area of the rotor.

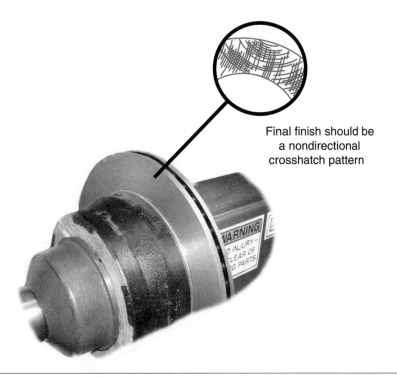

Final finish should be
a nondirectional
crosshatch pattern

Figure 7-67 A nondirectional finish on the surface of a rotor is a crosshatched pattern that does not follow the arc of rotor rotation. The surface helps pad break-in and reduces brake noise.

Figure 7-68 Use a sander to apply a nondirectional finish.

can also do the job with a sanding block with 150-grit aluminum oxide sandpaper. With the rotor turning at approximately 150 rpm, sand each rotor surface for at least 60 seconds with moderate pressure (Figure 7-68).

After the rotor has been sanded, clean each surface with hot water and detergent, which will float out the smallest particles of iron and abrasive materials. Then dry the rotors thoroughly with clean paper towels. If working in an area of high humidity or if the rotors will not be reinstalled immediately, wipe the friction surfaces with denatured alcohol and a clean cloth to be sure that all moisture is removed.

Machining to Remove Excessive Rotor Runout. Rotors with excessive runout may have been machined incorrectly previously, or the rotor may have become distorted by overtightening the wheel nuts. Such a rotor can be resurfaced if the amount and position of runout is marked when the rotor is mounted on the vehicle and these same conditions can be reproduced on the bench lathe. Proceed as follows:

1. Check the rotor lateral runout with the rotor on the vehicle as explained earlier in this chapter. The maximum clockwise rotation of the dial indicator needle indicates the place where the rotor has its maximum outward deflection. Mark this spot on the rotor hub with a plus (+) mark.

2. After marking the rotor, remove it from the vehicle and mount it on the lathe with the proper mounting adapters.

3. Install the dial indicator on the lathe and adjust the indicator pointer to the outside rotor surface.

4. Slowly turn the rotor clockwise and observe whether the runout amount and location are the same as they were on the car. If the readings are not the same as they were on the car, loosen the arbor nut and reposition the rotor and the adapters on the rotor. Repeat this procedure until the rotor runout matches the runout measured on the car as previously described.

5. When the conditions on the vehicle have been re-created on the lathe, machine off an amount of material equal to the maximum lateral runout.

If the rotor runout on the lathe cannot be made to match the runout found on the car, the rotor should be replaced.

To determine the runout of the rotor hub of a two-piece floating rotor, mount the dial indicator point against the hub flange. Rotate the axle and observe the movement of the dial indicator needle.

On-Vehicle Lathes

Excellent refinishing results can be achieved with an on-vehicle brake lathe (Figure 7-69). One advantage of an on-vehicle lathe is that the rotor does not have to be removed from the spindle or the hub. On-vehicle lathes also are ideal for rotors with excessive runout problems. The time and trouble needed to reproduce the exact runout condition on a bench lathe arbor is eliminated or reduced when you refinish the rotor on the vehicle.

To install the lathe, remove the wheel and then remove the caliper from its support and hang it out of the way with heavy wire. To hold a two-piece floating rotor to its hub, reinstall the wheel nuts with flat washers or adapters provided by the lathe manufacturer against the rotor. Torque the wheel nuts to specifications in the prescribed sequence to keep from introducing any runout into the rotor. Attach the lathe to the rotor using the hardware that comes with the lathe. Follow the manufacturer's mounting and operating instructions precisely.

As with the bench lathe, the cutting bits of the on-vehicle lathe straddle the rotor and depth-of-cut settings are made using adjustment knobs. The dial marks determine the depth of cut each tool is taking from the rotor surface.

The wheel bearings and spindle or axle serve as the lathe arbor, so they must be in good condition. Excessive bearing end play may prevent the use of an on-vehicle lathe or require bearing replacement before an on-vehicle lathe can be used. If any end play is present in an adjustable tapered roller bearing, carefully tighten the adjusting nut by hand just enough to remove the end play before installing the lathe. After turning the rotor, readjust the bearing as explained in Chapter 3 of this *Shop Manual.*

Figure 7-69 This on-car lathe has its own motor to rotate the rotor.

Some on-vehicle lathes mount on the caliper support, on the hub, or on a separate stand. Depending on lathe design, the rotor may be rotated by vehicle engine power or by the lathe's electric motor during refinishing.

> ⚠ **WARNING:** Using the engine to turn the rotor puts a lot of torque into the assembly even at idle speeds. Ensure that the caliper, hoses, and any wiring harness are tied back from the rotor and lathe. Injuries could occur from cut hoses or loose components.

If the engine is used to turn the rotor, the lathe can be used only on drive wheels. However, a problem exists because the differential gearing in the transaxle transmits the power to the opposite wheel, not to the rotor to be resurfaced. To prevent that opposite wheel from turning, that wheel can be lowered to the floor. This may reduce floor-to-lathe clearance to the point where it is difficult to run the lathe, however.

Another way to transfer drive power to the rotor is to lock the brakes at the opposite wheel. Ensure that the caliper on the side being machined is secured by a wire and clear of the lathe mechanism. Use a small C-clamp to slightly or completely push the piston into that caliper body. Leave the C-clamp in place until the rotor is machined. In this manner, the brakes can be applied using the brake depression tool from the alignment area. Otherwise, keep a second person in the driver's seat. Start the engine, shift the transmission into first gear or reverse (depending on location of the lathe), and idle the engine as slowly as possible.

The rotor should rotate as slowly as possible to provide a smooth surface. Spinning the rotor too fast will cause the tool bits to overheat and wear out faster. Excessive rotor speed during machining can damage the rotor.

During machining, the rotor must turn into the cutting edges of the bits. Depending on the design of the lathe, rotor rotation may change from one side of the car to the other. Therefore, the engine can drive the rotor in first gear or reverse to get the proper rotation.

Cross feed of an on-vehicle lathe may be automatic or manually controlled. If operated manually, advance the cutting bits slowly and steadily across the rotor. Set an automatic cross feed to 0.003 inch (0.08 mm) per revolution. Depth of cut should be shallower than on a bench lathe. Make successive cuts at a depth or 0.004 inch to 0.002 inch (0.10 mm to 0.05 mm) until the desired finish is achieved.

Self-powered lathes that drive the rotor with a motor mounted in the lathe have become more popular than lathes that use the engine to drive the rotor. Not only can a self-powered lathe be used on nondriving wheels, rotor speed can be controlled more exactly, and avoids dealing with engine exhaust inside the shop. Self-powered on-vehicle lathes produce almost as much torque as the engine at idle. Use extreme caution with this type of lathe.

As mentioned earlier, an on-car lathe may be mounted on the brake caliper support or on its own stand and indexed to the hub and the wheel studs. Each lathe has its own operating instructions, which you must follow carefully. All lathe operating procedures include most of the following steps:

1. Check for wheel bearing end play. If end play on an adjustable bearing is noted, adjust the bearing to remove the end play. If end play exceeds the carmaker's specifications on a nonadjustable bearing, replace the bearing.

2. Check the fluid level in the master cylinder and be sure it is not overly full. Remove fluid if necessary by connecting a flexible hose to the bleeder screw and running it to a catch container. This helps prevent corrosion and dirty fluid from reentering the master cylinder. The reservoir should be about half full when the caliper pistons are pushed back to demount the caliper in case fluid does not flow freely from the screw.

3. Place the transmission in neutral, release the parking brake, and raise the vehicle on a hoist to an appropriate working height.

4. Remove the wheel from the first rotor to be serviced. Remove any rust or corrosion from the axle or hub flange rotor mounting area, the rotor's matching surface to the flange, and around the wheel studs. This will help solve many runout problems.

5. Install spacers on the studs if required for the rotor drive adapter and reinstall the wheel nuts. Torque the nuts to specifications in the prescribed sequence. Even if the rotor is a one-piece rotor and if the vehicle engine is used to drive it for machining, it is a good idea to reinstall and properly torque the wheel nuts, ensuring that any runout created by wheel nut torque is removed during machining.

6. Remove the caliper from its support (push the piston back in its bore if necessary) and hang the caliper out of the way from the chassis or suspension, using a suitable support. Do not hang the caliper from the brake hose.

7. If the lathe is to be mounted on the caliper support, be sure that the area on the support around the mounting holes is free of dirt, rust, and gouges. Select the proper lathe-mounting adapters and mount the lathe on the caliper support. Tighten all fasteners securely.

8. If the lathe is self-powered, attach the wheel drive adapters to the wheel studs and align the drive motor stand with the hub axis. Lock the stand wheels and plug in the power cord.

9. Turn on the motor before moving the cutting bits close to the rotor to be sure the motor is turning in the right direction.

10. With the rotor turning in the right direction, operate the hand wheel to move the cutting bits until they are ½ inch in from the outer edge of the rotor. Then turn the depth-of-cut micrometer until the bits just lightly touch both rotor surfaces.

11. Perform a scratch test before machining the rotor.

12. Turn the hand wheel to move the bits outward and remove rust and dirt from the outer edge of the rotor.

13. Manually feed the bits inward until they are past the inner pad contact line. Then set the lathe stop for the inward feed, or in-feed.

14. Rotate both micrometer knobs clockwise for an initial cut of no more than 0.004 inch (0.10 mm).

15. Shift the lathe to automatic operation at the fast-feed rate. Switch the lathe to feed outward for the first rough cut.

16. After the lathe completes the first cut, turn off the lathe motor and check the uniformity of the cut. Make additional cuts if necessary, with the final cut at a depth of 0.002 inch (0.05 mm).

Rear Disc Brake Inspection and Replacement

Rear disc brakes are used on many vehicles. In most cases, the rear brakes are identical to the front disc brakes except for some type of parking brake mechanism. Figure 7-70 is an exploded view of a Honda rear disc brake that is typical of one with a parking brake mechanism. Photo Sequence 15 illustrates and explains pad replacement and caliper overhaul of this typical rear disc brake. All illustrations are provided by Honda Motor Co., Inc.

To complete this rear caliper service, install the caliper on the caliper bracket and tighten the caliper bolts to torque specifications. Reconnect the brake hose to the caliper with new sealing washers and tighten the banjo bolt to specifications. Then reconnect the parking brake cable to the arm on the caliper. Finally, reinstall the caliper shield.

Top off the master cylinder reservoir and bleed the brake system. Depress the brake pedal several times, then adjust the brake pedal. Before making adjustments, be sure that the parking brake arm on the caliper touches the pin.

 SERVICE TIP: Do not overlook the rear brakes when servicing a car with four-wheel disc brakes. Although the rear brakes do only 20 percent to 25 percent of the braking, they can run hotter than the front brakes on some cars because of they way the rear wheels are shrouded. Also, if the rear brakes are not up to full capacity, the front brakes will be overworked and wear out more quickly.

Classroom Manual
page 193

Special Tools

Service manual

Vehicle specific rear disc tools

Lift or jack with stands

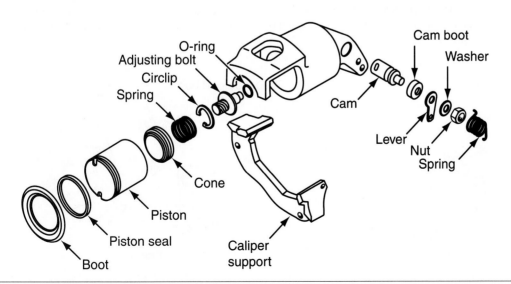

Figure 7-70 Service of this rear disc brake is explained in Photo Sequence 15.

Photo Sequence 15
Typical Procedure for Overhauling a Rear Brake Caliper

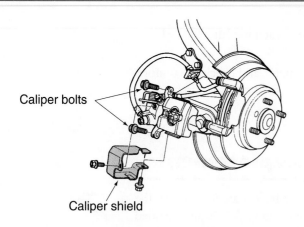

P15-1 Raise the vehicle on the lift and remove the rear wheels. The rear caliper may be protected by a plastic shield. Remove this shield and the bolts securing the caliper.

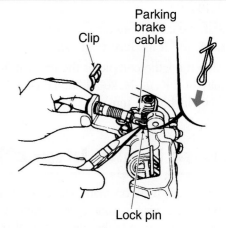

P15-2 Disconnect the parking brake cable from the lever on the caliper by pulling out the lock pin.

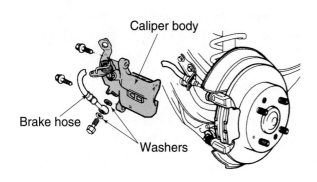

P15-3 Disconnect the brake hose from the caliper, remove the caliper mounting bolts, and lift the caliper off its support. To keep dirt from entering the caliper body, clean the outside of the caliper before disassembly.

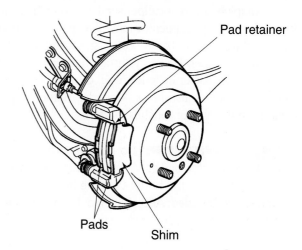

P15-4 Remove any pad shims and retainers and lift the pad spring out of the caliper. Then pull the pads off the caliper.

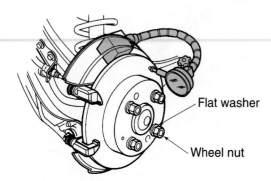

P15-5 Check the condition of the rotor as instructed earlier in this chapter. Thoroughly clean the rotor surface and inspect it for defects and damage. Measure rotor thickness and runout.

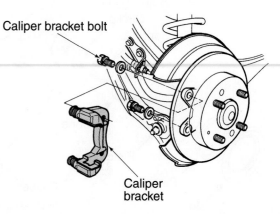

P15-6 Remove the bolts securing the caliper bracket and lift off the bracket. Thoroughly clean the caliper bracket.

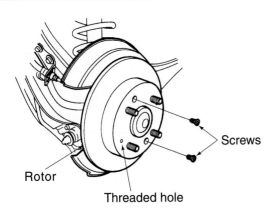

P15-7 This is a two-piece rotor that must be removed from the hub. Remove the screws and pull the rotor off the hub.

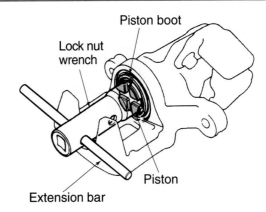

P15-8 To remove the caliper piston from the bore, use a proper size locknut wrench and extension bar. Turn counterclockwise to back the piston out of the bore. When the piston is free, remove the piston boot.

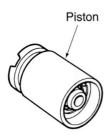

P15-9 Carefully inspect the piston for wear. Replace the piston if it is worn or damaged in any way.

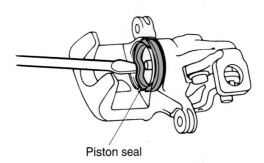

P15-10 Remove the piston seal from the caliper with the tip of a screwdriver or a wooden or plastic scraper. Do not scratch the bore.

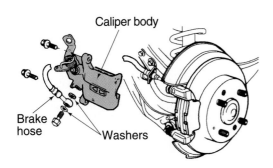

P15-11 Removing the brake spring and its related parts from the caliper requires several special tools. Install a rear caliper guide in the cylinder so the cutout on the guide aligns with a tab on the brake spring cover.

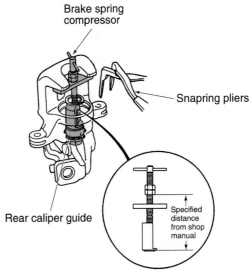

P15-12 Install a brake spring compressor between the caliper and the caliper guide. Turn the shaft of the compressor to compress the spring. Use snaping pliers to remove the circlip that holds the spring.

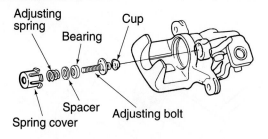

P15-13 After removing the circlip, remove the spring compressor from the caliper. Then remove the spring cover, adjusting spring, spacer, bearing, adjusting bolt, and cup.

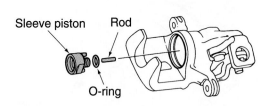

P15-14 Next remove the sleeve piston and O-ring. Then remove the rod from the can.

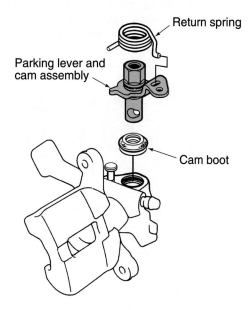

P15-15 Remove the return spring, the parking lever and cam, and the cam boot from the caliper body. Do not loosen the parking nut on the parking lever and cam with the cam installed in the caliper. If the lever and shaft must be separated, secure the lever in a vise before loosening the parking nut.

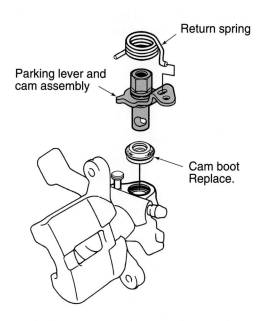

P15-16 Begin caliper reassembly by packing all cavities of the needle bearing with the specified lubricant. Coat the new cam boot with brake lubricant and install it in the caliper. Apply brake lubricant to the area on the pin that contacts the cam. Install the cam and lever into the caliper body. Then install the return spring. If the cam and lever were separated, reassemble them before installing the cam in the caliper.

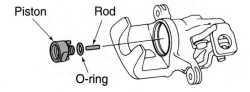

P15-17 Install the rod in the cam, followed by a new O-ring, lubricated with fresh brake fluid, on the sleeve piston. Then install the sleeve piston so the hole in the bottom of the piston is aligned with the rod in the cam and the two pins on the piston are aligned with the holes in the caliper.

Photo Sequence 15
Typical Procedure for Overhauling a Rear Brake Caliper (continued)

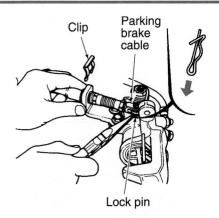

P15-18 Install a new cup with its groove facing the bearing side of the adjusting bolt. Fit the bearing, spacer, adjusting spring, and spring cover on the adjusting bolt; then install it in the caliper bore.

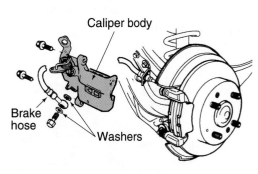

P15-19 Install the special caliper guide tool in the caliper bore, aligning the cutout on the tool with the tab on the spring cover. Adjust the special tool as shown in P15-18.

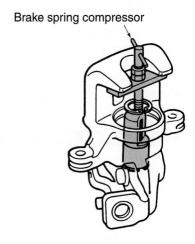

P15-20 Install the brake spring compressor and compress the spring until the tool bottoms out.

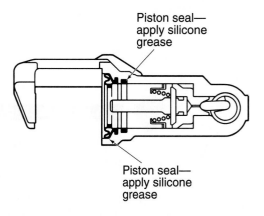

P15-21 Make sure that the flared end of the spring cover is below the circlip groove. Then install the circlip and remove the spring compressor. Ensure that the circlip is properly seated in the groove.

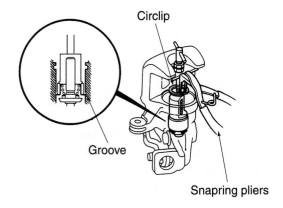

P15-22 Coat the new piston seal with fresh brake fluid and the piston boot with silicone lubricant and intall them in the caliper.

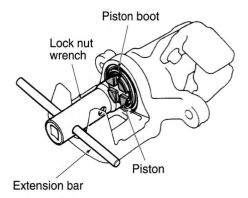

P15-23 Coat the outside of the piston with brake fluid and install it on the adjusting bolt while rotating it clockwise with the locknut wrench.

Photo Sequence 15
Typical Procedure for Overhauling a Rear Brake Caliper (continued)

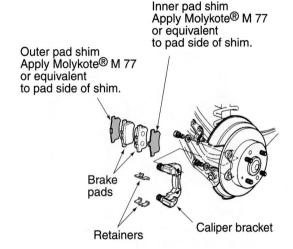

Outer pad shim
Apply Molykote® M 77
or equivalent
to pad side of shim.

Inner pad shim
Apply Molykote® M 77
or equivalent
to pad side of shim.

Brake pads

Retainers

Caliper bracket

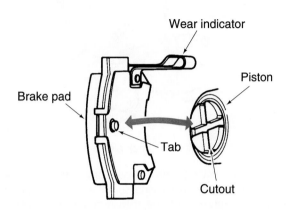

Wear indicator

Piston

Brake pad

Tab

Cutout

P15-24 Install the new or refinished rotor and the clean caliper bracket. Install the new brake pads, pad shims, retainers, and springs onto the caliper bracket.

P15-25 Lubricate the dust boot with a small amount of silicone lubricant before installing it to reduce the chances of twisting. On this rear caliper, rotate the piston clockwise to install it in the bore. Align the cutout in the piston with the tab on the inner brake pad by turning the piston back. The piston boot must fit properly without twisting. If the boot is twisted during installation, back it out until it is seated properly. Finally, reinstall the caliper and the splash shield.

CASE STUDY

It is usually not a good idea to buy into problems caused by someone else. Here is an example of why.

The owner of a truck wanted to save a couple bucks, so he changed his front brake pads himself. Shortly afterward, the truck started intermittently pulling to the right during braking. The owner took the truck to a shop and asked for help.

The brake techs at the shop found that the rotors had excessive runout and the calipers were in pretty sad shape, with some visible leakage. The techs surfaced the rotors and replaced the calipers. They also replaced the idler arm, which had excessive play. The brakes now worked fine on a test drive, so the truck was delivered to its owner. A week later, it was back.

This time, the truck randomly pulled both left and right. Again, the rotors got checked carefully, and the new calipers were exchanged on warranty. During the second service, one caliper mounting bolt was found to be loose and was tightened properly when the second set of calipers were installed. Again, the brakes were fine on a road test, and again the truck was returned to its owner.

Two weeks later, after a long trip, the truck was back at the shop with brake pull occurring again. This time the brake tech went through all the system lines and hoses. Pressure gauges installed on the calipers revealed erratic pressure at the front brakes. Apparently during the do-it-yourself brake bleeding after the first set of pads, dirt got into the system and partly clogged the hoses. If a qualified shop had done the job to begin with, the problem probably never would have developed.

Terms to Know

Aqueous	Discard dimension	Parallelism
Bearing end play	Heat checking	Rotor lateral runout
Bedding in	Loaded caliper	Vernier caliper
Burnishing	Nondirectional finish	
Cross-feed		

ASE-Style Review Questions

1. While servicing the front disc brakes on a FWD vehicle:
   Technician A determines the right wheel pad is worn and replaces both the right and left wheel pads.
   Technician B determines the pads are not worn but rotates their position to ensure even pad wear.
   Who is correct?
   A. A only   **C.** Both A and B
   B. B only   **D.** Neither A nor B

2. After servicing the disc brakes on a vehicle:
   Technician A reinstalls the wheel nuts using an impact wrench to ensure a tight fit.
   Technician B refills the master cylinder reservoirs to the proper level.
   Who is correct?
   A. A only   **C.** Both A and B
   B. B only   **D.** Neither A nor B

3. *Technician A* says that loaded calipers are replacement calipers that come with pads and hardware already installed.
   Technician B says that loaded calipers should be installed in axle sets.
   Who is correct?
   A. A only   **C.** Both A and B
   B. B only   **D.** Neither A nor B

4. *Technician A* says that it is very important that the rotor surface be made nondirectional during refinishing.
   Technician B says that rotors should always be refinished as part of routine disc brake service.
   Who is correct?
   A. A only   **C.** Both A and B
   B. B only   **D.** Neither A nor B

5. When performing disc brake work:
   Technician A works on one wheel at a time to avoid popping pistons out of the other caliper and to allow the other caliper to be used as a guide.

Technician B hangs the calipers from the brake hoses as a convenience to speed up the job.
   Who is correct?
   A. A only   **C.** Both A and B
   B. B only   **D.** Neither A nor B

6. When measuring rotor runout:
   Technician A uses a micrometer.
   Technician B uses a dial indicator.
   Who is correct?
   A. A only   **C.** Both A and B
   B. B only   **D.** Neither A nor B

7. *Technician A* says that newly installed wheel bearings are self-adjusting and require no special adjustment.
   Technician B says that bearing assemblies are interchangeable with one another.
   Who is correct?
   A. A only   **C.** Both A and B
   B. B only   **D.** Neither A nor B

8. *Technician A* says that the minimum wear thickness of a rotor is the discard thickness of the rotor.
   Technician B says that the refinishing dimension is cast into a rotor.
   Who is correct?
   A. A only   **C.** Both A and B
   B. B only   **D.** Neither A nor B

9. When refinishing a rotor on a lathe, the rotor wobbles excessively:
   Technician A says that the lathe arbor may be bent.
   Technician B say that the mounting adapters of the lathe may be distorted.
   Who is correct?
   A. A only   **C.** Both A and B
   B. B only   **D.** Neither A nor B

10. After servicing front disc brakes:

Technician A reinstalls the wheel and tire and tightens the nuts in the recommended tightening pattern.

Technician B pumps the brake pedal several times to position the new pads before test driving the vehicle.

Who is correct?

A. A only   **C.** Both A and B
B. B only   **D.** Neither A nor B

ASE Challenge Questions

1. Rear disc brakes are being discussed:
Technician A says that a stuck right rear caliper could cause a pull to the right when the brakes are applied.
Technician B says that an overheated rear rotor could be caused by misadjusted parking brake cable.
Who is correct?

A. A only   **C.** Both A and B
B. B only   **D.** Neither A nor B

2. The results of a road test are being discussed; the vehicle stops true but has a small vibration:
Technician A says that the vehicle probably has the wrong brake pads.
Technician B says that there is a problem with parallelism.
Who is correct?

A. A only   **C.** Both A and B
B. B only   **D.** Neither A nor B

3. A vehicle has a hard pedal but little braking effect:
Technician A says that the pads are glazed.
Technician B says that the piston seal is not allowing the brakes to self-adjust.
Who is correct?

A. A only   **C.** Both A and B
B. B only   **D.** Neither A nor B

4. There is a sharp odor coming from the front wheels just after pad and rotor replacement:
Technician A says that is normal for new composite pads and rotors.
Technician B says that the protective coating was not removed from the new rotors.
Who is correct?

A. A only   **C.** Both A and B
B. B only   **D.** Neither A nor B

5. Four-wheel disc parking brakes are being discussed:
Technician A says that some parking brakes could completely fail but not affect the service brakes.
Technician B says that a parking brake that was left on could destroy the pads and rotors.
Who is correct?

A. A only   **C.** Both A and B
B. B only   **D.** Neither A nor B

Job Sheet 22

22

Name: _____ Date: _____

Replace Brake Pads

NATEF Correlation

This job sheet is related to NATEF brake tasks: Select, handle, store, and fill brake fluids to proper level; remove caliper assembly from mountings; clean and inspect for leaks and damage to caliper housing; determine necessary action; clean and inspect caliper mounting and slides for wear and damage; determine necessary action; remove, clean, and inspect pads and retaining hardware; determine necessary action; reassemble, lubricate, and reinstall caliper, pads, and related hardware; seat pads, and inspect for leaks; clean, inspect, and measure rotor with a dial indicator and a micrometer; follow manufacturer's recommendations in determining need to machine or replace.

Objective

Upon completion and review of this job sheet, you should be able to replace the brake pads on a front wheel disc brake and inspect the disc brake components.

Tools and Materials

Lift or jack and jack stands

Impact tools

C-clamp

Disc brake silencer

Mechanic wire

Describe the Vehicle Being Worked On

Year _____ Make _____ Model _____

VIN _____ Engine type and size _____

ABS _____ yes _____ no _____ If yes, type _____

Procedure

Task Completed

1. Does the vehicle have ABS? _____

If yes, highlight the precautions required by the manufacturer concerning ABS and routine brake repairs. _____

_____

_____

Caliper/adapter torque _____ Wheel nut torque _____

> **CAUTION:** Before working on the brakes of a vehicle with an ABS, consult the service manual for precautions and procedures. Failure to follow procedures to protect ABS components during routine brake work could damage the components and cause expensive repairs.

2. Inspect the fluid level in the master cylinder. Adjust so the reservoir is about half full. ☐

3. Lift the vehicle and remove the wheel assembly.

⚠ **WARNING:** Wear safety glasses or face protection when using brake fluid. Injuries to the face or eyes could occur from spilled or splashed brake fluid.

4. Inspect the brake caliper mounting area for ABS components, caliper and adapter, and general condition of the steering, suspension, and brake components on each side of the vehicle. Results _____

_____

_____

_____

_____

☐

5. Position a catch basin and clean braking components. Dispose of the waste as required by law and shop policy.

■ **CAUTION:** Always clean around any lines or covers before removing or loosening them. Dirt and other contaminants will void the warranty and may damage the system components.

☐

6. Select the correct wrench and remove the caliper or adapter mounting fasteners.

☐

7. Slide the caliper/adapter from the rotor. Use a large flat screwdriver, if necessary, to pry the pads far enough to clear any ridge on the rotor.

■ **CAUTION:** Do not depress the brake pedal when the pads are being removed, installed, or have been removed from the caliper. Without the pads positioned properly, the piston could be forced from the bore when the pedal is depressed.

☐

8. Use mechanic wire to hang the caliper/adapter from a vehicle component.

☐

9. Remove the pads and antirattle (hardware) clips from the caliper.

10. Inspect the rotor for damage. Measure the thickness of the rotor, compare to specification, and determine serviceability. See Job Sheet 24 for rotor machining. Results _____

_____

_____

11. Inspect the caliper piston boot, slide pins and/or slide areas on the adapter and caliper. Results _____

_____

_____

☐

12. Use a C-clamp or similar tool to press the piston completely into the bore.

☐

13. Clean the slide pins or areas and lube with disc brake lubricant only.

☐

14. Coat the metal portion of each of the new pads with disc brake silencer. Allow to dry set (about 5 minutes).

NOTE: The next two steps may have to be adjusted depending on how the hardware and pads fit into the caliper and caliper support.

15. Install the inner pad and antirattle springs (hardware), if required, into the caliper. Ensure that the pad mates properly with the piston and the antirattle springs are installed properly. ☐

> **CAUTION:** Make certain the friction material, not the metal, faces the rotor. Serious damage to the rotor could occur if the pad is installed backwards.

16. If necessary, install the slide pins onto the caliper and caliper adapter. ☐

17. Slide the caliper/adapter with pads installed over the rotor and align the fastener holes. ☐

> **CAUTION:** Use the correct fastener in the correct manner. Replacement fasteners must meet the specifications for that application.

> **CAUTION:** Make certain that the fitting is not cross-threaded when reconnecting. This could damage the fitting or the component, or both.

> **CAUTION:** Use the correct fastener torque and tightening sequence when installing components. Incorrect torque or sequencing could damage the fastener(s) or component(s).

18. Install and torque the caliper/adapter fasteners. ☐

19. Install the wheel assembly and torque the lug nuts. ☐

20. Repeat step 2 through step 20 for the opposite wheel. ☐

21. Lower the vehicle when both wheel assemblies have been installed. ☐

22. Press the brake pedal several times to seat the pads to the rotor. ☐

> **CAUTION:** Before adding brake fluid, consult the vehicle service manual. Many manufacturers require a specific classification of brake fluid to be used.

> **CAUTION:** Prevent brake fluid from coming in contact with the vehicle's finish. Brake fluid damages paint and finish immediately on contact. If fluid contacts the finish, wash area thoroughly with running water using soap if possible.

> **CAUTION:** If possible, never use brake fluid from a previously opened container. Once opened, even tightly capped containers absorb moisture from the air.

> **CAUTION:** Do not reuse brake fluid. Discard old brake fluid according to EPA and local regulations.

23. Check the brake fluid level and top off as needed. ☐

> **WARNING:** Before moving the vehicle after a brake repair, pump the pedal several times to test the brake. Failure to do so may cause an accident with damage to vehicles, facility, or personal injury.

☐

☐

24. Perform a brake test to ensure the brakes will stop and hold the vehicle. Do this test before moving the vehicle from the bay.

☐

25. When the repair is complete, clean the area, store the tools, and complete the work order.

Problems Encountered _____

_____

Instructor's Response _____

_____

_____

Job Sheet 23

23

Name: _____ Date: _____

Measuring Rotor Runout

NATEF Correlation

This job sheet is related to NATEF brake tasks: Clean, inspect, and measure rotor with a dial indicator and a micrometer; follow manufacturer's recommendations in determining need to machine or replace; remove and reinstall rotor.

Objective

Upon completion and review of this job sheet, you should be able to measure a brake rotor for runout.

Tools and Materials

Dial indicator with clamping mount or magnetic mount

Brake rotor micrometer or outside micrometer

Service manual

Describe the Vehicle Being Worked On

Year _____ Make _____ Model _____

VIN _____ Engine type and size _____

ABS _____ yes _____ no _____ If yes, type _____

Procedure

Task Completed

NOTE: See Job Sheet 22 for gaining access to the rotor.

1. Rotor discard dimension _____

Rotor refinishing dimension _____

Type of rotor _____

Any special precautions to be followed for this rotor? _____

_____

_____

_____

2. Gain access to the rotor using Job Sheet 22. It is not necessary to remove pads from the caliper for this job sheet. ☐

3. Use the brake or outside micrometer to check the thickness of the rotor. Measure at four places. Give measurements in SAE and metric.

1 _____ 2 _____ 3 _____ 4 _____

Based on the measurements, is the rotor serviceable? _____ If not, what should be the next action? _____

_____

_____

_____

NOTE TO INSTRUCTORS: The above step is not needed for this job sheet, but it is included to show students there is no reason to measure runout if the rotor is unserviceable. Such measurement would be a waste of labor.

☐   **4.** Install the lug nuts to hold the rotor firmly against the hub if required.

☐   **5.** Check the wheel bearing for excessive end play. Correct as needed.

☐   **6.** Install the dial indicator mount onto a suspension member.

☐   **7.** Extend the mount's linkage over the rotor and install the dial indicator.

☐   **8.** Move the dial indicator so the plunger will contact the machined surface of the rotor about one-third of the way out from the center.

☐   **9.** Adjust the dial indicator until the plunger is against the rotor and retracted about halfway into the indicator.

☐   **10.** Set the dial face to zero.

     11. Rotate the rotor slowly while observing the needle on the dial indicator.

      Results _____

_____

_____

      Recommended action _____

_____

_____

☐   **12.** Move the dial indicator so the plunger will contact the machined surface of the rotor about two-thirds of the way out from the center.

     13. Repeat step 9 through step 11 and record the results and recommendation.

_____

_____

☐   **14.** Complete any actions necessary to replace, machine, and install the rotor onto the vehicle.

☐   **15.** When the repair is complete, clean the area, store the tools, and complete the repair order.

Problems Encountered _____

_____

Instructor's Response _____

_____

_____

Job Sheet 24

24

Name: _____ Date: _____

Machining Brake Rotors Off-Vehicle

NATEF Correlation

This job sheet is related to NATEF brake tasks: Remove and reinstall rotor; clean, inspect, and measure rotor with a dial indicator and a micrometer; follow manufacturer's recommendations in determining need to machine or replace; refinish rotor off-vehicle.

Objective

Upon completion and review of this job sheet, you should be able to measure and machine a brake rotor.

Tools and Materials

Bench-mounted brake lathe

Service manual

Brake disc (rotor) micrometer or outside micrometer

Describe the Vehicle Being Worked On

Year _____ Make _____ Model _____

VIN _____ Engine type and size _____

ABS _____ yes _____ no _____ If yes, type _____

Procedure

Task Completed

NOTE: See Job Sheet 22 for gaining access to the rotor.

1. Rotor discard dimension _____

Rotor refinishing dimension _____

Type of rotor _____

_____

Any special precautions to be followed for this rotor? _____

_____

_____

2. Remove the wheel assembly, caliper, and caliper support as necessary to gain access to the rotor (see Job Sheet 22). Remove the rotor. ☐

3. Inspect the rotor. Does its condition make it inadvisable to machine? _____

If yes, explain. _____

_____

4. Measure the thickness of the rotor (at atleast 12 points equally spaced around the rotor). Give measurements in SAE and metric sizes.

1 _____ 2 _____ 3 _____ 4 _____ 5 _____ 6 _____

7 _____ 8 _____ 9 _____ 10 _____ 11 _____ 12 _____

5. Measure the deepest groove, if any. Is it deep enough to affect machining? _____

If yes, explain and make a recommendation.

_____

NOTE TO INSTRUCTORS: The following steps are based generally on a bench brake lathe by Raybestos with one spindle and a floating rotor. Other lathes and/or rotors may require additional instruction for the student.

☐ 6. Select a centering cone that fits about halfway through the center hole on the rotor.

☐ 7. Select two identical clamps that fit the rotor without interfering with the machined surfaces of the rotor.

☐ 8. Slide one clamp onto the lathe shaft, open end out.

☐ 9. Slide on a spring, followed by the centering cone.

☐ 10. Slide on the rotor and outer clamp followed by the bushing, spacer (if needed), and the nut. Tighten; do not overtighten the nut.

☐ 11. Install the damping strap or pad as required.

☐ 12. Adjust the assembly inward toward the lathe body until it stops. Back out two turns.

☐ 13. Adjust the cutting head so the rotor fits about center between the cutting tips.

☐ 14. Move the cutting tips until they meet the rotor and reverse out about half a turn.

☐ 15. Adjust the cutting tips until they meet the rotor and reverse out about half a turn.

☐ 16. Ensure that the area around the lathe is clear and the lathe's drive mechanism is in neutral. Switch on the motor.

☐ 17. Adjust each cutting bit slowly until it comes in contact with the turning rotor. Hold in place and set the sliding scale to zero.

18. Reverse the bits away from the rotor and switch off the motor. Observe the scratch around the rotor. Is it even all the way around on each side of the rotor? If not, does the missed area on one side correspond with a scratch on the other side? Does the scratch indicate that the rotor is warped, or is there a problem with the setup or machine? Record your observations and make a recommendation on the next step.

_____

_____

☐ 19. Assuming that the rotor may be machined and the setup is correct, ensure that the cutting bits are away from the rotor. Switch on the motor and move the cutting bit toward the inside (rear) edge of the rotor's friction surface.

☐ 20. Adjust the cutting head until the cutting tips seem to be aligned with the rear (inward toward hub) of the machines surfaces.

☐ 21. Adjust the cutting tips until they contact the rotor.

22. Continue adjusting the tips until the scale is between 0.002 inch (0.5 mm) and 0.007 inch (0.6 mm). Both tips should cut into the surfaces. A 0.002-inch cut is the minimum, whereas a 0.007-inch cut is the maximum for a single cut. ☐

23. Engage fast speed on the lathe. ☐

24. Observe the rotor as it is being machined. Are there dark (uncut) areas? _____

    Explain the next step to be taken. _____

    _____

25. If the answer to step 23 is no, go to step 28. If the answer to step 23 is yes, go to step 25. ☐

26. When the cutting tips clear the rotor, disengage the drive and move the cutting tips back to the starting point. ☐

27. Adjust the cutting tips to 0.002 (0.5 mm) inch deeper and engage fast cut. ☐

28. Repeat step 19 through step 25. ☐

29. When the cutting tip clears the rotor, disengage the drive. ☐

30. Move the cutting tips to the starting point and set to cut 0.002 inch deeper, for semimetalic pads or 0.002 to 0.006 inch for organic pads. ☐

31. Engage the drive mechanism in slow speed. ☐

32. When the cutting bits clear the rotor, disengage the drive and stop the motor. ☐

33. When the rotor stops turning, remove it from the lathe. ☐

34. Wash the rotor in running hot, soapy water using a brush if possible. If a basin is required, use hot, soapy water and a brush to clean the machined surfaces. ☐

35. Rinse with clear water and blow dry with an OSHA-approved blowgun. ☐

36. Install the rotor onto the vehicle. ☐

37. Use Job Sheet 22 to install the caliper and other components. ☐

38. When repair is complete, clean the area and lathe, store the tools, and complete the repair order. ☐

Problems Encountered _____

_____

Instructor's Response _____

_____

_____

Job Sheet 25

25

Name: _____ Date: _____

Machining Brake Rotors On-Vehicle

NATEF Correlation

This job sheet addresses the following NATEF tasks: Clean, inspect, and measure rotor with a dial indicator and a micrometer; follow manufacturer's recommendations in determining need to machine or replace; refinish rotor on vehicle.

Objective

Upon completion of this job sheet, you should be able to machine a rotor mounted on the vehicle.

Tools and Materials

Basic hand tools

On-vehicle brake lathe

Describe the Vehicle Being Worked On

Year _____ Make _____ Model _____

VIN _____ Engine type and size _____

ABS _____ yes _____ no _____ If yes, type _____

Procedure

Task Completed

1. Any special precautions to be taken with machining rotors on-vehicle equipped with ABS? _____

   _____

   Rotor discard dimension (Do not remove the rotor if the dimensions are not readily visible.) _____ SAE _____ mm

   Rotor refinishing dimension _____ SAE _____ mm

   Type of rotor _____

   Any special precautions to be followed for this rotor? _____

   _____

2. Lift the vehicle to about waist level and remove the tire and wheel assembly. ☐

3. Remove the caliper, caliper adapter (mount), pads, and hardware. Use a wire to ☐ suspend the caliper from a suspension or body component, ensuring it is clear of the rotor and lathe during machining.

4. Use an outside (brake) micrometer to measure rotor thickness and parallelism. Note any ☐ wheel bearing movement and correct as necessary. Record your rotor measurements.

Minimum thickness by measurement _____ inch or _____ mm

Does the rotor meet the refinishing dimension? _____

Can the rotor be machined and still meet the discard specification? _____

Does any parallelism measurement exceed 0.002 inch (0.05 mm)? _____

Will machining the rotor to parallelism cause it not to meet minimum thickness specification? _____

Based on your inspection and measurement, is the rotor serviceable as is, does it need machining, or must it be replaced? _____

_____

CAUTION: Read and ensure you understand the operation of the lathe before proceeding. On-vehicle brake lathes generate a lot of rotational torque during operation. It can easily break a bone or cause other injury or property damage if operated improperly.

NOTE TO INSTRUCTORS: The following instructions are generally based on a Hunter Engineering on-vehicle brake lathe. Other lathes may use slightly different procedures and connections. Consult the lathe's operator manual before proceeding.

CAUTION: The vehicle transmission must be in neutral and the parking brake released if this is a rear rotor. Damage to the refinishing lathe or the vehicle may occur if the lathe cannot drive the rotor freely.

SERVICE TIP: It is best to leave the steering wheel unlocked. Depending on the front fender well opening, it may be necessary to steer the front wheels full left or right to properly install the cutting head over the rotor.

☐ 5. Remove the drive head from the lathe.

☐ 6. Loosen the lathe's stud guides; align and guide the guides onto the studs. Install the lug nuts finger tight.

☐ 7. Use the provided wrench to tighten the lug nuts and guide nuts firmly, lug nuts first.

☐ 8. Use a torque wrench to torque the lug nuts.

☐ 9. Move the lathe into alignment with the drive head. Rotate the rotor by hand and screw the drive head onto the lathe. If necessary, lift or lower the vehicle until the driving head and lathe are in near alignment.

CAUTION: If the lathe appears to bind or does bind, immediately switch off the motor. Do not try to make any adjustment or clear any obstacles with the lathe in operation. Serious injury could result, or damage to the lathe or vehicle could occur.

☐ 10. Clear everything from around the rotor and lathe machining head. Turn the lathe on and engage the drive.

☐ 11. Note the movement of the machining head and rotor. There should be a very slight up and down movement of the head. Are the machining head and rotor rotating smoothly without binding? _____

If yes, proceed to step 12. If not, disengage the drive, switch off the motor, loosen the stud and guide nuts, and reposition the head by realigning the guides. Retorque the lug nuts.

12. Disengage the drive and turn the lathe off. ☐

13. Manually move the cutting bits into place, with one bit to each side of the rotor. If necessary, spread the bits for clearance on each side of the rotor. ☐

14. Manually crank the bits over the rotor until each one is approximately half the distance between the outer and inner edges of the rotor's friction area. ☐

15. Move each bit in turn until it touches the rotor. Record the readings on the two depth adjustment knobs.

    Inside _____ Outside _____

16. Move the bits slightly away from the rotor and crank them back until they are clear of the rotor. ☐

17. Determine what depth of cut is to be made on the first cut. Add this amount to each of the readings recorded in step 15.

    Inside _____ Outside _____

18. Ensure that the area in and around the machining head is clear. Manually crank the bits until they are aligned with the inner edge of the friction material. ☐

19. Turn on the lathe. ☐

20. Set the adjustment knobs to the measurements from step 17. ☐

21. Engage the drive and allow the lathe to machine completely clear of the rotor before disengaging the drive and switching off the motor. ☐

22. Inspect the rotor on both sides. Does it need additional machining? _____

    If yes, return to step 20, except add another 0.002 inch (0.05 mm) to each adjustment knob's current reading and perform steps 21 and 22.
    If no, proceed to step 23. ☐

23. Disconnect the power supply from the lathe and remove the lathe from the rotor. ☐

24. Measure the rotor's thickness. Is the rotor still serviceable based on current thickness?

    _____ ☐

> **CAUTION:** Do not use compressed air. Bits of metal can be embedded in the skin, causing some injuries, or possibly can be embedded in an eye.

25. Clean the rotor's friction surface with hot water and a rag or other cleaning material. ☐

26. Install the caliper and the wheel assembly. Lower the vehicle and press the brake pedal several times to seat the pads. ☐

27. Conduct a test drive. ☐

Problems Encountered _____

_____

Instructor's Response _____

_____

_____

Drum Brake Service

Upon completion and review of this chapter, you should be able to:

- ❏ Diagnose drum brake problems, including poor stopping, pulling, grabbing, and dragging. Determine and perform needed repairs.

- ❏ Remove, clean, inspect, and service brake drums.

- ❏ Remove, clean, and inspect brake shoes and linings and related hardware, including all springs, pins, clips, levers, and adjusters. Determine and perform needed repairs.

- ❏ Clean and remove loose dirt, rust, and scale from the brake backing plates. Inspect and determine if backing plate replacement is needed.

- ❏ Disassemble, clean, inspect, and overhaul a wheel cylinder. Replace all cups, boots, and damaged hardware.

- ❏ Properly lubricate brake shoe support pads on backing plates, adjusters, and other drum brake hardware.

- ❏ Identify brake lining friction material from the edge code and select correct brake linings for a given vehicle.

- ❏ Identify primary and secondary shoes of a duo-servo brake and leading and trailing shoes of a leading-trailing brake and install them correctly.

- ❏ Refinish brake drums on a brake lathe.

- ❏ Adjust brake shoes and reinstall brake drums or wheel bearings.

- ❏ Reinstall wheels, torque wheel nuts, and make final adjustments.

Service Precautions

Specific CAUTIONS and WARNINGS are given throughout this chapter where needed to emphasize safety. The following general precautions apply to many different drum brake service operations and are presented at the beginning of this chapter to highlight their overall importance.

- ❏ When servicing drum brakes, never use an air hose or a dry brush to clean the brakes. Use OSHA-approved cleaning equipment to avoid breathing brake dust. See Chapter 1 in this *Shop Manual* for details on working safely around airborne asbestos fibers and brake dust.
- ❏ Do not spill brake fluid on the vehicle; it may damage the paint. If brake fluid does contact the paint, wash it off immediately. To keep fluid from spraying or running out of lines and hoses, cover the fittings with shop cloths when disconnecting them.
- ❏ Always use the DOT type of brake fluid specified by the vehicle manufacturer. Whenever possible, do not mix different brands of brake fluid.
- ❏ During servicing, keep grease, oil, brake fluid, or any other foreign material off the brake linings and drums. Handle brake shoes and drums carefully to avoid scratching the drums or nicking or scratching brake linings.

Basic Tools

Basic technician's tool set

Hydraulic floor jack

Diagnosing Drum Brake Problems

Many of the diagnostic procedures for drum brakes are similar to those listed in Chapter 7 on disc brakes. If problems are suspected in the drum brakes, road test the vehicle. Instructions for a safe, complete road test are given in Chapter 2 of this *Shop Manual*. Even driving across the driveway and into the service bay can reveal a lot about brake system condition.

Classroom Manual
pages 199–203

Figure 8-1 Apply the brake pedal and check for excessive travel and sponginess.

⚠ **WARNING:** Road test a vehicle under safe conditions and while obeying all traffic laws. Do not attempt any maneuvers that could jeopardize vehicle control. Failure to adhere to this precaution could lead to serious personal injury and vehicle damage.

Check for excessive travel and sponginess as the pedal is applied (Figure 8-1). Listen for noises: not just the obvious sounds of linings grinding on the drums, but mechanical clanks, clunks, and rattles.

For a complete inspection, the wheels must be removed for a clear view of the brake shoes, attaching hardware, and drum (Figure 8-2). Inspect the wheel and brake assembly for obvious damage that could affect brake system performance. Check the following:

1. Tires for excessive or unusual wear or improper inflation
2. Wheels for bent or warped rims
3. Wheel bearings for looseness or wear
4. Suspension system for worn or broken components
5. Brake fluid level in the master cylinder
6. Signs of leakage at the master cylinder, in brake lines or hoses, at all connections, and at each wheel cylinder
7. Damaged or improperly adjusted brake shoe
8. Loose, glazed, or worn lining
9. Weak or bad **return springs** and **hold-down springs.** Return springs pull the shoes to the released position after the pedal is released. Hold-down springs hold the shoes to the backing, but allow the shoes to move as needed during braking

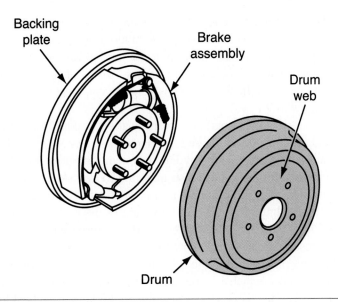

Figure 8-2 The first step of drum brake service is to inspect all the parts in the complete assembly.

10. Loose backing plate

11. Self-adjusters not operating

12. Oil, grease, or brake fluid on lining

A damaged or improperly adjusted brake shoe may cause brake drag or grab, pulling to one side, wheel lockup, hard pedal, excessive pedal travel, or noisy operation. A loose lining may cause pull to one side or a persistent brake chatter.

Glazed or worn linings are the most common cause of drum brake problems. Such problems include hard pedal, pulling to one side, wheel locking, brake chatter, excessive pedal travel, noisy brakes, grabbing brakes, or little or no braking power.

A broken or weakened return spring (Figure 8-3) will cause the brake to drag or the vehicle to pull to one side. A loose backing plate can cause brake drag, brake chatter, or a locked

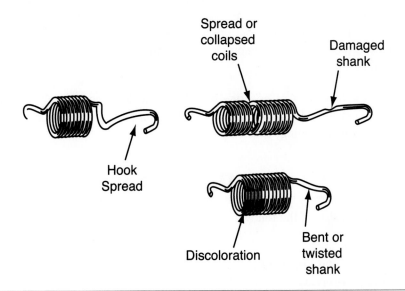

Figure 8-3 Inspect the shoe return springs for the damage noted. Also perform a bounce test. Drop the spring on a floor or steel table: if you hear a ring, it is a good spring; if you do not, the spring is defective.

wheel. If the self-adjuster operates incorrectly, the brake shoes will be improperly adjusted. The car may then pull to one side, have excessive pedal travel, or have a clicking sound in the shoes.

Oil, grease, or brake fluid on the linings can cause hard pedal, brake pull, wheel lock, brake chatter, uneven braking, noisy brakes, grabbing brakes, or loss of braking power. Grease or oil that is flung around the inside of the fender well is a sign of complete axle seal failure. The brakes must be replaced along with the seal.

Damaged or distorted brake drums also adversely affect braking. Excessive pedal travel may be a sign of a cracked drum. But as discussed in Chapter 4 of this *Shop Manual,* excessive pedal travel also may be caused by trouble in the hydraulic system. An out-of-round drum can cause the brake to drag and chatter, and the vehicle also may pull to one side. A scored drum can cause uneven braking with noisy or grabbing brake action. Drums that have become bell mouthed or distorted allow only partial lining contact and a twisting of the brake shoe that results in excessive pedal travel and hard pedal pressure. Threaded drums result when a dull or poorly shaped cutting bit is used to resurface the drums. These threads cause the brake shoes to move toward or away from the backing plate according to the direction of the thread spiral. This action creates a snapping or clicking action, which is especially noticeable on light brake applications at low speeds. A faulty wheel cylinder also may be the cause of brake drag, wheel lock, side pull, or a noisy and grabbing brake.

If any of these conditions are found during the test drive, remove the drums and inspect the brakes. Any wear in the shoes, shoe hold-down and retracting hardware, drums, or wheel cylinder will make a complete brake system overhaul necessary.

Disc brakes allow for a quick inspection every time the wheel is removed, but drum brake components remain covered by the drum when the wheel is removed. Fortunately, drum brakes wear about half as quickly as disc brakes, and the recommended inspection interval for many vehicles is now 30,000 miles or every 2 years.

Many state vehicle inspection laws specify that drums must be pulled at regular intervals and the system inspected. Always remove the drums for inspection whenever the customer has a brake-related complaint or concern or whenever you suspect a problem. Always service drum brakes and linings in axle sets.

✔ **SERVICE TIP:** In very cold weather, rear brake shoes can freeze to the drum surface. It takes the right (or wrong) combination of circumstances to cause this. It is most common when the parking brake is applied in damp weather or after driving through water and left applied when temperature drops below freezing. High metallic content in the brake linings further aggravates the problem, and some 1992–1993 GM brakes were particularly susceptible.

If the car has antilock brakes and one wheel stays locked momentarily, the difference in wheel speed may light the ABS warning lamp. If this is a persistent problem, replacing semi-metallic linings with organic linings is often a cure.

Drum Brake Service Operations

Classroom Manual
pages 203–219

Drum brake service consists of the following general operations:

❏ Drum removal
❏ Brake cleaning and inspection
❏ Wheel cylinder service
❏ Shoe replacement and hardware installation
❏ Drum refinishing and reinstallation
❏ Brake adjustment

Vehicle Preparation

For most drum brake service, the vehicle must be raised on a hoist or safety stands, wheels removed, and brakes disassembled to the extent necessary for the planned service. Follow these general preparation steps:

1. Disconnect the battery ground (negative) cable.

2. If the vehicle has electronically controlled suspension, turn the suspension service switch off.

3. Raise the vehicle on a hoist or safety stands and support it safely.

4. Remove the wheels from the brakes to be serviced. Brakes are always serviced in axle sets, so you may remove both front wheels, both rear wheels, or all four.

5. Vacuum or wet-clean the brake assembly to remove all dirt, dust, and fibers.

Special Tools

Lift or jacks with
  stands
Impact tools
Chalk

Brake Drum Removal

The brake drum must be removed for all brake service except bleeding the wheel cylinder and lines or a basic manual adjustment. Removal procedures are different for fixed and floating drums. In some cases, the parking brake may have to be adjusted (Figure 8-4). Self-adjusters (Figure 8-5) may have to be adjusted to gain shoe-to-drum clearance for drum removal. Wear on the drum friction surface creates a ridge at the unworn rim of the drum. As the self-adjusters move the shoes outward to take up clearance, the shoe diameter becomes larger than the ridge diameter. If adjustment is not retracted, the drum ridge may jam on the shoes and prevent drum removal (Figure 8-6). Trying to force the drum over the shoes may damage brake parts.

Before removing a drum, mark it "L" or "R" for left or right so that it gets reinstalled on the same side of the vehicle from which it was removed.

 CAUTION: Do not step on the brake pedal while a brake drum is off, or the wheel cylinder will pop apart.

Fixed (One-Piece) Drum Removal

Brake drums that are made as a one-piece unit with the wheel hub are common as rear drums on FWD cars and on the front wheels of older vehicles with four-wheel drum brakes. The hub

Special Tools

Dust cap pliers
Cotter pin puller
Brake adjusting tool
  (spoon)

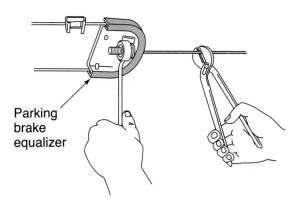

Parking
brake
equalizer

Figure 8-4 Loosen the parking brake cables to ensure enough shoe clearance for drum removal.

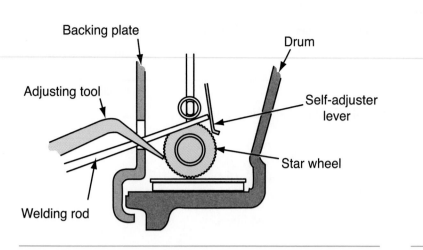

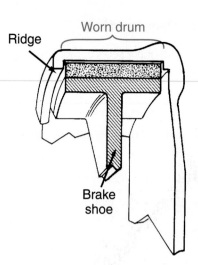

Figure 8-5 Use a heavy piece of wire, such as a piece of welding rod, to move the adjusting lever away from the star wheel. Then rotate the star wheel to release brake adjustment for drum removal.

Figure 8-6 These shoes have adjusted to fit a badly worn drum, and the drum will jam on the lining if you try to remove it without backing off the brake adjustment.

contains the wheel bearings and is held onto the spindle by a single large nut (Figure 8-7). This nut also is used to adjust bearing end play. Remove a fixed drum mounted on tapered roller bearings as follows:

1. Release the parking brake if removing a rear drum.

2. Use a pair of dust cap pliers or large slip-joint pliers to remove the dust cap from the hub.

3. Remove the cotter pin from the castellated nut or nut lock on the spindle. Then remove the nut lock, if used.

4. Remove the drum retaining nut and thrust washer.

5. Pull outward on the drum to slide it off the spindle (Figure 8-8). If the drum drags or catches on the brake shoes, slide it back onto the spindle and temporarily reinstall the

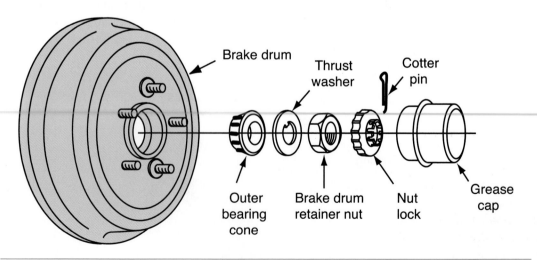

Figure 8-7 To remove a one-piece drum and hub, remove the single large retaining nut and slide the drum and hub off the spindle along with the bearings.

Figure 8-8 The drum brake assembly is covered by the drum.

spindle nut. Then retract the parking brake and the brake shoe adjustment. Brake adjustment is explained at the end of this chapter; parking brake adjustments are covered in Chapter 9 of this *Shop Manual.*

6. When removing the drum, be careful not to drop the outer wheel bearing on the floor and do not drag the inner bearing across the spindle, particularly the threads at the end of the spindle.

Occasionally the inner bearing race of a drum sticks on a spindle and prevents easy drum removal. In such a case, use a puller to remove the drum. After the drum is removed, use another puller or pair of small prybars to remove the bearing race from the spindle.

After any drum is removed, inspect the grease in the hub and on the bearings. If the grease is dirty or dried out and hard, it is a clue to possible bearing damage. Set the drum and all bearing parts aside for cleaning and close inspection. If the grease seems to be in good condition, place the drum on a bench with the open side down. Cover the outer bearing opening with a shop cloth to keep dirt out.

Floating (Two-Piece) Drum Removal

Floating drums are separate from the wheel hub or axle. On a rear-wheel-drive (RWD) vehicle, the drums are held in place on studs in the axle flange by the wheel and wheel nuts. Some FWD cars have rear hubs that contain sealed bearings and are mounted to the chassis on a spindle. These installations use a floating drum that is mounted on studs on the hub flange and held in place by the wheel and wheel nuts.

On many floating drums, you will find **push (speed) nuts** holding the drum onto two or three studs (Figure 8-9). Push nuts are made from thin stamped steel. They will not accept much torque. These push nuts are used to hold the drum and hub or axle together during vehicle assembly, before the wheels are installed. On most vehicles, the push nuts or speed nuts do not

Shops typically repack the wheel bearings and install new seals as a part of a brake repair.

Special Tools

Brake adjusting tool (spoon)

A **floating drum** is a brake drum that is separate from the wheel hub or axle.

Floating drums are the most common type used with RWD vehicles.

Figure 8-9 Push nuts (speed nuts) may be used to hold the drum in place when the wheel assembly is removed.

need to be reinstalled during service. Some Ford vehicles, however, have wheel studs with high shoulders that can catch the drum and hold it at an angle if you are not careful when installing the wheel. Save the push nuts or install new ones on these vehicles to hold the drum squarely against the axle or hub flange before installing the wheel.

Remove a fixed drum mounted on a hub or an axle flange as follows:

1. Release the parking brake.

2. Remove the push nuts, if installed, by grabbing each one with a pair of pliers and twisting it off the stud.

 **SERVICE TIP:** Diagonal wire cutters work very well in removing push nuts. An impact screwdriver is a very useful tool for loosening drum retaining screws.

3. Some drums are fastened to the axle flanges with small screws. Remove the screws with a suitable screwdriver or wrench to remove the drum.

4. Use a scribe or paint to make index marks on the drum and the hub so that the drum can be reinstalled in its original location.

5. If the drum is not rusted or stuck to its flange, lift it off and move it to a bench. If the drum drags or catches on the brake shoes, slide it back onto the flange or hub and retract the parking brake and the brake shoe adjustment. Brake adjustment is explained at the end of this chapter; parking brake adjustments are covered in Chapter 9 of this *Shop Manual.*

CAUTION: Do not attempt to hit directly on the drum between the studs. It is probable that the hammer will hit and damage at least one stud. Use the drift pin to deliver the hammer's force to the drum.

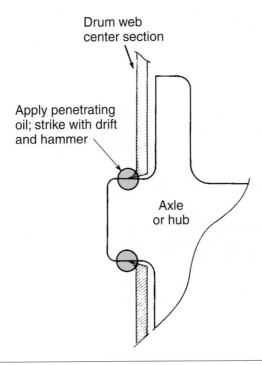

Drum web
center section

Apply penetrating
oil; strike with drift
and hammer

Axle
or hub

Figure 8-10 The center hole of a floating drum may catch on the axle flange or hub as you try to remove it.

✓ **SERVICE TIP:** The drums on some import vehicles have countersunk screws holding them to the hub or axle flange. An impact screwdriver may be required to loosen the screws. Other import drums may have two small (6 mm–8 mm) threaded holes through which bolts can be screwed to break the drum loose from the hub/axle flange. Both drum types will have the screws or holes between the wheel studs.

6. Penetrating oil may assist in loosening a stuck drum. If the drum is stuck to its flange or hub (Figure 8-10), use a large scribe or center punch to scribe around the joint at the drum and flange and break the surface tension. If the drum is still stuck, strike the edge of it with a dead-blow hammer at a 45-degree angle to loosen it. If a dead-blow hammer is not available, place a block of hard wood against the drum and strike it with a large ball-peen hammer. If the drum remains stuck, use a puller to separate it from the flange.

✓ **SERVICE TIP:** The use of a puller to remove a brake drum must be done with care. Tap the drum as the puller applies pulling force. The puller could crack, bend, or otherwise damage the drum.

Photo Sequence 16 shows most of the steps for typical brake drum removal from a rear axle.

Drum Brake Cleaning

Thorough cleaning of the brake assembly is a bigger job on drum brakes than on disc brakes. Dirt and dust created by drum brake lining wear stays inside the enclosure formed by the drum and backing plate. Fine metallic particles caused by drum wear combine with lining dust to create a unique brake grime, most of which accumulates on the backing plate, the shoes, the springs, and other parts of the brake assembly.

Photo Sequence 16
Typical Procedure for Removing a Brake Drum from a Rear Axle

P16-1 Release the parking brake and raise the car on a hoist. Mark the position of the wheel to the axle flange so the wheel can be reinstalled in the same position on the flange. This ensures proper wheel balance.

P16-2 Remove the wheel nuts holding the wheel to the hub and pull off the tire and wheel.

P16-3 Mark the position of the brake drum to the axle flange so the drum can be reinstalled in the same position on the flange.

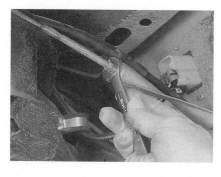

P16-4 Remove tension from the parking brake cables by loosening the adjusting nut at the equalizer.

P16-5 Remove the plastic or metal plug from the backing plate slot to expose the adjuster. On some backing plates, a lanced metal section must be punched or drilled out as shown here.

P16-6 Insert a small wire hook through the slot and pull/push the adjusting lever away from the star wheel about 1/16 inch.

P16-7 While holding the pawl away from the star wheel, use a screwdriver or brake adjuster to back off the star wheel about two dozen clicks or until the drum is loose on the wheel studs.

P16-8 Remove the drum from the studs and the axle flange. Some drums may be secured to the flange by two small screws. Remove them to remove the drum.

P16-9 If the drum is stuck to the flange, tap it lightly with a soft-faced mallet. If it is really stuck, free it as explained in the previous section of this chapter.

Photo Sequence 16
Typical Procedure for Removing a Brake Drum from a Rear Axle (continued)

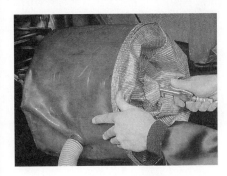

P16-10 Thoroughly clean the exposed brake assembly area and drum using a vacuum system with a HEPA filter.

P16-11 Inspect the brake assembly for broken springs, heat damage, leaking wheel cylinders, and other wear.

P16-12 Inspect the drum for scoring or other damage.

> ⚠️ **WARNING:** Do not blow dust and dirt off brake assemblies with compressed air outside of a brake cleaning enclosure. Airborne dust and possible asbestos fibers are an extreme respiratory hazard.

Special cleaning equipment is available for brake service and is discussed in Chapter 7 of this *Shop Manual*. Always ensure that cleaning equipment is in proper working condition and follow the manufacturer's instructions for the specific equipment in your shop.

Thoroughly clean the backing plates, struts, levers, and other metal parts to be reused. Examine the rear wheels for evidence of oil or grease leakage past the wheel bearing seals. Such leakage could cause brake failure and indicates the need for additional service work.

Brake dust waste must be collected, recycled, and disposed of in sealed, impermeable bags or other closed, impermeable containers. Any spills or release of brake waste material from inside of the enclosure or vacuum hose or vacuum filter should be immediately cleaned up using vacuuming or wet-cleaning methods. Review the asbestos safety instructions in Chapter 1 of this *Shop Manual*.

Cleaning with Vacuum-Enclosure Cleaning Systems

Vacuum-enclosure cleaning systems consist of a tightly sealed protective enclosure that covers and contains the brake assembly (Figure 8-11). The enclosure has built-in, impermeable sleeves and gloves that let you inspect and clean the brake parts while preventing the release of dust and potential asbestos fibers into the air. Examine the condition of the enclosure and its sleeves before beginning work. Inspect the enclosure for leaks and a tight seal.

Special Tools

Vacuum brake cleaner

Cleaning with Wet-Cleaning Systems

Low-pressure wet-cleaning systems wash dirt from the brake assembly and catch the cleaning solution in a basin (Figure 8-12). The cleaner reservoir contains water with an organic, nonpetroleum solvent or wetting agent. To prevent any asbestos-containing brake dust from becoming airborne, control the flow of liquid so that the brake assembly is gently flooded. Wash the backing plate, the shoes, and other brake parts used to attach the shoes before removing the old shoes. The cleaning solution must be treated as hazardous waste when changed. The filter on most cleaners with a filtration system is also treated as hazardous waste.

Special Tools

Brake cleaning tools/ equipment

Figure 8-11 This full-enclosure asbestos system traps brake dust and helps keep the shop's air free of dust.

Figure 8-12 Typical wet-cleaning equipment with a catch basin.

Figure 8-13 A HEPA-equipped vacuum cleaner is a good tool for cleaning brake components.

Cleaning with Vacuum Cleaning Equipment

Several types of vacuum cleaning systems are available to control brake dust in the shop. The vacuum system must have a HEPA filter to handle asbestos dust (Figure 8-13). A general-purpose shop vacuum is not an acceptable substitute for a special brake vacuum cleaner with a HEPA filter. After vacuum cleaning, wipe any remaining dust from components with a damp cloth.

Cleaning with Parts Washers

Special Tools

Parts washer

Almost all shops have a parts washer that is used to clean miscellaneous small and medium-size parts that are removed from vehicles. Traditionally, parts washers have used petroleum solvent, but many current models use water-based solvents and detergents. Do not use petroleum-based products to clean rotors and drums. Cast iron is porous and will hold the solvent and damage newly installed shoes. A parts washer that uses water-based solvents and detergents as the only solvent is safe. When the cleaning agent is changed, the old liquid is treated as hazardous waste.

Drum Brake Assembly Inspection

Classroom Manual
pages 203–219

Rear drum brakes wear much more slowly than front disc brakes because only 20 percent to 40 percent of the braking effort is provided by the rear brakes. On older four-wheel drum brake systems, front-to-rear wear is more equal, but front brakes still tend to wear more than rear brakes. Because most drum brakes today are used on the rear wheels with disc brakes on the front, it is important not to overlook rear brake inspection.

After the drum is removed, set it aside for inspection and measurement as explained later in this chapter. Then inspect the shoes and linings, the wheel cylinders, the springs, and other parts as explained below. Inspect each brake assembly before disassembly to help identify causes of problems. For example, if there is contamination on a brake lining, it could be brake fluid or gear oil. In either case, it indicates that the wheel cylinder and the axle oil seal should be checked very closely. Similarly, unusual wear on springs, struts, or levers can be a clue to incorrectly installed parts or abnormal operation. Such problems can be more easily identified while inspecting the complete assembly.

Lining and Shoe Inspection

Inspect three general conditions of the lining: thickness, wear pattern, and damage. Use a precision scale or a depth micrometer to measure lining thickness precisely. A tire depth gauge is also an excellent tool to measure lining thickness (Figure 8-14). Lining thickness is the first thing—but not the only thing—that determines the need for replacement. Most carmakers specify a minimum lining thickness of 1/32 inch (0.030 in. or 0.75 mm) above the shoe table or above the closest rivet head.

This thickness recommendation does not mean that a lining worn to 1/32 inch should continue in service; it means that 1/32 inch is the *absolute minimum* safe thickness at which the brake lining can perform to its minimum requirements. Worn linings cannot dissipate heat adequately, and the last 1/32 inch of lining will wear much faster than the first 1/32 inch of fresh linings. From a practical and safe standpoint, relined shoes should be installed before the lining has worn to the thickness of the shoe table at its thinnest point above the table or the closest rivet. This is approximately 1/16 inch. Use a depth gauge or scale graduated in 1/32-inch or 1/64-inch increments to measure lining thickness at several points.

Also inspect the lining for cracks, missing rivets, and looseness. Check for contamination from grease, oil, or brake fluid. A leaking wheel cylinder can deposit brake fluid on the linings. A leaking oil seal on a rear drive axle can let gear oil from the differential get onto the brakes. A less frequent, but possible, cause of lining contamination is a leaking grease seal for the wheel bearings in a hub. If linings are damaged or contaminated in any way, they must be replaced. Remember also that brake linings are serviced in axle sets so all linings for both wheels must be replaced if any are damaged.

Check the linings for unequal wear on any shoe of an axle set (Figure 8-15). Also look for uneven lining wear on any one shoe. If one lining on a duo-servo brake is worn more than the other,

Figure 8-14 Use a depth micrometer to measure lining thickness precisely.

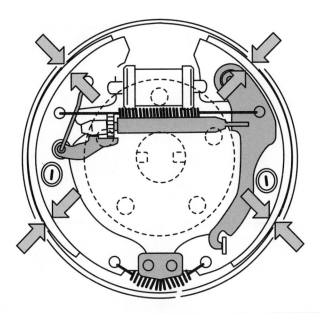

Figure 8-15 Inspect the linings for unequal wear, as well as for thickness, cracks, fluid or grease contamination, and other damage. Unequal wear is most common at the indicated points.

be sure the primary and secondary shoes are installed in the right locations. The primary shoe with the shorter lining should be the forward shoe. If the linings on one wheel are worn more than the other, that drum may be scored or rough. Uneven wear from side to side on any one set of shoes can be caused by a tapered drum. Check for parking brake cables to see that they are not sticking.

If brake shoes and linings have a slight blue coloring, it indicates overheating. In this case, the brake adjuster springs and hold-down springs should be replaced. Overheated springs lose their tension and could allow a new lining to drag and wear prematurely if not replaced.

If the lining of one shoe on one wheel is worn more than the other, check the less worn shoe for binding and incomplete application. A problem such as this is more likely to occur on a leading-trailing brake than on a duo-servo brake. If the lining is worn more in the center than the ends or vice versa, the lining may not have been arced properly to the drum diameter. If so, check the lining-to-drum fit closely when you install new shoes with a resurfaced drum. If the linings are worn more at the end where the parking brake applies the shoes, the parking brake may be adjusted too tightly. Linings worn badly at the toe or heel also may indicate an out-of-round drum.

> **☑ SERVICE TIP:** If a customer complains that the rear brakes lock easily or that the ABS activates at the rear during light to moderate braking, check the basics before condemning the expensive ABS hardware. Grabbing rear drum brakes are often caused by contaminated or cracked linings, fatigued return springs, or grossly uneven brake adjustment from one side to the other.

Wheel Cylinder and Axle Inspection

If left undetected and unrepaired, a leaking wheel cylinder can drain most of the fluid from half of the brake system, and the leaking fluid will contaminate and ruin the brake lining. Brake operation and safety are seriously compromised by a defective wheel cylinder.

Inspect the outside of the wheel cylinder for leakage. Then pull back the dust boots and look for fluid at the ends of the cylinder (Figure 8-16). Minor dampness or seepage and some staining is acceptable, but any noticeable fluid means that the cylinder should be overhauled or replaced. Also be sure that the pushrods engage the shoes properly.

Check the cylinder mounting on the backing plate for looseness and missing fasteners. Most cylinders are held to the backing plate by small cap screws, but some are secured with clips. Brake

Figure 8-16 Pull the boot slightly away from the cylinder and check for liquid or fluid within the boot.

Figure 8-17 The shoe support pads must be cleaned and lubricated with brake lubricant before installing the shoes.

spring tension and tight clearances in the mounting hole can make a clip-mounted cylinder seem secure even when the clip is missing. Inspect this type of cylinder mounting closely.

Inspect the brakes on a rear drive axle for contamination by gear oil leaking from the axle seal. Use a flashlight to inspect thoroughly behind the axle flange. If a leaking seal is found in the early stages, replace it before gear oil gets to the brake linings. Similarly, inspect the backing plate for grease leaking past the inner wheel bearing seal of a nondriving hub.

Backing Plate, Spring, and Hardware Inspection

Some vehicle manufacturers require that the hardware be replaced each time the shoes are changed.

Backing plates are rarely replaced unless they are damaged in an accident. Close inspection of the backing plate can solve problems with other brake parts, however. Inspect for a broken or bent plate or other obvious damage. Also look for uneven wear or scarring on the shoe support pads (Figure 8-17), which could indicate a bent shoe or incorrectly installed parts. Place a straightedge across two of the shoe support pads as far from each other as possible to check the straightness of the backing plate. If the straightedge does not contact both pads evenly and squarely, remove the backing plate for closer inspection and measurement. Check to see that deep grooves (scarring) have not been worn into the backing plates, which could cause the shoes to hang.

Inspect the return and hold-down springs for damage and unusual wear (Figure 8-18). Pry the brake shoes slightly away from the backing plate and release them. The hold-downs should pull the shoes sharply back to the plate.

✓ **SERVICE TIP:** If one piece of the brake hardware is damaged and needs replacing, replace all of the hardware on both sides. Hardware for drum brakes usually comes in kits that will include enough parts for both sides. Most brake hardware kits do not include hardware for the self-adjuster. It is not unusual to replace the brake hardware and not the self-adjuster or vice versa.

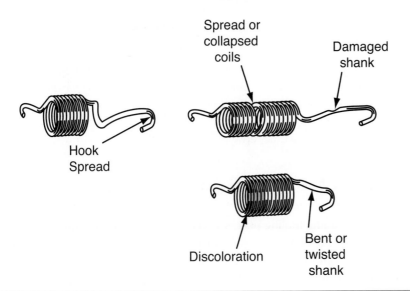

Figure 8-18 Inspect the springs for heat damage (discoloration), damaged or spread coils, and spread connecting hooks.

Inspect self-adjuster levers, pawls, and springs for wear and replace any defective parts (Figure 8-19). Pawls are the levers that actuate the self-adjuster (star wheels). Many shops make it standard practice to replace self-adjuster cables at each brake job, but at least inspect them for broken strands and obvious wear and stretching.

Check the parking brake linkage for damage and rust. Be sure that all parking brake levers and links are properly lubricated and free to move easily (Figure 8-20). A can of pressurized chain and cable lubricant can be used to lubricate a parking brake cable. Pull the cable out as far as possible and spray the visible cable into the end of the conduit. Work the cable back and forth several

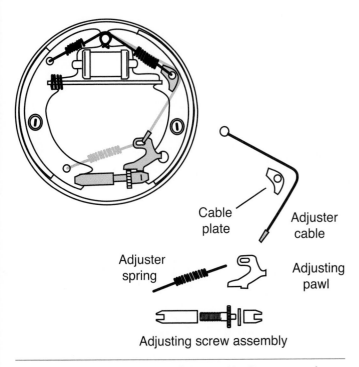

Figure 8-19 Inspect all of these self-adjuster parts for wear and damage.

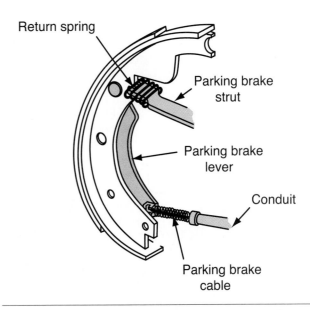

Figure 8-20 Inspect the parking brake linkage for wear and damage.

times and then clean any excess lubricant from the cable and conduit end. See Chapter 9 of the *Shop Manual* for further details on parking brake service.

✔ **SERVICE TIP:** Brake shoe return springs can be weakened by excessive heat and hard use but they may still look good. One traditional way to identify work-hardened and weak springs is to drop them on a concrete floor. If the springs land with a dull plop, they are good. If they land with a sharp ping and bounce off the floor, they have been overheated or work hardened and should be replaced.

Drum Brake Disassembly

Classroom Manual
pages 203–219

Duo-servo brakes and leading-trailing brakes all have a pair of shoes, return springs, hold-downs, and self-adjusters, but a lot of variety exists in the individual parts of drum brake assemblies. Figure 8-21 shows the components of a typical leading-trailing drum brake. Figure 8-22 show a duo-servo drum brake. Refer to these figures and Photo Sequence 17 for general guidance during disassembly and reassembly.

📋 **AUTHOR'S NOTE:** You will note some differences between the text and Photo Sequence 17. The text generally follows manufacturer recommended sequence, whereas the photo sequence is my way of doing it. Both will work, and you may even find a better method of doing it.

Special Tools

Lift or jack with stands

Impact tools

Return spring pliers

Hold-down spring removal tool

Service manual

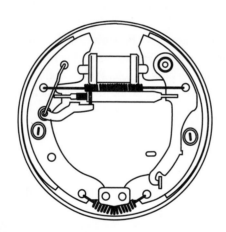

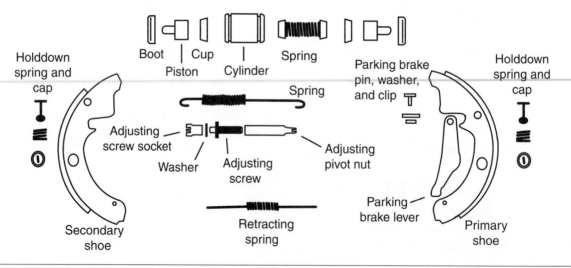

Figure 8-21 Exploded view of a typical leading-trailing drum brake.

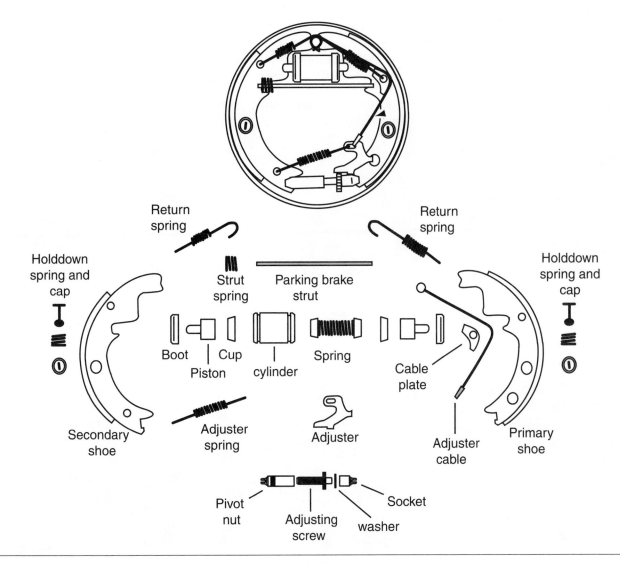

Figure 8-22 Exploded view of a typical duo-servo drum brake.

> **CAUTION:** Do not depress the brake pedal when the shoes are being removed, installed, or have been removed from the backing plate. Without the shoes positioned properly, the pistons could be forced from the wheel cylinder when the pedal is depressed.

If not thoroughly familiar with the brake design being serviced, obtain the vehicle service manual and refer to the procedures and illustrations. Remember that the brakes are mounted the same on both sides of most vehicles. After observing one side and moving to the opposite side remember that what *was* at the right hand *is* now at the left hand. It also may help to service the brake on one side of the vehicle at a time and use the brake on the opposite side as an assembly reference. Remember that left-hand and right-hand parts of many drum brakes look the same but are not. If parts, such as self-adjusters, are interchanged from one side to the other they will not work properly.

Raise the vehicle on a hoist or safety stands, remove the drum, and clean the brake assembly as explained earlier in this chapter. Then follow the general guidelines in the following sections, along with specific vehicle service procedures, to service drum brakes.

> **WARNING:** Wear safety glasses or face protection when using brake fluid. Injuries to the face or eyes could occur from spilled or splashed brake fluid. Brake fluid is irritating to the skin and eyes. In case of contact, wash skin with soap and water or rinse eyes thoroughly with water.

Typical Procedure for Disassembling a Drum Brake

P17-1 Remove the top return spring from the anchor and shoe. Remember which shoe return spring is on top. It must be reassembled in the same position.

P17-2 Remove the retaining spring from the secondary shoe. Do not remove the primary shoe retaining spring at this time.

P17-3 Rotate the secondary shoe forward and down. Catch the self-adjuster as the shoe releases the tension.

P17-4 Remove the parking brake strut and spring from the primary shoe. Withdraw it to the front. Remember which end holds the spring and which end faces forward.

P17-5 Remove the retaining spring on the primary shoe. Do not allow the shoe and self-adjuster mechanism to drop.

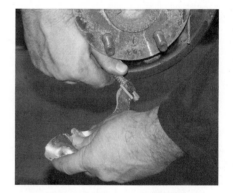

P17-6 There are two means to disconnect the parking brake from the primary shoe. One way is to remove the clip from the parking brake lever pin where it passes through the shoe. Do not lose the waved washer behind the shoe. The other way is to disconnect the cable from the lever as shown. The lever can be removed from the shoe later.

P17-7 Clean and lubricate the shoe support pads now. This is a common area that may be forgotten.

P17-8 Laying out the brake components in roughly the same position as they were on the vehicle will help in repositioning them correctly during assembly.

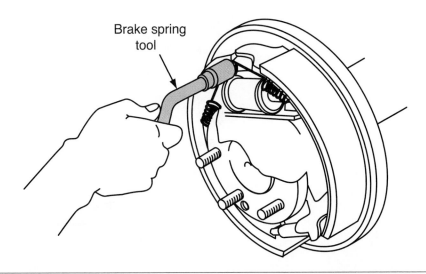

Brake spring tool

Figure 8-23 Use a brake spring tool to disengage the return springs from the anchor.

The shoe return springs are usually the first parts to be removed because their tension holds most of the other parts in place, but before removing the springs study their installation (see Figure 8-19). Keep fingers away from the return springs during removal and installation. If a spring slips off a tool or mounting point, its strong tension can cause injury or damage. Note how the springs are hooked over anchor posts and to which holes in the shoe webs they are attached.

Special brake spring pliers and other tools are available for spring removal and installation (Figure 8-23). The rear brakes on some older GM cars have a large horseshoe-shaped spring that serves as both a return spring and a hold-down spring (Figure 8-24). Special lever-type pliers are used to remove and install it (Figure 8-25).

Unispring

Figure 8-24 The rear brakes on some GM cars have a large horseshoe-shaped spring (Unispring) that serves as both a return spring and a hold-down spring.

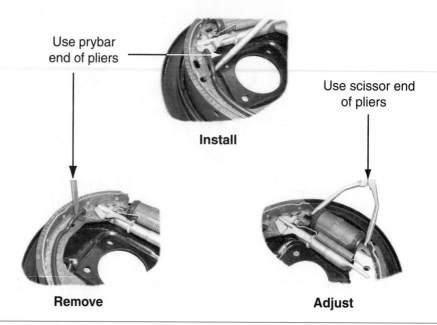

Use prybar end of pliers

Use scissor end of pliers

Install

Remove

Adjust

Figure 8-25 Use these special pliers to remove a GM Unispring. The pliers are also used to operate the self-adjusting ratchet mechanism prior to installing the brake drum.

After removing the return springs, remove the lighter hold-down springs and clips. Again, special tools are available to make the job easier and to avoid damage to parts. At this point, the shoes are loosened from the backing plate, and you can remove shoe-to-shoe springs and self-adjuster parts easily. Carefully note the positions in which the parts are installed so that you can reinstall them correctly.

CAUTION: Prevent dropping the self-adjusting screw assembly during disassembly and assembly. There is a small wave washer between the adjusting screw and the nearest end. This washer can be lost easily. Without it, the self-adjusting screw may bind and not adjust correctly. Normally, the entire screw assembly must be purchased to replace this washer.

Disconnect the parking brake cables and linkage from rear brakes with parking brake levers and struts (Figure 8-26 and Figure 8-27). It is often easiest to disconnect the parking brake linkage after the shoes are loosened from the backing plate to release tension on the cables.

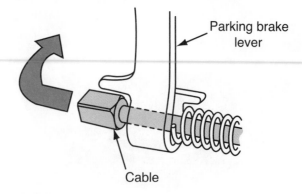

Parking brake lever

Cable

Figure 8-26 Pull the parking brake lever forward and disconnect the cable from the end of the lever.

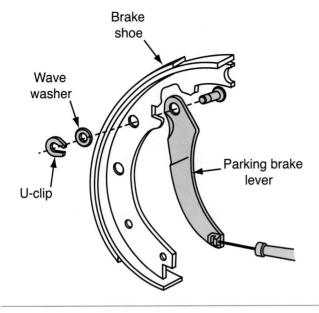

Figure 8-27 Then remove the clip and washer to separate the parking brake lever from the secondary or trailing brake shoe.

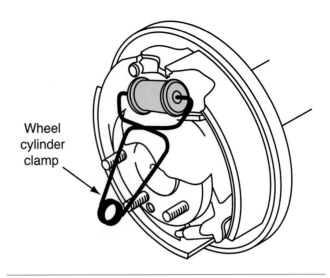

Figure 8-28 Install wheel cylinder clamps to keep the pistons from popping out of the cylinders when you disassemble the brakes.

On some brakes, sections of the shoe webs fit into the ends of the wheel cylinder pistons. On other brakes, short pushrods, or shoe links are installed between the cylinder pistons and the shoes. If the brakes being serviced have wheel cylinder pushrods, remove them for cleaning and reinstallation later.

If the wheel cylinders are not removed, install cylinder clamps (Figure 8-28) to keep the pistons from popping out of the cylinders. On some brake assemblies, wheel cylinder clamps can make disassembly and reassembly easier by holding the cylinder pistons in place while removing and reinstalling the shoes.

> **CAUTION:** Prevent brake fluid from coming in contact with the vehicle's finish. Brake fluid damages paint and finish immediately on contact. If fluid contacts the finish, wash the area thoroughly with running water and soap if possible.

> **CAUTION:** Always clean around the lines or covers before removing or loosening them. Dirt and other contaminants will void the warranty and may damage system components.

Although a wheel cylinder can be overhauled while installed on the backing plate, it really should be done with the cylinder removed. Dirt and abrasive particles from cleaning and honing can be removed from the cylinder more easily and kept out of the hydraulic lines and fittings. Remove the cylinder by disconnecting the brake line and removing its mounting screws or clip (Figure 8-29). Cap the brake line after the wheel cylinder is removed to keep dirt out of the hydraulic system.

After all parts are disassembled, inspect them again for damage that may have been hidden when they were installed. Clean individual parts in a parts washer and remove any remaining dirt from backing plates, parking brake linkage, and other parts still attached to the vehicle with brake cleaning equipment described previously.

The next major section of this chapter explains the options for wheel cylinder service or replacement. These paragraphs are then followed by procedures for drum brake reassembly.

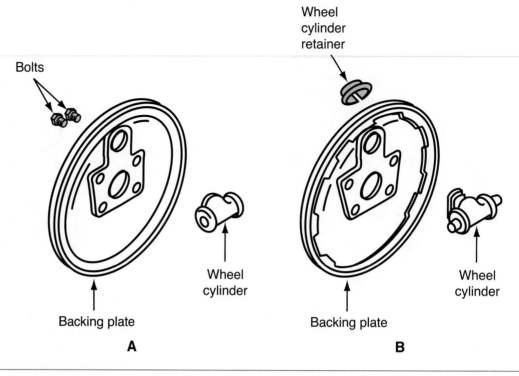

Figure 8-29 Two common ways to mount wheel cylinders to backing plates are with small bolts or cap screws (A) or with a spring-type retainer ring (B).

> ☑ **SERVICE TIP:** There is a mistaken notion in some shops that the pads on drum brake backing plates on which the shoes ride should be ground and sanded with a power sander to get them as smooth as possible. Do not do it.
>
> Yes, you should remove any big gouges or ridges with a file and lightly sand the pads by hand to clean them, but that is it. Grinding or power sanding can cut the pads down to different heights and cause the brake shoes to cock and bind and may also weaken the backing plate. Finish off any cleaning or light sanding with a light coat of brake lubricant, and that part of the brake job is completed.

Classroom Manual
pages 210–212

Special Tools

Denatured alcohol

Hone

Vise

A **piston stop** is a metal part on a brake backing plate that keeps the wheel cylinder pistons from moving completely out of the cylinder bore.

Wheel Cylinder Service

If external inspection reveals anything more than the slightest seepage or staining around a wheel cylinder, the cylinder should be disassembled for internal inspection, which can be done with the cylinder mounted on the backing plate. It is possible to overhaul some cylinders while they are installed. If the cylinder is recessed into the backing plate or if access is blocked by **piston stops,** parking brake linkage, or an axle flange, the cylinder must be removed for service. It is often faster to remove a wheel cylinder for overhaul than to struggle for clearance around other parts mounted on or near the backing plate. Moreover, dirt and abrasive particles from cleaning and honing can be removed from the cylinder more easily and completely on the bench than on the car.

> ☑ **SERVICE TIP:** On passenger cars and light trucks, it is more economical to the shop and customer to just replace the wheel cylinder. Always inspect the old wheel cylinder closely, however. The inspection may provide clues to the age and contamination of the brake fluid and the general condition of other hydraulic brake components.

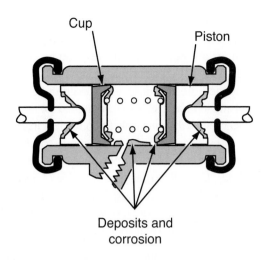

Cup

Piston

Deposits and corrosion

Figure 8-30 Corrosion and dirt can accumulate in wheel cylinders, which is another reason to service them every time the drum brakes are serviced.

✓ **SERVICE TIP:** On some cylinders the return spring is strong enough to push one or both pistons out. If you loosen the bleeder screw and push in on both pistons, some fluid is removed from the cylinder. Hold the pistons in place as the bleeder screw is tightened; this forms a low-pressure area that keeps the pistons in place. If the pistons are pushed outward by the return spring, they will draw fluid from the brake line and possibly allow air to enter the cylinder and line. If they do not move, air usually will not enter the bore. In either case, bleeding the brakes is shortened by a minute or two. Also, if you have done this much work to the brakes, there is no reason not to completely flush the system clear of all old brake fluid.

A four-wheel brake repair is an ideal time to flush the system. It is best to check for air anyway and a few minutes more will replace all of the old fluid.

Wheel cylinders are often overlooked during brake service. If the cylinder is not inspected closely and overhauled or replaced, its continued service life may well be shorter than the service life of new linings. Therefore, the next brake service may be needed sooner than necessary to fix damage from a leaking cylinder.

The brake shoe return springs hold the wheel cylinder pistons retracted in the cylinder bore. The pistons do not continually move outward as the linings wear as do disc brake caliper pistons. Nevertheless, over the life of drum brake linings, wheel cylinder pistons shift slightly in their bores. Corrosion and sludge can accumulate behind the piston seals (Figure 8-30), which can accelerate seal wear as the seals are pushed backward in the bore when new shoes are installed. This potential problem can be magnified if brake fluid is not changed regularly. The difference in piston seal position may be only several thousandths of an inch, but the possible problem can be compounded by long intervals between brake service due to very slow and gradual lining wear. For all of these reasons, many shops make it standard practice to overhaul or replace wheel cylinders whenever new shoes are installed. Some wheel cylinders cannot be rebuilt but must be replaced if there is leakage or damage.

Drum Brake Reassembly

Although drum brake reassembly is usually the opposite of disassembly, a few special steps are important to do the job correctly. Generally, do not use lubricated compressed air on brake parts because damage to rubber parts may result. Remember also that if any hydraulic part is removed or disconnected, to bleed all or part of the brake system.

Begin by installing new or overhauled wheel cylinders if the old ones were removed. Remove the caps from the brake line fittings and install the fittings into the cylinders, using a flare-nut wrench.

Classroom Manual
pages 203–219

It is very important to verify the locations of all components in a brake assembly. Compare the new brake shoes to the old ones to be sure that holes for springs and clips are in the same locations and that the new linings are positioned in the same locations on the shoes. This comparison of new and old brake shoes is particularly important for the primary shoes of duo-servo brakes. If a hardware kit is obtained, be sure it includes everything needed.

Also verify that right-hand and left-hand shoes are identified for installation on the correct wheel. Some shoes have pins for adjuster levers that will only fit correctly if installed on the specified right or left wheel (Figure 8-31). Some shoes have welded reinforcements to help support the web and table. The welds may interfere with mounting and brake operation if the shoes are installed on the wrong sides. A few brands mark the shoes "L" or "R" to designate the side of installation. Return springs, hold-down parts, and even parking brake levers may allow the wrong shoe to be installed on the wrong side of the car. Some part of the assembly will not fit, however, if the shoes are installed on the wrong sides. Photo Sequence 18 shows the general installation of a duo-servo brake assembly.

> ☑ **SERVICE TIP:** Brake shoe springs are a closed coil design in which the coils should touch each other when tension is released. You can identify a stretched spring by holding it up to the light to see if light shines through the coils. If it does, replace the spring. The coils are stretched.

> ▲ **WARNING:** When replacing the return springs with new ones or reinstalling the old ones, ensure that the correct spring is placed on the correct shoe. Sometimes the return springs for the front and rear shoes are different in shape and strength.

After determining the correct shoe locations, transfer any parking brake linkage parts from the old shoes to the new shoes. Any U-clips and wave washers that are bent for installation or that receive constant wear during operation should be replaced.

Remove nicks and rough spots from the raised shoe pads on the backing plate with emery cloth and then clean the area. Lightly coat the shoe pads with brake lubricant. Make sure that the

Special Tools

Lift or jack with stands

Impact tools

Return spring pliers

Hold-down spring removal tool

Service manual

Some return springs are color coded. When installing replacement springs, be sure they are a matching color.

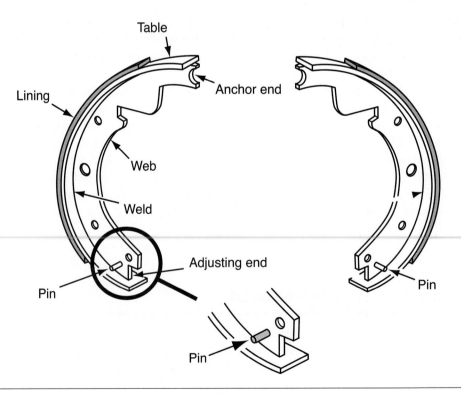

Figure 8-31 Verify that right-hand and left-hand shoes are identified for installation on the correct wheel.

Photo Sequence 18
Typical Procedure for Installing a Drum Brake Assembly

P18-1 Compare the new parts with the old to ensure that the correct ones are available.

P18-2 Depending on how the parking brake was disconnected, there are two means to connect it: slide the lever pin with the washer installed through the primary shoe and install a new retaining clip, or connect the cable to the lever as shown.

P18-3 Hold the self-adjuster mechanism in place on the primary shoe as the retaining spring is installed.

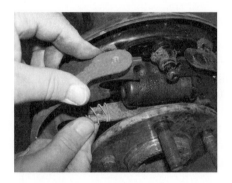

P18-4 Pull the top of the secondary shoe forward and fit the parking brake strut between the two shoes. The large slot in the strut goes to the rear and the spring to the front.

P18-5 Fit the holed end of the cable over the anchor and move the running end to the side. Install the shoes' return springs.

P18-6 Route the cable behind the primary return spring and down to the self-adjuster pawl position.

P18-7 Install the pawl and its spring(s).

P18-8 Pull the lower ends of the shoes apart and fit the self-adjuster between the two shoes. After releasing the shoes, check to make sure that the self-adjuster is correctly linked to the shoes and that the star wheel is to the rear.

P18-9 Hook the cable to the pawl and push up on the pawl. Route the cable over the cam. After releasing the pawl, ensure that the cable stayed in place on the cam.

P18-10 Check each component of the assembled drum brake for proper positioning and anchoring.

There are several brands of brake lubricant on the market. Use a well-known brand and ensure that it is labeled as brake grease.

backing plate bolts and bolted-on anchor pins are torqued to specifications. Ensure that riveted anchor pins are secure.

Install hold-down parts using the appropriate tools to mount the shoes on the backing plate. Connect parking brake linkage at the appropriate point during reassembly. This will vary for different brake designs. Lightly coat the surface of the parking brake pin with brake lubricant. Disc brake lubricant works well with drum brakes also. Install the lever on the pin with a new washer and clip (Figure 8-32). Attach the parking brake cables and be sure their movement is not restricted.

If the brakes you are servicing have wheel cylinder pushrods or links, install them between the wheel cylinder pistons and the shoes. If pushrods are not used, ensure that the shoe webs fit into the ends of the wheel cylinder pistons correctly.

Once again, ensure that left-hand and right-hand parts have not been interchanged from one side of the car to the other. This is particularly important for duo-servo star wheel adjusters, which

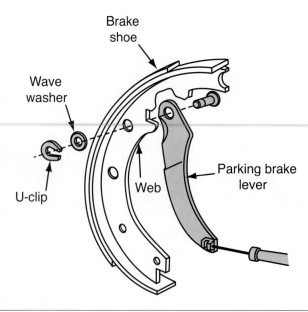

Figure 8-32 Installing the parking brake lever to the brake shoe.

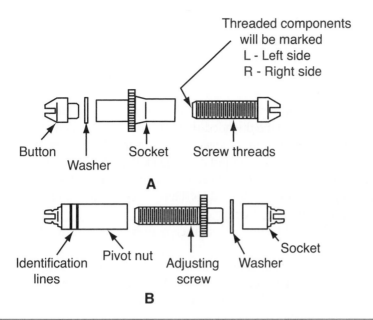

Figure 8-33 Look for left and right identification on adjuster assemblies. If an adjuster is installed on the wrong side of the car, it will not work. Typical DaimlerChrysler (A) and Ford (B) adjusters are shown here.

have left- and right-hand threads (Figure 8-33). The self-adjuster linkage for leading-trailing brakes also has definite left-hand and right-hand parts (Figure 8-34).

Disassemble the adjuster and clean the parts in denatured alcohol. Clean the threads with a fine wire brush. Make sure that the adjusting screw threads into the threaded sleeve over its

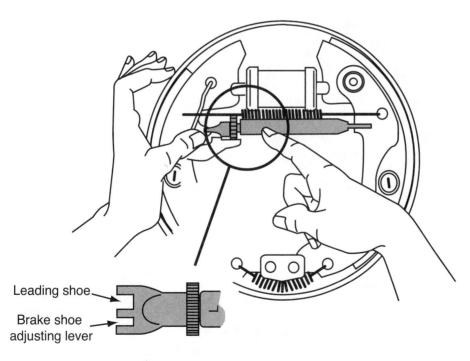

Note: Socket blade marked R and L. Install letter in upright position to ensure proper slot engagement to brake shoe adjusting lever.

Figure 8-34 This adjuster strut for a leading-trailing brake has specific right-hand and left-hand installation positions.

complete length without sticking or binding. Ensure that none of the star wheel teeth are damaged. Lubricate the screw threads with brake lubricant, being careful not to get any on the star wheel teeth. Also apply brake lubricant to the inside of the socket and the socket face. Finally, apply a continuous bead of lubricant to the open ends of the threaded sleeve and socket (end cap) when the threads are fully engaged.

When you install self-adjusters, set them close to their fully retracted positions. For example, screw a star wheel adjuster into the threaded sleeve until it bottoms and then turn it outward one or two turns. Setting the adjusters in a retracted position makes it easier to install the shoes and their return springs. After the shoes are installed, check the adjusters again and adjust them outward just enough to take up any slack between the adjusters and the shoes.

Before installing any of the adjuster cables, links, levers, guides, and other parts, inspect each carefully for the kinds of wear and damage shown in Figure 8-35. Many technicians and shop owners choose to replace all of the self-adjuster parts whenever drum brakes are serviced to ensure proper operation.

After the self-adjusters are installed, check their operation by prying the shoe to which the linkage is attached lightly away from its anchor or by pulling the cable or link to make sure the adjuster advances easily, one notch at a time. Adjuster cables tend to stretch, and star wheels and pawls can become blunted after a long period of use. Leave the adjusters close to a fully retracted position and the shoes at a minimum diameter. After brake assembly is complete, make the preliminary brake adjustment as explained in the following section.

⚠ **WARNING:** Wear safety glasses and keep your fingers away from the return springs during removal and installation. If a spring slips off a tool or mounting point, its strong tension can cause injury or damage.

Installing the return springs is usually the last step—or close to the last step—of brake reassembly. Use brake spring pliers and other special tools as required for spring installation (Figure 8-36).

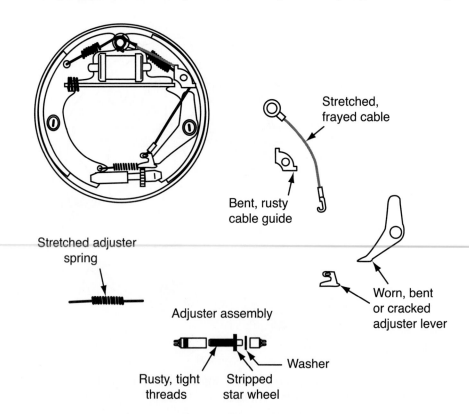

Figure 8-35 Check all self-adjuster parts closely for damage and wear.

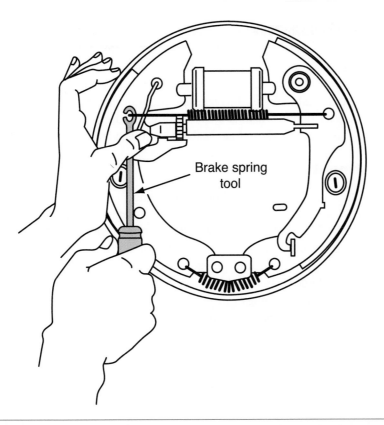

Figure 8-36 Use special tools and be very careful when installing brake shoe return springs.

Be sure to install each spring in the proper direction and in the proper holes in the shoe webs. Many return strings have longer straight sections at one end than at the other. If a spring is installed in the wrong hole on the shoe web, its operating tension will not be correct. Some vehicles require that a certain return spring be placed on top of the other. For instance, the return spring for the front shoe may have to be installed on the anchor first. This setup ensures the tip of the curve of the spring hook does not hang on parking brake components or other brake components. If a spring is stretched too far to install it in a wrong hole, damage to parts or personal injury can result. If the shoes seem not to fit at the upper anchor, recheck the parking brake and self-adjuster for proper installation.

✓ **SERVICE TIP:** If a brake shoe web has several holes close to each other, mark the one into which the end of a return spring is installed. Then match it to the corresponding hole in the replacement shoe. This saves time and hassle when installing new shoes, and it helps to verify that replacement shoes and return springs are the correct match for the vehicle.

Drum Installation

At this point, it is time to do the initial brake adjustment and install the drum. The next sections of this chapter cover initial brake adjustments and manual service adjustments of self-adjusters for duo-servo and leading-trailing brakes. Because brake drum inspection, measurement, and refinishing are separate operations that can be done independently of servicing the brake assembly, drum service is covered in the last sections of this chapter.

If the brake drum is a floating type that is installed over the wheel studs on an axle flange or hub, it can be installed after the initial brake adjustment. Speed nuts or push nuts that may have been installed on the drum when the vehicle was assembled can be reinstalled to help hold the drum in place while the wheel is installed. Their use is not essential, however.

If the drum is a one-piece assembly of drum and wheel hub, refer to Chapter 3 in this manual for instructions on cleaning, packing, installing, and adjusting wheel bearings.

Brake Adjustment

Unlike disc brakes in which self-adjustment is a basic design feature, drum brakes require manual adjustment when assembled. Furthermore, almost all drum brakes have self-adjuster mechanisms, which are parts that specifically readjust the brakes as the linings wear. With or without self-adjusters, drum brakes may require periodic manual adjustment to ensure proper lining-to-drum clearance. Correct brake adjustment ensures the proper brake pedal position and operation for safe braking.

Initial Adjustment

Special Tools

Brake tool adjuster
(spoon)

Brake shoe caliper

When you install the brake shoes and finish assembling all the brake components, the self-adjusters are retracted so that the shoes are at the minimum diameter. Before installing the drum, you should readjust the shoes to take up most of the clearance with the drum. This will make the final adjustment after the drum is installed faster and more accurate.

Before making the initial adjustment of rear drum brakes, check the parking brake cable adjustment. The parking brake may have been adjusted to remove slack with the old brake shoes installed. New shoes with fresh linings can reduce the lining-to-drum clearance allowed by the parking brake, even when the brake is released. If necessary, back off the parking brake adjustment before the initial manual adjustment of the brake shoes. The parking brake must operate freely with the brake shoes and linings centered on the backing plate. Check the parking brake and readjust it if necessary as a final step after manual adjustment of the service brakes. Refer to Chapter 9 of this *Shop Manual* for parking brake adjustment procedures.

Initial adjustment requires a brake shoe caliper (Figure 8-37), which is a measuring tool that gauges the inside diameter of the drum and the outside diameter of the installed shoes. Place the caliper into the drum as shown in Figure 8-37 and slide it back and forth to open the jaws to their widest point. Then tighten the lock screw to hold the caliper jaws in position.

Depending on the type of caliper used, the jaw opening on its opposite side is now set to equal the drum diameter or to a specific smaller diameter. The installed diameter of the brake shoes should be adjusted to approximately 0.020 inch to 0.040 inch (0.50 mm to 1.00 mm) smaller than the drum diameter. Check the vehicle service manual for exact specifications.

If the caliper opening used to gauge the brake shoes equals the drum diameter, readjust it undersized by the specified amount and then place it over the widest point of the brake shoes.

Figure 8-37 Measure the inside diameter of the drum with the brake drum micrometer and tighten the locknut.

As an alternative, you can leave the caliper set to the drum diameter and use a feeler gauge between the caliper and one shoe to adjust the shoes. If the caliper has built-in compensation for the drum and shoe diameters, simply place it over the widest point of the shoes.

With the caliper in place over the shoes (Figure 8-38), rotate the star wheel adjuster to expand the shoes until the caliper just slides over them without binding. Hold the self-adjuster pawl away from the star wheel with a screwdriver or a heavy wire hook while adjusting the star wheel. Some leading-trailing brakes have a separate star wheel or other adjuster for each shoe. For this type of brake assembly, adjust each shoe an equal amount.

After the initial adjustment, install the brake drum and apply the brakes several times to verify that the pedal is fairly high and firm. This will also center the shoes within the drum. Bleed the brakes, if necessary, to ensure that the lines are free of air. Then make a final, manual adjustment of the brakes as explained in the following paragraphs.

Manual Adjustment Precautions

Even with self-adjusters, drum brakes may need manual adjustment to compensate for incomplete self-adjustment at some time in their service life. Manual adjustment also may be required to verify that self-adjusters are working equally on each wheel of an axle set and to correct the adjustment if they are not.

Special Tools

Brake tool adjuster (spoon)

Some technicians rely on the self-adjusters to make the final adjustment automatically after new shoes are installed. This is not the best practice, however, because initial adjustment is only a rough adjustment and may not be equal on both wheels of an axle set. As a result, self-adjustment then may remain unequal, particularly on lightly used rear drum brakes of an FWD car. A final manual adjustment verifies an equal and complete adjustment on both wheels.

Perform manual brake adjustments with the vehicle supported on a hoist or stands and off the ground so that the wheels can rotate during adjustment. Exact adjustment procedures are different for different brake designs but all are based on the principle of expanding the shoes until they contact the drum and then backing off the adjustment a specified amount. With the brakes adjusted, drum on, pump the pedal once or twice to center the shoes. Recheck the brake adjustment. The pedal will center the two shoes and provides a better adjustment. Generally duo-servo

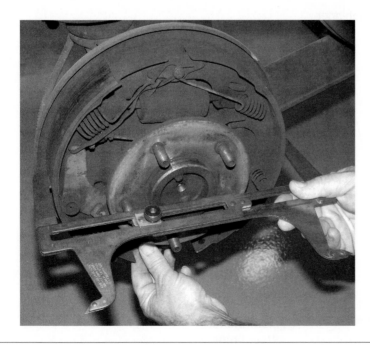

Figure 8-38 Flip the caliper over and place it over the widest point of the brake shoes. Adjust the brake shoes until they touch the caliper pads.

brake adjustments are backed off more than leading-trailing brake adjustments because duo-servo brakes need more clearance for the servo action to develop proper leverage.

On some vehicles, it may be necessary to tighten the adjuster quite a bit after the linings first contact the drum. Then when the adjustment is backed off, it may feel as if excessive retraction of the shoes is needed to get free wheel rotation. In addition, the pedal may feel springy or spongy when it is applied. All these symptoms are signs that the shoes are not properly arced to the drum. Excessive adjuster travel in either direction indicates that the shoes are bending due to incomplete lining contact with the drum. The only satisfactory long-term solution to this problem is to measure the shoe arc, remove the shoes, and either replace them with properly arced shoes or have the shoes arced to match the drum diameter and brake installation. Figure 8-39 shows three common kinds of lining arcing profiles. The arc profile must match the original design of the brake assembly.

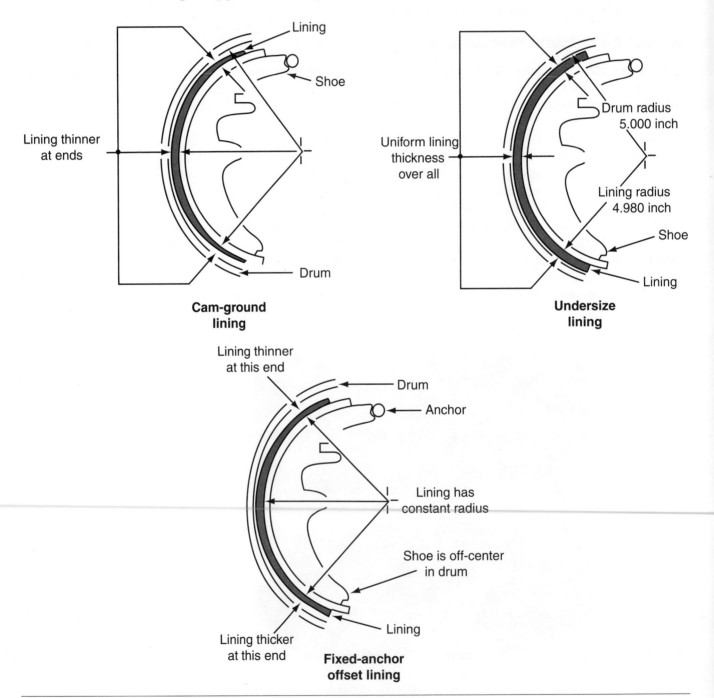

Figure 8-39 Three common profiles for the lining arc of new brake shoes.

Before performing any of the following brake adjustments, be sure that the parking brake is fully released. If the parking brake is holding the brake shoes off their anchors or is improperly adjusted in any way, back off the parking brake adjustment. Some noticeable slack should be present in the cables, and the linkage should not bind. Readjust the parking brake as explained in Chapter 9 of this *Shop Manual* after adjusting the service brakes.

Duo-Servo Star Wheel Adjustment

Duo-servo brakes use a single star wheel adjuster in the link that connects the bottoms of the brake shoes. Rotating the star wheel moves both shoes at the same time and adjusts the clearance with the drum as an assembly. Most duo-servo brakes are adjusted manually through a hold in the backing plate (Figure 8-40), but some cars have an adjustment hole in the outboard web of the drum (Figure 8-41). This latter style requires wheel and tire removal for brake adjustment.

The adjusting hole, in either the backing plate or the drum, is usually closed with a rubber, plastic, or metal plug that can be removed easily with a punch. After adjustment, the plug should be reinstalled to keep dirt and water out of the brakes. Some brakes are built without the hole in the backing plate or drum, but the location for the hole is scored or lanced. At the time of the first brake adjustment, the scored area (or knock out) must be removed with a drill or by knocking it

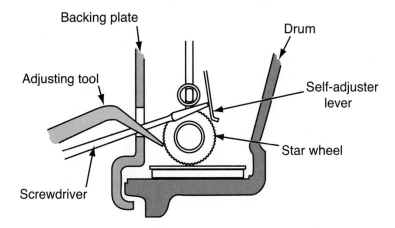

Figure 8-40 If the adjustment opening is in the backing plate, use a screwdriver to push the self-adjuster lever away from the star wheel during manual adjustment.

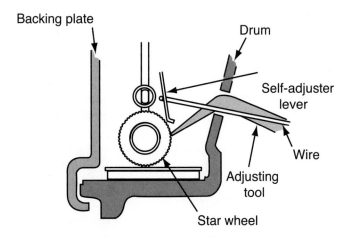

Figure 8-41 If the adjustment opening is in the drum, use a wire hook or bent piece of welding rod to pull the self-adjuster lever away from the star wheel during manual adjustment.

out with a small chisel or punch. After the brakes are adjusted, the hole should be closed with a plug as described above.

To adjust duo-servo brakes, insert a brake adjusting tool or a flat-blade screwdriver through the adjusting hole to engage the notches of the star wheel. With the other hand, use another small screwdriver or a wire hook to push or pull the self-adjuster pawl away from the star wheel. The method of disengaging the self-adjuster is different depending on brake design and on whether the adjuster is reached through the backing plate or through the drum.

☑ **SERVICE TIP:** A thin piece of plain brazing rod can be used to hold off the self-adjuster pawl during brake adjusting. The rod is strong enough to do the job and is small enough not to interfere with the adjusting tool.

It is possible to force the star wheel against the self-adjuster without disengaging it. However, if this is done, it will quickly wear down the teeth of the star wheel and the edge of the pawl. The self-adjuster will not operate properly then and may never adjust the brakes during their service life. Always disengage the self-adjuster before manually turning the star wheel.

To expand the adjuster link and decrease lining-to-drum clearance, the handle of the adjusting tool is usually moved upward to turn the star wheel. To retract the adjuster link and increase lining-to-drum clearance, move the tool handle downward. This general rule can vary for different brake designs, however, and even from one side of the car to the other. Check the vehicle service manual for instructions.

Rotate the vehicle wheel by hand as the brakes are adjusted. For duo-servo brakes, rotate the wheel in the direction of forward rotation while expanding the shoes. The self-energizing operation and servo action of the shoes will help to center the shoes in the drum during adjustment.

Exact adjustment instructions vary from one carmaker and one brake design to another. Some instructions say to adjust the brakes outward until a light drag is felt on the drum as the wheel is turned. Other instructions call for a heavy drag or locking the wheel. The instructions then direct that the shoes be backed off or retracted a specific number of notches or clicks of the star wheel. Generally, retract the shoes more on a duo-servo brake than on a leading-trailing brake.

It is very important to follow exactly service manual instructions for a given brake design. It is equally important that the adjustment be equal on both sides of the vehicle to avoid brake pull and to ensure equal lining wear.

☑ **SERVICE TIP:** Most brake adjustments are made through a star wheel adjuster, but if you ever have to work on an old 1976–1979 Chevette you may be momentarily stumped. Brake adjustment is made by turning bolts on each backing plate, one on each side of the axle. Turn the top of each bolt *toward* the axle to back off the adjustment for drum removal.

Leading-Trailing Star Wheel Adjustment

Special Tools

Brake tool adjuster
(spoon)

Most leading-trailing brakes have a single star wheel adjuster similar to the adjuster for duo-servo brakes. The adjuster linkage is mounted higher on the shoe webs (Figure 8-42) than the adjuster link used for duo-servo brakes, because leading shoes and trailing shoes are rigidly anchored to the bottom of the backing plate. Rotating the star wheel expands and retracts the shoes, just as it does for duo-servo brakes, and the self-adjuster mechanism usually must be retracted or released for manual adjustment.

Just as for duo-servo brakes, the adjusters for leading-trailing brakes have definite left-hand and right-hand parts (Figure 8-43). Verify that the adjusters are installed correctly before trying to adjust the brakes.

Some leading-trailing brakes have a separate star wheel or other adjuster for each shoe. For this kind of brake assembly, adjust each shoe an equal amount. Leading-trailing brakes with quadrant adjusters are not adjusted by turning a star wheel but by manually moving the toothed quadrant. Refer to the vehicle service manual for special procedures and tool requirements.

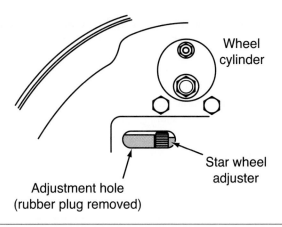

Wheel
cylinder

Star wheel
adjuster

Adjustment hole
(rubber plug removed)

Figure 8-42 This star wheel for a leading-trailing brake is mounted higher on the brake assembly than the adjuster for a duo-servo brake.

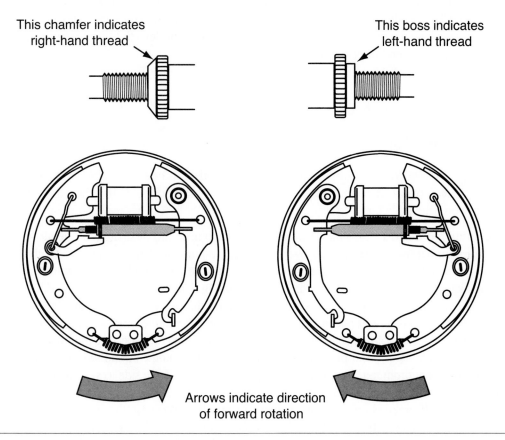

This chamfer indicates
right-hand thread

This boss indicates
left-hand thread

Arrows indicate direction
of forward rotation

Figure 8-43 Note the identification of the left-hand and right-hand adjuster parts on this leading-trailing brake.

Like duo-servo brakes, rotate the vehicle wheel by hand while adjusting leading-trailing brakes. Rotation methods are a bit different, however. If the brake has a single star wheel, start by rotating the wheel forward until specified drum contact is made. Then rotate the wheel in reverse to verify adjustment. If the brake has separate star wheels for each shoe, rotate the wheel forward while adjusting the forward shoe; then rotate the wheel in reverse as the rear shoe is adjusted.

Again, exact adjustment instructions vary from one carmaker and one brake design to another. Generally, retract the shoes less on a leading-trailing brake than on a duo-servo brake.

Follow service manual instructions exactly for a given brake design to obtain equal adjustment on both sides of the vehicle. After adjustment, install the plug in the adjustment access hole to keep dirt and water out of the brakes.

● **CUSTOMER CARE:** If you have a customer who is a very cautious and careful driver and always avoids panic stops or any kind of hard braking, recommend a brake inspection and adjustment check at every oil change. This kind of driver may never apply the brakes hard enough to activate the self-adjusters. Also remind those drivers whose vehicle brakes are adjusted by parking brake action to always apply the parking brake before exiting the vehicle.

Brake Drum Service

Classroom Manual
pages 203–207

Brake drums are resurfaced more often than disc brake rotors. In fact, most shops make drum refinishing or turning a standard part of complete drum brake service. The drums must be removed from the brakes anyway for any kind of service except simple adjustment, and many drum problems cannot be identified completely until the drum is mounted on a lathe. Maximum braking performance cannot be obtained with a drum that is damaged or defective. Complete drum service consists of these general operations:

❑ Drum inspection
❑ Drum measurement
❑ Drum resurfacing or turning

Drum Inspection

⚠ **WARNING:** Never use a drum that exceeds or is near its discard limits. Thin drums can overheat or shatter. This may cause a vehicle accident. The shop and technician may be held liable.

Brake drums work like a heat sink. Their job is to absorb heat and dissipate it to the air. As drums wear from normal use or are thinned by refinishing, the amount of metal available to absorb and release heat is reduced. As a result, the drums operate at increasingly higher temperatures. The drum structural strength also is weakened by the loss of metal. Braking forces can distort the drum's shape or lead to cracking or other problems. Drum inspection is the process of identifying these problems and deciding which can be serviced by refinishing and which require drum replacement.

An earlier section of this chapter explained the importance of inspecting the brake shoes and linings for indications of drum problems. For example, if the linings on one wheel are worn more than the other, that drum may be scored or rough. Uneven wear from side to side on any one set of shoes can be caused by a tapered drum. Linings worn badly at the toe or heel may indicate an out-of-round drum. If brake shoes and linings have a slight blue coloring, that indicates overheating, which is a clue to check the drum for the same condition. If the lining is worn more in the center than at the ends or vice versa, the lining may not have been arced properly to the drum diameter. If so, check the lining-to-drum fit closely when installing new shoes with a resurfaced drum.

Continue by inspecting the drum itself for the following conditions:

❑ *Scored Drum Surface.* Inspect the drum braking surface for scoring by running your fingernail across the surface (Figure 8-44). Any large score marks mean that the drum must be resurfaced or replaced. The most common cause of drum scoring (Figure 8-45A) is when road grit or brake dust becomes trapped between the brake lining and drum. Glazed brake linings that have been hardened by high temperature, or inferior linings that are very hard, also can groove the drum surface. Excessive lining wear that exposes the rivet heads or shoe steel will score the drum surface. If the scoring is not too deep, the drum can be refinished.

To see what a bell-mouthed drum looks like, take a small cone paper cup, cut off the closed end, and open the cup. The cup is now bell-mouthed. The smaller end represents the closed end of the brake drum.

Figure 8-44 To evaluate the amount of scoring in a drum, inspect it and run your fingernail across the drum surface.

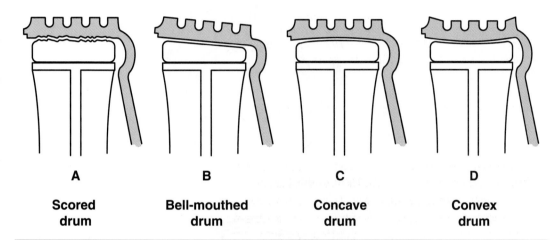

A	**B**	**C**	**D**
Scored drum	**Bell-mouthed drum**	**Concave drum**	**Convex drum**

Figure 8-45 Typical brake drum wear patterns and defects.

❑ *Bell-Mouthed Drum.* Bell mouthing is shape distortion caused by extreme heat and braking pressure (Figure 8-45B). It is most common on wide drums that are weakly supported at the outside of the drum. Bell mouthing makes full drum-to-lining contact impossible, so braking power is reduced. Drums must be refinished or replaced.

❑ *Concave Drum.* A concave wear pattern (Figure 8-45C) is caused by a distorted shoe that concentrates braking pressure on the center of the drum.

❑ *Convex Drum.* A convex wear pattern (Figure 8-45D) is caused by excessive heat or an oversized drum, which allows the open end of the drum to distort.

❑ *Hard Spots.* **Hard spots,** or chilled spots, in the cast-iron surface result from a change in metallurgy caused by heat. They appear as small raised areas (Figure 8-46A). Brake chatter, pulling, rapid wear, hard pedal, and noise may occur. Hard spots can be ground out of the drum, but since only the raised surfaces are removed, they can reappear when heat is reapplied. Drums with hard spots should be replaced.

❑ *Threading.* An extremely sharp or chipped tool bit or a lathe that turns too fast can literally cut a thread into the drum surface. During brake application, the shoes ride outward on the thread, then snap back with a loud crack. **Threading** can also cause faster lining wear and interfere with shoe alignment during braking. To avoid threading the drum surface, use a properly shaped bit and a moderate-to-slow lathe speed.

❑ *Heat Checks.* Unlike hard spots, **heat checks** are visible on the drum surface (Figure 8-46B). They are caused by high temperatures. Heat-checked drums may have a bluish-gold tint, which is another sign of high operating temperatures. Hardened carbide lathe bits or special grinding attachments are available to service these conditions.

Hard spots are circular blue-gold glazed areas on drum or rotor surfaces where extreme heat has changed the molecular structure of the metal.

Hard spots can cause chattering during drum machining, leaving a very rough finish.

Heat checks are small cracks on drum or rotor surfaces that usually can be machined away.

Drums that have severe heat checks should be replaced unless there is an economic or availability problem.

A Hard or chill spots

B Heat checks

C Cracked drum web

Figure 8-46 Hard spots, heat checks, and cracks are common drum problems.

The **drum web** is the closed side of a brake drum.

❏ *Cracked Drum.* Cracks in the drum are caused by excessive stress. They may appear anywhere, but they are most common near the bolt circle or at the outside of the **drum web** (Figure 8-46C). Cracks also may appear at the open edge of the braking surface. Fine cracks often are hard to see and often do not appear until after machining. Any crack, no matter how small, means the drum must be replaced.

✔ **SERVICE TIP:** Tapping the drum lightly with a hammer will give an indication if it is cracked. Set the drum, open side up, and tap the outside with a hammer. A good drum will produce a bell-like or ringing sound, whereas a cracked drum will have a flat, dull sound without any ring.

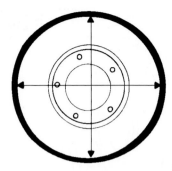

Figure 8-47 An out-of-round drum will have unequal diameter measurements.

❏ *Out-of-Round Drums.* Slightly out-of-round drums usually appear good to the eye, but the problem causes pulling, grabbing, and pedal vibration or pulsation. An out-of-round or egg-shaped condition (Figure 8-47) is often caused by heating and cooling during normal brake operation. An out-of-round drum can also be caused by setting the parking brake while the drums are hot. A severely out-of-round drum will cause pedal pulsation and/or vibration during braking. To test for an out-of-round drum before the drum is removed, adjust the brake to a light drag and feel the rotation of the drum by hand. Any areas of heavy drag or no drag may indicate a problem. Remove the drum and measure it at several points to determine the amount of distortion. A brake drum that is out of round enough to cause vehicle vibration or roughness when braking should be refinished. Remove only enough stock to return the brake drum to roundness.

❏ *Grease or Oil Contamination.* If the drums have been exposed to leaking oil or grease, thoroughly clean them with a nonpetroleum solvent such as denatured alcohol or brake cleaner. Locate the source of the oil or grease leak and fix the problem before reinstalling new or refinished drums.

Drum Measurements

During brake service, every drum must be measured with a drum micrometer or gauge to make sure that it is within the safe size limits. An old rule of thumb for drum refinishing was that a drum could be turned to 0.060 inch beyond its original diameter and should be replaced if its diameter was 0.090 inch or more beyond its original size. Thus, a 10-inch drum could be turned to 10.060 inches and should be discarded at 10.090 inches or more. Life is not that simple any more, and drum measurement requires more thought and precision.

The first step of drum measurement is to check the **discard diameter,** or maximum inside diameter, that is cast or stamped on the outside of the drum (Figure 8-48). The discard diameter is the allowable wear dimension, not the allowable machining dimension. There must be 0.030 inch (0.75 mm) left for wear *after* machining. That is, the refinished drum diameter must be at least 0.030 inch less than the discard diameter. If this dimension is exceeded, the drum will wear beyond its maximum allowable diameter during normal operation.

Ford Motor Company specifically says that a drum must be replaced if its diameter exceeds the discard dimension at any point. If you are unsure about any drum dimensions for refinishing or replacement, refer to the vehicle service manual for specifications. An old drum made before 1971 will not have a discard diameter marked on its surface, so check a service manual for exact specifications.

Special Tools

Brake drum
  micrometer
Service manual
Inside micrometer

Discard diameter
is the maximum
inside diameter that
is cast or stamped
on a brake drum; it is
the allowable wear
dimension, not the
allowable machining
dimension.

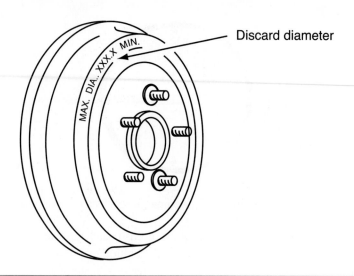

Discard diameter

Figure 8-48 The discard diameter is cast or stamped into every brake drum made since the early 1970s.

Begin measuring a drum by setting the drum micrometer to the nominal drum diameter such as 11.375 inches or 276 mm (Figure 8-49). Refer to a service manual if unsure about the original diameter. Then insert the micrometer into the drum. Insert the end with the movable plunger first and hold it against the drum as you insert the fixed end of the micrometer.

Hold the anvil steady against the drum surface, and move the dial end of the micrometer back and forth until the highest reading on the dial indicator is obtained (Figure 8-50). Add the dial indicator reading to the nominal drum diameter on the micrometer. For example, if the micrometer was

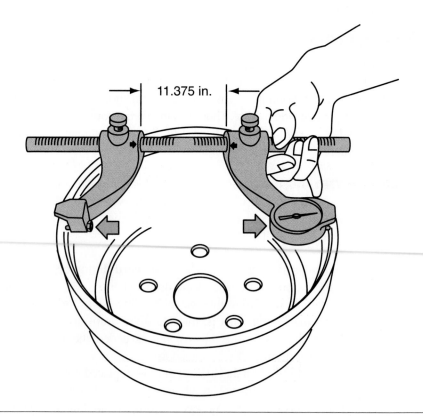

11.375 in.

Figure 8-49 Begin drum measurement by setting the micrometer to the nominal drum diameter.

Dial indicator

Figure 8-50 Insert the micrometer in the drum, hold the fixed anvil firmly against the inside surface, and carefully rock the dial indicator end until the highest reading is achieved. Take four measurements, 45 degrees apart.

set to 11.375 inches and the dial indicator reads 0.015 inch, the diameter is 11.375 plus 0.015, which equals 11.390 inches (Figure 8-51).

Compare this reading to the drum discard dimension and remember that it must be 0.030 inch (0.75 mm) or more *under* the discard dimension to allow for refinishing. Take at least four measurements, 45 degrees apart around the drum opening.

Measuring the drum at four locations around its opening checks for an out-of-round condition as well as the overall diameter. If the highest and lowest diameter measurements vary by 0.006 inch (0.15 mm) or more, machine the drum to correct the out-of-round condition. This guideline applies even if the drum is otherwise in good condition and within the maximum diameter limits. An out-of-round drum can cause brake chatter, grabbing, and pedal pulsation if not corrected.

The basic drum micrometer measurement will tell you the drum diameter and whether it can be refinished. If the drum is deeply scored, however, you must measure the diameter at the bottom of the deepest groove. To do this, a drum micrometer adapter with pointed anvils or a machinist's inside micrometer with a long extension to span the drum diameter is needed. Again, the diameter measured at the deepest groove must be 0.030 inch (0.75 mm) or more *under* the discard dimension.

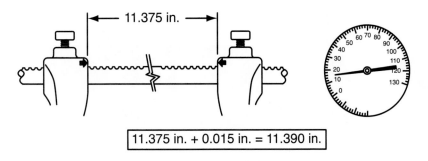

Figure 8-51 Add the dial indicator reading to the original micrometer setting to get the drum diameter.

Severe bell-mouth, concave, or convex wear may be visible to your eye or indicated by abnormal lining wear, but an inside micrometer should be used to measure precisely for these conditions. To check for any kind of wear or distortion across the drum surface, take several measurements in a straight line from the inside to the outside of the drum. Then repeat the measurements at about four positions, 45 degrees apart around the drum opening. If the highest and lowest measurements taken at any straight-line position across the drum vary by 0.006 inch (0.15 mm) or more, machine the drum to correct the problem.

If the drums are smooth and true and within safe limits, any slight scores can be removed by polishing with fine emery cloth. If scoring or light grooves cannot be removed by hand, the drum must be refinished or replaced. Even slightly rough surfaces should be turned to ensure a true drum surface and to remove any possible contamination on the surface from previous brake linings and road dust.

Many shops now use a newer version of the drum micrometer (Figure 8-52). The measurements are done in the same manner as with an older drum micrometer, but it is easier and faster with the new tool. Snap-on Tools and others have an electronic brake drum micrometer (Figure 8-53). This tool is even easier and faster than the tool mentioned above, primarily because the measurements can be converted from inches to metric with a push of a button. The tools shown in Figure 8-52 and Figure 8-53 can measure all light vehicle drums up to about 14 inches or 365 mm.

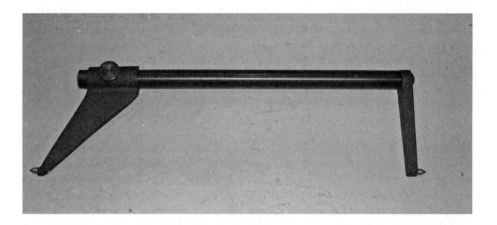

Figure 8-52 This drum caliper eliminates the need to set up the caliper and then adjust the reading according to the drum diameter. The diameter can be measured directly from this caliper.

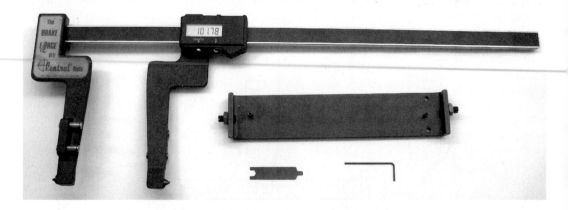

Figure 8-53 An electronic drum caliper is quick, easy to read, and can convert between measuring systems.

Refinishing Brake Drums

Brake drums with moderate to severe scoring or other defects can be refinished by either turning (cutting) or grinding on a brake lathe (Figure 8-54). Only enough metal should be removed to obtain a true, smooth friction surface. If too much metal is removed from a drum, the following unsafe conditions can result:

- ❏ Brake fade caused by the thin drum being unable to absorb heat during braking
- ❏ Poor and erratic braking due to distortion of the drum
- ❏ Noise caused by a thin drum vibrating during operation
- ❏ Drums cracking or breaking during a very hard brake application

If one drum must be machined, the other drum on the same axle also must be machined to the same diameter so that braking will be equal at both wheels. Drum diameters must be within 0.005 inch of one another.

> ✔ **SERVICE TIP:** It is not always necessary to refinish a drum to remove minor score marks. You can remove them with sandpaper or emery cloth.

Brake Lathes

Brake drums are refinished on bench lathes made specifically for drum machining. Many models of drum lathes exist, but regardless of the lathe used, it should be serviced regularly according to the manufacturer's maintenance procedures.

Different cutting assemblies are used for rotors and for drums. The attaching adapters, tool holders, vibration dampers, and cutting bits must be in good condition. Make sure mounting adapters are clean and free of nicks. Always use sharp cutting tools or bits and use only replacement cutting bits recommended by the equipment manufacturer. Dull or worn bits leave a poor surface finish, which will affect braking performance.

Classroom Manual
pages 203–207

Special Tools

Bench brake lathe

Catch basin

Hot soapy water

Figure 8-54 A typical bench brake lathe for machining (turning) drums and rotors.

The tip of the cutting bit should be slightly rounded, not razor sharp. This is even more important for turning a drum than for a rotor. A sharply pointed bit can cut a spiral groove into the drum that will cause noisy and erratic brake operation.

Mounting a Drum on a Lathe

The mounting procedure for a drum depends on whether the drum has wheel bearings mounted in its hub. For one-piece drums and hubs with bearings installed, remove the inner bearing and grease seal before mounting the drum on the lathe arbor. Refer to Chapter 3 in the *Shop Manual* for bearing removal and installation instructions.

A one-piece drum with bearing races in the hub mounts to the lathe arbor with tapered or spherical cones. A two-piece drum removed from its hub is centered on the lathe arbor with a spring-loaded cone and clamped in place by two large cup-shaped adapters.

Remove all grease and dirt from the bearing races before mounting the drum. It may be necessary to steam clean the grease out of the hub. Index the drum on the wheel bearing races to ensure that the machining is accurately indexed to the drum axis. Use the appropriate cones and spacers to lock the drum firmly to the arbor shaft. For floating drums without bearings, clean all rust and corrosion from the hub area with emery cloth or 120-grit sandpaper and use the proper cones and spacers to mount the drum to the arbor shaft.

When mounting a one-piece drum and hub, check the inner bearing races (cones) to be sure they are secure in the hub. If either race is loose, replace the drum and all bearings.

When the drum is on the lathe, install a rubber or spring-type vibration damper on the outer diameter of the drum (Figure 8-55) to prevent the cutting bits from chattering during refinishing. Use of the vibration damper results in a smoother finished surface. The damper also helps reduce unwanted noise. Photo Sequence 19 shows the general procedures for mounting a one-piece drum/hub on a brake lathe. For a floating drum, refer to Photo Sequence 14 for general mounting procedures.

Machining a Drum on a Lathe

Before removing any metal from the drum, verify that it is centered on the lathe arbor and that extra runout has not been created by the lathe mounting. If the drum is not centered and square

Figure 8-55 Install a vibration damper to keep the drum from chattering while being refinished.

404

Photo Sequence 19
Mounting a One-Piece Drum/Hub on a Bench Brake Lathe

P19-1 Ensure that the hub cavity is completely clean and that the bearing races are firmly secure within the hub.

P19-2 Select a tapered cone that fits into each bearing race. Normally the outer bearing race requires a slightly smaller cone than the inner race.

P19-3 After sliding the inner tapered cone onto the arbor, fit the hub onto the cone.

P19-4 Hold the drum/hub onto the inner cone as the outer tapered cone is fitted onto the arbor and into the outer bearing race.

P19-5 Hold everything in place on the arbor as the necessary spacer and/or bushing is installed onto the arbor.

P19-6 Screw on the arbor nut. It is a reverse (left-hand) thread. The interior of one end of the nut is not threaded. This end of the nut is facing the spacer/ bushing when installed on the arbor.

P19-7 Install the vibration damper around the outer circumference of the drum.

P19-8 The drum/hub is now properly mounted and ready for the final checks before machining.

with the arbor, machining can actually add runout. To check drum mounting, make a small scratch on one surface of the drum as follows:

1. Begin by backing the cutting assembly away from the drum and turning the drum through one complete revolution to be sure there is no interference with rotation.

2. Start the lathe and advance the cutting bit until it just touches the drum surface near midpoint of the drum's friction area.

CAUTION: Do not attempt to use a brake lathe without training. The minimum training is studying the lathe's operator manual. Serious injury or damage could occur if the lathe is improperly set up.

WARNING: A brake lathe can produce a lot of torque. Do not wear loose clothing or unrolled long-sleeved shirts while machining or setting up the lathe.

WARNING: Do not attempt to make adjustments or perform other actions in or near the cutting head. Allow the drum to come to a complete stop before loosening the nut. The lathe produces sufficient torque to break a bone or cause other injuries.

WARNING: Do not allow the lathe to operate without close supervision. Do not allow other persons near the lathe until it stops. Do not leave the lathe until it stops running. Inattention or lack of monitoring could cause an accident.

3. Let the cutting bit lightly scratch the drum, approximately 0.001 inch (0.025 mm) deep (Figure 8-56).

4. Move the cutting bit away from the drum and stop the lathe. If the scratch is all the way around the drum, the drum is centered and you can proceed with resurfacing.

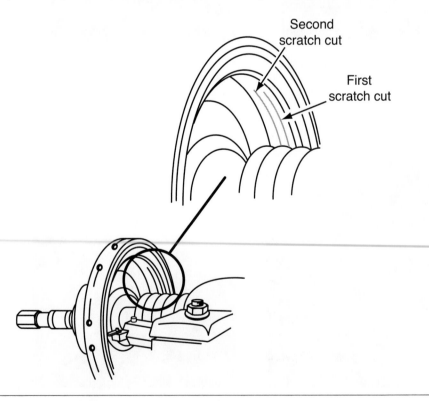

Second scratch cut

First scratch cut

Figure 8-56 A pair of scratch cuts will help you check for an out-of-round condition, as well as for the drum mounting on the lathe.

5. If the scratch appears intermittently, either the drum is out of round or it is not centered on the arbor. In this case, loosen the arbor nut and rotate the drum 180 degrees on the arbor; then retighten the nut.

6. Repeat step 2 through step 4 to make another scratch about ¼-inch away from the first (Figure 8-57).

7. If the second scratch appears intermittently, but at or near the first scratch test, the drum is significantly out of round, but it is properly centered on the lathe and you can proceed with machining.

8. If the second scratch appears opposite the first on the drum surface, remove the drum from the lathe arbor and recheck the mounting.

In extreme cases, the lathe arbor shaft may be bent. To determine if the arbor is bent, mount a dial indicator on the lathe and disconnect lathe power. Release the pulley belt tension by moving the controlling lever; then rotate the arbor slowly by turning the drive pulleys. Observe the dial

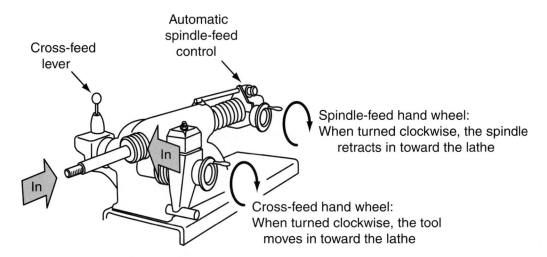

Cross-feed lever

Automatic spindle-feed control

In

In

Spindle-feed hand wheel: When turned clockwise, the spindle retracts in toward the lathe

Cross-feed hand wheel: When turned clockwise, the tool moves in toward the lathe

Clockwise rotation

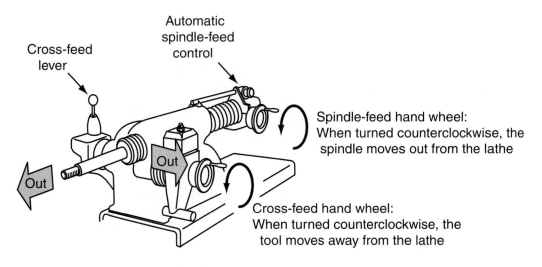

Cross-feed lever

Automatic spindle-feed control

Out

Out

Spindle-feed hand wheel: When turned counterclockwise, the spindle moves out from the lathe

Cross-feed hand wheel: When turned counterclockwise, the tool moves away from the lathe

Counterclockwise rotation

Figure 8-57 On most lathes, the spindle-feed hand wheel controls the movement of the cutter in and out of the drum, and the cross-feed hand wheel controls the depth of cut. Note that the lathe can be operated in either clockwise or counterclockwise rotation.

indicator needle. Movement of the needle more than one division (0.001 in.) indicates a bent arbor. Contact the lathe manufacturer for lathe service information.

A distorted mounting adapter sometimes can be corrected by installing it on a precision metal-working lathe and machining it. It is usually more practical to replace a defective adapter, however.

Adjusting Lathe Settings. Before starting to machine the drum, consider and adjust three lathe settings: lathe speed (rpm); cross-feed (depth of cut), and spindle feed (speed of travel across the drum surface). The lathe speed usually stays constant throughout the machining operations, but most lathes have at least two or three speed settings. Select the best speed for the drum being machined according to the lathe manufacturer's instructions. Most drums can be refinished and sanded satisfactorily at 150 rpm.

The spindle feed is the distance the cutting bit moves across the friction surface during each lathe revolution. A spindle feed of 0.010 inch to 0.020 inch (0.25 mm to 0.50 mm) per revolution is good for rough cuts on most drums. Make the finish cut at a slower spindle feed of about 0.002 inch to 0.005 inch (0.05 mm to 0.15 mm) per revolution. Most lathes have SAE and metric measurements.

The cross-feed, or depth of cut, is the amount of metal removed by the cutting tool in each pass across the drum. A setting of up to 0.015 inch (about 0.40 mm) can be used for a rough cut but only about 0.005 inch (0.15 mm) for the finish cut. Figure 8-57 shows the cross-feed and spindle-feed controls for a typical drum lathe. Note that the lathe can be operated clockwise or counterclockwise, depending on the desired spindle-feed direction.

Machining the Drum. When certain that the drum is securely mounted and properly centered, turn on the lathe and adjust it to the desired speed. Then advance the cutting bit to the open edge of the drum and remove the ridge of rust and metal that has formed there. Use several light cuts of 0.010 inch to 0.020 inch rather than one heavy cut.

Next, move the cutting bit to the closed (inner) edge of the drum and remove the ridge that also is present there. As the ridges are removed, note the point of the smallest drum diameter. Position the cutting bit at this point and adjust the hand wheel that controls the depth of cut to zero. This is the starting point for further depth-of-cut adjustments.

Reposition the cutting bit to the closed (inner) edge of the drum and adjust it for a rough cut as specified previously. The hand wheel micrometer is graduated to indicate the amount of metal removed from the complete diameter. For example, if the hand wheel indicates 0.010 inch, the lathe has made a cut 0.005 inch deep in the drum surface.

Adjust the cross-feed for a rough cut and engage the cross-feed mechanism. The lathe will automatically move the cutting bit from the inner to the outer edge of the drum. Make as many rough cuts as necessary to remove defects but stay within the dimension limits of the drum. If the cutting bit chatters as it passes over hard spots in the drum surface, grind the surface or discard the drum. Complete the turning operation with a finish cut as specified previously.

After turning, keep the drum mounted on the lathe, and deburr it with 80-grit sandpaper (do not use emery cloth) to remove all minute rough and jagged surfaces. If the drum does not clean up when turned to its maximum machining diameter, it must be replaced.

Some replacement brake drums are semifinished. A semifinished drum may require additional machining to obtain the proper dimensional specifications and surface finish. Fully finished drums do not require additional machining unless it is needed to match the diameter of an old drum on the same axle set or has been stored incorrectly. A quick, light scratch test on a new drum may prevent a comeback. New drums are also protected with a rustproofing coating that must be thoroughly cleaned off the friction surfaces. Use a nonpetroleum solvent such as brake cleaner or lacquer thinner to remove the coating.

Cleaning a Refinished Drum. The surface of a freshly refinished drum contains millions of tiny particles. If these metal particles are not cleaned away, they will become embedded in the brake lining. When the brake lining becomes contaminated with metal shavings, it becomes a fine grinding stone and soon scores the drum.

Figure 8-58 Cleaning a drum after refinishing.

The metal particles can be removed from the drum by washing thoroughly with hot, soapy water and wiping with a lint-free rag. Use compressed air to thoroughly dry the clean drum. After washing, wipe the inside of the drum (especially the newly finished surface) with a lint-free white cloth dipped in denatured alcohol or brake cleaning solvent. Use a cleaner that does not leave a residue. This operation should be repeated until dirt is no longer visible on the wiping cloth (Figure 8-58). Allow the drum to dry before reinstalling it on the vehicle.

Drum Installation

If the drum is a two-piece floating drum, make sure all mounting surfaces are clean. Apply a small amount of silicone dielectric compound to the pilot diameter of the drum before installing the drum on the hub or axle flange. If the drum has an alignment tang or a hole for a locating screw, make sure it is lined up with the hole in the hub or axle flange. If push nuts or speed nuts were used to hold the drum in position, they may be reinstalled. Their use is not mandatory, however.

Install the wheel and tire on the drum and torque the wheel nuts to specifications, following the recommended tightening pattern (Figure 8-59). Failure to tighten in the correct pattern may result in increased lateral runout, brake roughness, or pulsation.

Torque stick

Figure 8-59 You should not install wheel nuts directly with an impact wrench, but using a torque stick of the correct specification is acceptable.

If the drum is a fixed, one-piece assembly with the hub that contains the wheel bearings, clean and repack the bearings and install the drum as explained in Chapter 3 of this *Shop Manual*.

After lowering the vehicle to the ground, pump the brake pedal several times before moving the vehicle. This positions the brake linings against the drum and verifies that pedal operation is correct. If the vehicle is so equipped, turn the air suspension service switch back on.

● **CUSTOMER CARE:** Advise your customers to avoid unnecessary hard braking during the first 500 miles after brakes are relined. This is a critical break-in period for linings. A little early care can dramatically increase the life of the linings.

 C A S E   S T U D Y

The brake job was almost finished on the Hyundai Excel, but the new rear wheel cylinders would not hold their pushrods out against the shoes. When the brake pedal was released, the pistons retracted fully into the wheel cylinder bores. Consequently, the first time the pedal was pressed, it would go to the floor.

The new aftermarket wheel cylinders did not have internal springs, but the tech working on the job thought that might be the problem. He checked with a dealer service department and found that the original Hyundai wheel cylinders *did* have springs. A set of OEM cylinders solved the problem. The moral of the story is to always make certain that aftermarket parts truly are equivalent to original equipment parts.

Terms to Know

Discard diameter	Heat checks	Return spring
Drum web	Hold-down springs	Threading
Floaring drum	Piston stop	
Hard spots	Push (speed) nut	

ASE-Style Review Questions

1. Before trying to remove a brake drum for service:
   Technician A backs off the brake shoe adjuster.
   Technician B takes up all slack in the parking brake cable.
   Who is correct?
   A. A only
   B. B only
   C. Both A and B
   D. Neither A nor B

2. When inspecting a wheel cylinder:
   Technician A finds liquid brake fluid behind the piston boot and rebuilds the wheel cylinder based on this fact.
   Technician B does not rebuild the wheel cylinder if only dampness is found in the boot.
   Who is correct?
   A. A only
   B. B only
   C. Both A and B
   D. Neither A nor B

3. When adjusting the brake shoes on a car with self-adjusting brakes:
   Technician A moves the self-adjusting lever away from the star wheel.
   Technician B cranks down on the star wheel to expedite the job.
   Who is correct?
   A. A only
   B. B only
   C. Both A and B
   D. Neither A nor B

4. Drum linings are badly worn at the toe and heel areas of the linings:
   Technician A says that the problem is an out-of-round drum.
   Technician B says that the problem is a tapered drum.
   Who is correct?
   A. A only
   B. B only
   C. Both A and B
   D. Neither A nor B

5. *Technician A* says that the drum discard dimension is the maximum diameter to which the drums can be refinished.
   Technician B says that the drum discard diameter is the maximum allowable wear dimension and not the allowable machining diameter.
   Who is correct?
   A. A only
   B. B only
   C. Both A and B
   D. Neither A nor B

6. When machining a drum on a brake lathe:
   Technician A uses a spindle speed of approximately 150 rpm.
   Technician B makes a series of shallow cuts to obtain the final drum diameter.
   Who is correct?
   A. A only
   B. B only
   C. Both A and B
   D. Neither A nor B

7. *Technician A* says that as long as the drum-to-lining adjustment is correct, the diameters of the two drums on an axle set do not matter as long as they do not exceed the discard dimension.
   Technician B says that the drum diameters on given axles must be exactly the same.
   Who is correct?
   A. A only
   B. B only
   C. Both A and B
   D. Neither A nor B

8. *Technician A* says that new drums must be cleaned to remove the rustproofing compound from the drum surface.
   Technician B says that refinished drums must be cleaned to remove all metal particles from the drum surface.
   Who is correct?
   A. A only
   B. B only
   C. Both A and B
   D. Neither A nor B

9. *Technician A* says that brakes with linings with more wear at the wheel cylinder end indicate a normal wear condition.
   Technician B says that if one lining on a duo-servo brake is worn more than the other, the shoes may be installed incorrectly.
   Who is correct?
   A. A only
   B. B only
   C. Both A and B
   D. Neither A nor B

10. *Technician A* says that weak or broken return springs can cause brake drag or pulling to one side.
    Technician B says that the same problems can be caused by a loose backing plate or an inoperative self-adjuster.
    Who is correct?
    A. A only
    B. B only
    C. Both A and B
    D. Neither A nor B

ASE Challenge Questions

1. *Technician A* says that a misadjusted left rear brake could make the vehicle drift or pull to the right during brake application.
   Technician B says that a stuck piston in one wheel cylinder would have the same effect as a misadjusted brake.
   Who is correct?
   A. A only
   B. B only
   C. Both A and B
   D. Neither A nor B

2. Brake drums are being discussed:
   Technician A says that hard spots in the drums may cause brake chattering.
   Technician B says that chattering is usually caused by fluid-soaked brake pads.
   Who is correct?
   A. A only
   B. B only
   C. Both A and B
   D. Neither A nor B

3. *Technician A* says that measuring the drum with a drum micrometer will give the discard dimensions.
   Technician B says that a drum caliper is used to measure the interior and exterior of the drum.
   Who is correct?
   A. A only
   B. B only
   C. Both A and B
   D. Neither A nor B

4. The rear brakes lock on one side during moderate braking:
Technician A says that this is caused by glazed shoes and drums.
Technician B says that one leaking axle seal could cause this problem.
Who is correct?

A. A only

B. B only

C. Both A and B

D. Neither A nor B

5. Wheel cylinders are being discussed:
Technician A says that a leaking wheel cylinder may cause the wheel to grab or lock.
Technician B says that dampness or seepage found inside the wheel cylinder dust boot is not considered a cause for replacing the wheel cylinder.
Who is correct?

A. A only

B. B only

C. Both A and B

D. Neither A nor B

Job Sheet 26

26

Name: _____ Date: _____

Replace Brake Shoes

NATEF Correlation

This job sheet is related to NATEF brake tasks: Research applicable vehicle and service information, such as brake system operation, vehicle service history, service precautions, and technical service bulletins; select, handle, store, and fill brake fluids to proper level; remove, clean, and inspect brake shoes, springs, pins, clips, levers, adjusters/self-adjusters, other related brake hardware, and backing support plates; lubricate and reassemble; preadjust brake shoes and parking brake before installing brake drums or drum/hub assemblies and wheel bearings; install wheel, torque lug nuts, and make final checks and adjustments.

Objective

Upon completion and review of this job sheet, you should be able to replace the brake shoes on a rear-wheel drum brake and inspect the drum brake components.

Tools and Materials

Lift or jack and jack stands

Impact tools

Return spring tool

Describe the Vehicle Being Worked On

Year _____ Make _____ Model _____

VIN _____ Engine type and size _____

ABS _____ yes _____ no _____ If yes, type _____

Procedure

Task Completed

1. Highlight the precautions required by the manufacturer concerning ABS and routine brake repairs. _____

   _____

   _____

   Wheel nut torque _____

   CAUTION: Before working on the brakes of a vehicle with an ABS, consult the service manual for precautions and procedures. Failure to follow procedures to protect ABS components during routine brake work could damage the components and cause expensive repairs.

2. Inspect the fluid level in the master cylinder. Adjust so the reservoir is about half full. ☐

3. Lift the vehicle and remove the wheel assembly. ☐

⚠ **WARNING:** Wear safety glasses or face protection when using brake fluid. Injuries to the face or eyes could occur from spilled or splashed brake fluid.

4. Inspect the brake drum mounting. Are there fasteners holding the drum, is the drum rusty, or are there any other item(s) that may affect drum removal? Explain and make a recommendation. _____

_____

_____

_____

_____

☐   5. Remove drum.

6. Inspect the drum for damage. Measure the diameter of the drum, compare to specification, and determine serviceability. See Job Sheet 27 for drum machining.

Results _____

_____

☐   7. Position a catch basin and clean braking components. Dispose of the waste as required by law and shop policy.

NOTE TO INSTRUCTORS: The following steps are based on a floating drum and a duo-servo brake system with a cable-operated self-adjuster. Service on other type drums or brake systems may require additional training or guidance for the student.

■ **CAUTION:** Do not depress the brake pedal when the shoes are being removed, installed, or have been removed from the system. Without the shoes installed properly, the wheel cylinder pistons could be forced from the bore when the pedal is depressed.

☐   8. Select the return-spring tool (pliers) and remove the top return spring from the anchor. Mark or lay out in order of removal.

☐   9. Remove the second return spring and self-adjuster cable. Mark or lay out in order of removal.

☐   10. Use the retainer-spring tool to remove the spring from front (secondary) shoe. Mark or lay out in order of removal.

☐   11. Remove the shoe and detach it from the self-adjuster and self-adjuster spring.

☐   12. Do not drop self-adjuster components. Mark or lay out in order of removal.

☐   13. Use the retainer-spring tool to remove the spring from the rear (primary) shoe. Mark or lay out in order of removal.

☐   14. Remove the shoe and detach it from the self-adjuster mechanism and parking brake cable or lever. Mark or lay out in order of removal.

☐   15. Remove parking brake lever from rear shoe if necessary.

16. Inspect the wheel cylinder dust boot. Check under edge of boot for brake fluid.

Results and recommendation. _____

_____

_____

_____

17. Clean the slide areas on the backing plate and lube with brake lubricant only. Disc brake lubricant is sufficient.

☐

NOTE: The next two steps may have to be adjusted depending on how the hardware and shoes fit onto the backing plate and self-adjuster components.

CAUTION: Compare the new shoes to the old shoes. They should be exactly alike except the new shoes *may* have additional lining material. If in doubt, consult the instructor, service manual, or parts supplier to ensure the new shoes are correct. Installing incorrect shoes will affect braking.

18. Install the parking brake lever onto the rear shoe if necessary. ☐
19. Connect the parking brake lever to the parking brake cable as needed. ☐
20. Install the rear shoe and self-adjuster actuator onto the backing plate. ☐
21. Install the retainer spring. ☐
22. Install the front shoe and retaining spring. ☐
23. Install the top end of the cable and then install the return springs in reverse order of removal. ☐
24. Install the self-adjuster spring. ☐
25. Disassemble the self-adjuster, clean, lube, and reassemble if not done previously. Screw the self-adjuster to the minor position. ☐
26. Install the self-adjuster by pulling the front shoe forward and inserting the self-adjuster between the two shoes. ☐
27. Install the spring for the self-adjuster mechanism on the rear shoe. ☐
28. Use the brake gauge to set the initial adjustment of the brakes. ☐
29. Install the drum and complete the brake adjustment as needed. ☐
30. Install the wheel assembly and torque lug nuts. ☐
31. Lower the vehicle almost to the floor when both brakes are completed and the wheel assemblies have been installed. ☐
32. Press the brake pedal several times to center the shoes. ☐
33. Check brake adjustment and make changes as required. When completed, lower the vehicle to the floor. ☐

CAUTION: Before adding brake fluid, consult the vehicle service manual. Most manufacturers require a specific classification of brake fluid to be used.

CAUTION: Never use brake fluid from a previously opened container. Once opened, even tightly capped containers will absorb moisture from the air.

34. Check the brake fluid level and top off as necessary. ☐

WARNING: Before moving the vehicle after a brake repair, pump the pedal several times to test the brake. Failure to do so may cause an accident with damage to vehicles or facility, or personal injury.

35. Perform a brake test to ensure that the brakes will stop and hold the vehicle. Do this test before moving the vehicle from the bay.

☐ **36.** When the repair is complete, clean the area, store the tools, and complete the work order.

Problems Encountered _____

_____

Instructor's Response _____

_____

Job Sheet 27

27

Name: _____ Date: _____

Machining Brake Drums

NATEF Correlation

This job sheet is related to NATEF brake tasks: Research applicable vehicle and service information such as brake system operation, vehicle service history, service precautions, and technical service bulletins; remove, clean (using proper safety procedures), inspect, and measure brake drums; determine necessary action; refinish brake drum.

Objective

Upon completion and review of this job sheet, you should be able to measure and machine a brake drum.

Tools and Materials

Bench-mounted brake lathe

Service manual

Brake drum micrometer

Describe the Vehicle Being Worked On

Year _____ Make _____ Model _____

VIN _____ Engine type and size _____

ABS _____ yes _____ no _____ If yes, type _____

Procedure

Task Completed

1. Determine the drum discard diameter _____

   Type of drum _____

   _____

   Any special precautions to be followed for this drum? _____

   _____

   _____

2. Remove wheel assembly to gain access to the drum (Job Sheet 26). Remove the drum. ☐

3. Inspect the drum. Does its condition make it inadvisable to machine? _____ If yes, explain. _____

   _____

4. Measure the internal diameter of the drum (at least two points equally spaced around the drum)   1 _____ 2 _____

5. Measure the deepest groove, if any. Is it enough to affect machining? _____ If yes, explain and make recommendations. _____

   _____

   _____

NOTE TO INSTRUCTORS: The following steps are based generally on a bench-brake lathe by Raybestos with one spindle and a floating brake drum. Other lathes or drums may require additional instruction for the student.

☐   **6.** Select a centering cone that fits about halfway through the center hole of the drum.

☐   **7.** Select two identical clamps that fit the drum without interfering with the cutting head of the lathe.

☐   **8.** Slide one clamp onto the lathe shaft, open end out.

☐   **9.** Slide on a spring followed by the centering cone.

☐   **10.** Slide on the drum followed by the outer clamp, bushing, spacer (if needed), and the nut. Tighten, but do not overtighten, the nut.

☐   **11.** Install the damping strap.

☐   **12.** Adjust the assembly inward toward the lathe body until it stops. Reverse out two turns.

☐   **13.** Adjust the cutting head so it will reach the inner edge of the machined surface of the drum.

☐   **14.** Move the drum out (away from the lathe) until the cutting tip is about halfway through the machined surface.

☐   **15.** Adjust the cutting tip until it meets the drum and reverse out about half a turn.

☐   **16.** Ensure the area around the area is clear and the lathe's drive mechanism is in neutral. Switch on the motor.

☐   **17.** Adjust the cutting tip slowly until it comes in contact with the turning drum. Hold in place and set the sliding scale to zero. Move cutting bit away from drum.

☐   **18.** Adjust the cutting head until the cutting tip is aligned with the rear edge of the machined surfaces.

☐   **19.** Adjust the cutting tip until it contacts the drum.

☐   **20.** Continue adjusting the tip until the scale is set between 0.002 inch and 0.015 inch.

☐   **21.** Engage the fast speed on the lathe.

    22. Observe the drum as it is being machined. Are there dark (uncut) areas? _____

       Explain the next step to be taken. _____

       _____

       _____

☐   **23.** If the answer to step 22 is no, go to step 27. If the answer to step 22 is yes, go to step 24.

☐   **24.** When the cutting tip clears the drum, disengage the drive and move the cutting tip back to the starting point.

☐   **25.** Adjust the cutting tip to cut 0.002 inch deeper and engage fast cut.

☐   **26.** Repeat step 18 through step 25 as needed.

☐   **27.** When the cutting tip clears the drum, disengage the drive.

☐   **28.** Move the cutting tip to the starting point and set to cut between 0.002 inch and 0.005 inch deeper.

☐   **29.** Engage the drive mechanism in slow speed.

☐   **30.** When the cutting bit clears the drum, disengage the drive and stop the motor.

☐   **31.** When the drum stops turning, remove it from the lathe.

32. Wash the drum in running hot, soapy water and with a brush if possible. If a basin is required, use hot soapy water and a brush to clean the machined surfaces. ☐

33. Rinse with clear water and blow dry with an OSHA-approved blowgun. ☐

34. Install the drum on the vehicle. ☐

35. Use Job Sheet 26 to install shoes and other components. ☐

36. When the repair is complete, clean the area and lathe, store the tools and adapters, and complete the repair order. ☐

Problems Encountered _____

_____

Instructor's Response _____

_____

_____

Parking Brake Service

Upon completion and review of this chapter, you should be able to:

- ❏ Diagnose parking brake problems.
- ❏ Inspect the parking brake system for wear, rust, binding, and corrosion.
- ❏ Clean or replace all system parts as needed.
- ❏ Lubricate the parking brake system.

- ❏ Adjust calipers with integrated parking brakes.
- ❏ Adjust the parking brake and check system operation.
- ❏ Service electric parking brake systems.

Parking Brake Tests

Basic Tools

Basic technician's set
Hydraulic lift or safety stands

Parking brake service primarily consists of testing the system operation, adjusting the cables and linkage, and replacing components when necessary. In addition, auxiliary drum parking brakes used with some rear disc brakes may require adjustment, and vacuum-release mechanisms and warning lamp switches may require service.

Checking the service brake pedal travel is the first step in diagnosing a potential parking brake problem. The next step is to inspect and test the parking brake control and the linkage. Finally, a performance test will verify that the parking brake can hold the vehicle stationary as required by motor vehicle safety standards.

Rear Drum Brake Pedal Travel

Because the service brake shoes are the parking brakes for vehicles with rear drum brakes, service brake adjustment directly affects parking brake operation. If lining-to-drum clearance is excessive, the parking brake linkage may not have enough travel to apply the parking brakes completely.

If the parking brake lever or pedal must be applied to the full limit of its travel to engage the parking brakes, excessive clearance or slack exists somewhere in the system. The technician cannot tell immediately, however, if that clearance or slack is in the cables and linkage or in the shoe adjustment. To isolate the looseness in the system, press the service brake pedal and note its travel. If brake pedal travel seems excessive, the rear drum service brakes may need adjustment. Always check and adjust the service brakes before adjusting the parking brake linkage.

Rear Disc Brake Pedal Travel

To apply the rear calipers for parking brake use, the rear disc brakes on GM and Ford cars have lever-operated mechanisms inside the caliper pistons. These mechanisms include self-adjusters that keep the caliper piston from retracting too far in its bore. If the self-adjusters do not work properly, both the service brake pedal and the parking brake control linkage may have excessive travel.

GM rear calipers on cars built before the mid-1980s were known for parking brake adjuster problems. GM calipers were redesigned in the mid-1980s, however, and the self-adjusters on most older models were modified. Figure 9-1 is a cross section of the parking brake mechanism in the later-model GM rear caliper.

Owners of GM cars with rear disc brakes—of any year—should be reminded to use the parking brake regularly and not rely on the transmission park position to hold the car stationary. Applying and releasing the parking brakes keeps the parking brakes adjusted and keeps the mechanism from seizing. The most practical remedy for defective self-adjusters is to replace the calipers. The self-adjusters can be rebuilt, however, if replacement calipers are not available.

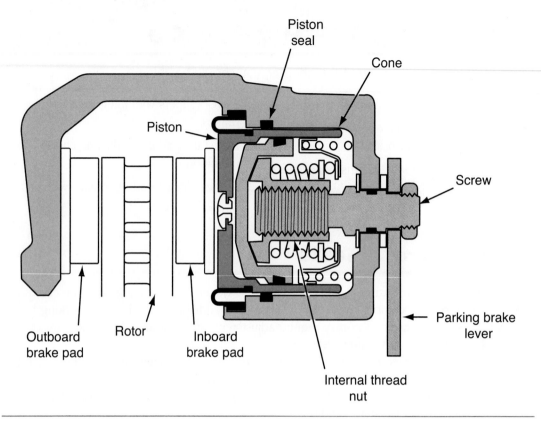

Figure 9-1 A GM screw-and-nut parking brake mechanism for rear disc brakes.

To check parking brake operation on some Ford vehicles with rear disc brakes (Figure 9-2), push the parking brake lever on the caliper forward by hand. If it moves more than 20 degrees, the self-adjuster needs readjustment. To do this, start the engine and apply the service brakes forty to fifty times with moderate pressure. Wait about 1 second between pedal applications. Then stop the engine and apply the pedal another thirty times with heavy foot pressure. If the travel of both the service brake pedal and the parking brake pedal or lever does not decrease, the rear calipers must be rebuilt or replaced.

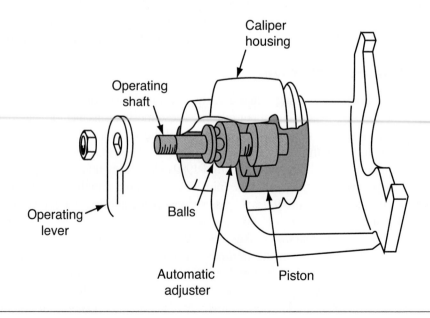

Figure 9-2 A Ford ball-and-ramp parking brake mechanism for rear disc brakes.

Figure 9-3 shows an exploded view of a Honda rear caliper. It uses a cam-and-bolt mechanism to make the brake adjustment, and it uses the cam to turn the bolt much like the self-adjusters on drum brakes. The cam is moved by application of the arm attached to the parking brake cables. If the parking brake is used properly, there is usually no requirement to adjust the parking brake for pad wear.

When the parking brake is applied the arm is moved to force the caliper piston outward. If the piston moves far enough, the arm moves the cam and cam lever, which, in turn, turn the bolt and prevent the piston from returning to its previous position within its bore. Although a manual adjustment of this portion of the parking brake is not normally needed, there is an adjustment available at the cable equalizer (Figure 9-4). Normally this adjustment is used when new components of the braking system are installed, if the parking brake needs to be backed off during caliper rebuild or replacement, and sometimes when new pads are installed. The adjustment is simple: Remove the protective plate from the bottom of the vehicle body; use the appropriate wrench to turn the adjusting nut clockwise to tighten or counterclockwise to slacken the cables. Hold the front rod with pliers while turning the adjusting nut. This is the most common type of parking adjustment.

Linkage Inspection and Test

Remind the customer that the parking brake is for holding the vehicle and not for stopping it. Many vehicles use the parking brakes to self-adjust the service brakes. If the service brake pedal is firm

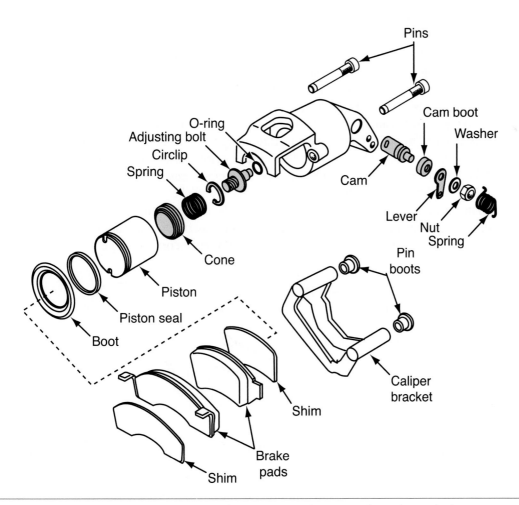

Figure 9-3 When the cable is pulled, the arm moves the cam and cam lever. The lever rotates the sleeve piston, which is attached to the adjusting bolt. If the movement is enough, the rod prevents the piston from fully returning to its previous position within the bore.

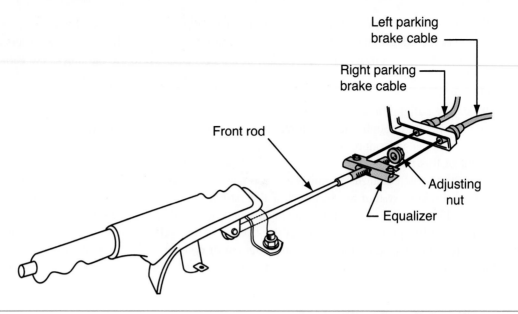

Figure 9-4 It is best to clean and lubricate the adjusting nut and front rod threads before attempting to turn the adjusting nut. Hold the rod on the unthreaded portion with pliers during movement of adjusting nut.

and has normal travel, the parking brakes should be fully applied when the lever or pedal moves one-third to two-thirds of its travel. Some carmakers say that the parking brakes should be applied completely when the ratchet mechanism moves a specified number of notches (Figure 9-5).

If the parking brake lever or pedal travels more than specified, the linkage adjustment is too loose. In this case, the brakes may not be applied well enough to hold the vehicle stationary, especially on a hill. If the parking brake lever or pedal travels less than specified, the linkage adjustment is too tight, which can cause the brake linings to drag on the drum or rotor when the vehicle is moving and lead to brake fade and premature lining wear. Overly tight parking brake linkage for rear drum brakes also can interfere with accurate service brake adjustment.

If the lever or pedal travel is within the specified range, raise the vehicle on a hoist with the parking brakes released. Rotate the rear wheels by hand and check for brake drag. Have an assistant operate the parking brake control while you check the movement of the cables and linkage. The pedal or lever should apply smoothly and return to its released position. The parking brake cables should move smoothly without any binding or slack. Protective **conduit** should be in good shape.

A **conduit** is a flexible metal housing or jacket that houses the parking brake cables to protect them from dirt, rust, abrasion, and other damage.

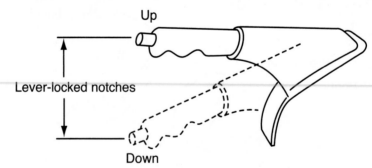

Lever-locked notches:

Cars with rear disc brakes: 7–11
Cars with rear drum brakes: 4–8

Figure 9-5 Some carmakers specify a certain amount of travel for the parking brake lever.

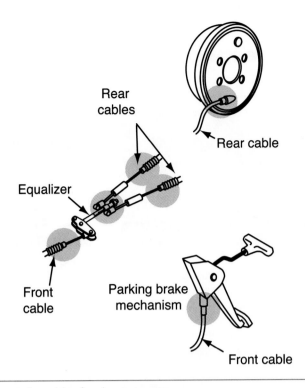

Rear
cables

Rear cable

Equalizer

Front
cable

Parking brake
mechanism

Front cable

Figure 9-6 Inspect the cables for damage.

Also inspect the cables (Figure 9-6) for broken strands, corrosion, and kinks or bends. If the brakes drag or the linkage binds in any way, the parking brakes must be adjusted or repaired as necessary. Clean and lubricate the parking brake and noncoated metal cables with a brake lubricant. See Photo Sequence 20 for a typical inspection and adjustment procedure. Damaged cables and linkage must be replaced.

The parking brake cables on some vehicles are coated with a plastic material. This plastic coating helps the cables slide smoothly against the nylon seals inside the conduit end fittings. It also protects the cable against corrosion damage. Plastic-coated cables do not need periodic lubrication, but these cables should be handled carefully during service. Avoid contact with sharp-edged tools or sharp surfaces on the vehicle underbody. Damage to the plastic coating will impair the smooth operation of the system and can also lead to corrosion.

A small amount of drag is normal for rear disc brakes, but heavy drag usually indicates an overadjusted parking brake mechanism. To determine if the misadjustment is in the linkage or in the caliper self-adjusters, disconnect the cables at a convenient point so that all tension is removed from the caliper levers. If brake drag is reduced, inspect the external linkage and lubricate, adjust, or repair it as necessary. If brake drag is still excessive with the parking brake linkage disconnected, check the caliper levers to be sure that they are returning fully against their stops. If the caliper levers are operating correctly and brake drag remains excessive, the calipers must be repaired or replaced.

Performance Test

FMVSS 105 requires that the parking brakes hold the vehicle stationary for 5 minutes on a 30 percent grade in both the forward and reverse directions (Figure 9-7). This standard can be the basis of a performance test for the parking brakes.

Stop the vehicle on a grade of approximately 30 percent, put the transmission in neutral, apply the parking brakes with moderate force, and release the service brakes. It is not necessary to wait a full 5 minutes, but the vehicle should remain stationary for a reasonable amount of time. Perform this test with the vehicle facing in both directions on the grade.

Photo Sequence 20
Typical Procedure for Inspecting and Adjusting Rear Drum Parking Brakes

P20-1 Proper adjustment of the parking brake begins with setting the parking brake to a near fully applied position.

P20-2 Raise the car on the hoist and make sure it is secure. Position the vehicle so you can rotate the rear wheels freely. (If the parking brake is applied and adjusted properly, you should be unable to rotate the wheels.)

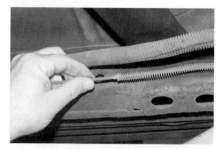

P20-3 Carefully inspect the entire length of the parking brake cable. Look for fraying, breakage, and deterioration.

P20-4 Spray all exposed metal areas of the cable assembly with penetrating oil. This ensures a free-moving system.

P20-5 Inspect the adjustment mechanism. Clean the threaded areas and make sure the adjusting nuts are not damaged.

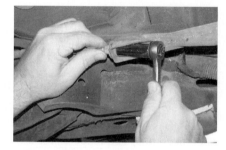

P20-6 If the parking brake needs adjustment when checked in P20-2, complete the following adjustment procedures: Loosen the adjustment locknut, then adjust the parking brake by tightening the adjusting nut.

P20-7 When you can no longer turn the wheels by hand, stop tightening the adjusting nut.

P20-8 Lower the vehicle and release the parking brake.

P20-9 Raise the vehicle and rotate the wheels. If the wheels turn with only slight drag, the parking brake is properly adjusted.

P20-10 After the proper adjustment is made, tighten the adjusting locknut. Apply a coat of white grease to all contacting surfaces of the adjustment assembly.

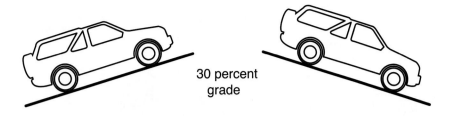

30 percent
grade

Figure 9-7 The parking brake must hold the vehicle on a 30 percent grade for 5 minutes in both the forward and the reverse directions.

 SERVICE TIP: Failure to use the parking brakes regularly is a leading cause of complaints about a low brake pedal. This is particularly true on some leading-trailing drum brakes that use the parking brake linkage to adjust the service brakes, and it is also a common problem with rear disc brakes. Advise your customers to use the parking brakes and do not overlook failure to use the parking brakes as the cause of a low-pedal problem.

Cable and Linkage Adjustment

Parking brake cable and linkage adjustment should be checked in accordance with the specific vehicle maintenance schedule. Most carmakers call for parking brake inspection at 7,500-mile or 15,000-mile intervals (12,000 km to 24,000 km). Parking brake adjustment also should be checked as part of every brake service job. Parking brake linkage often requires adjustment after rear shoes or pads are replaced.

Adjustment locations vary from one linkage design to another. Most pedal-operated parking brakes have the adjustment at a point under the center of the vehicle (Figure 9-8). Lever-operated brakes may have the adjustment under the vehicle, but it is more often located at the point where the cable, or cables, attach to the lever and is accessible from inside the car (Figure 9-9). Part of the center console or floor trim may have to be removed for adjustment access.

The service brakes must always be adjusted correctly before adjusting the parking brake linkage. If drum brake shoe adjustment is too loose, trying to adjust the parking brake may pull the shoes off their anchors or away from the wheel cylinder. The parking brakes may appear to be adjusted correctly when actually they are not. Service brake operation and further adjustment also may be adversely affected.

If service brake adjustment is unequal from side to side and one pair of shoes is looser than the other, the parking brake linkage may engage one wheel brake more tightly than the other. Only one wheel will provide stationary holding power, and parking brake effectiveness will be

Classroom Manual
pages 234–238

Special Tools

Service manual

Brake adjusting tool (spoon)

Lift or jack with stands

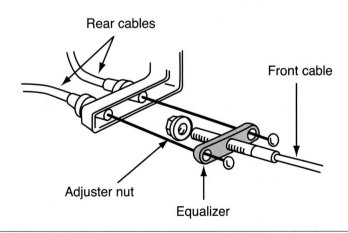

Rear cables

Front cable

Adjuster nut

Equalizer

Figure 9-8 Typical undercar adjustment for pedal-operated parking brakes.

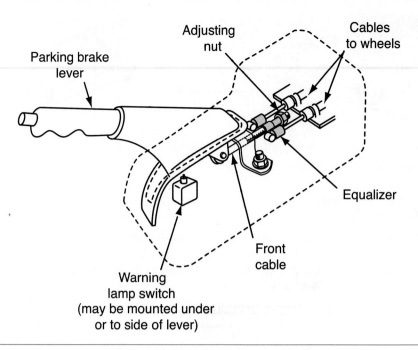

Figure 9-9 Adjustment points for lever-operated parking brakes usually are accessible from inside the car.

reduced. The problem of unequal service brake adjustment is most critical on lever-operated parking brakes that have a separate cable for each wheel. The lever moves each cable equally; but if one wheel brake adjustment is loose, full cable travel will not fully apply that wheel brake.

A typical cable adjuster has a threaded rod at the end of a cable and two jam nuts on the rod where it passes through an equalizer or a cable anchor. If the nuts are rusted or corroded, soak them with penetrating oil before trying to adjust the linkage.

To prevent damage to the threaded adjusting rod, clean the grease and dirt from the threads on either side of the adjusting nut before trying to turn the nut. Forcing the nut over dirty threads may damage the threads on the nut or the rod. Also lubricate the threads of the adjusting rod before turning the nut.

To make the adjustment, hold the adjuster nut with one wrench while loosening the locknut with another wrench. Then turn the adjuster nut to lengthen or shorten the cable as required. If the cable tries to twist as the adjuster nut is turned, grip an unthreaded section of the rod with locking pliers and hold the cable as you turn the adjuster nut. Finally, hold the adjuster nut in place with one wrench and retighten the locknut with another. On linkage with a separate adjustment for each cable, adjust each cable equally.

Drum Brake Cable Adjustment

Special Tools

Service manual

Lift or jack with stands

If the service manual is available for the vehicle, follow the adjustment instructions in the manual. Photo Sequence 20 shows a typical parking brake linkage adjustment for rear drum brakes. Most adjustment procedures include these eight basic steps:

1. Block the wheels, place the transmission in neutral, and release the parking brakes.
2. If the vehicle is equipped with automatic ride control, turn it off. Raise the vehicle on a hoist so that the wheels are off the ground.
3. If required, adjust the service brakes according to the carmaker's instructions before adjusting the parking brakes.
4. Verify that the parking brake linkage is clean, properly lubricated, and operating freely.
5. Tighten the linkage until the brakes drag as the wheel is rotated by hand.

6. Then loosen the adjustment until the wheel just turns freely.

7. Apply the parking brakes until the wheel locks and check the pedal or lever travel. The parking brakes should be fully applied when the pedal or lever moves one-third to two-thirds of its full travel or as specified by the manufacturer.

8. Apply and release the parking brakes several times and verify that the wheels turn freely when the brakes are released and are locked when the brakes are applied.

Some carmakers specify that parking brakes should be partially applied during adjustment. This is intended to apply a specific amount of slack in the linkage when the brakes are released. Instructions usually require the parking brake lever or pedal be applied a specific number of notches on the ratchet mechanism. Then adjust the brakes following the steps above.

Disc Brake Cable Adjustment

Special Tools

Service manual

The parking brake cables for the rear disc brakes on most vehicles are adjusted similarly to the drum brake cables described previously. The caliper levers that apply the caliper pistons are an added adjustment point, however. After servicing a rear brake caliper, operate the service brakes to position the brake pads before adjusting the parking brake. Begin by loosening the parking brake adjusting nut. Then start the engine and press the brake pedal several times to position the pads for normal operation.

Before tightening the cables in any way, leave them loose and be sure that they exert no tension on the caliper levers. Move the caliper levers by hand to verify that they move freely and that they return fully to their unapplied positions. If the calipers have stop lugs, be sure that the levers return completely to these stops. Tighten the cable adjustment to remove all slack and the levers just start to move. Then back off the adjustment until the levers are just released but the cables are not slack. Apply and release the parking brakes several times to be sure that the levers return to the fully released positions each time.

As another example, Figure 9-10 shows the rear disc brake parking brakes for a midsize Honda. On this design, the lever of the rear brake caliper must contact the brake caliper pin. To make the adjustment, pull the parking brake hand lever up one notch and then tighten the adjusting nut (see Figure 9-9) until the rear wheels drag slightly when turned. Next, release the

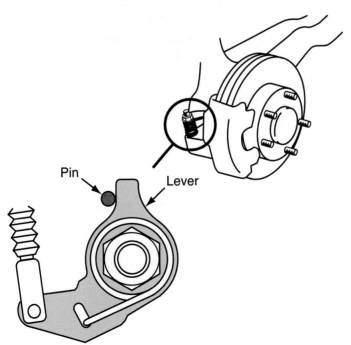

Figure 9-10 The rear disc brake parking brake for a midsize Honda.

parking brake lever and verify that the rear wheels do not drag when turned. Make any readjustments necessary. When the equalizer is properly adjusted, the rear brakes should be fully applied when the parking brake lever is pulled up five to ten notches.

If the parking brake pedal or lever does not engage the brakes completely at one-third to two-thirds of its travel, the self-adjusters in the calipers are not working properly. Try to adjust the self-adjusters by operating them as explained previously under the pedal travel test instructions. It is more likely, however, that the caliper will require repair or replacement.

Auxiliary Drum Brake Shoe Adjustment and Replacement

The auxiliary drum parking brakes used with some rear disc brakes require two separate adjustments. The cable and linkage adjustment is based on the same principles explained previously for drum brake cable adjustment. The parking brake shoes must be adjusted separately, however, before any adjustments are made to the cables and linkage.

Most auxiliary drum parking brakes have manual star wheel adjusters similar to the adjusters on drum service brakes (Figure 9-11). The adjuster is usually accessible through a hole in the outer surface in the drum portion of the rotor (Figure 9-12). On some cars, the adjuster is accessible through the hole for one of the wheel bolts. If the adjuster is accessible through a hole in the rotor, remove the wheel and then reinstall two or three wheel nuts to hold the rotor on the hub. If the adjuster is accessible through a wheel bolt hole, remove one wheel bolt. You can leave the wheel installed. Some adjusters are accessible through an opening in the inboard side of the backing plate (Figure 9-13). The star wheel adjuster on the parking brake is a simple device that can be operated with a screwdriver. Tighten the star wheel until the brakes lock and then back off the adjustment the specified number of notches or clicks.

The shoes and linings of auxiliary drum parking brakes usually outlive the vehicle on which they are installed. If they require replacement, the procedure is similar to the procedures for drum service brakes. The drum portion of the brake rotor cannot be resurfaced, but resurfacing should never be necessary because the parking brakes are applied and released when the vehicle is stationary. Rotating friction should never be applied to the drum area of the parking brake rotor.

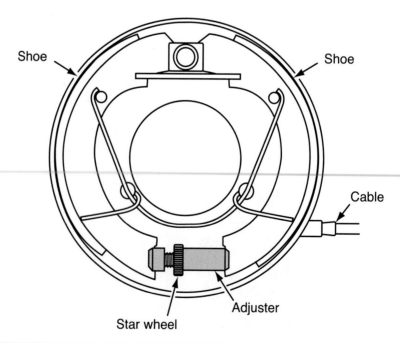

Figure 9-11 Auxiliary drum parking brake adjustment location.

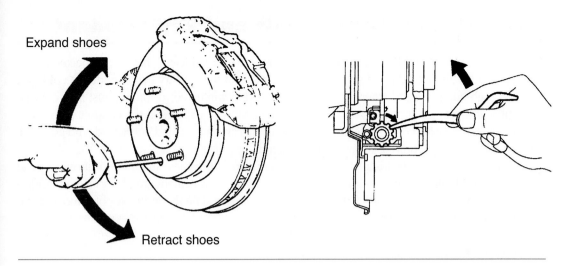

Expand shoes

Retract shoes

Figure 9-12 The adjusters for some auxiliary drum parking brakes are accessible through the outboard drum surface.

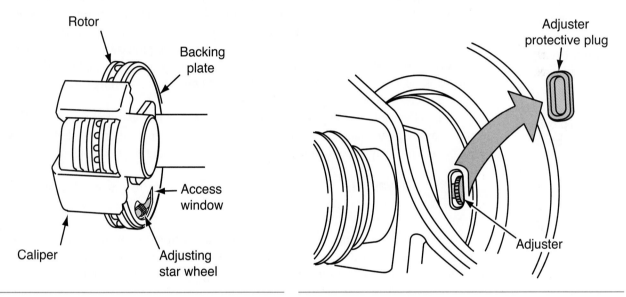

Figure 9-13 Some auxiliary drum parking brakes are adjusted through the backing plate.

Figure 9-14 Remove the protective plug to gain access to the parking brake adjuster.

Some newer vehicles use a similar method to adjust the drum-in-hat parking brakes but with some cautions that may save the technician time. When adjusting the parking brakes on a DaimlerChrysler 300 Series, the star wheel is accessed through the backing plate just like many other vehicles (Figure 9-14). The manufacturer directs that the parking brakes be adjusted until the wheel locks. The adjuster is then backed off six detents or teeth. At this point the wheel should be free to turn. If not, back off the adjuster one detent or tooth at a time. Daimler-Chrysler states that the star adjuster should not be backed off more than seventeen detents, however, because the adjuster screw will screw out of its shell and separate. Then the entire disc brake assembly will have to be removed along with at least some portion of the parking brakes to reconnect the star wheel assembly. A 10- to 15-minute job could become a 1-hour job and additional labor costs for which the customer should not be held responsible.

This particular vehicle series does not require any adjustments on the parking brake cable(s). Instead all adjustments are done at the shoes in a manner very similar to standard drum brakes. It is not an uncommon procedure, however. Study the service manual if not experienced with the type of vehicle and parking brake system being serviced.

Cable and Linkage Repair and Replacement

Classroom Manual
pages 234–238

Special Tools

Penetrating oil

Cleaning solvent

Catch basin

Wire brush

Propane torch

A good alternative to penetrating oil is chain and cable lubricant. This type of lubricant works its way into the cable or chain and provides long-term protection on parking brake cables.

Parking brake cables and linkage can easily last the life of a vehicle if they are inspected, lubricated, and adjusted according to specifications. In areas of the country that experience cold, wet weather, high humidity, and where the roads are salted in winter, corrosion and rust may damage the cables and prevent proper operation. The cable may stick or seize in the conduit, or individual strands may fray and rust. A sticking or seized cable can be loosened, lubricated, and adjusted. Because of the labor required, however, it is often more practical to replace the cable.

Parking brake linkage installations may have one to four cables, and their locations depend on vehicle and brake system designs. The details of cable replacement are always slightly different from one vehicle to another, but the following sections outline the principles of removing and installing parking brake cables.

Freeing Seized Cables

If a cable is only moderately corroded or sticking slightly, or if a replacement is unavailable, you can free the cable by working penetrating oil into the ends of the conduit. Let the oil soak in for a few minutes and then try to move the cable in and out of the conduit. Several applications of oil may be needed. A good alternative to penetrating oil is a chain and cable lubricant. This type of lubricant works its way into the cable or chain and provides long-term protection on parking brake cables.

Most corrosion and damage is confined to the first few inches of cable inside the conduit or to the cable sections outside of the conduit. You may need to disconnect one or both ends of the cable for access to corroded sections, however.

When the cable starts to move in the conduit, slide it in and out as far as it will go and continue to apply penetrating oil. Remove corrosion with solvent and emery cloth, steel wool, or a wire brush. You may need to apply heat to a seized cable to free it completely.

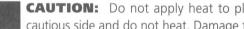

CAUTION: Do not apply heat to plastic-lined conduit. If uncertain, err on the cautious side and do not heat. Damage to the conduit may result.

In this case, disconnect and lower the cable away from the vehicle or remove it completely if possible. When the cable moves freely and dirt and corrosion have been removed, lightly lubricate the exposed sections of the cable with brake grease to avoid future problems.

Front (Control) Cable Replacement

WARNING: Disable the supplemental inflatable restraint system (SIRS) or air bag system when working on or near the system wiring. Accidental discharge of the air bag could occur, causing injury or damage.

SERVICE TIP: The following sections cover some simple cable replacement procedures. Some newer vehicles require that the front seat, portions of the dash and steering column cover, and the door sill be removed to gain access to the complete cable system. Still others require that rubber sealing grommets through the floor and/or bulkhead be replaced when new cable(s) are installed. Some cable replacement jobs can take several hours compared to older vehicles in which most parking brake cables could be replaced in an hour. Check the service manual and labor guide before providing an estimate to the customer.

If the parking brakes are operated by a lever, the control cable attachment is usually accessible from inside the car. A few lever-operated cables are accessible from under the car, however. If the

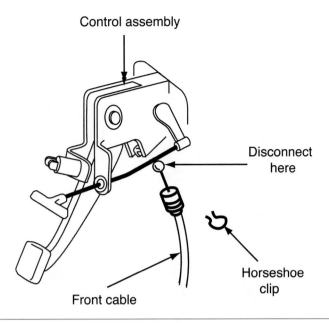

Control assembly

Disconnect
here

Horseshoe
clip

Front cable

Figure 9-15 Disconnect the upper end of the cable from the control pedal or lever.

parking brakes are operated by a pedal, the control cable attachment is usually under the instrument panel. Because of the awkward access to most such installations, it is a good idea to disconnect the battery ground cable to avoid damaging any electrical system components before working under the instrument panel. Also review the precautions in Chapter 1 of this *Shop Manual* about working around air bag installations.

To start replacing the control cable, release the parking brakes and raise the vehicle on a hoist. Disconnect the lower end of the cable from the equalizer before trying to disconnect the upper end. The lower end is more accessible, and disconnecting it first provides slack to help disconnect the upper end and any mounting clips or grommets.

Disconnect the upper end of the cable from the control pedal or lever (Figure 9-15) but do not pull it out of the vehicle yet. First, fasten a length of cord or flexible wire to the brake cable inside the car. Then pull the cable through the floor or fire wall from under the vehicle. Disconnect the cable from the cord or wire and leave the cord or wire in place through the cable opening. Fasten the cord or wire to the replacement cable and use it to help draw the cable through the opening in the floor or fire wall.

Connect the upper end of the new cable to the pedal or lever and reinstall all mounting clips, brackets, and grommets. If any attaching hardware is damaged, install new parts. Then connect the lower end of the cable to the equalizer or to an in-line connector on the rear cable. Adjust the parking brakes as explained previously.

The equalizer may be referred to as the parking brake adjuster.

Rear Cable Replacement

Almost all vehicles have two rear parking brake cables that are connected to the equalizer or adjuster. With the vehicle on a hoist, remove the cable adjusting nuts. Then disconnect the front end of the rear cable from the equalizer or from the front brake cable.

To remove the cable from a drum brake, first remove the wheel and brake drum. Disconnect the end of the rear cable from the rear parking brake lever on the rear shoe (Figure 9-16). Use the proper size offset box wrench or screwdriver to depress the conduit retaining prongs and slide the pronged fitting out through the hole in the backing plate (Figure 9-17).

Special Tools

Service manual

Lift or jack with
  stands

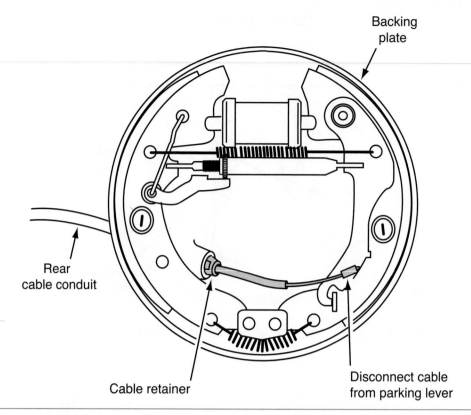

Figure 9-16 Disconnect the end of the cable from the brake shoe.

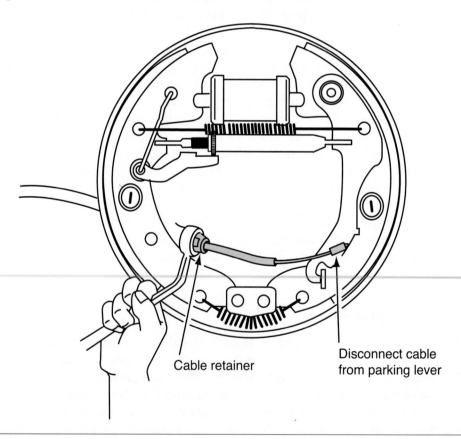

Figure 9-17 Use a box wrench to retract the conduit retaining prongs. A 13 mm box wrench is the common size used for this task.

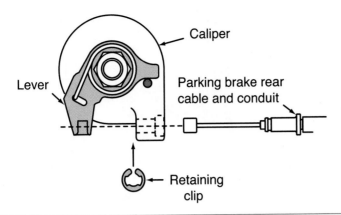

Figure 9-18 Disconnect the end of the cable from the parking brake lever on rear disc brakes.

On a typical rear disc brake, the cable conduit is secured at the rear disc brake by a clip or pin that connects the end of the cable to the parking brake lever on the caliper (Figure 9-18). Remove the clip and pin to free the cable. It may be necessary to remove grommets and pronged retainers before the rear cable can be freed. The routing of left-hand and right-hand rear parking brake cables may be different. Refer to the vehicle service manual for details and specific instructions.

Attach the new rear cable to the equalizer or adjuster. Install all grommets and clips used to secure the cable. Follow the original routing pattern.

On drum brakes, insert the cable and conduit into the hole in the backing plate. Ensure that the retaining prongs lock the conduit in place where it passes through the backing plate. Hold the brake shoes in place on the backing plate and engage the brake cable into the parking brake lever. Install the brake drum and the wheel. Install the brake cable adjusting nut. Adjust the parking brakes as explained previously.

On disc brakes, insert the parking brake rear cable and conduit end into the rear disc brake caliper and install the retaining clip or pin that secures the cable to the lever. Adjust the parking brakes as explained previously.

Vacuum-Release Parking Brake Service

⚠️ **WARNING:** Most late-model cars and light trucks have supplemental inflatable restraint systems (SIRS), known as air bags. To avoid accidental deployment of the air bag and possible injury or vehicle damage, always disconnect the battery ground (negative) cable, then the positive battery cable and wait 2 minutes before working near any of the impact sensors, steering column, or instrument panel. Do not use any electrical test equipment on any of the air bag system wires or tamper with them in any way unless specifically directed by your instructor.

Classroom Manual
pages 232–233

When working near the steering wheel, steering column, or other parts of the air bag system, disarm the system by first turning the ignition switch off. Then disconnect the battery ground (negative) cable from the battery; then disconnect the positive cable. Wait 2 minutes for the electronic module capacitor to be depleted. Review the air bag safety information in Chapter 1 of this *Shop Manual.*

Automatic (Vacuum) Release Systems

Some pedal-operated parking brakes have a vacuum release diaphragm or motor (Figure 9-19). Typically, a rod connects the vacuum-actuated diaphragm inside the motor to the parking brake release handle. A hose runs from the vacuum motor to the engine intake manifold. The vacuum hose is routed through a vacuum release switch on the steering column or floor shift console. The vacuum release switch supplies engine vacuum to the parking brake release motor when

Special Tools

Vacuum gauge/
  pump
Service manual

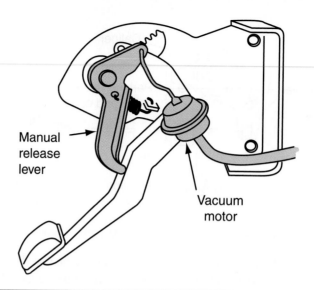

Figure 9-19 A typical foot-operated parking brake with a vacuum release mechanism.

the engine is running and the shift lever is placed in gear. This vacuum provides the power the control motor uses to release the parking brake.

The release switch vents motor vacuum when the shift lever is placed in park or neutral (and reverse on some vehicles). The release switch is operated by a cam on the shift linkage that opens and closes the switch as the driver moves the gear selector (Figure 9-20). The vacuum release system also can be released manually at any time by pulling on the manual release lever similar to ordinary parking brakes.

The vacuum required at the control motor is usually 10 in. Hg to 12 in. Hg (35 kPa to 40 kPa). Refer to the vehicle service manual for exact specifications and check system vacuum with a vacuum gauge. Perform all vacuum checks with the engine running at idle. Inspect the lines between all connecting points. Low vacuum is often caused by a loose hose connection or a leak in the hose. To detect a leak in the hoses, listen for a hissing sound along the hose route. Make certain lines are not crossed with another hose, connected to the wrong connection, kinked, or otherwise damaged.

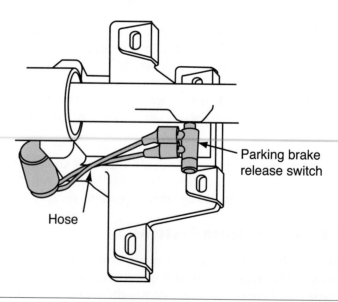

Figure 9-20 Vacuum hose connections and a vacuum release switch buried under the steering column.

CAUTION: Do not apply air pressure to the vacuum system of the parking brake release. Air pressure from the shop compressed air supply or other source will damage the actuator diaphragm in the control motor.

To test the operation of the system, run the engine at idle with the transmission gear selector in neutral. Apply the parking brakes and move the gear selector to drive. Watch the parking brake pedal to see if it returns to its unapplied position when the parking brakes release. If the parking brakes release, then the vacuum control is working properly.

If the parking brakes do not release, test for vacuum at the brake-release vacuum hose that connects to the release motor (see Figure 9-19). Remove the vacuum hose from the motor and tee the vacuum gauge into the line. Vacuum may be sent to the control motor only when the transmission is in a drive gear range. Refer to the service manual for operational details and vacuum specifications. Normally, at least 10 in. Hg (35 kPa) of vacuum is required to release the parking brake. If minimum vacuum is not present, check for a damaged component and replace it. On some vehicles the release switch can be adjusted to maintain contact with the shift lever tang. Ensure that the release switch has not moved out of place.

Parking Brake Release Switch Replacement

A faulty parking brake release switch will prevent engine vacuum from reaching the release motor. If vacuum is present at the inlet to the switch but does not exit the switch when the shift control is in a forward gear, replace the switch. To replace a parking brake release switch mounted on the steering column, begin by disconnecting the battery ground cable. Remove the necessary moldings (trim) from the instrument panel for access to the steering column cover. Remove the steering column cover and any reinforcement piece that may be present (Figure 9-21).

Disconnect the gear selector cable from its lever and remove the retaining hardware that secures the steering column tube (Figure 9-22). Carefully lower the steering column tube and disconnect the vacuum hose extension from the parking brake release switch. Remove the retaining screws that secure the parking brake release switch to the steering column tube and remove the switch (see Figure 9-20).

Begin installation of the new switch by shifting the system into neutral. Position the parking brake release switch over the column mounting bosses and push the switch against the turn signal canceling cam. Fasten the parking brake release switch with the retaining screws.

The new switch may have a plunger retainer that must be removed and discarded. After this retainer is removed, connect the vacuum hose extension to the parking brake release switch.

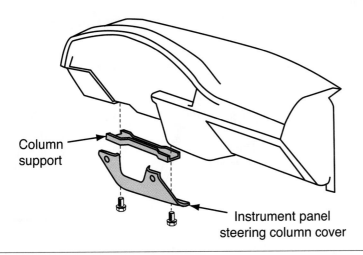

Column support

Instrument panel steering column cover

Figure 9-21 A typical steering column cover to be removed for switch access.

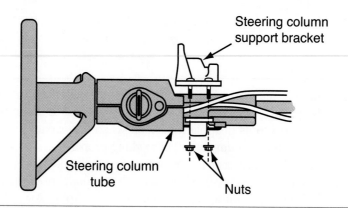

Figure 9-22 Drop the steering column from its bracket for switch access.

Reassemble the steering column tube, connect the gear selector cable to the steering column lever, and reinstall the instrument panel moldings (trim). Complete the installation by reconnecting the air bag backup power supply and battery ground cable.

Parking Brake Lamp Switch Test

Classroom Manual
page 234

With the ignition on and the parking brakes applied, the switch should close to light the BRAKE warning lamp on the instrument panel and remind the driver that the parking brakes are applied.

To test a typical parking brake switch, first gain access to the switch, which is often the hardest part of the job. For lever-operated parking brakes, it may be necessary to remove all or part of the center console. For pedal-operated parking brakes, locate the switch on the pedal bracket under the instrument panel.

After locating the switch, disconnect the electrical harness connector and apply the brake pedal or lever. Then use an ohmmeter or continuity tester to check the switch. The switch should be closed and continuity should exist with the pedal or lever applied. The switch should be open, with no continuity, with the pedal or lever released.

With the brake lever pulled up, use a continuity checker to ensure continuity exists between the positive terminal and a good ground (Figure 9-23). Continuity should be broken when the brake lever is in the down position.

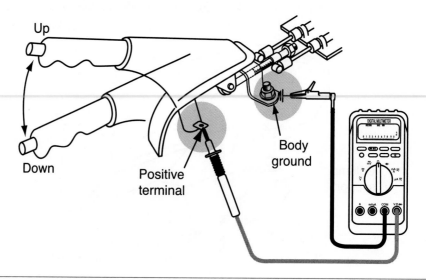

Figure 9-23 Testing the parking brake switch on a lever-operated installation.

Electric Parking Brake Service

Electric parking brakes may have electrical, mechanical, or hydraulic problems. The mechanical and hydraulic portions involve the rear service brakes, and problems may also be apparent in the operation of the service brakes. This section assumes that the mechanical and hydraulic components are good and only the electrical side is discussed.

Electro-Lok

This is a simple system to check. If the parking brakes partially hold, the problem is usually in the service brake system. Correct that before proceeding. If the parking brakes do not hold at all, first check the fuse for the system. Assuming the fuse is good, raise the vehicle if necessary to gain access to the system solenoid. Ensure that the parking brake switch is set to "on." Unplug the electrical harness from the solenoid and set a DMM to DC voltage. Probe the harness with the leads from a DMM (Figure 9-24). There should be voltage within 1 volt of battery voltage. If not, repair the two conductors, ground and positive, and/or the switch as needed and retest. If sufficient voltage is present, disconnect the DMM from the harness and set it to ohms. Probe the electrical connection on the solenoid (Figure 9-25). The resistance should be within the manufacturer's specification, which is usually about 40 ohms. If the resistance is too high or too low, replace the solenoid.

Continental Teves

This is an electric motor or solenoid that pulls on two parking brake cables similar to the standard manual parking brake control. This unit could have mechanical or electrical problems. The

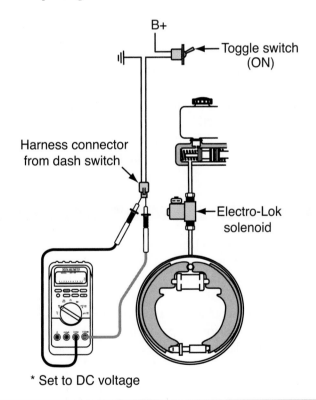

Figure 9-24 Ensure that the parking brake switch is on before doing this test. The voltage to the solenoid should be within 1 volt of battery voltage.

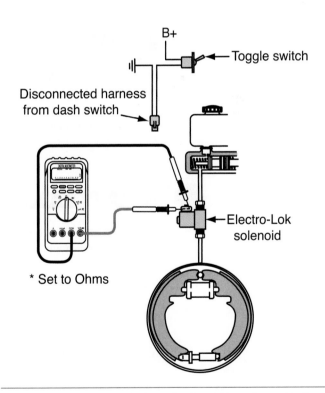

Figure 9-25 A typical good solenoid has about 40 ohms of resistance, but always check the manufacturer's specifications. An electrical motor may have much more resistance, and in some cases a resistance measurement cannot be made at all with the motor assembled.

cable(s) are adjusted similarly to standard cables. If the parking brakes partially apply, check the cable(s) and parking brakes adjustment first. If they do not apply at all, check the fuse for the motor. If the fuse is good, locate and gain access to the motor's connection harness. Test the harness and then the motor in the same manner as that performed on the Electro-Lok system. If sufficient voltage is present and the motor meets the manufacturer's specifications, then the motor will probably have to be replaced. The Continental Teves is a self-contained unit and internal repairs should not be attempted.

Electrically Direct Apply

This, too, is an electric motor. In fact, there are two, one at each rear wheel. Whereas the Continental Teves system could be used with drum brakes and some rear disc, this system is used with rear disc using the caliper to apply the parking brake force. Generally, the motors work like the screw-and-nut or the ball-and-ramp rear disc parking brakes. This system could function electrically and still not apply the parking brakes directly, however.

Check and repair the electrical side of the system in the same manner as the two previous systems. If the electrical system checks out, make sure the motor is mounted to the caliper correctly before condemning it. If the motor is loose on the caliper or its driver is not positioned on the caliper correctly, the motor will function correctly but still not be able to properly apply the caliper enough to hold the vehicle in place. A good point to remember about all three systems is that if installed correctly the first time, none of the electrical components noted will move out of position and will function almost perfectly during the life of the vehicle. Service is usually needed only if some portion of the electrical system or a component fails and that is seldom.

CUSTOMER CARE: Parking brakes are not designed for use in place of the service brakes and should be applied only after the vehicle is brought to a complete stop, except in an emergency. If you suspect a customer is using the parking brakes for other than their intended job, remind him or her of the dangers of this practice and stress the importance of keeping the service brakes in good working order.

On some ABSs, applying the parking brakes while the vehicle is moving will light the ABS warning lamp and set a diagnostic trouble code in the ABS computer.

CASE STUDY

The 1993 sedan had the pads for its rear disc brakes replaced as part of a routine brake job. Soon after, the owner complained that the rear brakes seemed to be dragging.

The tech who worked on the car originally got the comeback. He inspected and lubed the rear caliper slides to ensure that the calipers would move freely. Of most importance, he carefully adjusted the parking brake cables and double-checked to be sure that the pistons and pads retracted completely. That was the cure for the problem. When you service rear brakes—discs or drums—always check and adjust the parking brakes to be sure that the rear brakes do not drag.

Term to Know

Conduit

ASE-Style Review Questions

1. *Technician A* says that auxiliary drum parking brakes on a car with rear disc brakes are adjusted with a star wheel adjuster.
   Technician B says that mechanically actuated rear disc parking brakes have self-adjusters.
   Who is correct?
   A. A only
   B. B only
   C. Both A and B
   D. Neither A nor B

2. *Technician A* says that the vacuum release switch used with vacuum-operated parking brakes supplies engine vacuum to the parking brake vacuum motor when the transmission is placed in a forward gear.
   Technician B says that if a vacuum release switch fails, the cable must be disconnected to release the parking brakes.
   Who is correct?
   A. A only
   B. B only
   C. Both A and B
   D. Neither A nor B

3. While lubricating the parking brake system:
   Technician A applies graphite to the contact areas of plastic-coated cables.
   Technician B applies a quality grease to metal cables.
   Who is correct?
   A. A only
   B. B only
   C. Both A and B
   D. Neither A nor B

4. During parking brake adjustment:
   Technician A checks and adjusts the drum-to-lining clearance before adjusting the parking brakes.
   Technician B fully applies the parking brake lever before making the adjustment.
   Who is correct?
   A. A only
   B. B only
   C. Both A and B
   D. Neither A nor B

5. The parking brakes do not release:
   Technician A checks the drum-to-lining clearance for proper adjustment.
   Technician B checks for proper engine vacuum to the vacuum release system.
   Who is correct?
   A. A only
   B. B only
   C. Both A and B
   D. Neither A nor B

6. *Technician A* says that the vacuum release switch used with vacuum-operated parking brakes is located on the parking brake foot pedal or hand lever mechanism.
   Technician B says that the vacuum motor is mechanically linked to the release handle by a metal rod.
   Who is correct?
   A. A only
   B. B only
   C. Both A and B
   D. Neither A nor B

7. When adjusting the parking brake linkage:
   Technician A cleans and lubricates the threads of the adjusting mechanism bolt to avoid damaging it.
   Technician B makes certain the rear wheels cannot be rotated forward with the parking brakes fully applied.
   Who is correct?
   A. A only
   B. B only
   C. Both A and B
   D. Neither A nor B

8. When testing a vacuum control motor:
   Technician A checks for proper vacuum at the motor using a vacuum gauge installed with proper fittings.
   Technician B applies 5 psi to 10 psi of air pressure to the control motor to test the release action of the mechanical linkage.
   Who is correct?
   A. A only
   B. B only
   C. Both A and B
   D. Neither A nor B

9. Adjustment to rear disc brake parking brakes is being discussed:
   Technician A says that on some systems, a rear disc parking brake hand lever is pulled up one notch and the adjusting nut is tightened until the rear wheels drag slightly when turned.
   Technician B says that when typical pedal-activated rear disc parking brakes are fully released, the adjusting nut is tightened against the rear parking brake cable equalizer or adjuster until there is less than $\frac{1}{16}$ inch of movement of either rear parking brake lever at the rear disc brake caliper.
   Who is correct?
   A. A only
   B. B only
   C. Both A and B
   D. Neither A nor B

10. *Technician A* says that the cable running from the parking brake control to the equalizer or adjuster is commonly referred to as the front cable.
    Technician B says that there may be slight routing and length differences between the left and right rear cables in a parking brake system.
    Who is correct?
    A. A only
    B. B only
    C. Both A and B
    D. Neither A nor B

ASE Challenge Questions

1. A vehicle with rear disc brakes is being discussed:
   Technician A says that a misadjusted parking cable may cause the drum parking brakes to glaze over.
   Technician B says that the parking brakes may be automatically adjusted by the service brakes.
   Who is correct?
   - **A.** A only
   - **B.** B only
   - **C.** Both A and B
   - **D.** Neither A nor B

2. The parking brake control mechanism is being discussed:
   Technician A says that corrosion on the exposed portion of the cable should not cause the brakes to hang.
   Technician B says that a cable guide that is too close to the exhaust system could cause the parking brakes not to release.
   Who is correct?
   - **A.** A only
   - **B.** B only
   - **C.** Both A and B
   - **D.** Neither A nor B

3. *Technician A* says that the parking brakes are dragging because the rear disc brakes are misadjusted.
   Technician B says that a parking brake that is continuously hanging may be traced to the vacuum release switch under the dash.
   Who is correct?
   - **A.** A only
   - **B.** B only
   - **C.** Both A and B
   - **D.** Neither A nor B

4. Drum parking brakes are being discussed:
   Technician A says that a vehicle that can be rolled back and forth with the parking brakes set could have a backing plate loose on the axle housing.
   Technician B says that the drum parking brakes on the rear disc will not directly affect the service brakes.
   Who is correct?
   - **A.** A only
   - **B.** B only
   - **C.** Both A and B
   - **D.** Neither A nor B

5. *Technician A* says that a vehicle with the parking brakes applied will cause all wheels to slide if pulled by a tow bar.
   Technician B says that a vehicle's parking brakes that do not hold on a 30 percent slope do not meet FMVSS 105.
   Who is correct?
   - **A.** A only
   - **B.** B only
   - **C.** Both A and B
   - **D.** Neither A nor B

Job Sheet 28

28

Name: _____ Date: _____

Adjusting Parking Brake Cable

NATEF Correlation

This job sheet is related to NATEF brake tasks: Check parking brake cables and components for wear, rusting, binding, and corrosion; clean, lubricate, or replace as needed; check parking brake operation; determine necessary action.

Objective

Upon completion and review of this job sheet, you should be able to adjust the parking brake cable.

Tools and Materials

Lift or jack and stands
Service manual
Impact tools
Brake adjusting tool

Describe the Vehicle Being Worked On

Year _____ Make _____ Model _____

VIN _____ Engine type and size _____

ABS _____ yes _____ no _____ If yes, type _____

Procedure

Task Completed

1. Wheel nut torque _____

   Method of measuring and adjusting parking brake _____

   _____

   _____

   _____

2. Operate the parking brakes and test holding ability. Measure, if necessary, following instructions in the service manual. ☐

3. Make certain that the parking brakes are released. ☐

4. Lift the vehicle and remove the wheel assembly. ☐

5. Check service brake adjustment. Correct as necessary. ☐

NOTE TO INSTRUCTORS: This job sheet is set up based on a duo-servo brake system using a foot-operated lever and cables with the parking adjustment at the equalizer under the vehicle.

6. Loosen the locknut. ☐

☐   **7.** Turn the adjusting nut or bolt to remove excessive slack from the brake cable or to set cable to manufacturer's specifications.

☐   **8.** Operate the parking brakes and check against the manufacturer's test data.

☐   **9.** If the parking brake meets manufacturer's specifications, install any components removed and ensure that the wheels rotate smoothly with the parking brake off.

☐   **10.** Lower the vehicle to the floor.

☐   **11.** Test the parking brake.

☐   **12.** When the repair is complete, clean the area, store the tools, and complete the repair order.

Problems Encountered   _____

_____

Instructor's Response   _____

_____

Job Sheet 29

29

Name: _____ Date: _____

Testing Parking Brake Warning Light Circuit

NATEF Correlation

This job sheet is related to NATEF brake tasks: Check parking brake operation; determine necessary action; check operation of parking brake indicator light system.

Objective

Upon completion and review of this job sheet, you should be able to inspect and test the parking brake warning light circuit.

Tools and Materials

Service manual
Wiring diagram
Component locator
Digital multimeter
Jumper wire

Describe the Vehicle Being Worked On

Year _____ Make _____ Model _____

VIN _____ Engine type and size _____

ABS _____ yes _____ no _____ If yes, type _____

Procedure

Task Completed

1. Location of the parking brake warning light switch _____

   _____

   Special procedures from service manual _____

   _____

   NOTE: KOEO is key on, engine off.

2. Perform an operational check on the warning light.

   A. KOEO, parking brakes on. Is the light lit? If the answer is yes, test is complete and system is functional. If the answer is no, go to step 4.

   Results _____

   B. KOEO, parking brakes off. Is the light off? If the answer is yes, test is complete and system is functional. If the answer is no, go to step 4.

   Results _____

☐
☐
☐

3. KOEO, light is off, parking brakes are on.

    A.   Turn key off.

    B.   Locate parking brake warning light switch and disconnect the harness.

    C.   Connect a jumper wire from the wire to good ground.

    D.   KOEO. Is the light on? If the answer is no, go to step E. If the answer is yes, check adjustment of switch and/or replace switch.

        Results _____

☐

    E.   Key off. Connect the multimeter between the wire and ground, black lead to ground.

    F.   KOEO. Is battery voltage showing? If the answer is yes, replace warning lamp. If the answer is no, replace the warning lamp. If the light is now lit, the circuit is good. If the lamp still is not lit, then the circuit between the ignition switch, warning lamp, and the ground circuit must be checked. Consult the instructor and service manual for diagnostic routine.

        Results _____

4. KOEO, parking brakes are off, warning light is on.

    A.   Check master cylinder reservoir for low-level warning light. If equipped, top off brake fluid. Did the light go out? If the answer is yes, test is complete and system is functional; reconnect harness and perform operational test again. If the answer is no, go to step B.

        Results _____

☐

    B.   Locate parking brake warning light switch. Disconnect.

    C.   KOEO, parking brakes off, did the light go out? If the answer is no, the circuit between the ignition switch, warning lamp, and the ground circuit must be checked. Consult the instructor and service manual for diagnostic routine. If the answer is yes, go to step D.

        Results _____

☐

    D.   Adjust or replace switch. Perform operational test again.

Problems Encountered _____

_____

Instructor's Response _____

_____

Electrical Braking Systems Service

Upon completion and review of this chapter, you should be able to:

- ❏ Inspect and test the brake system to determine if a complaint is related to the base brakes system or the antilock system.
- ❏ Relieve high pressure from an ABS hydraulic accumulator.
- ❏ Bleed an ABS.
- ❏ Explain the purposes and major features of diagnostic trouble codes.
- ❏ Explain the differences between hard and soft trouble codes.
- ❏ Use a computer pin voltage, or pin out, chart to perform voltage and resistance tests on computer control circuits.

- ❏ Perform resistance and voltage waveform tests on a speed sensor and its circuit.
- ❏ Set up and test a wheel speed sensor with an oscilloscope.
- ❏ Interpret ac and dc signal waveforms.
- ❏ Remove and replace a wheel speed sensor.
- ❏ Remove and replace an ABS computer (control module).
- ❏ Test and replace Delphi DBS-7 system components.
- ❏ Test and replace Teves Mark 20 and Mark 20E components.

Introduction

The basic concept of ABSs dates back to the 1950s, but digital electronic control was not available until the 1980s. ABSs were installed on slightly more than 3 percent of domestic vehicles produced in 1987, on more than 50 percent by 1995, and almost all new vehicles built for the year 2000 had ABSs. As ABS use grows, the variations in designs will continue. For this reason, the manufacturer's specifications, service instructions, and electrical circuit diagrams for the system being serviced must be available and utilized.

ABSs are more likely to suffer from basic brake system problems than from problems in the ABS control circuitry or components, which means the technician should be familiar with the normal braking system and how to diagnose its problems. Some ABS installations require special testers to read trouble codes and perform some system tests. If the required tester is not available, many of the circuits can be tested with ordinary test equipment, but the diagnosis will not be as fast or as efficient. This chapter contains basic ABS troubleshooting information, as well as general test and repair guidelines that apply to most systems and two specific systems.

Basic Tools

Basic technician's tool set

Appropriate service manual

DMM

Brake System Troubleshooting

An ABS is an electrically controlled hydraulic system. The ABS function will not work if the base brakes do not work properly in their non-antilock mode. Some ABS problems will be electrical, but many will result from a hydraulic system malfunction. When such problems are detected by the ABS control module, it will disable the antilock function and light the antilock indicator lamp on the instrument panel. The cause can be as simple as a low hydraulic fluid level or a leaking hose, line, or connection.

Classroom Manual
pages 247–252

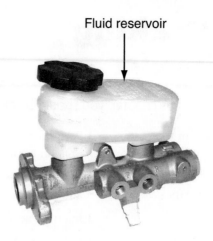

Fluid reservoir

Figure 10-1 On many vehicles, you can check the brake fluid level through the translucent master cylinder reservoir.

Brake System Check

Special Tools

Coworker

Use the following quick and simple brake system check to help determine if the base brake system is working properly and whether the problem is electrical or hydraulic. Several of these tests were discussed earlier in this manual as checks for the base brake system.

To check the base brake system:

1. Shut the engine off and check the master cylinder fluid level. On most late-model cars, this can be done by cleaning the translucent reservoir, and visually noting the fluid level in relation to the embossed line or mark on the reservoir (Figure 10-1). To evaluate fluid condition, remove the reservoir cap or cover.

2. Start the engine and note the brake and ABS indicator lamps. When the ignition is switched on, the BRAKE lamp should light with the parking brake off. The ABS indicator lamp should light for a few seconds, then go out (Figure 10-2).

3. Apply and release the parking brake. The BRAKE lamp should light when the parking brake is applied and go out when it is released. The ABS indicator lamp should light

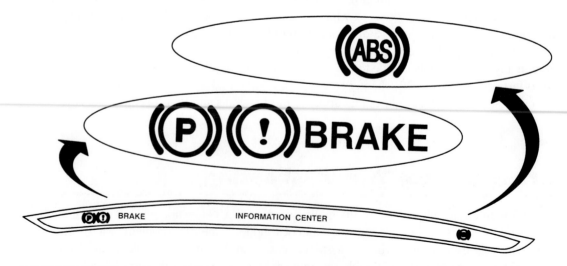

Figure 10-2 Both the BRAKE and the ABS indicator lamps should light when the ignition is in the RUN position.

while the engine is cranking and go out when it starts. On some systems, it may remain on for a few seconds after the engine starts, then go out.

4. Check the operation of the stoplamps by pressing the brake pedal while an assistant watches both fender mounted lamps and the CHMSL to make sure they light when the pedal is applied and go off when the pedal is released.

5. With the engine running, pump the brake pedal rapidly several times. The pedal height should remain about the same. If it increases during pumping, or if a very spongy feeling during pedal application is noted, air may be trapped in the hydraulic lines. It is common for the brake pedal on a car with antilock brakes to feel slightly softer than a car without them. On many systems the pedal can be forced to the floor with firm pressure. The important thing to check for is the uniformity or repeatability of the brake pedal operation. Pedal height and feel should remain constant as the pedal is applied and released. If they change with repeated pedal applications, air may be trapped in the brake lines.

6. If the car has vacuum-assisted power brakes, start the engine and run it at medium speed for a short time, then shut it off. Wait about 90 seconds and then apply the brake pedal moderately several times. The brake pedal should feel firmer with each stroke. When the vacuum reserve is exhausted, depress the brake pedal firmly and restart the engine. The pedal should drop slightly and then hold as vacuum is applied to the booster.

7. Check the condition of the vacuum hose between the power booster and intake manifold. Start the engine and listen for a hissing noise from the hose or hose connection. Such a noise indicates a vacuum leak. Replace the hose if it is deteriorated; replace the hose clamps if the hose connections are not tight.

8. Inspect all brake hoses, lines, and fittings for damage, deterioration, leakage, or chassis interference. This inspection requires that the vehicle be raised on a hoist. Fix any defects found.

> ⚠ **WARNING:** Use caution and obey all highway laws during the road test. Inattention to traffic while attempting to read a scan tool could result in an accident and serious injury.

9. Finally, test drive the vehicle to evaluate brake system performance. Accelerate to about 20 mph and use normal braking to stop the car. If it stops smoothly within 25 feet without swerving, the brakes are probably in good working order. Under hard braking, beginning from 25 mph or 30 mph, the brake pedal may pulsate and the ABS light may come on for a moment if an ABS event occurs.

Troubleshooting with the Brake Pedal

The brake pedal is a helpful diagnostic tool that is easily used by a trained technician. The technician can determine the probable cause of a wide variety of brake system problems by identifying apparent symptoms as the pedal is applied. Figure 10-3 lists symptoms and probable causes that can be determined in this way.

> ● **CUSTOMER CARE:** Advise your customers that pumping the brake pedal on a vehicle with antilock brakes actually defeats the operation of the antilock system. In fact, rapidly pumping the brake pedal may set a DTC and turn the ABS warning lamp on.

Troubleshooting at the Brake Pedal

SYMPTOM	PROBABLE CAUSE
Pedal surging, brake chatter, vehicle surge during braking	Front discs out of round; excessive disc thickness variation; bearings out of adjustment
	Rear drums out of round; hard spots caused by overheating
Brakes grab	Hard spots on front discs or rear drums; cracked pads or shoe linings
Car pulls to one side	Misaligned front end; drum brake components malfunctioning; frozen caliper pistons or contaminated front brake pads; pinched lines or leaking seals
Excessive pedal effort	Insufficient engine vacuum; defective booster; vacuum leak; frozen piston; contaminated or glazed linings
Rear brakes drag	Misadjusted parking brake; rear brakes out of adjustment; weak shoe return springs; frozen wheel cylinder pistons
All brakes drag	Frozen brake pistons; misadjusted stoplamp switch; restricted pedal return; defective master cylinder; contaminated brake fluid
Low-speed disc brake squeak	Worn pad lining
Scraping noise when brakes are applied	Brake linings completely worn out
Intermittent chirp when drum brakes are applied or released	Insufficient backing plate pad lubricant
Intermittent clunk when drum brakes are applied	Threaded drums
Rear-wheel lockup	Contaminated linings; front calipers frozen; defective combination valve
Pedal low and spongy with excessive pedal travel	Insufficient fluid in system; air in hydraulic system
Pedal low and firm with excessive pedal travel	Brakes out of adjustment
Brakes release slowly and pedal does not fully return	Frozen caliper or wheel cylinder pistons; defective drum brake return springs; binding pedal linkage; backing plate grooved
Brakes drag and pedal does not fully return	Contaminated brake fluid; defective master cylinder; defective vacuum booster or vacuum check valve; binding pedal linkage or lack of lubrication

Figure 10-3 An experienced technician can tell a great deal about the condition of a brake system just by the pedal feel.

ABS Hydraulic System Service

Classroom Manual
page 259

✏️ **AUTHOR'S NOTE:** Certain precautions must be taken when servicing the base brakes of some ABS installations. The technician must understand the system components and their operation before trying to bleed the brakes or perform other service(s) that requires opening the hydraulic system.

An integrated ABS installation using a pressurized accumulator is similar to the hydro-boost power brake system that uses the power steering fluid to operate both the steering gear and the brake booster. In both designs, the hydraulic fluid is under very high pressure. For example, 2,600 psi of hydraulic pressure exists in the DaimlerChrysler ABS-3 system whenever the ignition switch is on. Opening a hydraulic line in this system without first discharging the pressure can result in a messy and potentially dangerous situation.

⚠️ **WARNING:** Before opening the hydraulic system of an ABS installation, review the manufacturer's service instructions, with particular attention to safety precautions that relate to hydraulic pressure. Do not loosen any fittings or otherwise open hydraulic lines where pressure may be present. Failure to comply may result in vehicle damage or injury.

Depressurizing the System (Relieving Accumulator Pressure)

The technician must refer to the carmaker's service procedures if it is necessary to check brake fluid level, bleed the brakes, or open a hydraulic line to replace a component. These service procedures determine if it is necessary to depressurize the hydraulic accumulator first.

For example, the DaimlerChrysler ABS-3 system and the Teves ABS both use a pressurized hydraulic accumulator. The DaimlerChrysler ABS-3 system must be depressurized to bleed the brakes, but the Teves ABS requires that the accumulator be fully pressurized when bleeding the rear brakes. The Teves front brakes can be bled with or without accumulator pressure.

When the automaker's procedure requires that the system be depressurized, the ignition switch should be off. Pump the brake pedal at least twenty-five times with about 50 pounds of pedal force. As the accumulator pressure discharges, a change in pedal feel will be noticed. When an increase in pedal effort is felt, pump the pedal a few more times. This will remove all hydraulic pressure from the system. Some systems may require up to fifty pedal applications to relieve accumulator pressure completely. It is always better to proceed on the side of caution, so apply the pedal a few more times after the pressure seems to have been relieved.

Special Tools
Service manual

Fluid Level Check and Refill

Some integrated antilock systems require different procedures to check and refill the fluid reservoir. The DaimlerChrysler-Bosch ABS-3 system must be depressurized before the reservoir cap is removed. The fluid level should be at the top of the white screen in the filter-and-strainer assembly.

To check fluid level in an early Teves Mark II system, turn the ignition on and pump the brake pedal until the hydraulic pump motor starts. When the pump shuts off, you can visually check the level through the translucent reservoir. Under certain conditions, the fluid level may be above the MAX fill line on the reservoir. If this is the case, shut the ignition off and then turn it back on. Pump the brake pedal again to start the hydraulic pump. When the pump stops, the fluid level should be accurate. If the fluid is below the MAX fill line (Figure 10-4), remove the cap and add enough fluid to bring it to the correct level.

To check the fluid level on later Teves systems, depressurize the system and look at the fluid through the translucent reservoir. The level should be at the FULL mark. With the system pressurized, the level will be somewhere below the FULL mark.

Special Tools
Service manual

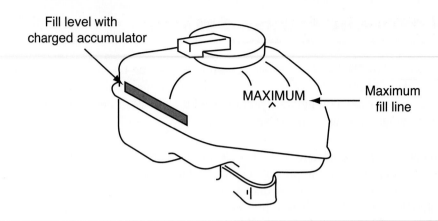

Fill level with
charged accumulator

MAXIMUM

Maximum
fill line

Figure 10-4 The hydraulic reservoir used with the early Ford Teves Mark II ABS has two fill marks. One indicates the normal level with a charged accumulator; the other is a MAX fill level mark.

Do not overfill the reservoir in either system. This will cause the fluid to overflow when the accumulator discharges during its normal operation.

Bleeding the System

Special Tools

Service manual

Tubing with
  transparent
  container

Scan tool with ABS
  cartridge

Coworker

On most ABSs a coworker can speed up the bleeding process, especially if the vehicle is raised.

 CAUTION: Brake fluid is corrosive to body finish and the paint on the vehicle lifts. Brake fluid is also irritating to the skin and the eyes. Always use a capture container when bleeding brakes to reduce human and equipment exposure.

✓ **SERVICE TIP:** Some vehicles require the manufacturer's scan tool to properly bleed their ABSs. Many aftermarket scanners cannot perform this task correctly.

Bleeding an ABS is fundamentally the same as bleeding a non-ABS hydraulic system. Some variety exists in extra steps that may be required for different systems, however. On one hand, some nonintegral ABS installations can be bled in the same way as a non-ABS hydraulic system, with no extra steps. On the other hand, on some integral systems, such as the Teves Mark II ABS, the rear brakes must be bled with a fully pressurized accumulator. The ignition must be on, and the brake pedal must be applied while the rear bleed screws are opened one at a time. Because of the procedure variations that exist, check the manufacturer's instructions before bleeding a system. The details of bleeding procedures can change from one model year to the next, so reviewing the carmaker's instructions is always a good idea.

⚠ **WARNING:** If a system requires that a high-pressure accumulator be charged to bleed the brakes, follow the manufacturer's instructions and precautions exactly when working with high-pressure hydraulic components.

Some manufacturers state that either pressure bleeding or manual bleeding is acceptable for their various ABS installations. Others specify either one method or the other.

CAUTION: Do not use DOT 5 silicone brake fluid in an ABS. ABS operation will be degraded.

Use only the DOT 3, DOT 4, or synthetic brake fluid specified by the vehicle manufacturer. When bleeding an ABS hydraulic system, it is good practice to flush the system completely to ensure that all old fluid and possible contamination are removed.

✓ **SERVICE TIP:** If a brake-pull problem develops *after* bleeding an ABS, completely flush and bleed the system again. If sludge or dirt of any kind gets trapped in antilock brake solenoids or valves, it can unbalance hydraulic pressure at the wheels—even for non-ABS braking. Flushing the dirt out of the system often solves an antilock brake pulling problem.

The following paragraphs summarize the special instructions for bleeding some of the most popular systems. These sections do not cover all possible ABS installations, however, and should not be substituted for manufacturers' specific procedures. They do provide general guidelines and examples of ABS bleeding methods. Before bleeding any antilock system, repair any conditions that would set diagnostic trouble codes and clear all codes from computer memory.

Bendix 9 (Jeep). The Bendix 9 system can be bled with pressure equipment or manually. The entire system, including the accumulator, pump, and master cylinder, must be bled if any hydraulic connection is opened. To bleed these hydraulic control components, loosen the fittings on the side of the hydraulic modulator one at a time while an assistant holds steady pressure on the brake pedal. Bleed the accumulator, pump, and master cylinder before bleeding the wheel brakes.

To bleed the wheel brakes, turn the ignition off and pump the brake pedal thirty to forty times to exhaust accumulator pressure. Leave the ignition off and the accumulator depressurized while bleeding the brakes. The usual bleeding sequence for the Bendix 9 system is: modulator, RR, LR, RF, LF.

Bendix 10. The Bendix 10 system can be bled with pressure equipment or manually. Before bleeding the wheel brakes, turn the ignition off and pump the brake pedal thirty to forty times to exhaust accumulator pressure. Leave the ignition off and the accumulator depressurized while bleeding the brakes. The usual bleeding sequence for the Bendix 9 system is: LR, RR, LF, RF.

Bendix 6. If the hydraulic modulator is not removed or otherwise exposed to air, the bleeding procedure for the Bendix 6 system is the same as for the Bendix 10 system. If air has entered the modulator, special procedures beyond the scope of this manual are needed to purge the air from the system. Refer to a Bendix-DaimlerChrysler shop manual or aftermarket service manual for the specific year and model of vehicle being serviced.

Bendix LC4. The Bendix LC4 system can be bled only manually. Pressure-bleeding equipment does not produce pressure high enough to remove all air from the system. The usual bleeding sequence is: RR, LF, LR, RF.

Bleeding the LC4 system is a two-stage process that requires the use of a **scan tool.** Several aftermarket scan tools provide the necessary capabilities. Follow the instructions for a particular service equipment.

Bendix ABX-4. Unless air has entered the modulator, the technician can bleed the Bendix ABX-4 system as he or she would a non-antilock system. Either manual or pressure-bleeding methods can be used, but Bendix 4 and ABX-4 systems are best bled manually. The usual bleeding sequence is: LR, RF, RR, LF.

If air has entered the hydraulic modulator, bleed the wheel brakes first and then bleed the modulator, using a scan tool to control system operation. Follow the instructions for the particular service equipment. After bleeding the modulator, repeat the bleeding operations at all four wheels. To thoroughly purge these systems, however, certain solenoid valves must be cycled while holding down the brake pedal. Do this only when a Bendix 4 and ABX-4 brake bleeding procedure in a DaimlerChrysler service manual gives specific instructions to do so.

Bendix Mecatronic. The Bendix Mecatronic system can be bled using either manual or pressure-bleeding methods as would be done with a non-antilock system. The usual bleeding sequence is: RR, LF LR, RF.

A **scan tool** is a diagnostic tool that allows the technician to communicate with the vehicle's onboard computers. It can be used to command some vehicle actuators to perform certain tasks.

Bosch 3. Before bleeding the wheel brakes, turn the ignition off and pump the brake pedal thirty to forty times to exhaust accumulator pressure. Leave the ignition off and the accumulator depressurized while bleeding the brakes. Bleed the system using either manual or pressure-bleeding methods the same as would be done with a non-antilock system. The usual bleeding sequence is: LR, RR, LF, RF.

To bleed the master cylinder and hydraulic booster, relieve accumulator pressure and top off the reservoir. Connect a transparent hose to the bleeder screw on the side of the hydraulic assembly and place the other end in a clear container filled with a few ounces of brake fluid. Open the bleeder screw one-half to three-quarters turn and turn the ignition on. The pump will run and force fluid from the hydraulic assembly into the bleeding container. Close the bleeder screw when the fluid flowing into the container is free of air bubbles. Disconnect the bleeder hose, turn the ignition off, and top off the fluid reservoir.

Bosch 2, 2E, 2S, 2U, Micro, and ABS-ASR. The Bosch 2 system versions can be bled using either manual or pressure-bleeding methods generally the same as would be done with a non-antilock system. All of these Bosch 2 ABS versions have been used by many carmakers for many years. Although the basic bleeding procedures are similar, it is important to check manufacturers' instructions for specific years and models. The bleeding sequence usually begins with the wheel farthest from the master cylinder, but different sequences are specified by different manufacturers.

Bosch 5.3. Before bleeding the wheel brakes, pump the brake pedal several times to purge the vacuum reserve for the power brake system. Ensure that the master cylinder is kept at least half full during the bleeding process. Raise the vehicle with an assistant inside. Do not start the engine or turn on the ignition key. Manually bleed the system starting with the right rear. Have the assistant slowly depress the brake pedal and hold before opening the bleeder screw. After tightening the screw, the pedal must be slowly returned to the rest position. Wait at least 15 seconds before pressing the pedal again. Repeat, including the waiting period, until fresh, air-free fluid is forced from the screw. Continue with the other wheels in sequence: LR, RF, LF.

Delco Moraine III. Before bleeding the wheel brakes, turn the ignition off and pump the brake pedal thirty to forty times to exhaust accumulator pressure. Leave the ignition off and the accumulator depressurized while manually bleeding the front brakes. Bleed the RF caliper first, then the LF caliper.

To bleed the rear brakes, turn the ignition on to let the pump recharge the accumulator. Have an assistant press the brake pedal slowly while the RR bleeder screw is slowly opened. Do not press the brake pedal to the limit of its travel. Doing so discharges the accumulator too fast.

Hold the bleeder screw open for about 15 seconds while the assistant maintains pedal pressure. Then close the bleeder screw, release pedal pressure, refill the fluid reservoir, and repeat the procedure at the LR wheel.

Delco Powermaster III. The Powermaster III must be bled using a two-step routine. A coworker will be needed. The front brakes are bled first. With the ignition in OFF, pump the brake pedal twenty-five to forty times to depressurize the accumulator. Manually bleed the front brakes.

With the front brakes bled, turn the ignition to RUN (not START) and allow the pump to pressurize the accumulator. Leave the ignition in the RUN position as the rear wheels are bled.

Start with the right rear wheel. Open the bleeder valve and have the coworker depress the brake pedal about halfway and hold for 15 seconds. The accumulator pressure will flush the old fluid and air from the brakes. After the 15 seconds have expired, close the bleeder screw, top off the reservoir as needed, and repeat the procedures on the left rear wheel.

The Powermaster unit is bled next. Turn the ignition to RUN and apply light force to the brake pedal. Open the bleeder screw closest to the engine. When clear, air-free fluid flows from the screw, close it and open the second bleeder screw. Once the fluid from the second screw is clean, close the bleeder screw. Turn the ignition to OFF and depressurize the accumulator by pumping the brake pedal twenty-five to forty times. Allow time for the fluid to de-aerate, about 2 minutes, and refill the reservoir with DOT 3 brake fluid. The next step is bleeding the boost section of the Powermaster.

The boost section can be bled manually or with a scan tool. The scan tool is used to cycle the valves about ten times to purge all air. This is done manually by applying light pressure to the brake pedal and turning the ignition to RUN for 3 seconds. Turn the ignition to OFF and then repeat the 3 seconds in the RUN position. Repeat this cycle ten times. This will clear the boost of any air. The last step is to depressurize the accumulator as done before, wait 2 minutes, top off the reservoir, and finally test the brake pedal for height and feel.

Delphi Chassis (Delco Moraine) ABS-VI. To bleed the ABS-VI system, the pistons in the front and rear modulators must be in their upper positions to unseat the check balls in the hydraulic circuits. GM recommends using a scan tool with bidirectional control to do this. Although a scan tool makes the job faster and easier, the pistons can be positioned manually.

To manually position the modulator pistons, start the engine and run it for 10 seconds. Verify that the ABS lamp is off and listen for several sharp clicks that indicate the pistons being driven to their upper positions. If these distinct clicks are not heard, drive the car at 5 mph and again listen for the clicking of the ABS control motors. When the clicks become audible, carefully drive the car to the service bay without activating ABS operation.

Before bleeding the wheel brakes, attach a hose to the rear bleeder screw on the hydraulic modulator (closest to the master cylinder) and manually bleed all air from the bleeder port (Figure 10-5). Then repeat this procedure at the forward bleeder screw. After bleeding the modulator, loosen all of the brake line fittings on the outboard side of the modulator and manually bleed air from the upper ends of the brake lines (Figure 10-6).

The last stage of bleeding an ABS-VI system is to bleed the wheel brakes, either manually or with pressure equipment. Because of the variations in ABS-VI installations and because the system has been split either diagonally or front to rear on different vehicles, consult the carmaker's service instructions for the specific wheel bleeding sequence.

Delphi DBC-7. The DBC-7 can be bled using routine pressure, suction, or manual bleeding provided no air has entered the system. Like all brake systems, any time components other than the pads or shoes are replaced, assume that air has entered the system. To bleed an air-contaminated DBC-7 system a pressure bleeder and scan tool must be used following the guidelines described next.

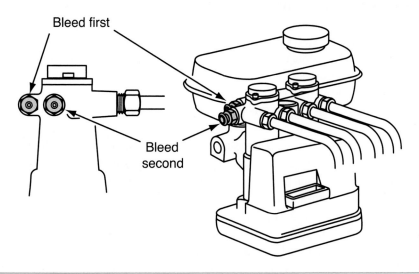

Bleed first

Bleed second

Figure 10-5 Bleed these two points on the ABS-VI modulator first.

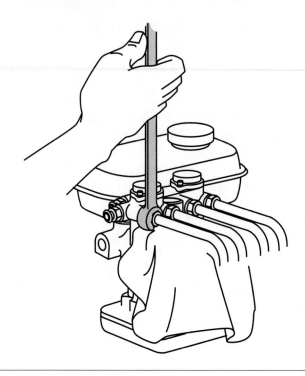

Figure 10-6 Bleed air from the brake line fittings at the outboard side of the ABS-VI modulator.

Connect the pressure bleeder to the reservoir of the master cylinder and the scan tool to the OBD-II 16-pin connector under the dash. Turn the ignition switch to RUN with the bleeder screws closed. Use the pressure bleeder to pressurize the system to 35 psi. Set the scan tool to "automatic bleed procedure." The valves and solenoid in the BPMV will cycle on/off for 1 minute. The scan tool will now indicate that each wheel be bled. The pump will operate during the bleeding process. The valves for the wheel to be bled will cycle for 1 minute. This procedure will repeat for each wheel. After the last wheel is bled, the scan tool will cycle the solenoids for 20 seconds to purge the system of any remaining air. Depressurize the pressure bleeder and disconnect it from the master cylinder. The final step is to apply the pedal several times to check for height and feel. If the pedal is still low or spongy, repeat the entire process.

Kelsey–Hayes EBC2 and 325 (RABS and RWAL). The Kelsey–Hayes rear-wheel antilock systems, used on light trucks and SUVs, can be bled manually or with pressure equipment in generally the same way as a non-antilock system. On most systems, the combination valve must be held open for pressure bleeding. The modulator must be bled separately if air has entered it. Some modulators have bleeder screws; others require that a brake line fitting must be loosened for bleeding.

Because of the wide variety of trucks and SUVs on which this system has been used, refer to the vehicle manufacturer's instructions for the wheel bleeding sequence.

Kelsey–Hayes EBC4. Either manual or pressure-bleeding methods can be used for this system. If no air has entered the hydraulic system, standard non-ABS methods can be used. The usual bleeding sequence is: RR, LR, RF, LF. If air has entered the modulator, special bleeding procedures must be used. Refer to the manufacturer's service instructions for details on a specific year, make, and model.

Kelsey–Hayes EBC410. Standard non-ABS manual or pressure-bleeding methods can be used for this system if no air has entered the hydraulic lines. The usual bleeding sequence is: RR, LR, RF, LF. If air has entered the ABS valve assembly, special bleeding procedures that require a scan tool must be used. Refer to the manufacturer's service instructions for details on a specific year, make, and model.

Teves Mark II. The Teves Mark II system can be bled manually or with pressure equipment. The front brakes can be bled with or without system accumulator pressure, but the accumulator must be fully charged to bleed the rear brakes manually. The ignition must be on, and the brake pedal must be applied while the rear bleed screws are opened one at a time. Because of the high pressure at the bleed screws, the technician must be very careful when opening a screw to bleed the line. Within 10 seconds after opening the screw, the fluid flow should be free of air bubbles and the screw can be closed. Refill the reservoir after bleeding each wheel. The pump will run periodically during the bleeding procedure to maintain accumulator pressure.

Teves Mark IV. Standard non-ABS manual or pressure-bleeding methods can be used for this system. Unlike the Teves Mark II system, the accumulator of the Mark IV system must be depressurized before bleeding the rear brakes. The bleeding sequence varies for different years, makes, and models.

Proper bleeding for these systems consists of first manually bleeding the base brakes, then bleeding the hydraulic control unit (HCU), and finally repeating a manual base brake bleeding sequence. Refer to manufacturers' recommended brake-bleeding procedures for these systems.

Teves Mark 20 and Mark 20E. The Teves Mark 20 requires that the two systems, base and ABS, be bled independently of each other. Bleed the base system first, using manual or pressure bleeding. The sequence of bleeding is LR, RF, RR, LF. The ABS must be bled using Chrysler's DRB-III scan tool. After connecting the scan tool to the vehicle, check for any stored DTCs. Run the DRB-III "bleed brakes" program. The valves will be cycled to purge any air still in the lines. Bleed the base system a second time after bleeding the ICU.

> ☑ **SERVICE TIP:** Before forcing a caliper piston back into its bore to remove the caliper or change the pad, attach a bleeder hose to the caliper bleeder screw, put the other end in a container of clean brake fluid, and open the bleeder.
>
> That is only half the preparation, however. Next you have to keep old fluid from flowing back up the brake lines when you push the piston back in its bore. The best method to prevent contaminated brake fluid from returning to the master cylinder is to lightly pressurize the system. To do this use the alignment's brake depressor to apply the brake pedal about a half inch. This will block the ports in the master cylinder and hold moderate pressure in the lines.
>
> *Then,* push the piston slowly and steadily back in its bore. Any sediment in the caliper bore will be forced out through the bleeder, not back into the brake system where it can damage expensive ABS components.

General ABS Troubleshooting

The following sections outline the troubleshooting principles for ABS control systems and for individual components. These guidelines apply to all antilock systems. The most benefit gained from any diagnostic guideline is when it is used as part of organized, systematic troubleshooting. The four major steps for accurate troubleshooting are:

Classroom Manual
pages 252–260

1. Basic inspection and vehicle checkout
2. Control system diagnosis—trouble codes, self-tests, and data readings
3. Operating symptom diagnosis
4. Intermittent problem diagnosis

The details of these steps are nothing new or revolutionary. These steps are simply up-to-date applications of proven test principles. The following sections explain how to use these principles effectively.

Basic Inspection and Vehicle Checkout

The system computer does not monitor or control all parts of the ABS. The mechanical components, vacuum and hydraulic lines, wiring, and mechanical parts should be inspected first. The technician also should check for body damage, mechanical damage or tampering, and newly installed accessories. The following steps will make the inspection and checkout easier.

Sometimes the best source of information about a vehicle is ignored: the owner/driver.

Verify the Customer's Complaint. Get the customer to describe the problem in as much detail as possible. Ask these kinds of questions:

- ❏ Does the problem exist all the time or some of the time? Does it occur regularly or at random?
- ❏ Does the problem occur at certain times or temperatures?
- ❏ What are *all* the symptoms: noises, vibrations, smells, vehicle performance, or any combination?
- ❏ Has the problem occurred before and what was done to fix it?
- ❏ When was the vehicle last serviced and what was done to it?

Listen to the customer's complaint carefully and get as much information as possible. Avoid asking closed-end questions such as: "Does it happen when the engine is hot?" Ask open-ended questions such as: "What temperature is the car when this happens?"

Finally, check the vehicle to verify that the problem exists as described. Try to re-create the conditions that the customer describes. It may not always be possible to duplicate the conditions exactly, but try to come as close as possible. A road test of the vehicle, or leaving the vehicle standing overnight to re-create a cold-operating problem may be necessary. If the problem caused a soft code, try to get the code to recur during testing.

Before getting all wrapped up in all the possibilities, check the brake fluid level. Some ABSs will switch on both warning lights if the fluid is low. The simple things sometimes cause the greatest problems.

Inspect and Check Out the Vehicle. Look for obvious faults and try to eliminate simple problems first. Look for loose or broken wires, connectors, and hydraulic lines or hoses. Check for leaks. Check for mechanical and electrical tampering and collision damage. After completing the basic inspection and checkout of a system, begin the control system diagnosis.

Test the Control System

Test from the general to the specific. If an immediate look at the computer system for the cause of a problem is done, other possible causes may be overlooked. The basic inspections and tests described in other sections of this *Shop Manual* begin with general checks or area tests. Testing the computer control system also should begin with general area tests before moving on to pinpoint tests.

Check the System Warning Lamp. Vehicles with ABSs have an amber or yellow instrument panel lamp that lights to indicate major system problems (Figure 10-7). Any such indicator lamp should light when the ignition key is turned on without starting the engine. This is a basic bulb check similar to the bulb check for alternator or other brake system warning lamps. The ABS lamp may go out within 1 or 2 seconds of turning on the ignition, or it may stay lit for a longer time.

If the lamp does not light when the ignition is turned on, the computer probably will not go into the diagnostic self-test mode. The problem may be as simple as a burned out bulb, or the problem may be with the computer itself. Begin by checking the fuse and the bulb. Then test the lamp circuit for correct power and ground. Follow the specific carmaker's instructions for checking the lamp to verify that the system can perform its self-test functions.

If the warning lamp lights steadily with the engine running, it indicates that a system problem exists. The following section on trouble codes explains the differences between hard and soft codes and permanent and intermittent faults. On many vehicles, some system problems will set a trouble code in the computer memory but will not light the warning lamp.

Figure 10-7 The ABS warning lamp will light to indicate system problems.

If the warning lamp does not light with the key on or if the computer will not go into a diagnostic mode, it may be necessary to make some voltage tests at a diagnostic connector. Almost all diagnostic connectors have a ground terminal that is used for one or more test modes. Use a voltmeter to check the voltage drop between the diagnostic ground terminal and the battery negative terminal. High ground resistance or an open circuit can keep the computer out of the self-test mode and may be a clue to other system problems.

Various other terminals on the diagnostic connector may have other levels of voltage applied to them at different times. Some may have battery (system) voltage present under certain conditions. Others may have 5 volts, 7 volts, or a variable voltage applied for specific test conditions.

Self-test programs and diagnostic modes operate differently on different vehicles, but all provide the same basic kind of information. Most ABS computers with any kind of self-test capability allow a scan tool or special tester to read system trouble codes. The following paragraphs outline the common features and principles of trouble code diagnosis.

 SERVICE TIP: Dirty or damaged wheel speed sensors and damaged sensor wiring harnesses are a leading trigger to turn on ABS warning lamps. Do not rush to condemn the ABS computer or hydraulic module before checking the speed sensors and their signals carefully.

SERVICE TIP: A poor ground connection for the ABS controller will generally lead to the storage of multiple, nonrelated codes.

Troubleshooting Trouble Codes. Automotive computers can test their own operation and the operation of input and output circuits. Most computers have one or more of the following self-test capabilities:

1. They can recognize the absence of an input or output signal or a signal that is continuously high or low when it should not be.

2. They can recognize a signal that is unusual or out of limits for a period of time or a signal from one sensor that is abnormal when compared to the signal from another sensor.

3. They can send a test voltage signal to a sensor or actuator to check a circuit, or they can operate an actuator and check the response of a sensor.

If the computer recognizes a condition that is not right, it records a **diagnostic trouble code (DTC).** A DTC is a 2-, 3-, 4-, or 5-digit numeric or alphanumeric code that indicates a particular system problem. Most systems will light the ABS warning lamp for many—but not all—codes. In addition, most ABS computers will store the code in long-term memory.

A DTC can indicate a problem in a particular circuit or subsystem, but it does not always pinpoint the exact cause of the problem. Checking DTCs is an overall or area test of the system.

Carmakers have different names for codes and categorize them in different ways, for example:

- ❏ GM: trouble codes
- ❏ Ford: service codes
- ❏ DaimlerChrysler: fault codes

With the introduction of the second-generation on-board diagnostic systems (OBD-II) for engine control systems in 1995, the term *diagnostic trouble code* has been widely applied to codes for all automotive control systems. Two general terms used for all codes are hard codes and soft codes.

Hard Codes. A **hard code** indicates a failure that is present at the time of testing and permanent until it is fixed. If the ignition is turned off and the codes cleared, a hard code will reappear immediately or within a few minutes because the problem still exists until corrected. Hard codes indicate full-time problems that are not too difficult to diagnose. Ford refers to hard codes as on-demand codes because they are detected by the computer immediately on demand when it runs a self-test.

A hard code enables the technician to go right to a certain area or areas and begin pinpoint testing. Carmakers' diagnostic charts or pinpoint test procedures are designed to troubleshoot hard codes. The procedures assume that the problem is present at the time of testing.

Soft Codes. A **soft code,** or memory code, indicates an intermittent problem: one that comes and goes. Soft codes are the computer's way of remembering a problem that occurred sometime in the past before testing, but that is not present now. The problem may not reappear if the codes are cleared, and a retest of the system is done. It may have happened at a certain speed or temperature or under some other conditions that cannot be re-created in the shop.

Ford refers to soft codes as continuous memory codes because they are stored continuously in the computer's memory until cleared. Some GM divisions call soft codes on late-model vehicles history codes and identify them as such on the computer's data stream.

Because soft codes indicate intermittent problems, diagnostic charts and pinpoint tests usually do not isolate the problem immediately. The special intermittent test procedures later in this section will help troubleshoot soft codes accurately. To find the problems that cause soft codes, electrical connectors should not be opened or disconnected until they have been checked in normal operation or by doing a wiggle test. Disconnecting and reconnecting a connector may temporarily solve a problem without revealing the root cause.

Special Tools

Service manual

Scan tool with ABS
   cartridge

Digital multimeter

Determine Whether Codes Are Hard or Soft. After checking trouble codes, write down any codes that may be present. Remember that if the codes are cleared, soft codes will not reappear right away. Some antilock systems will display only one or up to three codes at a time, even when more faults exist. For such systems, each code must be repaired and cleared in sequence and then the system must be retested until no more codes appear.

If a code is a hard code, the technician can go to the manufacturer's test or troubleshooting chart for that code number. If it is determined that a code is a soft code, use the intermittent diagnostic procedures outlined later in the chapter to help pinpoint the problem.

All carmakers advise that trouble codes should be diagnosed and serviced in a basic order: hard codes first followed by soft codes.

Operating Range Tests

The signal from an analog sensor can drift out of range as the sensor ages or wears. Some sensors can develop an erratic signal, or dropout, at one point in the signal range. A loose or corroded ground connection for a sensor also can force the signal out of limits.

These sensor problems and similar problems can cause definite malfunctions without setting a code. The operation of many sensors can be checked, however, by using the operating range charts provided by the carmakers. These charts list signal-range specifications for voltage, resistance, frequency, or temperature that the sensor provides under varying conditions.

Use a DVOM, a frequency counter, or other appropriate instrument to test the sensor signal at the sensor connector and, if necessary, at the main connector to the computer. Back probing many sensor connectors or installing jumper wires will provide connection points for the meter. A breakout box or harness may be needed to check sensor signals at the main computer connector. If possible, operate the sensor through its full range and check the signal at several points.

Use Computer Pin Voltage Charts. A computer pin voltage or pin out chart identifies all the connector terminals at the main computer connector by number, circuit name, and function. The voltage or resistance levels that should be present under various conditions also are often listed. Some circuits have different voltage specifications with the key on and the engine off, during cranking, and when the engine is running. Use the pin voltage charts to check input and output signals at the computer. Figure 10-8 is an example of part of a pin out chart for the main connector at an ABS control module. Checking signals at the computer is closely related to sensor operating range tests.

Check Ground Continuity. With the key on, circuit energized, and current flowing, use a digital voltmeter to check the voltage drop across the main computer ground connection and across the ground connection of any sensor that you think may be causing a problem (Figure 10-9). Low-resistance ground connections are critical for electronic control circuits.

Antilock brake control module connector (end view)

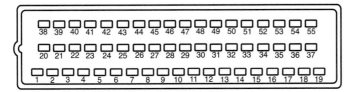

EXAMPLES

Pin number	Circuit	Circuit function
1	530 (LG/Y)	Ground
2	498 (PK)	ABS valve assembly
3	532 (O/Y)	ABS power relay
4	—	Not used
5	549 (BR/W)	ABS pedal sensor switch
6	—	Not used
26	535 (LB/R)	ABS switch No. 2
27	524 (PK/BK)	RR brake sensor-LO
28	519 (LG/BK)	LR brake sensor-LO
29	516 (Y/BK)	RF brake sensor-LO
30	522 (T/BK)	LF brake sensor-LO
31	462 (P)	Pump motor speed

Figure 10-8 Part of a pin out chart for testing circuit functions at an ABS control module.

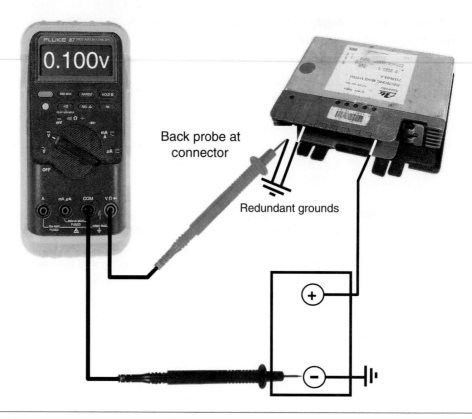

Figure 10-9 With any circuit closed and current flowing, voltage drop across any ground connection and the battery negative (–) terminal must be no more than 0.1 volt (100 millivolts).

With the ignition on, voltage drop across the ground connection for any electronic circuit should be 0.1 volt or less. The voltage drop across a high-resistance ground connection in series with a sensor circuit reduces the signal voltage of the sensor. This ground resistance can offset the signal voltage enough to cause serious problems. For example, a 0.5-volt drop across the ground connection on a sensor that operates on a 5-volt reference equals a 10 percent measurement error.

Troubleshooting Intermittent Problems

Intermittent problems can be the hardest to diagnose and fix. If the technician is lucky, the intermittent problem will set a soft code in the computer memory. This gives a clue, at least, about the general area in which to start testing. Remember, however, that if the codes are cleared, the problem may not recur right away. The conditions that caused the problem may have to be simulated or the vehicle road tested to catch the intermittent fault. The following paragraphs outline some basic points that can help the technician troubleshoot intermittent problems.

Use Wiggle Tests and Actuator or Sensor Special Tests. Most control systems have long-term memory that will record soft codes for intermittent problems. Ford recommends wiggle tests in which the car computer and a scan tool are put in communication so that the scan tool will indicate when a soft code is set. Then tap or wiggle wiring and connectors to try to get the problem to occur. If the problem does occur, remember what was done when the code set. Use the scan tool to read the codes from the car again to verify the fault.

Many vehicles have special tests that let the technician command the computer to switch actuators on and off for testing. Some scan tools, for example, allow you to operate the pumps of the Delco Moraine ABS-VI system for bleeding.

Check Connectors for Damage. Many intermittent problems are caused by damaged connectors and terminals. Unplug the connectors in the problem circuit and inspect them carefully for:

❑ Bent or broken terminals
❑ Corrosion
❑ Terminals that have been forced back in the connector shell, causing an intermittent connection
❑ Loose, frayed, or broken wires in the connector shell

NOTE: Do not use spray electrical contact cleaner on a harness connector unless the wiring harness has been disconnected from the ABS controller. Electrostatic discharge (ESD) may damage the ABS controller.

Road Test and Record Data. If the vehicle transmits computer data in a road test or normal operating condition—and most ABS computers do—drive the vehicle and try to duplicate the problem. Use the snapshot or data recording function of a scan tool to record the computer data when the problem occurs. Then analyze the data in the shop to try to locate the cause.

Switch Testing

ABS control systems receive relatively simple input signals compared to powertrain control systems. Input signals for an ABS computer or control module come principally from switches and speed sensors. The following sections outline common troubleshooting guidelines for these common devices.

CAUTION: Do not reroute cables or use wheel sensors from other model vehicles during repairs. The speed signals could be wrong for the repaired vehicle or be distorted by **electromotive force (EMF)** from other magnetic/electrical fields.

The brake switch, the cruise control switch, and the brake fluid warning switch are examples of ABS input signals that come from simple switches. When used as a control system sensor, a switch provides a digital, on-off, high-low voltage signal. Such a signal indicates either one or the other of two operating conditions such as "brakes released" or "brakes applied."

To provide this kind of input signal, the switch may be installed between the battery and the computer or between the computer and ground. If the switch is installed between the computer and ground a reference voltage is applied to the switch circuit inside the computer, across a fixed resistor (Figure 10-10). A pull up resistor creates the on-off condition that can be recognized as a digital signal. The fixed resistor is often called a pull up resistor because it pulls the reference voltage up to the open-circuit level when the switch is open and drops the voltage when the switch is closed. The computer takes its input signal internally at a point between the resistor and the switch (Figure 10-10).

The reference voltage for the switch circuit is most often the 5-volt reference used for other computer circuits. For some circuits, it may be full system voltage (approximately 12 volts) or some other voltage level. The important thing to remember is that the voltage signal to the computer is either high or low, depending on the switch position, and indicates one of two operating states.

Testing a switch circuit is the basic process of placing the switch in a known operating position (open or closed) and using a voltmeter to check the voltage signal received by the computer. One common way to do this is to back probe the switch circuit wire terminal at the computer harness connector with the positive (+) lead of the voltmeter. Figure 10-11 shows this test method with the switch both open and closed.

Classroom Manual
pages 259–260

Electromotive force (EMF) is the proper terminology for voltage. EMF always creates a magnetic field around a conductor. This may interfere with EMF (or the voltage signal) in an adjacent conductor.

Reference voltage is a fixed voltage supplied to the sensor by a voltage regulator inside the computer or control module.

Different manufacturers may refer to reference voltage as VREF or REFV.

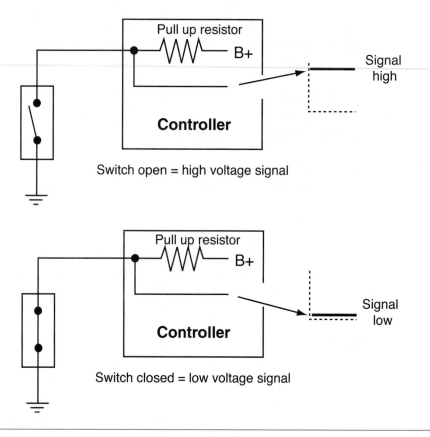

Switch open = high voltage signal

Switch closed = low voltage signal

Figure 10-10 The basic switch circuit used to provide an on-off input signal to a computer.

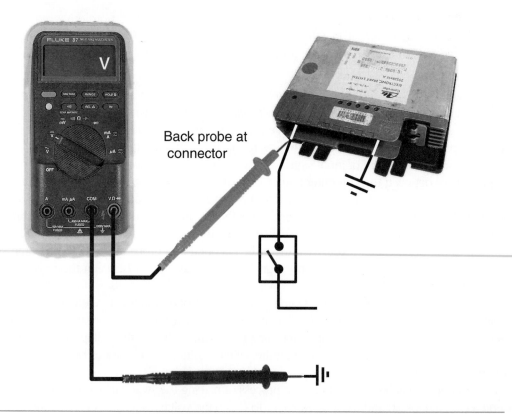

Figure 10-11 Back probe the switch circuit at the computer harness connector and operate the switch to test the input voltage signal.

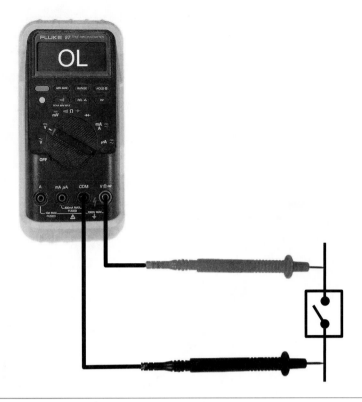

Figure 10-12 Use an ohmmeter or self-powered test lamp to check continuity with the switch disconnected from the circuit.

With the switch open, no current flows in the circuit so no voltage can be dropped across the fixed resistor. The input signal is open-circuit reference voltage, and that is what the test voltmeter should read. With the switch closed, current flows through the circuit, and the fixed resistor drops all of the reference voltage. The input signal is near zero volts, and that is what the test voltmeter should read.

If the switch is easily accessible, you also can remove or disconnect it from the circuit and check continuity with an ohmmeter or self-powered test lamp (Figure 10-12).

Wheel Speed Sensor Testing

Sensor Resistance Testing. All pickup coil sensors have resistance specifications. One basic test for any speed sensor is to disconnect it from its circuit and measure its resistance by connecting an ohmmeter across the two terminals of the sensor wiring harness (Figure 10-13). If resistance is out of limits, either high or low, replace the sensor.

A resistance test for a speed sensor is only a starting point, however. Sensor resistance often can be within limits, but the sensor can produce a faulty signal. Damage to the sensor trigger wheel (tone ring), for example, can produce an uneven signal even when the pickup coil is electrically in good condition.

Testing a WSS with an Oscilloscope. The best way to test a wheel speed sensor is with an **oscilloscope.** An **oscilloscope** is a test instrument that can display voltage or amperage over a period of time with a graph display. The primary difference between oscilloscopes is the labeling of the controls and which control to activate to achieve a satisfactory graph. Photo Sequence 21 shows a typical sequence for setting up an oscilloscope.

Figure 10-13 Use an ohmmeter to measure the resistance of the pickup coil winding.

A good PM wheel sensor will create a waveform similar to the one shown in Figure 10-14. Note that the highest voltage is between 1.5 volts and 2 volts. This signal is typical of a wheel sensor signal when the wheel is rotated by hand at about 20 to 30 revolutions per minute. Also note that the slope of ascending and descending voltage is mostly smooth and uniform. Figure 10-15 and Figure 10-16 show a too high voltage reading and a too low voltage, respectively, at the same wheel speed. The voltage slopes are ragged and uneven. Even if the voltage signal value were close to correct, this "wobbly" graph would point to a problem with the sensor or possibly the wheel and/or axle assembly.

A magnetoresistive sensor will produce a positive voltage signal similar to the one shown in Figure 10-17, but the digital signal may be positive or negative polarity. The depicted signal would be commonly referred to as a square wave. Figure 10-18 and Figure 10-19 show a typical signal for a slow wheel and a fast wheel, respectively. The voltage is the same but the frequency of the signals is different.

Most oscilloscopes have sample "perfect" waveform patterns for each type of sensor they are capable of testing. They can also store patterns for later use. While a signal is present on the display, press the FREEZE button on the oscilloscope. The pattern will be frozen and can be saved. Follow the oscilloscope's instructions to save this waveform. Once saved, locate the example or test pattern and compare it to the one just saved. Although there will be some minor differences between the two, they should be reasonably similar. The use of the oscilloscope on specific ABS/TCS brands is discussed later.

Checking Speed Sensor Bias Voltage. A speed sensor with a pickup coil can generate a signal voltage through the electromagnetic action of the rotating trigger wheel and the magnetic field surrounding the coil winding. It does not need a reference voltage provided by the computer, as a resistive sensor does.

Photo Sequence 21
Setting an Oscilloscope for Use

P21-1 Switch on the oscilloscope's power. Some oscilloscope instructions recommend that the instrument be connected to the vehicle's power supply.

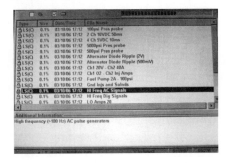

P21-2 Select the SENSOR item from the oscilloscope's menu. Some instruments will allow the technician to select the type of sensor being tested.

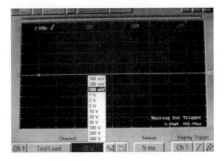

P21-3 Once the display screen shows the graphing chart, set the voltage to 0.5 volt. This means each horizontal grid line is valued at one-half volt.

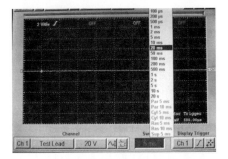

P21-4 Set the time scale to 20 milliseconds (ms). Each vertical line now represents 20 ms of time.

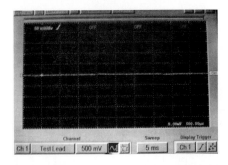

P21-5 Select either ac or dc voltage from the menu. Ac should be used for PM speed sensors and dc for a magnetoresistive type.

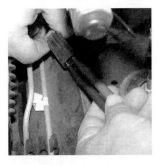

P21-6 Unplug the harness from the sensor. Connect one oscilloscope lead to each terminal of the sensor. The sensor can now be tested for a return voltage signal.

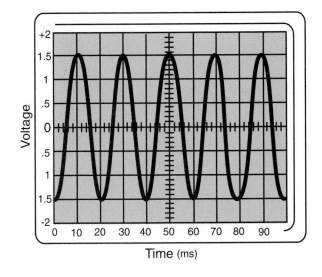

Good wheel speed sensor

Figure 10-14 A good analog or sine waveform will form a positive voltage value to an equal negative voltage value in smooth, even slopes. A typical PM WSS will produce less than 2 volts.

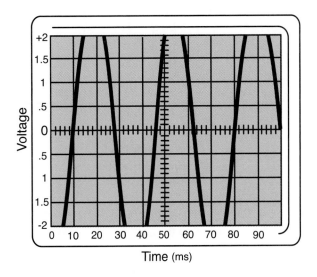

High-voltage wheel speed sensor

Figure 10-15 This PM WSS is generating over 2 volts, which is considered too high a voltage output.

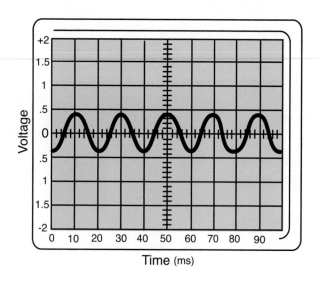

Low-voltage wheel speed sensor

Figure 10-16 This PM WSS is hardly generating any voltage. The controller will probably not be able to read a voltage this low.

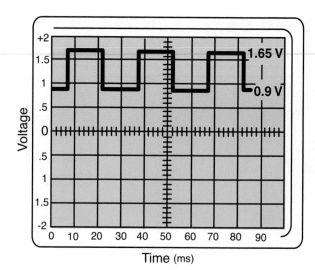

Digital waveform

Figure 10-17 A typical digital (dc) waveform generated by a wheel speed sensor like the magnetoresistive type. This is commonly called a square waveform and may be either positive or negative in polarity.

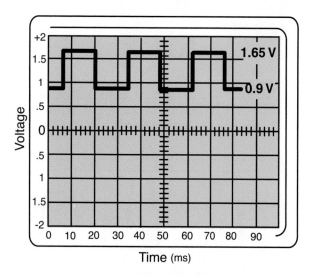

Magnetoresistive wheel speed sensor for slow wheel

Figure 10-18 Compare this typical dc waveform for a slow turning wheel to the one depicted in Figure 10-19.

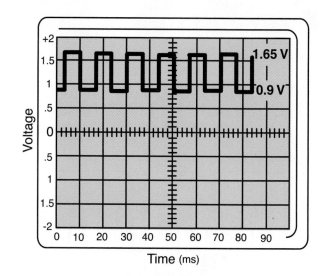

Magnetoresistive wheel speed sensor for fast wheel

Figure 10-19 This dc waveform is typical of a fast turning wheel. Compare it to Figure 10-18.

However, most ABS speed sensors receive a bias voltage from the system computer for two reasons:

❏ The bias voltage lets the system computer detect an open or a short circuit for the sensor before the wheel turns.
❏ The bias voltage elevates the sensor signal off the common ground plane of the vehicle electrical system to reduce signal interference.

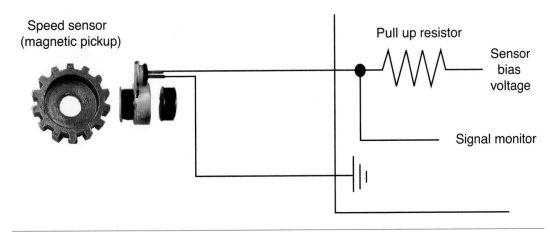

Figure 10-20 Simplified speed sensor circuit showing bias voltage and the signal monitor point.

Figure 10-20 shows a simple speed sensor bias voltage circuit that uses a pull up resistor inside the computer. The bias voltage varies from manufacturer to manufacturer. It may be the 5 volts used for other computer circuits, or it may be a different value such as 1.5 volts or 1.8 volts. Check the carmaker's test procedures and system specifications to determine the required bias voltage when troubleshooting a speed sensor circuit.

The computer monitors the sensor signal at a point between the fixed pull up resistor and the pickup coil (see Figure 10-20). When power is applied to the circuit, current flows through the pull up resistor and through the pickup coil to ground. The voltage drop at the signal monitor point is a predetermined portion of the reference voltage and a known value that is part of the computer program.

If an open circuit exists, no current flows through the circuit, and no voltage is dropped across the pull up resistor. The signal monitor voltage will be high: equal to open-circuit bias voltage (Figure 10-21). In this case, the computer will immediately set a trouble code for an open circuit fault.

If a shorted or grounded circuit exists, all—or almost all—of the bias voltage is dropped across the pull up resistor. The signal monitor voltage will be lower than the programmed signal monitor voltage (Figure 10-22). In this case, the computer will immediately set a trouble code for a grounded or shorted circuit.

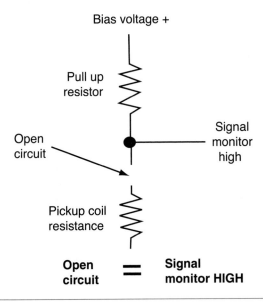

Figure 10-21 If the sensor circuit is open, signal monitor voltage will be high.

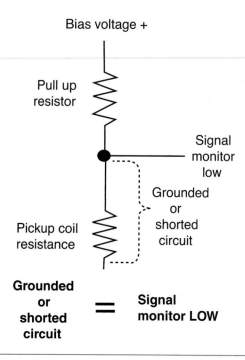

Bias voltage +

Pull up resistor

Signal monitor low

Pickup coil resistance

Grounded or shorted circuit

Grounded or shorted circuit **=** **Signal monitor LOW**

Figure 10-22 If the sensor circuit is grounded or shorted, signal monitor voltage will be low.

The simple voltage divider circuit shown in Figure 10-21 and Figure 10-22 lets the computer detect an electrical fault as soon as the ignition is turned on. The wheel does not need to turn even one revolution. A technician can verify an open- or short-circuit fault by connecting a voltmeter between the high-voltage side of the pickup coil circuit and ground. Depending on the circuit fault, the meter should read close to full bias voltage or close to zero volts with the ignition on.

Some speed sensors receive a bias voltage to raise the signal above the common ground plane of the vehicle electrical system, as well as to detect open- and short-circuit problems. Ford Motor Company particularly favors this type of signal biasing.

Figure 10-23 and Figure 10-24 are signal waveforms from a rear antilock brake system (RABS) speed sensor. Note that the waveforms in both figures are evenly shaped sine waves. High resistance in the circuit of Figure 10-23 created a trouble code by offsetting the signal voltage too high above the zero-voltage point (shown by the small rectangle at the right side of the waveform) and reducing the signal amplitude.

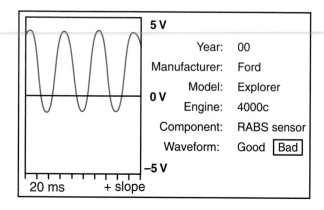

Year:	00
Manufacturer:	Ford
Model:	Explorer
Engine:	4000c
Component:	RABS sensor
Waveform:	Good [Bad]

Figure 10-23 Faulty RABS speed sensor signal.

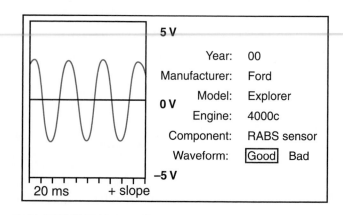

Year:	00
Manufacturer:	Ford
Model:	Explorer
Engine:	4000c
Component:	RABS sensor
Waveform:	[Good] Bad

Figure 10-24 Good RABS speed sensor signal.

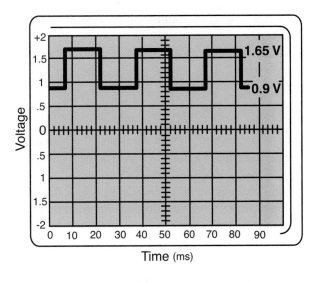

Digital waveform

Figure 10-25 This positive square waveform has a low of about 0.9 volt and a high of about 1.6 volts.

The programs for many Ford control systems require that a biased ac signal voltage still must cross a certain amount below the zero point of the voltage scale. The signal in Figure 10-23 barely drops below zero volts. The signal in Figure 10-24, however, clearly drops below zero volts to the negative side of the scale and has a greater amplitude than the signal in Figure 10-23.

The circuit conditions shown in these two illustrations can be seen only when bias voltage is applied to the circuit and the sensor is operating. An oscilloscope or a graphing multimeter is an important tool for troubleshooting these kinds of problems in any ABS speed sensor circuit.

Testing a Magnetoresistive WSS

The return signal will be approximately 0.9 volt or 1.65 volts at 7 mA and 14 mA, respectively. A digital multimeter (DMM) and an oscilloscope are needed to properly test the magnetoresistive sensor. Before testing the sensor ensure that the remaining sensor circuit is good and that there are 12 volts from the controller to the sensor as measured with a DMM. If the circuitry is correct, connect the oscilloscope to the sensor by back probing into the sensor/harness connection. Set the oscilloscope to voltage and adjust the scale to the 0.5 volt reading and 20 milliseconds (ms). Refer back to Photo Sequence 21 for general steps to set up the oscilloscope.

The initial voltage reading will be approximately 1.6 volts or 0.9 volt depending on the location of a tooth to the sensor head. Rotate the wheel slowly by hand while observing the oscilloscope display. The signal voltage should form a square wave digital pattern (Figure 10-25). The repeating waveform should be consistent with a high voltage of about 1.6 volts and a low voltage of 0.9 volt.

ABS Component Replacement

Troubleshooting and diagnosis are larger factors in ABS service than is component replacement. Preceding sections of this chapter provide general guidelines for the most common troubleshooting requirements. Vehicle manufacturers also provide specific test procedures for specific

Special Tools

Service manual

Wiring diagram

Electronic component locator

Lift or jack with stands

components in their service manuals. Look for this information in carmakers' service manuals or aftermarket information sources.

When a component does require replacement, the methods are generally straightforward mechanical procedures. Some special tools may be needed to service some ABS components. You can find this information, along with replacement procedures, in manufacturers' and aftermarket service manuals.

⚠️ **WARNING:** ABS service may require opening the hydraulic system. ABS hydraulic systems may operate with pressures of 2,000 psi or higher. The system must be completely depressurized before opening any hydraulic connection. In most cases, the system can be depressurized by applying and releasing the brake pedal at least twenty-five times. Follow the vehicle manufacturer's instructions for complete information on hydraulic system service and safety.

Photo Sequence 22 shows the key steps of a typical ABS pump and motor removal. This sequence outlines basic pump service for a GM Teves system. These or similar steps are examples of common component replacement methods.

✔️ **SERVICE TIP:** When doing any kind of service work on an ABS, let the system be the guide. Many ABS components have decals (usually yellow) with important service directions. For example, many ABS accumulators have decals with instructions on how to depressurize the system before opening the hydraulic lines or checking fluid level. The decals are there for the technician's benefit. Pay attention to them.

Wheel Speed Sensor Replacement

A problem in a wheel speed sensor may require replacement of the sensor pickup coil and its harness or the tone ring or both. Some wheel sensors have an adjustable air gap between the sensor head and the teeth on the tone ring. An equal number of sensors are nonadjustable.

If a sensor is adjustable, follow the carmaker's adjustment procedure exactly. Some nonadjustable sensors use a lightweight plastic or paper spacer on the mounting surface (Figure 10-26 and Figure 10-27). The spacer sets the air gap correctly. A new spacer of the correct thickness should be used whenever a sensor is reused.

Manufacturers differ in their requirements for servicing the wiring harnesses on speed sensors. The short, two-wire harness that is part of the sensor assembly on GM vehicles is made with very fine wire strands to provide maximum flexibility with minimum circuit resistance. GM specifies that the sensor harness should not be repaired by any method. If it is damaged, the complete sensor assembly must be replaced. DaimlerChrysler says that the harnesses on most of its wheel speed sensors can be repaired by soldering and reinsulating with heat-shrink tubing.

Because of the importance of signal accuracy from a wheel speed sensor, it is generally preferable to replace a sensor assembly, including the harness, instead of trying to repair the harness. Consult the manufacturer's specifications before deciding how to service these components.

Sensor tone rings that are pressed on the inside of the rotor or on the axle shaft often can be replaced. If the sensor ring is an integral part of the wheel bearing assembly, hub assembly, or the outer constant velocity joint on the axle, the entire component must be replaced if the sensor ring is damaged.

Observe these additional guidelines when servicing ABS wheel speed sensors:

❏ Unplug sensor electrical leads when replacing suspension components.

❏ If a sensor tone ring (trigger wheel) is replaceable, fit the new one in place by hand. Do not hammer or tap the sensor in place.

Photo Sequence 22
Pump and Motor Removal

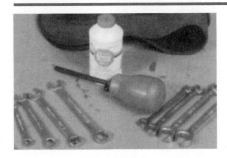

P22-1 You will need fender covers, a combination wrench set, a flare-nut wrench set, a syringe, and fresh brake fluid.

P22-2 Place the fender covers on the vehicle and disconnect the battery ground (negative) cable.

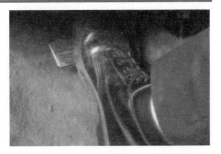

P22-3 Press and release the brake pedal thirty to forty times to depressurize the accumulator. The pedal should become firmer and travel less when the accumulator is depressurized.

P22-4 Disconnect the electrical connectors from the pressure switch and the motor.

P22-5 Use a clean syringe to remove about half of the brake fluid from the reservoir.

P22-6 Unscrew the accumulator from the hydraulic module. Then remove the O-ring from the accumulator.

P22-7 Disconnect the high-pressure hose from the pump.

P22-8 Disconnect the wire retaining clip. Then pull the return hose out of the pump body.

P22-9 Remove the bolt that attaches the pump and motor to the hydraulic module.

P22-10 Remove the pump and motor assembly by sliding it off the locating pin.

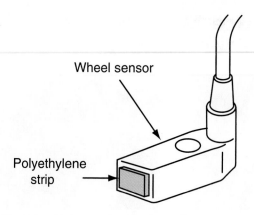

Figure 10-26 Some wheel speed sensors use a plastic spacer to set the air gap correctly.

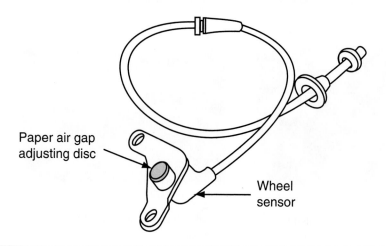

Figure 10-27 Other wheel speed sensors use a paper spacer. In either case, the spacer should not be removed during sensor installation.

❑ If a wheel sensor ring or tone wheel is replaceable and is pressed into place, do not remove the old ring or install a new one by hammering or prying. Use a hydraulic press with the proper special tools.

❑ Remove the vehicle wheel when replacing a wheel sensor.

❑ Some wheel sensors require an anticorrosion coating before installation to prevent galvanic corrosion. Never substitute grease unless the carmaker specifies its use.

Sensor assemblies that are a permanent part of the wheel bearing and hub assembly are used on many late-model GM cars. They need no adjustment and plug directly into the vehicle wiring harness.

Computer (Control Module) Replacement

Vehicle computers do fail but not with great frequency or regularity. Too many computers have been replaced because someone "thought it might be the problem," or "didn't know what else to do." Too often, computer replacement does not cure the problem. The shop has spent several hundred dollars and still has a car to fix. Before deciding to replace a computer, always check the manufacturer's technical service bulletins (TSBs) for specific information on revised computer part numbers and the problems they were designed to correct. Also check with dealership parts departments for the latest part number information. Computers and other electronic parts usually are special-order items. All electronic parts are absolutely, positively *nonreturnable*. Once purchased, they are the shop's. To avoid costly, unnecessary computer replacement, check these items:

1. *Battery voltage supply to the computer and the main system ground.* Be sure the battery is fully charged and provides at least 9.6 volts to 10 volts during cranking. Be sure the charging system is maintaining correct battery charge. Most computers receive battery voltage through a fuse or fusible link. Be sure that battery voltage is available at the specified terminals of the computer's main connector. Most computers are grounded remotely through several wires in the harness. Trace and check the ground connections to ensure good continuity.

2. *Operation of a system power relay.* Some computers receive power through a system power relay. If the vehicle is so equipped, check the relay operation. This relay may be remote mounted or underneath the electrohydraulic plastic cover.

3. *Sensor reference voltage and ground circuits.* Many sensors share a common reference voltage supply from the computer and a common ground. Incorrect or erratic reference voltage or a bad common ground can affect operation of several sensors simultaneously. The symptoms may appear as if the computer has a major system problem. Repairing a wiring connection may correct the problem.

4. *Resistance and current flow through all computer-controlled solenoids and relays.* Every output device (solenoid or relay) controlled by a computer has a minimum resistance specification. The actuator resistance limits the current through the computer output control circuit. If the actuator is shorted, current can exceed the safe maximum and damage the computer. In most cases, current through a computer-controlled output device should not exceed 0.75 ampere (750 milliamperes). Before replacing a computer, check all output circuits for shorts or low resistance that could damage the computer.

▲ **WARNING:** Be sure that the ignition is off when removing and installing a control module or other electronic component.

As a general rule, the system computer should be at the bottom of the list of things to replace. Again, computers can fail, but a sensor or actuator problem, bad wiring, or a mechanical fault in the engine is a more likely cause of a problem.

☑ **SERVICE TIP:** Beware of **electrostatic discharge (ESD)** when handling electronic components. ESD is static electricity that can destroy the microscopic circuits of electronic integrated circuits. If some simple precautions are taken, however, ESD problems can be avoided.

☑ **SERVICE TIP:** The best method to prevent damage to electronic components from static is to always wear an antistatic strap during the repairs. The strap can be worn on the wrist or may be a waist belt. Some companies require their technicians to wear one as part of the uniform.

ESD occurs when two dissimilar materials are rubbed together or quickly separated from each other. Electrical charges build up on the surfaces and then discharge when a circuit path is available. If a person slides across the front seat of a car and then touches a metal surface, he or she can build up and then discharge several thousand volts of static electricity. ESD of 8,000 volts to 10,000 volts is very common and the person will not even feel or see it. When ESD generates a small spark from the fingertip to a metal surface and the snap is felt, a charge of 40,000 volts to 50,000 volts of ESD is involved.

High-voltage ESD does not hurt a person because it is moving only a few microamperes of current. That small amount of current, however, can destroy the microscopic interconnections of an integrated circuit. Those few microamps blow the circuit as does a low-current fuse. ESD problems can be avoided by observing a few simple precautions when handling electronic components:

❑ Do not remove an electronic part from its packing material until it is time for installation.

❑ Do not hold an electronic part by its connector pins.

❑ Before entering a vehicle to remove or replace an electronic part, touch an exposed metal part of the vehicle to discharge any static charge from your body.

❑ Avoid sliding across upholstery or carpeting when removing or installing an electronic part. If this is not possible, touch an exposed metal part of the vehicle with a free hand before installing a new component.

❑ When available, use an antistatic grounding strap attached to the wrist and clipped to a metal part of the vehicle body to prevent static charges from accumulating.

Testing Specific Manufacturers' Systems

Most of the following systems are very close to those discussed in previous sections, but these two systems, the Delphi DBC-1 and the Continental Teves Mark 20E, seem to be two of the most prominent ABSs/TCSs being used today. Also, the procedures for testing them are very similar to other brands except in some terminology and DTC numbering and definition.

Each system discussed here uses a scan tool as the primary tester. Older ABSs are not scan tool friendly and the DTCs will have to be retrieved manually. In addition, ABS/TCS solenoids or actuators cannot be activated for testing nor is data available electronically for diagnosing. In those systems, the DTC is the starting point for pinpoint or individual testing of each suspected device one by one.

Delphi DBC-7

Classroom Manual
pages 264–269

A scan tool is the quickest way to provide initial diagnosing and some testing of the DBC-7 system. The scan tool can retrieve DTCs and data, and it can activate certain devices within the system and measure their response to commands. Photo Sequence 23 shows the general steps to program and connect a scan tool to the DBC-7 data link connector under the dash.

But as with any malfunctioning system on a vehicle, a well-conducted test drive will tell the technician about the system's performance. If the ABS amber warning light is on, retrieve the DTC(s). If none are significant enough to be a safety issue and the service brakes seem to be satisfactory, clear the codes and conduct a test drive. Figure 10-28 shows the list of the types of possible codes for GM anti-lock brake systems.

All ABSs conduct a proof-out or test of their electronic circuits when the ignition switch is turned to RUN. The DBC-7 is no different. During key on, engine off, the dash warning lights will illuminate and the EBCM cycles the solenoid and pump on/off. It checks the system for proper operation and conducts an internal check as well. At speeds above 10 mph, the EBCM will conduct a dynamic test the first time the brakes are applied. The driver may feel some pedal movement and more slowing than anticipated. The WSSs are also monitored continuously while the vehicle is being operated. If a fault is detected, the ABS warning lamp will be illuminated and the ABS will be deactivated until a repair is accomplished. Should the dynamic rear proportioning be affected by the fault, then the red brake lamp will also be switched on.

Because the fault codes were cleared prior to the test drive, check to see if one or more of the same codes were set during the test drive. This gives the technician the most probable starting

Photo Sequence 23
Typical Procedure for Using a Scan Tool on the Delphi DBC-7

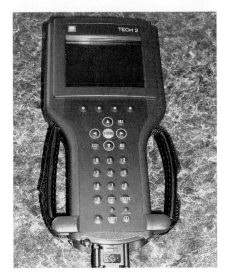

P23-1 Attach the communication cable to the scan tool.

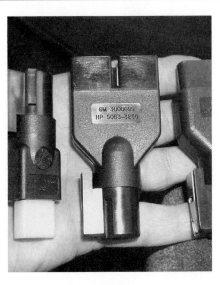

P23-2 Select and connect the correct adapter for the vehicle to the end of the communication cable.

P23-3 Connect the scan tool to the data link connector (DLC) with the ignition off.

P23-4 Turn on the scan tool by depressing the power button.

P23-5 Press enter to access the main menu.

P23-6 Select diagnostics.

P23-7 Select the correct year of the vehicle.

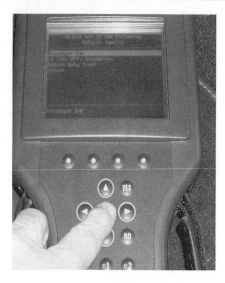

P23-8 Select the correct model of vehicle.

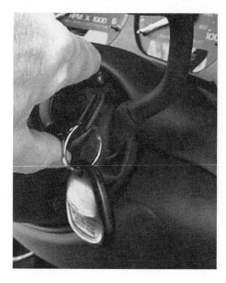

P23-9 Turn on the ignition but leave the engine off.

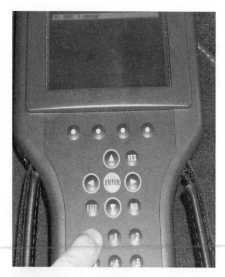

P23-10 Select the chassis, body style, and brake application. You may now choose data list mode to monitor system conditions such as wheel speed sensors and brake switch input. Trouble codes may be viewed, or, if required, the special functions mode will operate ABS components such as the pump and relays. The snapshot mode can be used to capture a fault as it occurs on a controlled test drive.

P23-11 After the problem has been isolated and repaired, clear all stored trouble codes, and test drive. Always turn the ignition off before disconnecting the scan tool from the DLC.

PUMP OPERATION CODES
0214 Relay contact or circuit open
0216 Motor shorted to ground
0217 Controller pump motor relay circuit open
0218 Motor circuit shorted to power
0243 Pump motor circuit open
0244 Pump motor stalled

WHEEL SENSORS
0321 LF wheel at zero speed
0322 RF wheel at zero speed
0323 LR wheel at zero speed
0324 RRr wheel at zero speed
0325 LF speed changes excessive
0326 RF speed changes excessive
0327 LR speed changes excessive
0328 RR speed changes excessive
0332 LF sensor circuit open or shorted
0333 RF sensor circuit open or shorted
0334 LR sensor circuit open or shorted
0335 RR sensor circuit open or shorted

CONTROLLER CODES
0436 Low system supply voltage
0437 High system supply voltage
0438 Brake thermal model exceeded (brakes hot)
0441 Variable effort steering circuit malfunction
0445 Tire pressure low
0447 Brake fluid low
0454 Shutdown detected
0455 Controller malfunction internally
0461 LF inlet valve solenoid open/shorted
0462 LF outlet valve solenoid open/shorted
0463 RF inlet valve solenoid open/shorted
0464 RF outlet valve solenoid open/shorted
0465 LR inlet valve solenoid open/shorted
0466 LR outlet valve solenoid open/shorted
0467 RR inlet valve solenoid open/shorted
0468 RR outlet valve solenoid open/shorted
0472 LF TCS valve solenoid open/shorted
0474 RR TCS valve solenoid open/shorted

DRP CODES
0548 DRP function disabled when combinations of other codes affect DTC operation.

TCS CODES
0675 Engine problem: PCM requested ETC/TCS be shut down
0676 Serial data problem: torque deliver signal incorrect
0677 Serial data problem: torque requested signal incorrect
0678 Engine problem: PCM inhibited TCS operation temporarily

SWITCH CODES
0791 Brake switch open during deceleration
0793 Code 0791 set in last ignition switch operation
0794 Brake switch circuit always working when ignition switch is on
0795 Brake lamp circuit always open

SPECIAL VEHICLE CODES
0815 Circuit malfunction in idle control
0857 YAW sensor malfunction
0858 4WD switch and circuit open/shorted

COMMUNICATION CODES
0998 Class 2 serial link inoperative

Figure 10-28 A list of DTC's definitions similar to the ones on GM and other brands. Note some special vehicles may have special codes.

point in the final diagnosis and repair. At times a code setting malfunction during the last drive cycle does not reappear during the next drive. Then that code is moved to the history file and the warning light will not be illuminated. After 100 drive cycles and no new codes set, all previously set codes will be cleared. A scan tool is required to clear all system DTCs if they remain set. It is also best to remember to clear the codes and, in some cases, reprogram or flash the controller once all repairs are completed.

A scan tool provides the technician with a means to retrieve DTCs and current operating data known as enhanced data. These data include the status of the brake switch, ABS activity, ETC/TCS activity, TCS switch, drive cycles since the DTC was set, and the vehicle's current speed.

Some standard specifications for the DBS-7 are: the ground resistance for the EBTSM is 5 ohms or less and the pump resistance is 2 ohms or less between the terminals. WSS on the DBS-7 should be at least 100 mV when the wheel is rotated at 1 revolution per second. One thing that is true for using DTCs as the diagnostic starting point is: Once the DTC is retrieved and defined, refer to the service manual for the diagnostic charts and follow the steps in the charts exactly. Trying a shortcut may make the repair take much longer and could create additional costs to the shop and technician.

EBCM Replacement

Replacement of parts is very similar to the procedures discussed under the component removal section. However, not all components are accessed easily. Read the service manual carefully, paying particular attention to CAUTIONS and WARNINGS. To replace the EBCM, perform the following actions first:

1. Turn the ignition switch to off.
2. On a 3.8L engine, remove the fuel injector sight shield.
3. Remove the fender upper diagonal brace.

4. Disconnect the accelerator and cruise control cables at the bracket.

5. Disconnect and move the cruise control module from the strut tower.

6. Remove the air intake duct.

The EBCM is now accessible. Start with unlocking the red tab and then sliding the connector cover to open (Figure 10-29). Clear the area around the EBCM before unplugging the harness connector. There are four retainers holding the EBCM and the BPMV together. Remove the retainer and gently pull the two sections apart. The installation is the reverse of the removal procedures. Be careful mating the BBCM and BPMV together and tighten the four retainers to 44 in-lb (5 N.m). Once the EBCM is in place and all connections and removed components are reinstalled, turn the ignition to RUN but do not START the engine. Perform the diagnostic system check covered in the first paragraphs of this section.

BPMV Replacement

Perform the first six steps in EBCM replacement. Disconnect, plug, and move the brake lines (pipes) from the BPMV body (Figure 10-30). Shift the lines to one side and then disconnect, plug, and shift the two master cylinder lines at the BPMV unit. Remove the two bolts and one nut holding the BPMV to the bracket and strut tower. Disconnect the ground strap at the BPMV end and remove the EBCM/BPMV assembly from the vehicle. If only the BPMV is being replaced, separate it from the EBCM as discussed above. The installation is the reverse of removal. Once the assembly has been reinstalled and all removed or disconnected components are in place, it is necessary to use the automated bleed process to clear the system of air.

Wheel Speed Sensor Replacement

To replace any wheel speed sensor, first lift the vehicle and remove the appropriate wheel and tire assembly. To remove a front-wheel speed sensor, first disconnect the jumper harness from the

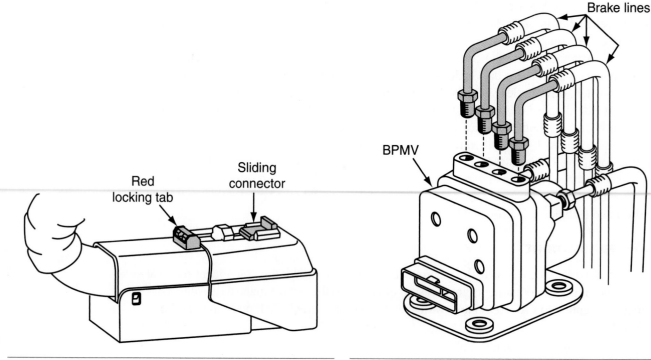

Figure 10-29 Unlock the red tab and slide the connector to open the EBCM housing.

Figure 10-30 Remove and plug all the brake lines (tubes) at the BPMV end.

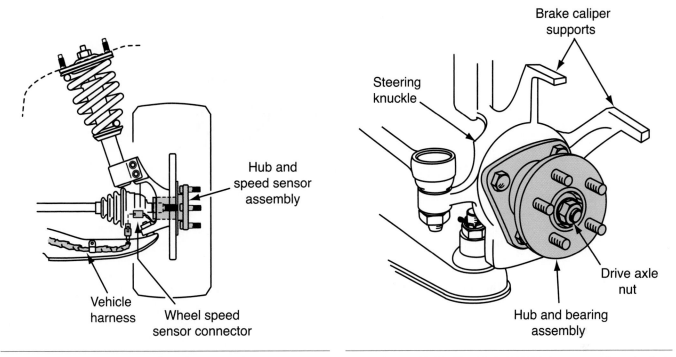

Figure 10-31 The coil of the harness is the jumper harness from the vehicle harness to the WSS connector lead.

Figure 10-32 The DBC-7 WSS is integrated into the hub and wheel speed sensor. The entire assembly is replaced as a unit.

speed sensor (Figure 10-31). Remove the brake caliper and rotor, followed by the hub and bearing assembly (Figure 10-32). The hub and bearing assembly is separated from the steering knuckle by removing the three fasteners and using a puller (Figure 10-33). The installation of the front-wheel speed sensor is the reverse of removal. With all components in place and the vehicle resting on its wheels, perform the diagnostic system check.

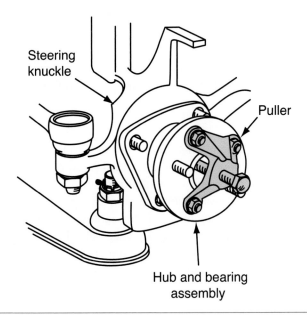

Figure 10-33 The hub and bearing assembly is removed from the steering knuckle using a puller like the one shown.

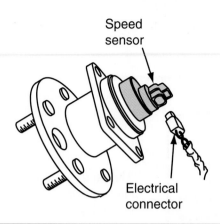

Speed
sensor

Electrical
connector

Figure 10-34 Unplug the rear harness from the rear-wheel sensor. Like its counterpart at the front, this sensor is incorporated into the hub and bearing assembly.

The removal of the rear sensor is similar to that of the front. After removing the tire and wheel assembly, disconnect the harness from the speed sensor (Figure 10-34). Remove the hub and bearing assembly. The installation, like always, is the reverse of the removal. Always perform the diagnostic system check any time a DBC-7 component is replaced.

Switches and Warning Lights

Replacing warning lights and TCS switches involves some disassembly of the instrument panel and/or the console. Although straightforward, it does require some expertise in the handling of plastic or flexible parts. This is best left to someone with instrument experience. If necessary, the new technician must closely follow the step-by-step instructions and use the special tools outlined in the vehicle's service manual. Pulling or prying in the wrong place can result in damage to some expensive plastic.

Teves Mark 20E

Classroom Manual
pages 269–272

The Mark 20E mostly uses the same test instruments and about the same procedures as any other ABS/TCS. But as mentioned earlier, the Mark 20E uses magnetoresistive wheel sensors and in some vehicles has acceleration switches with mercury. Testing a magnetoresistive sensor was discussed earlier so this section covers those other items that are particular to this system.

Initial Diagnosis

The startup cycle occurs when the ignition switch is switched to RUN. The ABS warning light will illuminate for up to 5 seconds. Popping sounds and slight pedal movement may be noticed. This is the controller cycling the solenoids and valves and is a normal occurrence. A humming sound may be heard during the initial drive-off at around 15 to 25 mph (25 to 40 kpm). Again the controller is testing the system. In this case, the pump and motor are temporarily activated.

The Mark 20 and Mark 20E will allow some wheel slip during braking to gain maximum stopping traction. This allowed slip may cause some tire chirping on some surfaces, but it is not a sign of total lockup. The system measures or allows wheel slippage based on the following slip values:

0 percent = free turning wheel
25 to 30 percent = allowable slippage
100 percent = total lockup

This means the system will allow up to 25 to 30 percent of slippage. The wheel in question is turning at a speed 25 to 30 percent slower than the other wheels at a given speed. The tire may chirp on some road surfaces. The easiest way to determine if the ABS is functioning properly is a road test on dry pavement. Accelerate the vehicle to about 20 to 25 mph and lock the brakes down, forcing the system into an ABS active mode. Once the vehicle has stopped, check behind the vehicle for black marks on the pavement. No black marks means the system is functioning properly.

Test Drive Diagnosis

During the test drive the technician should be listening and feeling for premature ABS functioning. Premature ABS operation may occur at any speed, with any braking force, and on any type of road surface. Some symptoms of premature ABS are clicking sounds from the ICU, pump operation, and/or pedal pulsations. Neither brake warning light will illuminate nor will any fault codes be stored.

To determine if there is premature ABS operation it may be necessary to use a high-end scan tool or DaimlerChrysler's DRB-III scan tool. Check the following components to isolate the root cause of the fault: damaged or incorrect tone wheels, damaged speed sensor mount, loose sensor fasteners, air gap too large between the sensor and tone wheel, or excessive tone wheel runout. Any of these faults could cause premature ABS functioning and should be examined closely during the diagnosis.

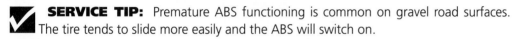 **SERVICE TIP:** Premature ABS functioning is common on gravel road surfaces. The tire tends to slide more easily and the ABS will switch on.

Code Retrieval

An aftermarket scan tool can usually retrieve DTCs from a Mark 20E. However, some Teves systems may not show a numerical code but instead display a text fault message similar to the ones shown in Figure 10-35. Also some aftermarket scan tools may not be able to access the ABS controller or activate some test sequences.

Text codes similar to DaimlerChrysler Teves ABS Applications
LF WSS failed EBCM continuity check
LF WSS signal failed EBCM check
RF WSS failed EBCM continuity check
RF WSS signal failed EBCM check
LR WSS failed EBCM continuity check
LR WSS signal failed EBCM check
RR WSS continuity failed EBCM check
RR Wheel signal failed EBCM check
Failure in valve power circuit
Internal Failure within EBCM
Amber warning lamp circuit failed test
Amber warning lamp relay failed test
System voltage too low
System voltage too high
Tone wheel fault detected
Communication circuits inoperative

Figure 10-35 A list of DTC text message similar to those for a Teves ABS/TCS on DaimlerChrysler vehicles as well as other similar manufacturers.

Front-Wheel Speed Sensor Replacement

Lift the vehicle and remove the tire and wheel assembly. Remove the cable routing bracket from the strut (Figure 10-36). Remove the inner fender grommet and unplug the cable from the vehicle harness. Remove the sensor retaining bolt and pull the sensor from the steering knuckle. If the sensor is stuck, do not use pliers to attempt to turn it. This will damage the sensor. Instead use a brass punch and hammer to tap the edge of the sensor (Figure 10-37). This should loosen it enough for removal. Installation of this sensor is not quite the reverse of removal.

The first installation steps are: connect the cable to the harness, install the grommet, and remount the cable bracket to the strut. Before installing the sensor into its cavity, coat the outside

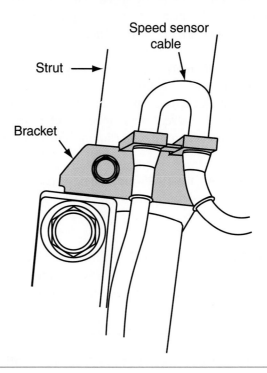

Figure 10-36 Remove the fastener and bracket from the strut.

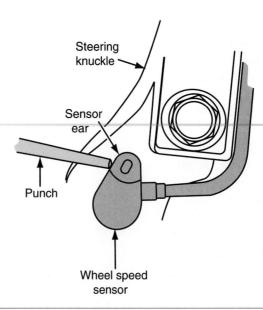

Figure 10-37 If the sensor is stuck, do not twist with pliers. Tap it gently with a brass punch at the point shown.

of the sensor body with high-temperature multipurpose grease. Install the sensor, tightening the retainer to 60 in-lb (7 N.m). Mount the tire and wheel assembly and road test the vehicle to ensure that the service brakes and the ABS are working properly.

Rear-Wheel Speed Sensor Replacement

Model year 2002 and later DaimlerChrysler vehicles equipped with Mark 20 or Mark 20E have the rear WSS mounted to the rear disc brake caliper, which is a little more difficult to replace than the front sensor. Before lifting the vehicle, it is necessary to remove the rear seat cushion and back to locate and disconnect the sensor cable from the vehicle's harness. Lift the vehicle and remove the tire and wheel assembly. Remove the grommet and cable from the floor pan, then remove the cable bracket from the strut. Unscrew the sensor fastener and pull the sensor from the disc caliper (Figure 10-38). The installation is the reverse of removal with the last step being the installation of the rear seat back and cushion.

Tone Wheel Inspection and Replacement

Replacement of the front and rear tone wheels is straightforward. The front tone wheel is a part of the drive axle so the drive axle will have to be replaced. The rear tone wheel is part of the hub and bearing assembly; therefore, it can be replaced only by replacing the entire hub and bearing assembly. But before condemning either tone wheel perform an inspection and check for the following faults: damaged or missing teeth, misalignment of sensor and tone wheel, incorrect air gap, tone wheel contacting sensor, tone wheel runout, or tone wheel loose on its mounting. Any of these faults could cause ABS fault codes, premature ABS functioning, and/or deactivation of the ABS.

ICU Replacement

The first step in this replacement procedure is to isolate the battery from the remainder of the vehicle. Disconnect the remote ground cable from the right strut tower. Place the nut back on the stud and lay the cable against the strut, ensuring that only the cable's insulation is actually touching the metal (Figure 10-39).

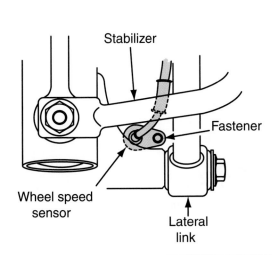

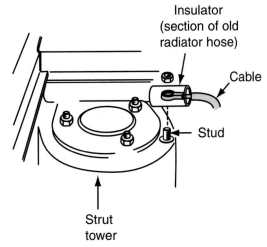

Figure 10-38 The fastener for this Mark 20E wheel sensor requires an extension, a socket, and a ratchet. The sensor is difficult to access, but do not pull on it with pliers.

Figure 10-39 Ensure that the metal end of the cable cannot come in contact with the vehicle body. An old radiator hose works well as a temporary insulator and holder.

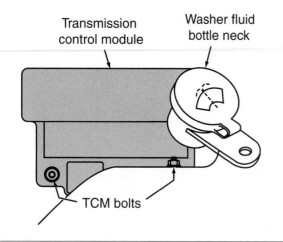

Transmission control module

Washer fluid bottle neck

TCM bolts

Figure 10-40 Shift the washer bottle neck before removing the transmission control module fasteners.

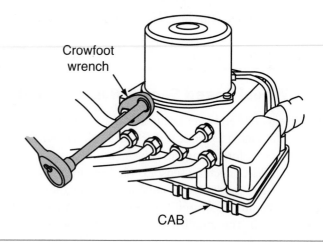

Crowfoot wrench

CAB

Figure 10-41 Using a crowfoot wrench is the only way to properly loosen and torque the brake line (tube) fittings.

☑ **SERVICE TIP:** A safer way is to disconnect the cable from the stud and reinstall the nut. Use a piece of old radiator hose that will slide over the cable end and insulation. This will effectively prevent any accidental contact between the cable and metal while the ICU is being replaced.

Use the brake pedal depressor from the alignment machine to depress and hold the brake pedal down about an inch. This isolates the master cylinder by blocking the reservoir ports. Under the hood remove the speed control servo from its mount and disconnect the electrical harness. Move the servo to the side. Shift the washer bottle filler neck to the side by removing the fastener. Do not remove the neck from the washer reservoir. After removing the fasteners, lift the transmission control module (TCM) up and out of the way (Figure 10-40).

Clean all around the ICU and the brake lines thoroughly. Using a crowfoot wrench, loosen and disconnect the master cylinder lines and brake tubes at the ICU ends (Figure 10-41). Disconnect the ICU 24-pin electrical connector by moving the lock up as far as possible. This removes the lock and disconnects the plug (Figure 10-42).

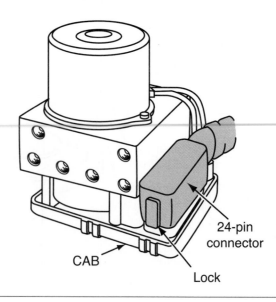

24-pin connector

CAB

Lock

Figure 10-42 Once the lock is unsnapped and moved, the 24-pin connector can be removed from the CAB.

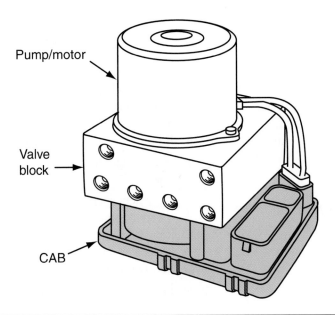

Figure 10-43 The ICU can be disassembled by removing the pump/motor then flipping the unit over to access the fasteners that hold the valve block and CAB together.

Lift the vehicle and remove the left front tire and wheel assembly and the inner splash shield. Remove the ICU fasteners. Lift the ICU out of the bracket, around it, and out through the wheel opening.

Installation is the reverse, closing with the following steps. If necessary, reprogram or initialize using a DRB-II scan tool. Bleed the service brakes and ABS and road test. The ICU can be separated into three units if desired: pump and motor, CAB, and valve block (body) (Figure 10-43).

Like other ABSs/TCSs, replacement of the warning lights and switches in the dash is best left to someone with experience in instrument panel and console work. Some supposedly simple removable of dash and console panels can cause breakage or damage to plastic components.

CASE STUDY

A student brought his Dakota in with the ABS warning light on. The retrieved DTC indicated a problem with the wheel speed sensor. The ABS was a Kelsey-Hayes RWAL with the sensor mounted on the differential housing. Upon inspection it was found that the sensor harness connector was loose and covered in mud. However, after unplugging and cleaning the connector plug on the harness and sensor the problem remained. A voltage and resistance check of the controller-to-sensor circuit revealed an open somewhere. The student pressure-washed the undercarriage and an inspection soon revealed the problem. Apparently while mud-bogging, the harness just forward of the rear axle met with an obstacle and was partially pulled apart. Soldering the ends of the conductor back together solved the problem and the student headed for the mud hole again the following weekend.

Terms to Know

Diagnostic trouble code (DTC)	Hard code	Scan tool
Electromotive force (EMF)	Oscilloscope	Soft code
Electrostatic discharge (ESD)		

ASE-Style Review Questions

1. Brake bleeding with an ABS installed is being discussed:
   Technician A says that the Delphi DBC-7 must have the service brakes bled three times in sequence.
   Technician B says that some ABSs require a scan tool to properly bleed the brakes.
   Who is correct?
   - **A.** A only
   - **B.** B only
   - **C.** Both A and B
   - **D.** Neither A nor B

2. When multiple trouble codes are present in an ABS, look for
   - **A.** a weak connection at a common ground.
   - **B.** an open circuit.
   - **C.** low-voltage signals.
   - **D.** high-voltage signals.

3. *Technician A* says that some antilock systems display trouble code numbers on a scan tool.
   Technician B says that some systems display codes by flashing the amber ABS warning lamp and red brake system warning lamp in a given sequence.
   Who is correct?
   - **A.** A only
   - **B.** B only
   - **C.** Both A and B
   - **D.** Neither A nor B

4. *Technician A* says that the amber ABS warning lamp signals an ABS malfunction.
   Technician B says that the red BRAKE warning lamp also can signal an ABS problem.
   Who is correct?
   - **A.** A only
   - **B.** B only
   - **C.** Both A and B
   - **D.** Neither A nor B

5. *Technician A* says that some lateral acceleration sensors may contain mercury.
   Technician B says that some lateral acceleration sensors must be handled as hazardous material or waste.
   Who is correct?
   - **A.** A only
   - **B.** B only
   - **C.** Both A and B
   - **D.** Neither A nor B

6. The Teves Mark 20 and Mark 20E are being discussed:
   Technician A says that the Mark 20E uses a speed sensor with a supplied voltage of 12 volts.

   Technician B says that the Mark 20 uses a magneto-resistive wheel speed sensor.
   Who is correct?
   - **A.** A only
   - **B.** B only
   - **C.** Both A and B
   - **D.** Neither A nor B

7. The DBC-7 ABS/TCS has all of the following features EXCEPT
   - **A.** PM wheel speed sensors.
   - **B.** it may be three- or four-channel.
   - **C.** it provides codes in text message format.
   - **D.** it has a speed sensor integrated in the hub/bearing assembly.

8. *Technician A* says that some ABSs must be depressurized before bleeding the brakes or removing components.
   Technician B says that some ABSs require a scan tool to bleed the brakes.
   Who is correct?
   - **A.** A only
   - **B.** B only
   - **C.** Both A and B
   - **D.** Neither A nor B

9. *Technician A* says that TCSs add several new components to a vehicle and require testing through their own diagnostic connectors.
   Technician B says that TCSs are examples of adding new features to a vehicle primarily through computer software.
   Who is correct?
   - **A.** A only
   - **B.** B only
   - **C.** Both A and B
   - **D.** Neither A nor B

10. Diagnosing a DBC-7 system is being discussed:
    Technician A says that the system must be depressurized before retrieving DTCs.
    Technician B says that thoroughly inspecting the vehicle and speed sensor harness is one of the first steps in proper diagnosis.
    Who is correct?
    - **A.** A only
    - **B.** B only
    - **C.** Both A and B
    - **D.** Neither A nor B

ASE Challenge Questions

1. *Technician A* says that a Teves system can be tested without driving the vehicle.
   Technician B says that the DRB-III is the only scan tool that can be used on a DBC-7 system.
   Who is correct?
   A. A only
   B. B only
   C. Both A and B
   D. Neither A nor B

2. The Teves Mark 20E is being discussed:
   Technician A says that pedal movement is normal when the ignition switch is first switched to RUN.
   Technician B says that a humming sound at about 25–40 kph is normal during the first driveaway of the day.
   Who is correct?
   A. A only
   B. B only
   C. Both A and B
   D. Neither A nor B

3. *Technician A* says that some ABSs require some pressure to be bled from the accumulator so the scan tool can accurately measure startup pressures.
   Technician B says that some ABSs require tools other than a scan tool to retrieve trouble codes.
   Who is correct?
   A. A only
   B. B only
   C. Both A and B
   D. Neither A nor B

4. *Technician A* says that the Teves Mark 20E is bled in the following sequence: service brakes, ABS unit, service brakes.
   Technician B says that the DBC-7 system requires the use of DOT-3 brake fluid.
   Who is correct?
   A. A only
   B. B only
   C. Both A and B
   D. Neither A nor B

5. *Technician A* says that early RABS used a jumper wire to supply short-term power to the control module memory.
   Technician B says that on DaimlerChrysler vehicles only the amber light will be lit for any ABS problems.
   Who is correct?
   A. A only
   B. B only
   C. Both A and B
   D. Neither A nor B

Job Sheet 30

30

Name: _____ Date: _____

ABS Warning Lamp Check

NATEF Correlation

This job sheet is related to NATEF brake tasks: Diagnose antilock brake system (ABS) electronic control(s) and components using self-diagnosis and/or recommended test equipment; determine necessary action.

Objective

Upon completion of this job sheet, you will have tested the ABS warning lamp circuit and its ability to detect a fault.

Tools and Materials

Scan tool

Describe the Vehicle Being Worked On

Year _____ Make _____ Model _____

VIN _____ Engine type and size _____

ABS type _____

Procedure

Task Completed

1. Set the parking brake. ☐

2. While observing the ABS warning lamp, turn the ignition on and then back off. ☐

3. Did the ABS warning lamp stay on for a few seconds and then go out?

Yes   No

4. Disconnect the ABS control module connector. ☐

5. Close the connector release on the module connector. Do not reconnect the connector to the module. ☐

6. Is the ABS warning lamp on?

Yes   No

Should it be on?

Yes   No

7. Reconnect the ABS connector.

8. Verify if a code has been set.

Yes   No

Why? _____

Problems Encountered _____

_____

Instructor's Response _____

_____

Job Sheet 31

31

Name: _____ Date: _____

Use Scan Tool to Scan ABS for Codes

NATEF Correlation

This job sheet is related to NATEF brake tasks: Diagnose antilock brake system (ABS) electronic control(s) and components using self-diagnosis and/or recommended test equipment; determine necessary action.

Objective

Upon completion and review of this job sheet, you should be able to use a scan tool to retrieve ABS trouble codes.

Tools and Materials

Service manual

Scan tool

Describe the Vehicle Being Worked On

Year _____ Make _____ Model _____

VIN _____ Engine type and size _____

ABS type _____

> **NOTE TO INSTRUCTORS: The following procedures are based on an MT2500 from Snap-on Tools using a primary domestic and the domestic transmission cartridges. The vehicle is an OBD-II-equipped vehicle. The use of other scan tools or vehicles may require additional instruction for the student.**

Procedure

Task Completed

1. Remember that some scan tools require that vehicle ID be entered before making a connection to the DLC.

   VIN _____

2. Enter the vehicle identification into the scan tool by:

   A. Pressing and holding the left button on the scan tool. ☐

   B. Enter the tenth digit of the VIN using the scroll knob on the right side of the tool. When the correct digit is visible, press the Y (yes) button. The tool should display the year and make.

   Results _____

   C. Enter the eighth digit of the VIN using the scroll knob on the right side of the tool.

   _____

   When the correct digit is visible, press the Y (yes) button. The tool should display the engine type and size.

   Results _____

D. Enter the fifth digit of the VIN using the scroll knob on the right side of the tool. When the correct digit is visible, press the Y (yes) button. The tool should display the model.

Results _____

☐   **E.** Answer the automatic transmission questions using the Y (yes) or N (no) button.

☐   **F.** Answer the air-conditioning questions using the Y (yes) or N (no) button.

     G. The scan tool should state that the vehicle ID is stored and the button can be released. Is the vehicle identified correctly? _____

     H. The scan tool should state the location of the OBD-II connection on the vehicle and which personality key to use.

Results _____

☐   **I.** Release the left button if all is correct.

☐   **3.** Select and connect the scan tool OBD-II connector to the scan tool. Install the personality key into the OBD-II connector.

☐   **4.** Connect the scan tool to the vehicle's OBD-II connector.

     5. The scan tool should switch on, identify itself, and show the identification for the vehicle entered in step 2. If correct, press the Y (yes) button.

Results _____

☐   **6.** At the next screen, select "Codes and Data."

☐   **7.** Switch ignition to KOEO.

     8. The scan tool should show five-digit codes if present. Various data should also be displayed. Record all codes or data displayed on this screen. An example could be a code shown as P0403.

Results _____

     9. If any other information is displayed, correct the data or consult the service manual for the vehicle and scan tool to resolve conflicts. Example of an error message may be: NO COMMUNICATION. IS KEY ON? ENSURE THE CABLE IS PROPERLY CONNECTED, ETC
☐   (or words to that effect).

     10. Once all trouble codes have been recorded, switch the ignition to off. Unplug the scan
☐   tool and store.

☐   **11.** When the task is complete, store the tools and complete the repair order.

Problems Encountered _____

_____

Instructor's Response _____

_____

Job Sheet 32

32

Name: _____ Date: _____

Testing an ABS Wheel Speed Sensor

NATEF CORRELATION

This job sheet is related to NATEF brake tasks: Test, diagnose and service ABS speed sensors, toothed ring (tone wheel), and circuits using a graphing multimeter (GMM)/digital storage oscilloscope (DSO) (includes output signal, resistance, shorts to voltage/ground, and frequency data).

Objective

Upon completion and review of this job sheet, you should be able to inspect and test an ABS wheel speed sensor with an oscilloscope.

Tools and Materials

Service manual
Wiring diagram
Component locator
Oscilloscope
Lift or jacks with stands

Describe the Vehicle Being Worked On

Year _____ Make _____ Model _____

VIN _____ Engine type and size _____

ABS _____ yes _____ no _____ If yes, type _____

Procedure

Task Completed

1. If vehicle is FWD, set transaxle to neutral and ignition to accessory position. ☐

2. Lift the vehicle until one of the front wheel sensors is accessible. ☐

3. Turn wheels to left or right for better access to the sensor. ☐

4. Locate and disconnect the speed sensor. ☐

5. Program the oscilloscope according to its operator's manual. ☐

6. Connect the oscilloscope leads to the terminals on the sensor. ☐

✓ **SERVICE TIP:** If the oscilloscope operator can stay clear and keep the leads clear of the wheel, a coworker can rotate the wheel using the engine.

7. Rotate the wheel at a constant speed while observing the graph.

General results _____

_____

_____

_____

☐ **8.** Freeze and save/print the graph.

9. Speed the wheel faster while observing the graph.

General results _____

_____

☐ **10.** Freeze and save/print the graph.

☐ **11.** Stop the wheel(s).

☐ **12.** After the wheel(s) have stopped, disconnect the oscilloscope and reconnect the harness to the sensor.

☐ **13.** Lower the vehicle.

☐ **14.** Shift the transaxle to park, switch the ignition to off, and set the parking brakes.

15. Compare the saved/printed graphs to the oscilloscope example file or the figures in the text.

16. Record the operational action of this speed sensor and make any recommendation.

_____

_____

_____

_____

Problems Encountered _____

_____

Instructor's Response _____

_____

Job Sheet 33

33

Name: _____ Date: _____

Replace an ABS Wheel Speed Sensor

NATEF Correlation

This job sheet is related to NATEF brake tasks: Remove and install antilock brake (ABS) electrical/electronic and hydraulic components.

Objective

Upon completion and review of this job sheet, you should be able to replace an ABS wheel speed sensor.

Tools and Materials

Service manual

Lift or jack with stands

Impact tools

Describe the Vehicle Being Worked On

Year _____ Make _____ Model _____

VIN _____ Engine type and size _____

ABS _____ yes _____ no _____ If yes, type _____

Procedure

Task Completed

1. Determine:

   Wheel sensor fastener torque _____

   Wheel lug nut torque _____

   Special tools and procedures for setting air gap _____

   _____

   _____

2. Lift vehicle to a good working height. ☐

3. Remove the wheel assembly. ☐

4. Inspect the area around the speed sensor. ☐

5. Remove any component blocking access to the speed sensor and/or fasteners. Items removed, if any.

   _____

6. Remove the speed sensor. ☐

7. Inspect the tone ring as much as possible.

Results _____

_____

☐   **8.** Remove the new speed sensor from its shipping carton.

☐   **9.** Install the speed sensor into its mounting.

☐   **10.** Secure the special tools and set the air gap to specifications.

☐   **11.** Install and torque the speed sensor fastener.

☐   **12.** Install the wheel assembly and torque the lug nuts to specifications.

☐   **13.** Lower the vehicle.

☐   **14.** Clear any ABS codes and road test.

☐   **15.** When the repair is complete, clean the area, store the tools, and complete the repair order.

Problems Encountered _____

_____

Instructor's Response _____

_____

Appendix

ASE Practice Examination

1. A vehicle has a drift to the left as the brakes are applied:
   Technician A says that the left front disc brake is grabbing.
   Technician B says that the right rear drum brake is adjusted too loose.
   Who is correct?
   A. A only   **C.** Both A and B
   B. B only   **D.** Neither A nor B

2. The customer complains of rear wheel lockup during medium braking on a pickup truck with RWAL:
   Technician A says that the metering valve may be defective.
   Technician B says that a frame-mounted proportioning valve could be stuck.
   Who is correct?
   A. A only   **C.** Both A and B
   B. B only   **D.** Neither A nor B

3. Four-wheel, three-channel antilock brakes are being discussed:
   Technician A says that the system has four speed sensors.
   Technician B says that the two rear wheels are independently controlled.
   Who is correct?
   A. A only   **C.** Both A and B
   B. B only   **D.** Neither A nor B

4. Proportioning valves are being discussed:
   Technician A says that a bad valve may cause the rear brakes to engage too quickly during initial brake application.
   Technician B says that this valve prevents excessive pressure from reaching the front brakes during normal braking.
   Who is correct?
   A. A only   **C.** Both A and B
   B. B only   **D.** Neither A nor B

5. *Technician A* says that a jerk of the steering to the right only as the brakes are applied could indicate improper wheel alignment.
   Technician B says that brake fluid could cause the steering wheel to shake as the brakes are applied.
   Who is correct?
   A. A only   **C.** Both A and B
   B. B only   **D.** Neither A nor B

6. The brake pedal drops slowly to the floor after the vehicle has stopped and the brake is kept applied. The fluid level does not drop.
   Technician A says that the dump valve on the RABS could be leaking.
   Technician B says that the master cylinder could have an internal leak.
   Who is correct?
   A. A only   **C.** Both A and B
   B. B only   **D.** Neither A nor B

7. There is a heavy shuddering and vibration as the brakes are heavily applied on a dry road:
   Technician A says that the rotors are not parallel.
   Technician B says that the ABS is probably functioning.
   Who is correct?
   A. A only   **C.** Both A and B
   B. B only   **D.** Neither A nor B

8. ABSs are being discussed:
   Technician A says that a bent or damaged tone ring could cause the wheel to lock up.
   Technician B says that a damaged tone ring could cause the ABS to deactivate.
   Who is correct?
   A. A only   **C.** Both A and B
   B. B only   **D.** Neither A nor B

9. A vehicle with four-wheel antilock brakes has a problem with the right rear wheel locking:
   Technician A says that this could be caused by a bad speed sensor mounted at the wheel.
   Technician B says that the speed sensor mounted at the differential could cause this problem.
   Who is correct?
   A. A only   **C.** Both A and B
   B. B only   **D.** Neither A nor B

10. The brake pedal pulsates during braking:
    Technician A says that this should cause the ABS to deactivate.
    Technician B says that this could be caused by warped drums.
    Who is correct?
    A. A only   **C.** Both A and B
    B. B only   **D.** Neither A nor B

11. A vehicle brake system is being bled:
    Technician A says that failure to build pressure at the left wheel could be caused by a bad master cylinder.
    Technician B says that the metering valve could cause low pressure in the rear brakes.
    Who is correct?
    A. A only
    B. B only
    C. Both A and B
    D. Neither A nor B

12. The right front wheel locks up during almost every brake application:
    Technician A says that a crimped line to the left wheel could be the cause.
    Technician B says that the interior of the right brake hose could be damaged.
    Who is correct?
    A. A only
    B. B only
    C. Both A and B
    D. Neither A nor B

13. Brake (stop) lights are being discussed:
    Technician A says that the loss of the CHMSL lamp will cause one or both of the other brake lamps not to work.
    Technician B says that the loss of all brake lights could be caused by a loose ground at one of the lights.
    Who is correct?
    A. A only
    B. B only
    C. Both A and B
    D. Neither A nor B

14. The red brake warning light is lit with the engine running:
    Technician A says that bleeding the brakes could cause this condition.
    Technician B says that a faulty metering valve could be at fault.
    Who is correct?
    A. A only
    B. B only
    C. Both A and B
    D. Neither A nor B

15. A scraping noise is heard as the brakes are applied:
    Technician A says that this could be normal for brakes with audible wear indicators.
    Technician B says that the wheel bearings may be the cause.
    Who is correct?
    A. A only
    B. B only
    C. Both A and B
    D. Neither A nor B

16. *Technician A* says that a stuck proportioning valve could prevent or reduce front brake application.
    Technician B says that a stuck proportioning valve could cause the vehicle to nosedive during initial brake application.
    Who is correct?
    A. A only
    B. B only
    C. Both A and B
    D. Neither A nor B

17. The vehicle's brake lights come on late (more pedal movement needed):
    Technician A says that the booster's pushrod may need adjusting.
    Technician B says that the brake (stop) light switch is sticking open.
    Who is correct?
    A. A only
    B. B only
    C. Both A and B
    D. Neither A nor B

18. There is a distinct odor of overheated brakes from the front wheels:
    Technician A says that the pushrod may be adjusted too short.
    Technician B says that apparently the driver did not completely release the parking brake.
    Who is correct?
    A. A only
    B. B only
    C. Both A and B
    D. Neither A nor B

19. One of the rear wheels tramps (hops) when the brakes are applied:
    Technician A says that the brake drum is out of round.
    Technician B says that the brake shoes are soaked with fluid.
    Who is correct?
    A. A only
    B. B only
    C. Both A and B
    D. Neither A nor B

20. The driver complains that sometimes the pedal goes almost to the floor with little braking effect:
    Technician A says that the master cylinder has an internal leak.
    Technician B says that the ABS may not have been properly bled or serviced.
    Who is correct?
    A. A only
    B. B only
    C. Both A and B
    D. Neither A nor B

21. An ABS control module has set a code 22. The service manual states that a code 22 indicates a broken tone ring.
Technician A says that the ABS circuit cannot directly detect a broken tone ring.
Technician B says that a working speed sensor circuit, but incorrect data, may cause the ABS control module to set a code 22.
Who is correct?
A. A only
B. B only
C. Both A and B
D. Neither A nor B

22. *Technician A* says that sliding rear wheels on a three-channel ABS can be caused by a bad speed sensor in the final drive or differential.
Technician B says that the speed sensors on a four-channel ABS are wired in parallel.
Who is correct?
A. A only
B. B only
C. Both A and B
D. Neither A nor B

23. The driver complains of a clicking noise every time the brakes are initially applied:
Technician A says that the pads' antirattle springs or clips may not be installed correctly.
Technician B says that the shoes' return springs are stretched.
Who is correct?
A. A only
B. B only
C. Both A and B
D. Neither A nor B

24. During drum brake inspection, it was found that two of the return springs had deep nicks in them from a previous repair.
Technician A says that so long as the nicks are less than one-half of the diameter of the spring wire they can be reused.
Technician B says to replace all of the hardware for that wheel.
Who is correct?
A. A only
B. B only
C. Both A and B
D. Neither A nor B

25. *Technician A* says that if only one wheel locks up during heavy braking then the ABS system is working correctly.
Technician B says that one wheel locking may indicate a problem with that brake or the brake on the opposite side.
Who is correct?
A. A only
B. B only
C. Both A and B
D. Neither A nor B

26. Tires sizes and ABSs are being discussed:
Technician A says that the size of the tires could affect the operation of the ABS.
Technician B says that rear tires that are larger in diameter than front tires may cause the ABS to function prematurely.
Who is correct?
A. A only
B. B only
C. Both A and B
D. Neither A nor B

27. *Technician A* says that to check the brake fluid level the ABS accumulator must be depressurized on the Teves Mark II ABS.
Technician B says that most ABS accumulators must be depressurized to bleed the front brakes.
Who is correct?
A. A only
B. B only
C. Both A and B
D. Neither A nor B

28. A Delphi DBC-7 has a code C1278 and the TCS is temporarily disabled (engine problem):
Technician A says that the ABS is also disabled.
Technician B says that the throttle position sensor may be at fault.
Who is correct?
A. A only
B. B only
C. Both A and B
D. Neither A nor B

29. A vehicle equipped with the Delphi DBC-7 system is being discussed:
Technician A says that the vehicle must be traveling at least 10 miles per hour before the ECBM can perform a dynamic test.
Technician B says that the red brake light and the amber ABS light may be illuminated if dynamic rear proportioning is affected by an ABS fault.
Who is correct?
A. A only
B. B only
C. Both A and B
D. Neither A nor B

30. Replacement of a Delphi DBC-7 BPMV is being discussed:
Technician A says that the BPMV can be bled manually.
Technician B says that the engine must be running with a scan tool connected to bleed the BPMV.
Who is correct?
A. A only
B. B only
C. Both A and B
D. Neither A nor B

31. Wheel speed sensors are being discussed:
Technician A says that a standard feeler gauge may be used to adjust the air gap.
Technician B says that the air gap is automatically set on all vehicles when the sensor is bolted in place.
Who is correct?
A. A only
B. B only
C. Both A and B
D. Neither A nor B

32. A vehicle has the amber and red brake lights on with the engine running.
Technician A says that the proportioning valve has moved off center.
Technician B says that a leaking brake line could set the fault.
Who is correct?
A. A only
B. B only
C. Both A and B
D. Neither A nor B

33. The rear brakes are being serviced:
Technician A says that the small drum-within-disc for the parking brake should be machined in the same manner and as frequently as a regular drum.
Technician B says that if the piston will not screw back into the caliper the piston may be screwed off the screw.
Who is correct?
A. A only
B. B only
C. Both A and B
D. Neither A nor B

34. A wheel speed sensor on a Delphi DBC-7 is being replaced. All of the following are true *EXCEPT*:
A. Remove the caliper and rotor.
B. Remove the hub assembly from the steering knuckle.
C. Disconnect the sensor wiring from the sensor.
D. Remove the sensor from the hub assembly.

35. A vehicle equipped with a Teves Mark 20E ABS has popping sounds when the ignition key is first switched to the RUN position:
Technician A says that this is a normal operation for this ABS.
Technician B says that the brake pedal may move at this time.
Who is correct?
A. A only
B. B only
C. Both A and B
D. Neither A nor B

36. A wheel speed sensor on a Teves Mark 20E is being tested:
Technician A says that a DMM is the best instrument for testing this sensor.
Technician B says that the sensor should have a high voltage of 1.6 volts and a low voltage of 0.90 volt.
Who is correct?
A. A only
B. B only
C. Both A and B
D. Neither A nor B

37. Drum brakes are being discussed:
Technician A says that a grabbing brake could be traced to a leaking axle seal.
Technician B says that a wheel cylinder is checked for leaks by pulling the cup to one side.
Who is correct?
A. A only
B. B only
C. Both A and B
D. Neither A nor B

38. A vehicle drifts toward one side on straight flat roads:
Technician A says that a stuck caliper could cause this problem.
Technician B says that the shoes on one rear wheel may be adjusted too tight.
Who is correct?
A. A only
B. B only
C. Both A and B
D. Neither A nor B

39. A vehicle is brought to the shop with a low but firm brake pedal:
Technician A says that this could be caused by low brake fluid in the master cylinder.
Technician B says that this could be caused by worn unadjusted brake pads.
Who is correct?
A. A only
B. B only
C. Both A and B
D. Neither A nor B

40. Replacement of a Teves Mark 20E rear speed sensor is being discussed:
Technician A says that the rear seat must be removed on some vehicles.
Technician B says that the sensor must be removed from the hub assembly.
Who is correct?
A. A only
B. B only
C. Both A and B
D. Neither A nor B

41. A Teves Mark 20E ICU is being replaced:
Technician A says to disconnect the negative cable directly from the battery negative terminal.
Technician B says to isolate the master cylinder by applying and holding the brake pedal down about an inch.
Who is correct?
A. A only
B. B only
C. Both A and B
D. Neither A nor B

42. Technician A says to replace a double-flare fitting with an ISO-type fitting as new brake lines are required.
Technician B says that failure of both rear brakes to apply could be traced to one flexible hose.
Who is correct?
A. A only
B. B only
C. Both A and B
D. Neither A nor B

43. The red brake warning light does not go out during KOER conditions:
Technician A says to bypass the low-fluid switch with a jumper to make the light go out.
Technician B says that the metering valve should be the first thing to check.
Who is correct?
A. A only
B. B only
C. Both A and B
D. Neither A nor B

44. The amber brake warning light does not come on during KOEO conditions:
Technician A says to make sure the switch in the combination valve is properly connected.
Technician B says to check the lamp.
Who is correct?
A. A only
B. B only
C. Both A and B
D. Neither A nor B

45. Technician A says that a failure of one set of contacts within a brake (stop) light switch could cause a failure of just the CHMSL.
Technician B says that an audible alarm could indicate worn disc brake pads.
Who is correct?
A. A only
B. B only
C. Both A and B
D. Neither A nor B

46. Technician A says that a failure of a brake (stop) lamp could cause a failure of the turn signal for that corner.
Technician B says that a dual-filament lamp can be used to replace a single-filament brake lamp.
Who is correct?
A. A only
B. B only
C. Both A and B
D. Neither A nor B

47. A set of drum brakes is being reinstalled:
Technician A says to adjust the parking brake cables before adjusting the service brakes.
Technician B says to use a brake micrometer to adjust the shoes before installing the drum.
Who is correct?
A. A only
B. B only
C. Both A and B
D. Neither A nor B

48. Technician A says to always check and adjust the service brakes before adjusting the parking brake.
Technician B says that adjusting the front parking brake cable may be done at the lever between the seats.
Who is correct?
A. A only
B. B only
C. Both A and B
D. Neither A nor B

49. Traction control is being discussed:
Technician A says that a lit amber light during KOER indicates that the traction controls are disabled.
Technician B says that a failure within the TCS does not necessarily disable the ABS.
Who is correct?
A. A only
B. B only
C. Both A and B
D. Neither A nor B

50. Non-ABS brake systems are being discussed:
Technician A says that rear wheel lockup during moderate braking could be corrected by replacing the combination valve.
Technician B says that rear wheel lockup during moderate braking could be corrected with new tires.
Who is correct?
A. A only
B. B only
C. Both A and B
D. Neither A nor B

Glossary
Glosario

accumulator   A container that stores hydraulic fluid under pressure. It can be used as a fluid shock absorber or as an alternate pressure source. A spring or compressed gas behind a sealed diaphragm provides the accumulator pressure.

acumulador   Recipiente que contiene líquido hidráulico bajo presión. Se puede utilizar como líquido amortiguador o como fuente de presión alternativa. Un resorte o gas comprimido detrás de un diafragma sellado proporciona presión al acumulador.

American Wire Gauge (AWG)   A system for specifying wire size (conductor cross-sectional area) by a series of gauge numbers. The lower the number, the larger the wire cross section.

Calibre americano de cables (AWG)   Sistema para especificar el tamaño del cable (área de la sección transversal del conductor) por medio de una serie de números de calibre. Mientras menor sea el número, mayor será la sección transversal del cable.

ampere (A)   The unit for measuring electric current. One ampere equals a current flow of 6.28×1018 electrons per second.

amperio (A)   Unidad de medida de la corriente eléctrica. Un amperio equivale a un flujo de corriente de $6,28 \times 1018$ electrones por segundo.

amplitude   Signal strength or the maximum measured value of a signal.

amplitud   Fuerza de la señal, o máximo valor medido de una señal.

analog   A signal that varies proportionally with the information that it measures. In a computer, an analog signal is voltage that fluctuates over a range from high to low.

analógica   Señal que varía proporcionalmente con la información que mide. En un computador, una señal analógica es la tensión que fluctúa en un margen de alto a bajo.

anodized finish   A protective metal surface finish formed by an electrolytic coating of a special oxide.

pulido anodizado   Pulido de superficie protectora metálica formada por un revestimiento electrolítico de un óxido especial.

aqueous   Water based.

acuoso   Con agua como base.

asbestos   The generic name for a silicate compound that is very resistant to heat and corrosion. Its excellent heat dissipation abilities and coefficient of friction make it ideal for automotive friction materials such as clutch and brake linings. Asbestos fibers are a serious health hazard if inhaled, however.

amianto   Nombre genérico de un compuesto de silicatos muy resistente al calor y a la corrosión. Su excelente capacidad de disipación de calor y su coeficiente de fricción lo hacen ideal para materiales de fricción para automoción, como el embrague y los forros de frenos. Sin embargo, las fibras de amianto constituyen un serio peligro para la salud si se inhalan.

asbestosis   A progressive and disabling lung disease caused by inhaling asbestos fibers over a long period of time.

asbestosis   Enfermedad pulmonar progresiva que incapacita a la persona causada por la inhalación de fibras de amianto durante un periodo de tiempo prolongado.

aspect ratio   The ratio of the cross-sectional height to the cross-sectional width of a tire expressed as a percentage.

relación entre dimensiones   Relación entre la altura y la anchura de la sección transversal de un neumático, expresada en un porcentaje.

atmospheric pressure   The weight of the air that makes up the Earth's atmosphere.

presión atmosférica   Peso del aire que constituye la atmósfera de la Tierra.

autoranging   A feature of an electrical test meter that lets it shift automatically from one measurement range to another for a given value such as voltage or resistance.

autodeterminación de escalas   Característica de un medidor de prueba eléctrico que le permite cambiar automáticamente de una escala de medida a otra para un valor determinado, como por ejemplo, la tensión o la resistencia.

backing plate   The mounting surface for all other parts of a drum brake assembly except the drum.

placa de refuerzo   Superficie en que se montan todas las partes de un freno de tambor, excepto el tambor.

banjo fitting   A round, banjo-shaped tubing connector with a hollow bolt through its center that enables a brake line to be connected to a hydraulic component at a right angle.

ajuste de banjo   Conector redondo, en forma de banjo, atravesado en el centro por un perno hueco, que permite conectar en ángulo recto una línea de frenos a un componente hidráulico.

bearing end play   The designed looseness in a bearing assembly.

extremo del cojinete   Soltura diseñada para el conjunto de cojinetes.

bedding-in   See burnishing.

lustre   Véase satinado.

bench bleeding   The process of filling a master cylinder with brake fluid, clamping it in a vise, and bleeding air from it before installing it on a vehicle.

extracción de vapor del banco   Proceso de relleno de un cilindro maestro con líquido de frenos sujetándolo con un tornillo y de extracción del aire antes de instalarlo en un vehículo.

bleeder screw   A screw that opens and closes a bleeding port in a caliper or a wheel cylinder.

tornillo del extractor de vapor   Tornillo que abre y cierra la válvula de extracción de vapor en un calibre o un cilindro de la rueda.

brake bleeding   A procedure that pumps fresh brake fluid into the brake hydraulic system and forces out air bubbles and the old aerated fluid through bleeding ports in calipers and wheel cylinders.

extracción de vapor de los frenos   Procedimiento que bombea líquido de frenos nuevo al sistema hidráulico de frenos y hace salir las burbujas de aire y el líquido viejo, aireado, por las válvulas de extracción de vapor en calibres y cilindros de las ruedas.

brake caliper   The part of a disc brake system that converts hydraulic pressure back to mechanical force that applies the pads to the rotor. The caliper is mounted on the suspension or axle housing and contains a hydraulic piston and the brake pads.

calibre del freno   Parte de un sistema de frenos de disco que vuelve a convertir la presión hidráulica en fuerza mecánica que aplica las pastillas al rotor. El calibre va montado en el alojamiento del eje o la suspensión, y contiene un pistón hidráulico y las pastillas de freno.

brake pad   The part of a disc brake assembly that holds the lining friction material that is forced against the rotor to create friction to stop the vehicle.

pastilla de freno   Parte de un conjunto de frenos de disco que aloja el material de fricción del forro que se fuerza contra el rotor para crear la fricción que detendrá el vehículo.

brake shoes   The curved metal parts of a drum brake assembly that carry the friction material lining.

zapatas de freno   Partes metálicas curvas de un conjunto de frenos de tambor que llevan el forro de material de fricción.

brush hone   A cylinder hone made of abrasive modules attached to flexible nylon cords that are attached to the hone shaft. A brush hone removes peaks or high spots on the surface of a cylinder bore and produces a flattened or plateaued finish. Also called a flexhone.

cepillo pulimentador   Cilindro hecho de módulos abrasivos unidos a cables flexibles de nylon que van fijados al eje del pulimentador. El cepillo de pulimentar elimina picos o puntos altos en la superficie del hueco de un cilindro y produce un acabado plano o liso. También llamado pulimentador flexible.

burnishing   The process of applying friction materials to each other to create a desired wear pattern.

satinado   Proceso de aplicación y fricción de materiales entre ellos para crear un patrón de desgaste deseado.

caliper support   The bracket or anchor that holds the brake caliper.

soporte del calibre   Mordaza o ancla que aloja el calibre de freno.

camber   The inward or outward tilt of the wheel measured from top to bottom and viewed from the front of the car.

inclinación   Inclinación de la rueda hacia dentro o hacia fuera, medida de arriba abajo y vista desde la parte frontal del vehículo.

Canadian Center for Occupational Health and Safety (CCOHS)   A Canadian federal agency similar to the U.S. OSHA with a similar mandate, responsibility, and authority.

Centro Canadiense de Salud y Seguridad Ocupacional (CCOHS)   Agencia federal canadiense similar a la OSHA de EE.UU. con mando, responsabilidad y autoridad similares.

carbon monoxide   A poisonous gas that is a by-product of incomplete combustion; also an air pollutant.

monóxido de carbono   Gas venenoso, subproducto de una combustión incompleta; también es un agente contaminante del aire.

caster   The backward or forward angle of the steering axis viewed from the side of the car.

inclinación del eje   Inclinación hacia delante o hacia atrás del eje de dirección visto desde el costado del auto.

center high-mounted stoplamp (CHMSL)   A third stoplamp on vehicles built since 1986, located on the vehicle centerline no lower than 3 inches below the rear window (6 inches on convertibles).

luz de frenado trasera superior (CHMSL)   Tercera luz de frenado, en vehículos fabricados después de 1986, ubicada en la línea central del vehículo no más baja que a 3 pulgadas por debajo de la ventanilla trasera (6 pulgadas en los convertibles).

cold inflation pressure   The tire inflation pressure after a tire has been standing for 3 hours or driven less than 1 mile after standing for 3 hours.

presión de inflado en frío   Presión de inflado del neumático después de haber estado en reposo durante tres horas o haber recorrido menos de una milla después del mismo tiempo de reposo.

combination valve   A hydraulic control valve with two or three valve functions in one valve body.

válvula de combinación   Válvula de control hidráulico con dos o tres funciones de paso en el mismo cuerpo de válvula.

conduit   A flexible metal housing or jacket that houses the parking brake cables to protect them from dirt, rust, abrasion, and other damage.

conducto   Alojamiento de metal flexible o forro que recubre los cables del freno de estacionamiento para protegerlos de la suciedad, el polvo, la abrasión y otros daños.

crocus cloth   Fine polishing cloth or emery cloth.

trapo púrpura de hierro   Tela fina de pulir o tela de esmeril.

cross-feed   The distance the cutting bit of a brake lathe moves across the friction surface during each lathe revolution.

alimentación transversal   Distancia que la fresa de un torno de frenos se mueve a través de la superficie de fricción durante cada revolución del torno.

crosshatch   A crisscross pattern formed by a cylinder hone.

trama   Un patrón entrecruzado formado por un pulimentador de cilindros.

cup expander   A metal disc that bears against the inner sides of wheel cylinder seals to hold the seal lips against the cylinder bore when the brakes are released. This keeps air from entering the cylinder past the retracting pistons and seals.

cubeta de expansión   Disco metálico que se ajusta a los costados internos de las juntas del cilindro de las ruedas para sujetar los bordes de la junta contra el hueco del cilindro al aplicar los frenos. Así se impide que el aire que entra en el cilindro pase más allá de los pistones retráctiles y las juntas.

cup seal   A circular rubber seal with a depressed center section surrounded by a raised sealing lip to form a cup. Cup seals often are used on the front ends of hydraulic cylinder pistons because they seal high pressure in the forward direction of travel but not in the reverse.

cubeta de obturación   Junta de goma circular con una sección central hundida rodeada por un borde de junta saliente que forma una copa. Las cubetas de obturación se suelen usar en los extremos frontales de los pistones de los cilindros hidráulicos porque impiden el paso de alta presión hacia delante pero no en el sentido contrario.

cylinder hone   An abrasive tool used to remove dirt, light corrosion, and machining irregularities from a cylinder bore.

pulimentador de cilindros   Herramienta abrasiva usada para eliminar la suciedad, la corrosión leve y las irregularidades de la superficie interior de un cilindro.

Department of Transportation (DOT)   The U.S. government executive department that establishes and enforces safety regulations for motor vehicles and for federal highway safety.

Departamento de Transportes (DOT)   El departamento ejecutivo del Gobierno de los EE.UU. que establece y hace cumplir las normas de seguridad para vehículos a motor y para la seguridad vial federal.

diagnostic energy reserve module (DERM)   Used to test the air bag system, store codes if a fault is detected, and supply the system with a prime voltage.

módulo de diagnóstico de reserva de energía (DERM)   Se usa para probar el sistema de air bag, almacenar códigos si se detecta un fallo, y proporcionar tensión principal al sistema.

diagnostic trouble code (DTC)   A numerical code generated by an electronic control system that has self-diagnostic capabilities as the result of a system self-test or monitoring to indicate a problem in a circuit or subsystem or to indicate a general condition that is out of limits; also called trouble codes, service codes, or fault codes by various carmakers.

código de problema de diagnóstico (DTC)   Código numérico generado por un sistema electrónico de control que posee capacidades de autodiagnóstico como resultado de una prueba o control propio del sistema para indicar un problema en un circuito o subsistema, o para indicar una condición general que se sale de los límites; los diversos fabricantes de automóviles también los denominan códigos de problemas, códigos de servicio o códigos de fallos.

diaphragm   A flexible membrane, usually made of rubber, that isolates two substances or areas from each other. A rubber diaphragm isolates brake fluid in the master cylinder reservoir from the air. A diaphragm separates the two chambers of a power brake vacuum booster.

diafragma   Membrana flexible, por lo común de goma, que aisla una sustancia o una zona de otra. Un diafragma de goma aísla del aire el líquido de frenos en el depósito del cilindro maestro. Un diafragma separa las dos cámaras de un reforzador de vacío en frenos de potencia.

digital   A signal that is either on or off and that is translated into the binary digits zero and one. In a computer, a digital signal is voltage that is low or high or current flow that is on or off.

digital   Señal que está activada o desactivada y que se traduce por los dígitos binarios cero y uno. En un computador, una señal digital es una tensión alta o baja, o flujo de corriente que está abierto o cerrado.

digital multimeter (DMM)   A volt, ohm , milliamp meter—also called a volt-ohm-milliamp (VOM) meter or a digital volt-ohmmeter (DVOM)—that tests voltage, resistance, and current.

multímetro digital (DMM)   Medidor de voltios, ohmios, miliamperios—también llamado medidor VOM o medidor digital de voltios-ohmios (DVOM)—que prueba la tensión, la resistencia y la corriente.

disc brake   A brake in which friction is generated by brake pads rubbing against the friction surfaces on both sides of a brake disc or rotor attached to the wheel.

freno de disco   Freno en el que la fricción se genera al rozar las pastillas de freno contra las superficies de fricción a ambos lados de un disco o rotor de freno que está unido a la rueda.

discard diameter   The maximum inside diameter that is cast or stamped on a brake drum. The discard diameter is the allowable wear dimension, not the allowable machining dimension.

diámetro de descarte   El diámetro interior máximo que se funde o estampa en un tambor de frenos. El diámetro de descarte es la dimensión de desgaste tolerable, no la dimensión de moldeo permisible.

discard dimension   The minimum thickness of a brake rotor or the maximum diameter of a drum. If disc or drum wear exceeds the discard dimensions, the disc or drum must be replaced. A disc or drum should never be machined to the discard dimension.

dimensión de descarte   Grosor mínimo de un rotor de frenos o diámetro máximo de un tambor. Si el desgaste del disco o el tambor excede las dimensiones de descarte, el disco o el tambor deben reemplazarse. Nunca se debe disponer un disco o tambor hasta la dimensión de descarte.

DOT 3, DOT 4, and DOT 5   U.S. Department of Transportation specification numbers for hydraulic brake fluids.

DOT 3, DOT 4 y DOT 5   Números de especificación del Departamento de Transportes de los EE.UU. para los líquidos de frenos hidráulicos.

double flare   A type of tubing flare connection in which the end of the tubing is flared out, then is formed back on to itself.

doble ensanche   Tipo de conexión de tubos en la que el extremo del tubo se acampana y luego se vuelve a doblar.

drum brake   A brake in which friction is generated by brake shoes rubbing against the inside surface of a brake drum attached to the wheel.

freno de tambor   Freno en que la fricción la generan zapatas que rozan contra la superficie interior de un tambor de freno unido a la rueda.

drum web   The closed side of a brake drum.

membrana del tambor   Lado cerrado de un tambor de freno.

duo-servo brake   A drum brake that develops self-energizing action on the primary shoe, which in turn applies servo action to the secondary shoe to increase its application force. Brake application force is interrelated for the primary and the secondary shoes. Also called a dual-servo or a full-servo brake.

freno servoduo   Freno de tambor que desarrolla acción autónoma sobre la zapata primaria, que a su vez aplica servoacción a la zapata secundaria para aumentar su fuerza de aplicación. La fuerza de aplicación de freno es interrelacionada para las zapatas primarias y las secundarias. También se conoce como freno servo dual o totalmente asistido.

electromotive force (EMF)   The proper terminology for voltage. EMF always creates a magnetic field around a conductor. This may interfere with EMF (or the voltage signal) in an adjacent conductor.

fuerza electromotriz (EMF)   La terminología apropiada para el voltaje. EMF crea siempre un campo magnético alrededor de un conductor. Esto puede interferir con EMF (o la señal del voltaje) en un conductor adyacente.

Environmental Canada   The Canadian version of the U.S. EPA with requirements that relate to Canada's more northern environment and citizens.

Environmental Canada   La versión canadiense de la EPA de EE.UU. con requisitos que se relacionan a los medioambientes y ciudadanos del norte de Canadá.

Environmental Protection Agency (EPA)   The U.S. government executive department that establishes and enforces regulations to protect and preserve the physical environment, best known for regulations relating to air quality.

Agencia Protectora del Medio Ambiente (EPA)   Departamento gubernamental de EE.UU. que establece y hace cumplir las normas que protegen y preservan el entorno físico, más conocido por las normas que regulan la calidad del aire.

equalizer   Part of the parking brake linkage that balances application force and applies it equally to each wheel. The equalizer often contains the linkage adjustment point.

compensador   Parte del acoplamiento del freno de estacionamiento que equilibra la fuerza ejercida y la aplica por igual en ambas ruedas. Con frecuencia, el compensador incluye el punto de ajuste del acoplamiento.

Federal Motor Vehicle Safety Standards (FMVSS)   U.S. government regulations that prescribe safety requirements for various vehicles, including passenger cars and light trucks. The FMVSS regulations are administered by the U.S. Department of Transportation (DOT).

Normas federales de seguridad de vehículos a motor (FMVSS) Normas gubernamentales de EE.UU. que dictan los requisitos de seguridad para diversos vehículos, incluyendo los automóviles de pasajeros y los camiones ligeros. El Departamento de Transportes de los EE.UU. (DOT) es el que administra las normas FMVSS.

fixed caliper brake A brake caliper that is bolted to its support and does not move when the brakes are applied. A fixed caliper must have pistons on both the inboard and the outboard sides.

freno de calibre fijo Calibre de freno que se fija al soporte con un perno y no se mueve al aplicar los frenos. Un calibre fijo debe tener pistones tanto en el lado exterior como en el interior.

floating caliper A caliper that is mounted to its support on two locating pins or guide pins. The caliper slides on the pin in a sleeve or bushing. Because of its flexibility, this kind of caliper is said to float on its guide pins.

calibre flotante Calibre que se monta al soporte mediante dos pasadores de posición o pasadores guía. El calibre se desliza dentro del pasador a través de un manguito o casquillo. Debido a su flexibilidad, se dice que este tipo de calibre flota en los pasadores guías.

floating drum A brake drum that is separate from the wheel hub or axle. A floating drum usually is held in place on studs in the axle flange or hub by the wheel and wheel nuts.

tambor flotante Tambor de freno que está separado del buje o eje de la rueda. Un tambor flotante suele ir sujeto en los pasadores de la pestaña del eje o buje mediante la rueda y las tuercas de la rueda.

floating rotor A rotor and hub assembly made of two separate parts.

rotor flotante Conjunto de rotor y buje en dos piezas distintas.

force Power working against resistance to cause motion.

fuerza Energía que trabaja contra la resistencia para producir movimiento.

friction The force that resists motion between the surfaces of two objects or forms of matter.

fricción Fuerza que se opone al movimiento entre las superficies de dos objetos o formas de materia.

fusible link A wire of smaller gauge that is connected into a circuit to act as a fuse. A fusible link will melt when exposed to excessive current to protect other circuit components.

enlace fusible Alambre de menor medida que se conecta a un circuito para actuar como fusible. Los enlaces fusibles se funden al ser expuestos a una corriente excesiva para proteger otros componentes del circuito.

graphing meters Electronic diagnostic instruments that come in different sizes and capabilities and that can graph the flow of electrical values; commonly known as scopes or oscilloscopes.

Medidores gráficos Herramientas de diagnóstico electrónico que son de diferentes tamaños y capacidades, y que pueden graficar el flujo de los valores eléctricos; se les conoce comúnmente como ámbitos u osciloscopios.

gravity bleeding The process of letting old brake fluid and air drain from the brake hydraulic system through a wheel bleeder screw.

extracción de vapor de la gravedad Proceso que permite drenar el líquido de frenos sucio y el aire del sistema hidráulico de frenos a través de un tornillo extractor de vapor de la rueda.

hard code A diagnostic trouble code from a vehicle computer that indicates a problem that is permanently present at the time of testing; may or may not keep the system from operating.

código duro Código de problema de diagnóstico del computador de un vehículo que indica que existe un problema permanente en el momento de la prueba; puede o no dejar inoperante el sistema.

hard spots Circular blue-gold glazed areas on drum or rotor surfaces where extreme heat has changed the molecular structure of the metal.

puntos duros Zonas circulares vidriadas de color oro azulado en las superficies del tambor o del rotor, que aparecen donde el calor excesivo ha cambiado la estructura molecular del metal.

heat checks Small cracks on drum or rotor surfaces that usually can be machined away.

grietas térmicas Pequeñas fisuras en las superficies del tambor o del rotor que normalmente se pueden hacer desaparecer con una máquina.

heat-shrink tubing Plastic tubing that shrinks in diameter when exposed to heat; used to insulate electrical wiring and repairs.

tuberías que se contraen con calor Cañerías de plástico que encogen al ser expuestas al calor; se usan para aislar cables eléctricos y en reparaciones.

height-sensing proportioning valve A proportioning valve in which hydraulic pressure is adjusted automatically according to the vertical movement of the chassis in relation to the rear axle during braking; sometimes also called a weight-sensing proportioning valve.

válvula dosificadora de detección de altura Válvula dosificadora en la que la presión hidráulica se ajusta automáticamente según el movimiento vertical del chasis en relación con el eje trasero durante el frenado; a veces también se la denomina válvula dosificadora de detección de peso.

high-efficiency particulate air (HEPA) filter A filter that removes the smallest particulates from air.

filtro de macropartículas de aire de gran eficiencia (HEPA) Filtro de aire que retiene las partículas más pequeñas.

high impedance High input resistance provided by a digital meter.

alta impedancia Gran resistencia de entrada proporcionada por un medidor digital.

hold-down springs Small springs that hold drum brake shoes in position against the backing plate while providing flexibility for shoe application and release.

resortes de sujeción Pequeños resortes que mantienen las zapatas de frenos de tambor en posición contra la placa de refuerzo, a la vez que dan flexibilidad para aplicar y soltar la zapata.

hydraulic control unit A housing that contains the valves and solenoids to control individual wheel braking pressures. Its function is similar to the control valve assembly on RABS/RWAL systems.

unidad de control hidráulica Una cubierta que contiene las válvulas y los solenoides para controlar presiones de los frenos individuales de la rueda. Su función es similar al montaje de válvula de control en sistemas de RABS/RWAL.

hydro-boost A hydraulic power brake system that uses the power steering hydraulic system to provide boost for the brake system.

reforzador hidráulico Sistema hidráulico de frenos de potencia que utiliza el sistema hidráulico de dirección de potencia para reforzar el sistema de frenado.

hydroplane The action of a tire rolling on a layer of water on the road surface instead of staying in contact with the pavement. Hydroplaning occurs when water cannot be displaced from between the tread and the road.

aquaplaning　Acción de un neumático que rueda sobre una capa de agua sobre la superficie vial en lugar de mantenerse en contacto con el pavimento. El "aquaplaning" se produce cuando no se puede desplazar el agua entre los dibujos del neumático y la calle.

hygroscopic　The chemical property or characteristic of attracting and absorbing water, particularly out of the air. Polyglycol brake fluids are hygroscopic.

higroscópico　Que posee la propiedad química o característica de atraer y absorber agua, en especial del aire. Los líquidos de frenos con poliglicol son higroscópicos.

international system of units or metric system　The modern international metric system used by the automotive industry and other industries.

sistema métrico o sistema internacional de unidades　Sistema métrico internacional moderno que utiliza, entre otras, la industria automotriz.

ISO flare　A type of tubing flare connection in which a bubble-shaped end is formed on the tubing; also called a bubble flare.

ensanche ISO　Tipo de conexión con ensanche del tubo en el cual un extremo toma forma de burbuja; también se le llama ensanche de burbuja.

lateral runout　A side-to-side variation or wobble as the tire and wheel are rotated.

desviación lateral　Variación u oscilación de un lado a otro cuando se hacen girar el neumático y la rueda.

leading shoe　The first shoe in the direction of drum rotation in a leading-trailing brake. When the vehicle is going forward, the forward shoe is the leading shoe, but the leading shoe can be the front or the rear shoe depending on whether the drum is rotating forward or in reverse and whether the wheel cylinder is at the top or the bottom of the backing plate. The leading shoe is self-energizing.

zapata tractora　Primera zapata en la dirección de giro del tambor en un freno de tracción-remolque. Cuando el vehículo avanza, la zapata delantera es la tractora, pero la zapata tractora puede ser la delantera o la trasera dependiendo de si el tambor está girando hacia delante o hacia atrás y de si el cilindro de la rueda está en la parte superior de la placa de refuerzo o en la inferior. La zapata tractora es autónoma en cuanto a energía.

leading-trailing brake　A drum brake that develops self-energizing action only on the leading shoe. Brake application force is separate for the leading and the trailing shoes. Also called a partial-servo or a nonservo brake.

freno tracción-remolque　Freno de tambor que desarrolla una acción autónoma sólo sobre la zapata tractora. La fuerza de aplicación de freno es independiente para las zapatas tractoras y para las de remolque. También se conoce como freno parcialmente asistido o freno no asistido.

linear　In a straight line.

lineal　En línea recta.

loaded calipers　Pairs of calipers sold as service replacements that have been overhauled and are loaded with new pads. The complete caliper assembly is ready to install when purchased.

calibres cargados　Pares de calibres vendidos como repuestos revisados y a los que se han colocado nuevas pastillas. Al comprarlo, el conjunto está listo para ser instalado.

manual bleeding　The process of using the brake pedal and master cylinder as a hydraulic pump to expel air and brake fluid from the system. Manual bleeding is a two-person operation.

extracción de vapor manual　Uso del pedal del freno y el cilindro maestro como bomba hidráulica como procedimiento para expulsar el aire y el líquido de frenos del sistema. La extracción de vapor manual es una operación para realizar entre dos.

material safety data sheets (MSDS)　Information sheets issued by the manufacturers of hazardous materials. An MSDS provides detailed information on dangerous ingredients, corrosiveness, reactivity, toxicity, fire and explosion data, health hazards, spill and leak procedures, and special precautions. Federal law requires that an MSDS be available for each hazardous material in your workplace.

hojas de datos de seguridad de materiales (MSDS)　Hojas de información proporcionadas por los fabricantes de materiales peligrosos. Una MSDS ofrece información detallada sobre ingredientes peligrosos, corrosión, capacidad de reacción, toxicidad, datos sobre incendios y explosiones, peligros para la salud, procedimientos en caso de derrame o goteo de estos materiales, y precauciones especiales a tomar. La ley federal exige que haya una MSDS para cada material peligroso en el lugar de trabajo.

maximum refinishing limit　A brake rotor measurement that determines the point at which a rotor should not be refinished even if the minimum thickness is correct.

Límite máximo de re-acabado　Medida del rotor del freno que determina el punto en el que al rotor no debe de dársele un reacabado aun si está correcto el grosor mínimo.

metering valve　A hydraulic control valve used primarily with front disc brakes on RWD vehicles. The metering valve delays pressure application to the front brakes until the rear drum brakes have started to operate.

válvula de dosificación　Válvula de control hidráulica que se usa principalmente con frenos delanteros de disco en vehículos RWD. La válvula de dosificación demora la aplicación de presión a los frenos delanteros hasta que hayan comenzado a funcionar los frenos traseros de tambor.

multiple　A metric measurement unit that is larger than the stem unit through multiplying by a power of ten.

múltiplo　Unidad del sistema métrico mayor que la unidad base, que resulta de multiplicar por una potencia de diez.

nondirectional finish　A finish on the surface of a rotor that will not start a premature wear pattern on the pads. The nondirectional sanded surface also helps pad break in and reduces brake noise.

pulido adireccional　Pulido de la superficie de un rotor que no muestra un patrón de desgaste prematuro en las pastillas. Una superficie lijada adireccional también favorece la aplicación de la pastilla y reduce el ruido de la frenada.

Occupational Safety and Health Administration (OSHA)　A division of the U.S. Department of Labor that establishes and enforces workplace safety regulations.

Administración de Seguridad y Salud en el Trabajo (OSHA)　División del Departamento de Trabajo de EE.UU. que establece y hace cumplir las normas de seguridad en el lugar de trabajo.

ohm　The unit used to measure the amount of electrical resistance in a circuit or an electrical device. One ohm is the amount of resistance present when one volt forces one ampere of current through a circuit or a device. Ohm is abbreviated with the Greek letter omega (Ω).

ohmio Unidad usada para medir la cantidad de resistencia eléctrica de un circuito o de un aparato eléctrico. Un ohmio es la cantidad de resistencia presente cuando un voltio hace pasar un amperio de corriente a través de un circuito o de un aparato. El ohmio se abrevia con la letra griega omega (Ω).

O-ring A circular rubber seal shaped like the letter "O."

toroide Junta de goma circular con la forma de la letra "O".

oscilloscope An electronic test instrument that can display voltage or amperage over a period of time as a graph display.

osciloscopio Instrumento de prueba electrónica que puede mostrar el voltaje o el amperaje por un período de tiempo en forma de gráfica.

pad hardware Miscellaneous small parts such as antirattle clips and support clips that hold brake pads in place and keep them from rattling.

hardware de pastillas Piezas pequeñas surtidas, como grapas antiresonancia y grapas de apoyo que mantienen las pastillas de los frenos en su lugar y evitan que suenen.

pad wear indicator A device that warns the driver when disc brake linings have worn to the point where they need replacement. Wear indicators may be mechanical (audible) or electrical.

indicadores del desgaste de las pastillas Dispositivo que advierte al conductor de que los forros de los frenos de disco se han gastado tanto que es necesario cambiarlos. Los indicadores de desgaste pueden ser mecánicos (acústicos) o eléctricos.

parallel circuit An electrical circuit in which all the loads are connected to form more than one current path with the power source.

circuito paralelo Circuito eléctrico en el que todas las cargas se conectan de manera que formen más de una ruta con la fuente de alimentación.

parallelism Thickness uniformity of a disc brake rotor. Both surfaces of a rotor must be parallel with each other within 0.001 inch or less.

paralelismo Uniformidad de grosor de un rotor de freno de disco. Ambas superficies de un rotor deben ser paralelas entre sí con una tolerancia de 0,001 pulgada o menos.

parking brake control The pedal or lever used to apply the parking brakes.

control del freno de estacionamiento Pedal o palanca que se usa para aplicar los frenos de estacionamiento.

pawl A hinged or pivoted component that engages a toothed wheel or rod to provide rotation or movement in one direction while preventing it in the opposite direction.

trinquete Componente articulado o embisagrado que se engrana con una rueda o varilla dentada para ofrecer rotación o movimiento en un sentido mientras que lo impide en el sentido contrario.

pedal free play The clearance between the brake pedal or booster pushrod and the primary piston in the master cylinder.

holgura de pedal Hlgura entre el pedal del freno o la varilla de refuerzo y el pistón primario en el cilindro maestro.

permanent magnet (PM) generator A reluctance sensor that generates a voltage signal by moving a conductor through a permanent magnetic field.

generador de magneto permanente (PM) Sensor de reluctancia que genera una señal de tensión al mover un conductor a través de un campo magnético permanente.

phenolic plastic Plastic made primarily from phenol, a compound derived from benzene and also called carbolic acid.

plástico fenólico Plástico hecho principalmente de fenol, compuesto derivado del benceno, llamado también ácido carbólico.

pickup coil sensor A reluctance sensor that generates a voltage signal by moving a conductor through a permanent magnetic field.

sensor de bobina captadora Sensor de reluctancia que genera una señal de tensión al mover un conductor a través de un campo magnético permanente.

piston stop A metal part on a brake backing plate that keeps the wheel cylinder pistons from moving completely out of the cylinder bore.

tope de pistón Parte metálica de la placa de refuerzo del freno que impide que los pistones del cilindro de la rueda se salgan del hueco del cilindro.

plateaued finish A flattened finish to a cylinder bore in which all of the high spots or peaks are removed.

acabado amesetado Acabado plano dado al hueco de un cilindro en el que se han eliminado todos los puntos altos o picos.

polyglycol Polyalkylene-glycol-ether brake fluids that meet specifications for DOT 3 and DOT 4 brake fluids.

poliglicol Líquidos de frenos de polialquilenglicoléter que están dentro de las especificaciones DOT 3 y DOT 4 para líquidos de frenos.

PowerMaster A self-contained hydraulic power brake system with its own hydraulic reservoir and independent electric pump; developed by General Motors for cars and trucks that could not economically use vacuum-assisted power brakes.

PowerMaster Sistema de frenos hidráulico autónomo con su propia reserva hidráulica y bomba eléctrica independiente; producido por General Motors para coches y camiones que no podían usar económicamente frenos de potencia asistidos por vacío.

pressure bleeding The process of using a tank filled with brake fluid and pressurized with compressed air to expel air and brake fluid from the system. Pressure bleeding is a one-person operation.

extracción de vapor a presión Proceso en el que se utiliza un tanque lleno de líquido de frenos y se ejerce presión con aire comprimido para expulsar el aire y el líquido de frenos del sistema. La extracción de vapor a presión es una operación para realizar una sola persona.

pressure differential The difference between two pressures on two surfaces or in two separate areas. The pressures can be either pneumatic (air) or hydraulic.

diferencial de presión Diferencia entre dos presiones en dos superficies o áreas distintas. La presión puede ser neumática (aire) o hidráulica.

pressure differential valve A hydraulic valve that reacts to a difference in pressure between the halves of a split brake system. When a pressure differential exists, the valve moves a plunger to close the brake warning lamp switch.

válvula de diferencial de presión Válvula hidráulica que reacciona ante una diferencia de presión entre las dos partes de un sistema de frenos dividido. Cuando hay una diferencia de presión, la válvula mueve un pistón que cierra el conmutador de la luz indicadora de freno.

primary shoe The leading shoe in a duo-servo brake. The primary shoe is self-energizing and applies servo action to the secondary shoe to increase its application force. Primary shoes have shorter linings than secondary shoes.

zapata primaria Zapata guía en un servofreno dual. La zapata primaria es autónoma y aplica una servoacción a la zapata secundaria para aumentar su fuerza de aplicación. Las zapatas primarias tienen forros más pequeños que las secundarias.

proportioning valve   A hydraulic control valve that controls the pressure applied to rear drum brakes. A proportioning valve decreases the rate of pressure application above its split point as the brake pedal is applied harder.

válvula de dosificación   Válvula de control hidráulico que controla la presión aplicada a los frenos de tambor traseros. Una válvula de dosificación disminuye la proporción de aplicación de presión por encima de su punto de separación cuando se presiona más fuerte el pedal del freno.

pull up resistor   A fixed resistor in a voltage divider circuit for a sensor input signal; a variable-resistor sensor changes its voltage drop proportionally to the quantity being measured, and the pull up resistor drops the rest of the reference voltage.

resistor de arranque   Resistor fijo en un circuito divisor de tensión que alimenta a un sensor de señal de entrada; un sensor del resistor variable cambia su caída de tensión proporcionalmente a la cantidad que ha medido y el resistor de arranque baja el resto de la tensión de referencia.

push (speed) nut   A lightweight, stamped steel retainer that pushes onto a stud to hold two parts together temporarily.

tuerca de empuje (velocidad)   Dispositivo de retención ligero, de acero estampado, que presiona contra un borne para juntar dos piezas temporalmente.

quick take-up master cylinder   A dual master cylinder that supplies a large volume of fluid to the front disc brakes on initial brake application, which takes up the clearance of low-drag calipers.

cilindro maestro de tensor rápido   Cilindro maestro doble que proporciona una gran cantidad de líquido a los frenos de disco delanteros en la primera frenada, compensando la holgura de los calibres de baja resistencia.

quick take-up valve   The part of the quick take-up master cylinder that controls fluid flow between the reservoir and the primary low-pressure chamber.

válvula de tensor rápido   Parte del cilindro maestro del tensor rápido que controla el flujo de líquido entre el depósito y la cámara de baja presión primaria.

radial ply tire   Tire construction in which the cords in the body plies of the carcass run at an angle of 90 degrees to the steel beads in the inner rim of the carcass. Each cord is parallel to the radius of the tire circle.

neumático de capas radiales   Fabricación de neumáticos en los que los cordones de las capas del cuerpo de la carcasa se fijan con un ángulo de 90 grados a las nervaduras de acero en el borde interior de la carcasa. Cada cordón es paralelo a los radios del círculo del neumático.

radial runout   An out-of-round condition in which the radius of the wheel or tire is not consistent from the wheel center to any point on the rim or the tread.

desgaste radial   Pérdida de redondez en la que el radio de la rueda o neumático no es consistente desde el centro de la rueda a cualquier punto de la llanta o de la rodadura.

reaction disc (or plate and levers)   The components in a vacuum power booster that provide pedal feel or feedback to the driver.

disco de reacción (o placa y palancas)   Componentes de un reforzador de vacío que proporcionan sensaciones en el pedal o respuesta al conductor.

reamer   A tool that is used to smooth the interior of a pipe or tube where the cutter broke through the metal wall

escariador   Herramienta que se usa para alisar el interior de un conducto o tubo donde la herramienta de cortar perforó la pared metálica.

reference voltage   A fixed voltage supplied to the sensor by a voltage regulator inside the computer or control module; as the sensor changes, the return voltage is altered and sent back to the computer for use. Most computer control systems operate with a 5-volt reference voltage.

tensión de referencia   Tensión fija suministrada al sensor por un regulador de tensión del interior del computador o módulo de control; al cambiar el sensor, la tensión de vuelta se altera y se envía de vuelta al computador para su uso. La mayoría de los sistemas de control por computador funcionan con una tensión de referencia de 5 voltios.

refractometer   A test instrument that measures the deflection, or bending, of a beam of light.

refractómetro   Instrumento de prueba que mide la deformación, o desviación, de un rayo de luz.

reluctance sensor   A magnetic pulse generator or pickup coil that sends a voltage signal in response to varying reluctance of a magnetic field.

sensor de reluctancia   Generador de impulsos magnéticos o bobina captadora que envía una señal de tensión como respuesta a una variación en la reluctancia de un campo magnético.

replenishing port   The rearward port in the master cylinder bore; also called other names.

puerto de abastecimiento   Puerto situado más atrás en el hueco del cilindro maestro; también recibe otros nombres.

return spring   A strong spring that retracts a drum brake shoe when hydraulic pressure is released.

resorte de vuelta   Resorte fuerte que retrae una zapata de freno de tambor cuando se libera la presión hidráulica.

rigid hone   Three or four abrasive stones mounted in small fixtures that are free to tilt inward and outward with spring tension holding them against the cylinder wall as the hone turns.

pulimentadora rígida   Tres o cuatro piedras abrasivas montadas en accesorios pequeños que están libres para inclinarse hacia dentro y hacia fuera con la tensión del resorte que las sujeta contra la pared del cilindro mientras gira la pulimentadora.

rolled finish   A manufacturing process designed to close the pores of porous metals such as aluminum. A hardened metal shaft is rolled over repeatedly with great pressure on the surface of the porous metal until the pores close. Honing this type of surface will reopen the metal pores.

acabado laminado   Proceso de fabricación destinado a cerrar los poros de los metales porosos como el aluminio. Un eje de metal endurecido se pasa como un rodillo, repetidamente y a gran presión, sobre la superficie del metal poroso hasta que se cierran los poros. Si se pule este tipo de superficie, se volverán a abrir los poros del metal.

rosin flux solder   Solder used for electrical repairs that does not contain acid flux or other corrosive materials.

soldadura de flujo de resina de trementina   Aleación para soldar usada en reparaciones eléctricas, que no contiene fundente ácido ni otros materiales corrosivos.

rotor   The rotating part of a disc brake that is mounted on the wheel hub and contacted by the pads to develop friction to stop the car. Also called a disc.

rotor Parte giratoria de un freno de disco que va montada en el buje de la rueda y entra en contacto con las pastillas para causar el rozamiento que detiene el vehículo. También llamado disco.

rotor lateral runout Rotor wobble from side to side as it rotates.

desviación lateral del rotor Oscilación del rotor de un lado a otro mientras gira.

scan tool A test computer that plugs into a diagnostic connector on a vehicle and reads trouble codes and other operating information from a vehicle onboard computer.

herramienta scan Computador de prueba que se enchufa a un conector de diagnóstico en un vehículo y lee los códigos de problema y otras información de funcionamiento en un computador de a bordo del vehículo.

secondary shoe The trailing shoe in a duo-servo brake. The secondary shoe receives servo action from the primary shoe to increase its application force. Secondary shoes provide the greater braking force in a duo-servo brake and have longer linings than primary shoes.

zapata secundaria Zapata de remolque en un servofreno dual. La zapata secundaria recibe acción asistida de la primaria para aumentar su fuerza de aplicación. Las zapatas secundarias proporcionan mayor potencia de frenado en un servofreno dual y tienen forros más grandes que las primarias.

self-adjuster A cable, lever, screw, strut, or other linkage part that provides automatic shoe adjustment and proper lining-to-drum clearance as a drum brake lining wears.

autoregulador Cable, palanca, tornillo, puntal u otro mecanismo de unión que proporciona un ajuste automático a la zapata y la holgura adecuada entre forro y tambor cuando se desgasta el forro de un freno de tambor.

self-energizing operation The action of a drum brake shoe when drum rotation increases the application force of the shoe by wedging it tightly against the drum.

operación autoactivada Acción de una zapata de freno de tambor cuando la rotación del tambor aumenta la fuerza de aplicación de la zapata al ajustarse como una cuña contra el tambor.

semimetallic lining Brake friction materials made from a mixture of organic or synthetic fibers and certain metals; these linings do not contain asbestos.

forro semimetálico Materiales de fricción del freno hechos con una mezcla de fibras orgánicas o sintéticas y ciertos metales; estos forros no contienen amianto.

sensors The discrimination and arming sensors that are pressure-sensitive switches that complete an electrical circuit during an impact of sufficient G-force to inflate the air bag.

sensores Sensores de discriminación y acción que son conmutadores sensibles a la presión y que completan un circuito eléctrico durante un impacto de suficiente fuerza G como para inflar el air bag.

series circuit An electrical circuit in which all the loads are connected in series with each other to form a single current path.

circuito en serie Circuito eléctrico en el cual todas las cargas están conectadas en serie entre sí formando una sola ruta para el flujo de corriente.

series-parallel circuit An electrical circuit in which some loads are in parallel with each other and one or more loads are in series with the power source and with the parallel branches.

circuito paralelo en serie Circuito eléctrico en el cual algunas cargas están conectadas en paralelo entre sí, y una o más cargas están en serie con la fuente de alimentación y con las ramas paralelas.

servo action The operation of a drum brake that uses the self-energizing operation of one shoe to apply mechanical force to the other shoe to assist its application. Broadly, servo action is any mechanical multiplication of force.

servoacción Acción de un freno de tambor en el que una zapata actúa de forma autónoma al aplicar una fuerza mecánica a la otra zapata y ayudarla en su funcionamiento. Más ampliamente, una servoacción es cualquier multiplicación mecánica de una fuerza.

setback A difference in wheelbase from one side of a vehicle to the other.

retroceso Diferencia en la distancia entre ejes en un lado y otro del vehículo.

sliding caliper A caliper that is mounted to its support on two fixed sliding surfaces, or ways. The caliper slides on the rigid ways and does not have the flexibility of a floating caliper.

calibre de desplazamiento Calibre que se monta al soporte sobre dos superficies o vías fijas de deslizamiento. El calibre se desliza por vías rígidas y no tiene la flexibilidad de un calibre flotante.

slope The numerical ratio, or proportion, of rear drum brake pressure to full system pressure that is applied through a proportioning valve. If half of the system pressure is applied to the rear brakes, the slope is 1:2, or 50 percent.

atenuación diferencial Razón numérica o proporción entre la presión del freno de tambor trasero y la presión total del sistema que se aplica a través de una válvula de dosificación. Si la mitad de la presión del sistema se aplica a los frenos traseros, la atenuación diferencial es 1:2 ó del 50 por ciento.

sodium azide The explosive compound used to inflate an air bag.

azida de sodio Compuesto explosivo que se emplea para inflar un air bag.

soft code A diagnostic trouble code from a vehicle computer that indicates a problem not present at the time of testing, indicating an intermittent problem that occurred sometime before testing; stored in long-term computer memory and usually erased after fifty to one hundred ignition cycles if the problem does not occur.

código blando Código de problema de diagnóstico originado en el computador del vehículo que indica un problema no existente en el momento de la prueba, señalando un problema intermitente que ocurrió antes de la prueba; queda almacenado en la memoria a largo plazo del computador y se suele borrar después de entre cincuenta y cien ciclos de encendido si la falla no vuelve a presentarse.

solderless connector A wiring repair connector that is joined to two wires by physical crimping.

conector sin soldadura Conector para reparar el cableado que se une a dos cables ajustándolo físicamente.

specific gravity The weight of a volume of any liquid divided by the weight of an equal volume of water at equal temperature and pressure. The ratio of the weight of any liquid to the weight of water, which has a specific gravity of 1.000.

gravedad específica Peso de un volumen de cualquier líquido dividido por el peso de un volumen igual de agua a la misma presión y temperatura. La relación entre el peso de cualquier líquido y el del agua, que tiene una gravedad específica de 1,000.

splice clip A special, noninsulated connector used with solder to connect two wires.

grapas de empalme Conector especial, sin aislamiento, que se utiliza con soldadura para conectar dos cables.

split point The pressure at which a proportioning valve closes during brake application and reduces the rate at which further pressure is applied to rear drum brakes.

punto de separación Presión a la que la válvula de dosificación se cierra durante la aplicación de los frenos y se reduce la relación con la que se aplica más presión a los frenos de tambor traseros.

spool valve A cylindrical sliding valve that uses lands and valleys around its circumference to control the flow of hydraulic fluid through the valve body.

válvula de carrete Válvula cilíndrica de deslizamiento que usa partes planas y hundimientos en su circunferencia para controlar el flujo del líquido hidráulico a través del cuerpo de la válvula.

square-cut seal A fixed seal for a caliper piston that has a square cross section.

junta cuadrada de pistón Junta fija para un pistón de calibre con sección transversal cuadrada.

squib coil A tungsten wire used to ignite the solid fuel of the inflator module

bobina detonadora Alambre de tungsteno usado para la ignición del combustible sólido del módulo inflador

star wheel A small wheel that is part of a drum brake adjusting link. Turning the star wheel lengthens or shortens the adjuster link to position the shoes for proper lining-to-drum clearance.

rueda en estrella Rueda pequeña que forma parte de un acoplamiento de ajuste de un freno de tambor. Al hacer girar la estrella se alarga o acorta el acoplamiento para poner en posición las zapatas y conseguir una tolerancia conveniente entre forro y tambor.

steering knuckle The outboard part of the front suspension that pivots on the ball joints and lets the wheels turn for steering control.

charnela de dirección Parte exterior de la suspensión delantera que pivota en las juntas de bola y permite que las ruedas giren para controlar la dirección.

stem unit Any metric unit to which a prefix can be added to indicate larger or smaller measurements to some power of ten. For example, kilometer is formed by adding the prefix "kilo" to the stem unit "meter."

unidad base Cualquier unidad métrica a la que se puede añadir un prefijo para indicar medidas mayores o menores en potencias de diez. Por ejemplo, kilómetro se forma añadiendo el prefijo "kilo" a la unidad base "metro".

stroking seal A seal, similar to a lip seal, that is installed at the rear of a caliper piston. Unlike a fixed seal that is installed in a groove in the caliper bore, a stroking seal is installed on the piston and moves with it.

junta del pistón Junta similar a la del borde, que se instala en la parte trasera de un pistón de calibre. A diferencia de la junta fija que se instala en una hendidura de la superficie interior del calibre, la junta del pistón se instala en el pistón y se mueve con éste.

submultiple A metric measurement unit that is smaller than the stem unit through dividing by a power of ten.

submúltiplo Unidad del sistema métrico menor que la unidad base que resulta de dividir por una potencia de diez.

supplemental inflatable restraint system (SIRS) The driver or driver/passenger side protection used during a forward direction collision.

sistema supletorio inflable de retención (SIRS) Protección de lado del conductor o del conductor/pasajero utilizada durante una colisión frontal.

surge bleeding A supplementary bleeding method in which one person rapidly pumps the brake pedal to dislodge air pockets while another person opens the bleeder screw. Surge bleeding is a two-person operation but should not be used as the only bleeding procedure for a brake system.

extracción de vapor por impulsos Método suplementario de extracción de vapor en el cual una persona bombea rápidamente el pedal del freno para desalojar las bolsas de aire mientras otra abre el tornillo extractor. La extracción de varpor por impulsos se hace entre dos personas, pero no se debe emplear como único procedimiento de extracción de vapor para un sistema de frenos.

table The outer surface of a brake shoe to which the lining is attached.

tabla Superficie exterior de una zapata de freno a la que se une el forro.

tandem booster A power brake vacuum booster with two small diaphragms in tandem to provide additive vacuum force.

reforzador en serir Reforzador de vacío para frenos con dos pequeños diafragmas en tándem que proporcionan más fuerza vacío.

tapered roller bearing A specific kind of bearing that is based on tapered steel rollers held together in a cage.

cojinete de bolillas cónicas Clase específica de cojinete que se basa en bolillas cónicas de acero encerradas en una jaula.

Technical Service Bulletins (TSB) A vehicle manufacturer bulletin that outlines the details for the diagnosis and repair of a specific fault in a specific range of vehicles. Is not considered a warranty or recall issue.

Boletines de Servicio Técnico (TSB) Boletín del fabricante del vehículo que resalta los detalles para el diagnóstico y reparación de una falla específica en un tipo de vehículo. No se considera como garantía o asunto de retiro.

toe angle The difference in the distance between the centerlines of the tires on either axle (front or rear) measured at the front and rear of the tires and at spindle height.

ángulo de separación Diferencia de la distancia entre las líneas centrales de los neumáticos en cada eje (delantero o trasero) medida en las partes delantera y trasera de los neumáticos y a la altura del husillo.

torque stick An extension for an impact wrench that includes the correct size socket for a wheel nut and that acts as a torsion bar to limit the torque applied by the impact wrench when installing a wheel.

varilla de momento de torsión Extensión para llave de impacto que incluye la boca de tamaño correcto para una tuerca de la rueda, que actúa como barra de torsión para limitar el momento de torsión aplicado por la llave cuando se instala una rueda.

trailing shoe The second shoe in the direction of drum rotation in a leading-trailing brake. When the vehicle is going forward, the rear shoe is the trailing shoe, but the trailing shoe can be the front or the rear shoe depending on whether the drum is rotating forward or in reverse and whether the wheel cylinder is at the top or the bottom of the backing plate. The trailing shoe is non-self-energizing, and drum rotation works against shoe application.

zapata de remolque Segunda zapata en la dirección de giro del tambor en un freno de tracción-remolque. Cuando el vehículo avanza, la zapata trasera es la remolcada, pero la zapata de remolque puede ser la delantera o la trasera dependiendo de si el tambor está girando hacia delante o hacia atrás y de si el cilindro de la rueda está en la parte superior de la placa de refuerzo o en la inferior. La zapata de remolque no es autónoma, y el giro del tambor funciona contra la aplicación de la zapata.

tread   The layer of rubber on a tire that contacts the road and contains a distinctive pattern to provide traction.

rodadura   Capa de goma de un neumático que está en contacto con la carretera y contiene un patrón distintivo que favorece la tracción.

tread wear indicator   A continuous bar that appears across a tire tread when the tread wears down to the last ⁄32 (⁄16) inch. When a tread wear indicator appears across two or more adjacent grooves, the tire should be replaced.

indicador de desgaste de la rodadura   Barra continua que aparece transversalmente en la rodadura del neumático cuando ésta se desgasta más de ⁄32 (⁄16) pulgadas. Cuando aparece el indicador de desgaste de rodadura cruzando dos o más hendiduras adyacentes, se debe cambiar el neumático.

tubing bender   A collection of interchangeable curved sections used to bend a tube or pipe to the correct radii without crimping.

acodador   Colección de secciones curvas intercambiables que se emplean para doblar un tubo o conducto hasta obtener el radio correcto sin estrecharlo.

tubing cutter   A cutter used to cut a pipe or tube at a flat angle.

cortatubos   Herramienta cortante empleada para cortar un conducto o tubo en ángulo recto.

vacuum   In automotive service, vacuum is generally considered to be air pressure lower than atmospheric pressure. A true vacuum is a complete absence of air.

vacío   En relación con los automóviles, el vacío se suele considerar como una presión de aire menor que la atmosférica. Un vacío verdadero es la completa ausencia de aire.

vacuum bleeding   The process of using a vacuum pump filled with brake fluid and attached to a wheel bleeder screw to draw old brake fluid and air from the system. Vacuum bleeding is a one-person operation.

extracción de vapor por vacío   Proceso en el que se usa una bomba de vacío llena de líquido de frenos y unida a un tornillo extractor de rueda para extraer del sistema el aire y el líquido de frenos viejo. La extracción de vapor por vacío es una operación para realizar una sola persona.

vacuum suspended   A term that describes a power brake vacuum booster in which vacuum is present on both sides of the diaphragm when the brakes are released. The most common kind of vacuum booster.

de suspensión de vacío   Término que describe un reforzador de vacío para frenos en el que existe vacío a ambos lados del diafragma cuando se sueltan los frenos. El tipo más común de reforzador de vacío.

vent port   The forward port in the master cylinder bore; also called other names.

válvula de ventilación   Válvula delantera de la parte interior del cilindro principal; también recibe otros nombres.

ventilated rotor   A rotor that has cooling fins cast between the braking surfaces to increase the cooling area of the rotor.

rotor ventilado   Rotor que tiene álabes refrigerantes entre las superficies de frenado para aumentar la superficie refrigerante del rotor.

vernier caliper   A measuring caliper with a vernier scale for very fine measurements. The scale may be graduated in U.S. customary or metric units.

calibre vernier   Calibre de medir con una escala vernier para mediciones afinadas. La escala puede estar graduada en unidades métricas o en las tradicionales americanas.

vernier scale   A fine auxiliary scale that indicates fractional parts of a larger scale.

escala vernier   Escala auxiliar afinada que indica fracciones de una escala mayor.

volt   The unit of electrical force or pressure.

voltio   Unidad de fuerza o presión eléctrica.

voltage   The electromotive force that causes current to flow. The potential force that exists between two points when one is positively charged and the other is negatively charged.

tensión   Fuerza electromotriz que hace que la corriente fluya. La fuerza potencial que existe entre dos puntos cuando uno está cargado positivamente y el otro negativamente.

ways   Machined surfaces on the caliper support on which a sliding caliper slides.

conductos   Superficies talladas en el soporte del calibre por las que se desliza un calibre de desplazamiento.

wear-indicating ball joint   A ball joint with a visual indicator to show the amount of wear on the joint.

juntas de bola de indicación de desgaste   Junta de bola con un indicador visual que muestra la cantidad de desgaste de la rótula.

web   The inner part of a brake shoe that is perpendicular to the table and to which all of the springs and other linkage parts attach.

membrana   Parte interior de una zapata de freno perpendicular a la tabla y en la que se fijan todos los resortes y otras partes de acoplamiento.

Index

Note: Page numbers in bold print reference non-text material.